T0203046

Lecture Notes in Computer Science 14489

Founding Editors

Gerhard Goos
Juris Hartmanis

Editorial Board Members

The series Lecture Notes in Computer Science (LNCS), including its subseries Lecture Notes in Artificial Intelligence (LNAI) and Lecture Notes in Bioinformatics (LNBI), has established itself as a medium for the publication of new developments in computer science and information technology research, teaching, and education.

LNCS enjoys close cooperation with the computer science R & D community, the series counts many renowned academics among its volume editors and paper authors, and collaborates with prestigious societies. Its mission is to serve this international community by providing an invaluable service, mainly focused on the publication of conference and workshop proceedings and postproceedings. LNCS commenced publication in 1973.

Zahir Tari · Keqiu Li · Hongyi Wu
Editors

Algorithms and Architectures for Parallel Processing

23rd International Conference, ICA3PP 2023
Tianjin, China, October 20–22, 2023
Proceedings, Part III

 Springer

Editors
Zahir Tari
Royal Melbourne Institute of Technology
Melbourne, VIC, Australia

Keqiu Li
Tianjin University
Tianjin, Tianjin, China

Hongyi Wu
University of Arizona
Tucson, AZ, USA

ISSN 0302-9743   ISSN 1611-3349 (electronic)
Lecture Notes in Computer Science
ISBN 978-981-97-0797-3   ISBN 978-981-97-0798-0 (eBook)
https://doi.org/10.1007/978-981-97-0798-0

This Springer imprint is published by the registered company Springer Nature Singapore Pte Ltd.
The registered company address is: 152 Beach Road, #21-01/04 Gateway East, Singapore 189721, Singapore

Paper in this product is recyclable.

Preface

On behalf of the Conference Committee, we welcome you to the proceedings of the 2023 International Conference on Algorithms and Architectures for Parallel Processing (ICA3PP 2023), which was held in Tianjin, China from October 20–22, 2023. ICA3PP2023 was the 23rd in this series of conferences (started in 1995) that are devoted to algorithms and architectures for parallel processing. ICA3PP is now recognized as the main regular international event that covers the many dimensions of parallel algorithms and architectures, encompassing fundamental theoretical approaches, practical experimental projects, and commercial components and systems. This conference provides a forum for academics and practitioners from countries around the world to exchange ideas for improving the efficiency, performance, reliability, security, and interoperability of computing systems and applications.

A successful conference would not be possible without the high-quality contributions made by the authors. This year, ICA3PP received a total of 503 submissions from authors in 21 countries and regions. Based on rigorous peer reviews by the Program Committee members and reviewers, 193 high-quality papers were accepted to be included in the conference proceedings and submitted for EI indexing. In addition to the contributed papers, six distinguished scholars, Lixin Gao, Baochun Li, Laurence T. Yang, Kun Tan, Ahmed Louri, and Hai Jin, were invited to give keynote lectures, providing us with the recent developments in diversified areas in algorithms and architectures for parallel processing and applications.

We would like to take this opportunity to express our sincere gratitude to the Program Committee members and 165 reviewers for their dedicated and professional service. We highly appreciate the twelve track chairs, Dezun Dong, Patrick P. C. Lee, Meng Shen, Ruidong Li, Li Chen, Wei Bao, Jun Li, Hang Qiu, Ang Li, Wei Yang, Yu Yang, and Zhibin Yu, for their hard work in promoting this conference and organizing the reviews for the papers submitted to their tracks. We are so grateful to the publication chairs, Heng Qi, Yulei Wu, Deze Zeng, and the publication assistants for their tedious work in editing the conference proceedings. We must also say "thank you" to all the volunteers who helped us at various stages of this conference. Moreover, we were so honored to have many renowned scholars be part of this conference. Finally, we would like to thank all speakers, authors, and participants for their great contribution to and support for the success of ICA3PP 2023!

October 2023

Jean-Luc Gaudiot
Hong Shen
Gudula Rünger
Zahir Tari
Keqiu Li
Hongyi Wu
Tian Wang

Organization

General Chairs

Jean-Luc Gaudiot	University of California, Irvine, USA
Hong Shen	University of Adelaide, Australia
Gudula Rünger	Chemnitz University of Technology, Germany

Program Chairs

Zahir Tari	Royal Melbourne Institute of Technology, Australia
Keqiu Li	Tianjin University, China
Hongyi Wu	University of Arizona, USA

Program Vice-chair

Wenxin Li	Tianjin University, China

Publicity Chairs

Hai Wang	Northwest University, China
Milos Stojmenovic	Singidunum University, Serbia
Chaofeng Zhang	Advanced Institute of Industrial Technology, Japan
Hao Wang	Louisiana State University, USA

Publication Chairs

Heng Qi	Dalian University of Technology, China
Yulei Wu	University of Exeter, UK
Deze Zeng	China University of Geosciences (Wuhan), China

Workshop Chairs

Laiping Zhao   Tianjin University, China
Pengfei Wang   Dalian University of Technology, China

Local Organization Chairs

Xiulong Liu   Tianjin University, China
Yitao Hu   Tianjin University, China

Web Chair

Chen Chen   Shanghai Jiao Tong University, China

Registration Chairs

Xinyu Tong   Tianjin University, China
Chaokun Zhang   Tianjin University, China

Steering Committee Chairs

Yang Xiang (Chair)   Swinburne University of Technology, Australia
Weijia Jia   Beijing Normal University and UIC, China
Yi Pan   Georgia State University, USA
Laurence T. Yang   St. Francis Xavier University, Canada
Wanlei Zhou   City University of Macau, China

Program Committee

Track 1: Parallel and Distributed Architectures

Dezun Dong (Chair)   National University of Defense Technology,
                               China
Chao Wang   University of Science and Technology of China,
                               China
Chentao Wu   Shanghai Jiao Tong University, China

Chi Lin	Dalian University of Technology, China
Deze Zeng	China University of Geosciences, China
En Shao	Institute of Computing Technology, Chinese Academy of Sciences, China
Fei Lei	National University of Defense Technology, China
Haikun Liu	Huazhong University of Science and Technology, China
Hailong Yang	Beihang University, China
Junlong Zhou	Nanjing University of Science and Technology, China
Kejiang Ye	Shenzhen Institute of Advanced Technology, Chinese Academy of Sciences, China
Lei Wang	National University of Defense Technology, China
Massimo Cafaro	University of Salento, Italy
Massimo Torquati	University of Pisa, Italy
Mengying Zhao	Shandong University, China
Roman Wyrzykowski	Czestochowa University of Technology, Poland
Rui Wang	Beihang University, China
Sheng Ma	National University of Defense Technology, China
Songwen Pei	University of Shanghai for Science and Technology, China
Susumu Matsumae	Saga University, Japan
Weihua Zhang	Fudan University, China
Weixing Ji	Beijing Institute of Technology, China
Xiaoli Gong	Nankai University, China
Youyou Lu	Tsinghua University, China
Yu Zhang	Huazhong University of Science and Technology, China
Zichen Xu	Nanchang University, China

Track 2: Software Systems and Programming Models

Patrick P. C. Lee (Chair)	Chinese University of Hong Kong, China
Erci Xu	Ohio State University, USA
Xiaolu Li	Huazhong University of Science and Technology, China
Shujie Han	Peking University, China
Mi Zhang	Institute of Computing Technology, Chinese Academy of Sciences, China

Jing Gong	KTH Royal Institute of Technology, Sweden
Radu Prodan	University of Klagenfurt, Austria
Wei Wang	Beijing Jiaotong University, China
Himansu Das	KIIT Deemed to be University, India
Rong Gu	Nanjing University, China
Yongkun Li	University of Science and Technology of China, China
Ladjel Bellatreche	National Engineering School for Mechanics and Aerotechnics, France

Track 3: Distributed and Network-Based Computing

Meng Shen (Chair)	Beijing Institute of Technology, China
Ruidong Li (Chair)	Kanazawa University, Japan
Bin Wu	Institute of Information Engineering, China
Chao Li	Beijing Jiaotong University, China
Chaokun Zhang	Tianjin University, China
Chuan Zhang	Beijing Institute of Technology, China
Chunpeng Ge	National University of Defense Technology, China
Fuliang Li	Northeastern University, China
Fuyuan Song	Nanjing University of Information Science and Technology, China
Gaopeng Gou	Institute of Information Engineering, China
Guangwu Hu	Shenzhen Institute of Information Technology, China
Guo Chen	Hunan University, China
Guozhu Meng	Chinese Academy of Sciences, China
Han Zhao	Shanghai Jiao Tong University, China
Hai Xue	University of Shanghai for Science and Technology, China
Haiping Huang	Nanjing University of Posts and Telecommunications, China
Hongwei Zhang	Tianjin University of Technology, China
Ioanna Kantzavelou	University of West Attica, Greece
Jiawen Kang	Guangdong University of Technology, China
Jie Li	Northeastern University, China
Jingwei Li	University of Electronic Science and Technology of China, China
Jinwen Xi	Beijing Zhongguancun Laboratory, China
Jun Liu	Tsinghua University, China

Kaiping Xue	University of Science and Technology of China, China
Laurent Lefevre	National Institute for Research in Digital Science and Technology, France
Lanju Kong	Shandong University, China
Lei Zhang	Henan University, China
Li Duan	Beijing Jiaotong University, China
Lin He	Tsinghua University, China
Lingling Wang	Qingdao University of Science and Technology, China
Lingjun Pu	Nankai University, China
Liu Yuling	Institute of Information Engineering, China
Meng Li	Hefei University of Technology, China
Minghui Xu	Shandong University, China
Minyu Feng	Southwest University, China
Ning Hu	Guangzhou University, China
Pengfei Liu	University of Electronic Science and Technology of China, China
Qi Li	Beijing University of Posts and Telecommunications, China
Qian Wang	Beijing University of Technology, China
Raymond Yep	University of Macau, China
Shaojing Fu	National University of Defense Technology, China
Shenglin Zhang	Nankai University, China
Shu Yang	Shenzhen University, China
Shuai Gao	Beijing Jiaotong University, China
Su Yao	Tsinghua University, China
Tao Yin	Beijing Zhongguancun Laboratory, China
Tingwen Liu	Institute of Information Engineering, China
Tong Wu	Beijing Institute of Technology, China
Wei Quan	Beijing Jiaotong University, China
Weihao Cui	Shanghai Jiao Tong University, China
Xiang Zhang	Nanjing University of Information Science and Technology, China
Xiangyu Kong	Dalian University of Technology, China
Xiangyun Tang	Minzu University of China, China
Xiaobo Ma	Xi'an Jiaotong University, China
Xiaofeng Hou	Shanghai Jiao Tong University, China
Xiaoyong Tang	Changsha University of Science and Technology, China
Xuezhou Ye	Dalian University of Technology, China
Yaoling Ding	Beijing Institute of Technology, China

Yi Zhao   Tsinghua University, China
Yifei Zhu   Shanghai Jiao Tong University, China
Yilei Xiao   Dalian University of Technology, China
Yiran Zhang   Beijing University of Posts and
                                 Telecommunications, China
Yizhi Zhou   Dalian University of Technology, China
Yongqian Sun   Nankai University, China
Yuchao Zhang   Beijing University of Posts and
                                 Telecommunications, China
Zhaoteng Yan   Institute of Information Engineering, China
Zhaoyan Shen   Shandong University, China
Zhen Ling   Southeast University, China
Zhiquan Liu   Jinan University, China
Zijun Li   Shanghai Jiao Tong University, China

Track 4: Big Data and Its Applications

Li Chen (Chair)   University of Louisiana at Lafayette, USA
Alfredo Cuzzocrea   University of Calabria, Italy
Heng Qi   Dalian University of Technology, China
Marc Frincu   Nottingham Trent University, UK
Mingwu Zhang   Hubei University of Technology, China
Qianhong Wu   Beihang University, China
Qiong Huang   South China Agricultural University, China
Rongxing Lu   University of New Brunswick, Canada
Shuo Yu   Dalian University of Technology, China
Weizhi Meng   Technical University of Denmark, Denmark
Wenbin Pei   Dalian University of Technology, China
Xiaoyi Tao   Dalian Maritime University, China
Xin Xie   Tianjin University, China
Yong Yu   Shaanxi Normal University, China
Yuan Cao   Ocean University of China, China
Zhiyang Li   Dalian Maritime University, China

Track 5: Parallel and Distributed Algorithms

Wei Bao (Chair)   University of Sydney, Australia
Jun Li (Chair)   City University of New York, USA
Dong Yuan   University of Sydney, Australia
Francesco Palmieri   University of Salerno, Italy

George Bosilca	University of Tennessee, USA
Humayun Kabir	Microsoft, USA
Jaya Prakash Champati	IMDEA Networks Institute, Spain
Peter Kropf	University of Neuchâtel, Switzerland
Pedro Soto	CUNY Graduate Center, USA
Wenjuan Li	Hong Kong Polytechnic University, China
Xiaojie Zhang	Hunan University of Technology and Business, China
Chuang Hu	Wuhan University, China

Track 6: Applications of Parallel and Distributed Computing

Hang Qiu (Chair)	Waymo, USA
Ang Li (Chair)	Qualcomm, USA
Daniel Andresen	Kansas State University, USA
Di Wu	University of Central Florida, USA
Fawad Ahmad	Rochester Institute of Technology, USA
Haonan Lu	University at Buffalo, USA
Silvio Barra	University of Naples Federico II, Italy
Weitian Tong	Georgia Southern University, USA
Xu Zhang	University of Exeter, UK
Yitao Hu	Tianjin University, China
Zhixin Zhao	Tianjin University, China

Track 7: Service Dependability and Security in Distributed and Parallel Systems

Wei Yang (Chair)	University of Texas at Dallas, USA
Dezhi Ran	Peking University, China
Hanlin Chen	Purdue University, USA
Jun Shao	Zhejiang Gongshang University, China
Jinguang Han	Southeast University, China
Mirazul Haque	University of Texas at Dallas, USA
Simin Chen	University of Texas at Dallas, USA
Wenyu Wang	University of Illinois at Urbana-Champaign, USA
Yitao Hu	Tianjin University, China
Yueming Wu	Nanyang Technological University, Singapore
Zhengkai Wu	University of Illinois at Urbana-Champaign, USA
Zhiqiang Li	University of Nebraska, USA
Zhixin Zhao	Tianjin University, China

Ze Zhang   University of Michigan/Cruise, USA
Ravishka Rathnasuriya   University of Texas at Dallas, USA

Track 8: Internet of Things and Cyber-Physical-Social Computing

Yu Yang (Chair)   Lehigh University, USA
Qun Song   Delft University of Technology, The Netherlands
Chenhan Xu   University at Buffalo, USA
Mahbubur Rahman   City University of New York, USA
Guang Wang   Florida State University, USA
Houcine Hassan   Universitat Politècnica de València, Spain
Hua Huang   UC Merced, USA
Junlong Zhou   Nanjing University of Science and Technology,
                                        China
Letian Zhang   Middle Tennessee State University, USA
Pengfei Wang   Dalian University of Technology, China
Philip Brown   University of Colorado Colorado Springs, USA
Roshan Ayyalasomayajula   University of California San Diego, USA
Shigeng Zhang   Central South University, China
Shuo Yu   Dalian University of Technology, China
Shuxin Zhong   Rutgers University, USA
Xiaoyang Xie   Meta, USA
Yi Ding   Massachusetts Institute of Technology, USA
Yin Zhang   University of Electronic Science and Technology
                                        of China, China
Yukun Yuan   University of Tennessee at Chattanooga, USA
Zhengxiong Li   University of Colorado Denver, USA
Zhihan Fang   Meta, USA
Zhou Qin   Rutgers University, USA
Zonghua Gu   Umeå University, Sweden
Geng Sun   Jilin University, China

Track 9: Performance Modeling and Evaluation

Zhibin Yu (Chair)   Shenzhen Institute of Advanced Technology,
                                        Chinese Academy of Sciences, China
Chao Li   Shanghai Jiao Tong University, China
Chuntao Jiang   Foshan University, China
Haozhe Wang   University of Exeter, UK
Laurence Muller   University of Greenwich, UK

Lei Liu   Beihang University, China
Lei Liu   Institute of Computing Technology, Chinese
                             Academy of Sciences, China
Jingwen Leng   Shanghai Jiao Tong University, China
Jordan Samhi   University of Luxembourg, Luxembourg
Sa Wang   Institute of Computing Technology, Chinese
                             Academy of Sciences, China
Shoaib Akram   Australian National University, Australia
Shuang Chen   Huawei, China
Tianyi Liu   Huawei, China
Vladimir Voevodin   Lomonosov Moscow State University, Russia
Xueqin Liang   Xidian University, China

Reviewers

Dezun Dong	Xiaolu Li
Chao Wang	Shujie Han
Chentao Wu	Mi Zhang
Chi Lin	Jing Gong
Deze Zeng	Radu Prodan
En Shao	Wei Wang
Fei Lei	Himansu Das
Haikun Liu	Rong Gu
Hailong Yang	Yongkun Li
Junlong Zhou	Ladjel Bellatreche
Kejiang Ye	Meng Shen
Lei Wang	Ruidong Li
Massimo Cafaro	Bin Wu
Massimo Torquati	Chao Li
Mengying Zhao	Chaokun Zhang
Roman Wyrzykowski	Chuan Zhang
Rui Wang	Chunpeng Ge
Sheng Ma	Fuliang Li
Songwen Pei	Fuyuan Song
Susumu Matsumae	Gaopeng Gou
Weihua Zhang	Guangwu Hu
Weixing Ji	Guo Chen
Xiaoli Gong	Guozhu Meng
Youyou Lu	Han Zhao
Yu Zhang	Hai Xue
Zichen Xu	Haiping Huang
Patrick P. C. Lee	Hongwei Zhang
Erci Xu	Ioanna Kantzavelou

Jiawen Kang

Jie Li

Jingwei Li

Jinwen Xi

Jun Liu

Kaiping Xue

Laurent Lefevre

Lanju Kong

Lei Zhang

Li Duan

Lin He

Lingling Wang

Lingjun Pu

Liu Yuling

Meng Li

Minghui Xu

Minyu Feng

Ning Hu

Pengfei Liu

Qi Li

Qian Wang

Raymond Yep

Shaojing Fu

Shenglin Zhang

Shu Yang

Shuai Gao

Su Yao

Tao Yin

Tingwen Liu

Tong Wu

Wei Quan

Weihao Cui

Xiang Zhang

Xiangyu Kong

Xiangyun Tang

Xiaobo Ma

Xiaofeng Hou

Xiaoyong Tang

Xuezhou Ye

Yaoling Ding

Yi Zhao

Yifei Zhu

Yilei Xiao

Yiran Zhang

Yizhi Zhou

Yongqian Sun

Yuchao Zhang

Zhaoteng Yan

Zhaoyan Shen

Zhen Ling

Zhiquan Liu

Zijun Li

Li Chen

Alfredo Cuzzocrea

Heng Qi

Marc Frincu

Mingwu Zhang

Qianhong Wu

Qiong Huang

Rongxing Lu

Shuo Yu

Weizhi Meng

Wenbin Pei

Xiaoyi Tao

Xin Xie

Yong Yu

Yuan Cao

Zhiyang Li

Wei Bao

Jun Li

Dong Yuan

Francesco Palmieri

George Bosilca

Humayun Kabir

Jaya Prakash Champati

Peter Kropf

Pedro Soto

Wenjuan Li

Xiaojie Zhang

Chuang Hu

Hang Qiu

Ang Li

Daniel Andresen

Di Wu

Fawad Ahmad

Haonan Lu

Silvio Barra

Weitian Tong

Xu Zhang

Yitao Hu

Zhixin Zhao
Wei Yang
Dezhi Ran
Hanlin Chen
Jun Shao
Jinguang Han
Mirazul Haque
Simin Chen
Wenyu Wang
Yitao Hu
Yueming Wu
Zhengkai Wu
Zhiqiang Li
Zhixin Zhao
Ze Zhang
Ravishka Rathnasuriya
Yu Yang
Qun Song
Chenhan Xu
Mahbubur Rahman
Guang Wang
Houcine Hassan
Hua Huang
Junlong Zhou
Letian Zhang
Pengfei Wang
Philip Brown
Roshan Ayyalasomayajula

Shigeng Zhang
Shuo Yu
Shuxin Zhong
Xiaoyang Xie
Yi Ding
Yin Zhang
Yukun Yuan
Zhengxiong Li
Zhihan Fang
Zhou Qin
Zonghua Gu
Geng Sun
Zhibin Yu
Chao Li
Chuntao Jiang
Haozhe Wang
Laurence Muller
Lei Liu
Lei Liu
Jingwen Leng
Jordan Samhi
Sa Wang
Shoaib Akram
Shuang Chen
Tianyi Liu
Vladimir Voevodin
Xueqin Liang

Contents – Part III

Addressing Coupled Constrained Reinforcement Learning via Interative Iteration Design

Wei Huang$^{(\boxtimes)}$ (ID) and Shichao Zhang$^{(\boxtimes)}$ (ID)

School of Computer Science and Engineering, Central South University,
Changsha 410083, China
{csu_hw,zhangsc}@csu.edu.cn

Abstract. Coupled constraints are a natural setting in many programming problems, such as edge computing, which makes agents more perplexed when updating policies. Existing primal-dual approaches often require additional conditions to be satisfied, and policy gradient methods easily fall into the inherent trade-off between maximizing task reward and satisfying constraints because the constraint information cannot be fed into the network. For such a setting, we design an Interactive Iteration (InIt) structure for policy gradient, which decouples high-dimensional coupled action space into multiple independent sub-action spaces to reduce the dimensionality of action space and the coupling of constraints. Then, we utilize policy gradient with strong convergence guarantees to sequentially solve the sub-actions and iteratively derive the optimal hybrid action to avoid the trade-off. Further, we propose a balancing loss for self-coupled actions, which enables the policy to pursue high task reward while complying with the objective constraints via the balancing feedback from the environment. Additionally, we conceive a notion of coupling compactibility to guide the decoupling of high-dimensional coupled action space to reduce model redundancy with minimal reward loss. Finally, we select two cases from the CCMDP scenario to evaluate our model, and the results demonstrate the superiority of our approach, especially for high-dimensional coupled action space.

Keywords: Coupled constraints · CMDP · Programming · Edge Computing · Policy gradient

1 Introduction

Many problems, especially in the real world, require that policies meet constraints. For instance, in edge computing, we want to strive for higher quality of service for individual tasks while also limiting the allocated resources to pursue overall performance [22]. Or we might want to minimize the energy usage of cooling data centers while ensuring that the temperature remains below some level [14]. Existing approaches typically formulate such constrained Reinforcement

© The Author(s), under exclusive license to Springer Nature Singapore Pte Ltd. 2024
Z. Tari et al. (Eds.): ICA3PP 2023, LNCS 14489, pp. 1–20, 2024.
https://doi.org/10.1007/978-981-97-0798-0_1

Learning (RL) problems as Constrained Markov Decision Processes (CMDP) [1,11,24]. In RL, most existing works mainly utilize primal-dual approaches to solve CMDP [3], which transforms the original problem into a dual problem and assumes that some conditions are satisfied. The objective is to maximize the expected total reward while satisfying a constraint on the expected total utility.

It is well known that the CMDP can be equivalently written as a programming [17,19]. However, to solve programming problems, we need to obtain the transition probabilities of the environment [19], which is often unavailable in realistic environment. RL characterizes that it utilizes the nonlinear mapping ability of neural networks to fit the transition probabilities of the environment and is capable to deal with continuous space [17,21]. For unconstrained Markov Decision Processes (MDP) problems, the agent can fit the transition probabilities of the environment well. For CMDP problems with simple constraints, the agent can only comprehend the boundaries of constraints by trial and error in the process of interacting with the environment. However, the agent appears less intelligent when handling CMDP problems that involve complex coupled constraints. The example in Fig. 1 illustrates this point well.

As shown in Fig. 1, at time epoch 1, three tasks are generated in the system, whose state is $S_n, n \in \{1,2,3\}$. The three tasks are uploaded to the server at the next time epoch. Subsequently, server $m \in \{1,2\}$ allocates computation resources $f_{n,m}$ to task n and enters the next state S'_n. $t(\cdot)$ is defined as the task processing delay function, and T_n is the maximum allowable delay associated with task n. The objective of the decision-maker is to minimize $t(y_n, f_{n,m})$ while satisfying $t(y_n, f_{n,m}) \leq T_n$. If only the task upload is taken into consideration, as shown by the blue dashed box, the delay constraint is denoted as $t(y_n) \leq T_n$. The objective of decision-maker is still to minimize $t(y_n)$. It is not difficult to observe that the action y_n is independent with each other, that is, adjusting y_n has no relation to minimizing $t(y_{\bar{n}}), \bar{n} \in \{1,2,3\}$ and $\bar{n} \neq n$. In other words, the presence of constraints does not conflict with the goal pursued by the agent, thus policy gradient methods can effectively handle such simple CMDPs. However, when task offloading is considered, the delay constraint is redefined as $t(y_n, f_{n,m}) \leq T_n$. Adjusting $f_{n,m}$ not affects the selection of y_n, but also the allocation of resources $f_{\bar{n},m}$. To clarify, the goal pursued by the agent may be prohibited by the constraints [21], which we refer to as the trade-off between the objective and the constraints. For CMDP with coupled action space like this, we refer to them as the Coupled Constraint Markov Decision Process (CCMDP), for which classic policy gradient methods are often helpless.

In fact, the majority of programming problems with coupled constraints are intrinsically unsolvable [3], and the earlier solutions primarily rely on the Lagrangian multiplier method [5,7,16]. Traditional Lagrangian multiplier methods sacrifice the search for the solving space to expedite decision-making, resulting in solutions that fall short of expectations. Existing RL-based approaches also inherit the primal-dual method from Lagrangian multiplier methods to solve CMDP [1,2,11,24]. These approaches all require some conditions to be satisfied, such as the Slater condition. Additionally, much of these works focus on multi-

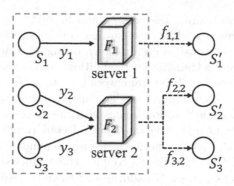

Fig. 1. A simple example of a CMDP with the coupled action space, where $S_1, S_2, S_3, S_1', S_2'$ and S_3' represent the instantaneous state. y_1, y_2 and y_3 indicate the selection decision. $f_{1,1}, f_{2,2}$ and $f_{3,2}$ express the resource allocation decision, and F_1 and F_2 depict the total resources of server 1 and server 2. Therefore, we have resource allocation constraints $f_{1,1} \leq F_1$ and $f_{2,2} + f_{3,2} \leq F_2$. In this example, there are three users who upload their tasks with the initial state $S_n, n \in \{1, 2, 3\}$ to the server, and then the server allocates resources to the tasks and enters the next state S_n'. Define $t(\cdot)$ as the task processing delay function and T_n as the maximum allowable delay associated with user n, and its corresponding constraint is $t(y_n, f_{n,m}) \leq T_n, m \in \{1, 2\}$. The objective is to reduce $t(y_n, f_{n,m})$ as much as possible.

objective and the non-linearity of constraints rather than the coupling. In this paper, we investigate the CCMDP, which pose two challenges:

(i) As the CCMDP can be considered a generalization of the CMDP, any difficulties existing in the CMDP will arise here too. For instance, in most reality environment, the preference function is usually non-linear. Ref [21] and Ref. [5] demonstrate that such optimization problems are non-convex;

(ii) Different from the CMDP, coupled constraints include inter-coupling and self-coupling between actions. These coupled constraints can be neither convex nor concave in the policy [24]. In particular, for the CCMDP with self-coupled constraints, we have proved that the Bellman optimality equation does not hold (see Lemma 2 for details). This presents a challenge for policy improvement and analysis.

Challenge (i) is inherent in the CMDP, and certain requirements on preference functions and constraints are inevitable. Both the traditional Lagrangian multiplier approaches [6,7,16] or the existing RL-based primal-dual approaches [2,8,24] require additional conditions to be satisfied. Although policy gradient methods can establish stronger convergence guarantees than action-value methods [21], which has contributed to the success of RL, their prerequisite is that the transition probabilities of the environment are known or easily fitted. However, constraint information in CMDP cannot be reflected in states and learned by neural networks, so agents can only indirectly perceive the boundaries of constraints through interactions with the environment. In high-dimensional action space with coupled constraints, the policy could be told to either focus entirely on maximizing task reward while disregarding constraint satisfaction, or focus entirely on satisfying the constraints (as shown in Fig. 1), leading to an inability to learn the environment's transition probabilities. To fix this deficiency, we propose an Interactive Iteration (InIt) framework to extend the existing policy gradient, which decouples high-dimensional action space with complex constraints into multiple low-dimensional action spaces with simple constraints. This design can exclude invalid regions that violate coupled constraints and compress the solving space. Although each sub-action space is independent and its constraints are also convex, the CCMDP with self-coupled constraints still does not satisfy the Bellman optimality equation. To address challenge (ii), we design a balancing loss for the self-coupled action network, which finds a single policy that achieves high task reward without violating the objective constraints. Specifically, the contributions of this paper are summarized as follows:

- We propose the InIt framework to extend the policy gradient method. InIt can reduce the dimensionality of action space and the coupling of constraints by decoupling high-dimensional coupled action space into multiple independent sub-action spaces. Since the constraints of each sub-action space are convex, it is easier to learn the transition probabilities of the environment. Further, it utilizes policy gradient with strong convergence guarantees to iteratively solve the sub-actions and derive the optimal hybrid actions to avoid the trade-off.
- We design a balancing loss for self-coupled constraints in the action network. Specifically, during interaction with the environment, we provide additional penalties for actions that exceed the objective constraint. Conversely, we provide differential reward for actions that do not meet the constraint. Meanwhile, the balancing loss is also incorporated into the backpropagation process to update self-coupled actions.
- Furthermore, we analyze the nature of action coupling, and conceive a notion, that is the coupling compactibility, to evaluate the degree of coupling correlation between two actions. Specifically, high coupling compactibility means that the decision-making of one action will significantly affect another action. Similarly, low coupling compactibility indicates that the decision-making of one action is less likely to affect another action. In other words, we can concatenate actions with low coupling compactibility to reduce redundancy introduced by the decoupling process with minimal reward loss.

2 Preliminaries

2.1 Coupled Constrained Markov Decision Process

A standard MDP is defined as $<\mathcal{S}, \mathcal{A}, \mathcal{P}, \mathcal{R}, \gamma, \mathcal{T}>$, where $\mathcal{S}$ is a state space, $\mathcal{A}$ is an action space, $\mathcal{P} = \{\mathcal{P}_t\}_{t=1}^T$ is a collection of transition kernels $\{\mathcal{P}_t\}$: $\mathcal{S} \times \mathcal{A} \to \Delta(\mathcal{S})$, $\mathcal{R} = \{r_t\}_{t=1}^T$ is a collection of reward functions $r_t : \mathcal{S} \times \mathcal{A} \to \mathbb{R}$, $\gamma \in [0, 1)$ is a discounted factor and $\mathcal{T} \in \mathbb{N}_+$ is the horizon. The agent interacts with MDP by performing its policy $\pi : \mathcal{S} \to \mathcal{A}$. The objective of the agent is to optimize its policy to maximize the expected discounted cumulative reward $\mathcal{J}(\pi) = \mathbb{E}_\pi[\sum_{t=0}^T \gamma^t r_t]$, where $s_0 \sim \mathcal{S}$ is the initial state. $a_t \sim \pi(s_t)$ is the action taken by the agent. $s_{t+1} \sim \mathcal{P}(s_{t+1}, r_t \mid s_t, a_t)$ is the next state. The state-action value function Q_π at the t-th step is defined as $Q_\pi = \mathbb{E}_\pi[\sum_{i=t}^T \gamma^t r_t \mid s_i = s, a_i = a]$.

Coupled constrained RL is typically formulated as CCMDP. Before defining CCMDP, we first introduce the definitions of coupling and self-coupling in detail. Please refer to Definition 1 and Definition 2, respectively.

Definition 1 (Coupling). *Given a function $f(x, y)$, and constraints c_x, c_y and $c(x, y)$, where c_x is the constraint associated only with variable x, c_y is the constraint associated only with variable y, and $c(x, y)$ is the constraint associated with the both. Due to the existence of $c(x, y)$, the value ranges of variables x and y are constrained by new constraints c'_x and c'_y, where $c'_x \subseteq c_x$ and $c'_y \subseteq c_y$. When variable x changes, c'_y changes with it. Similarly, when variable y changes, c'_x changes too. Then, we call $c(x, y)$ a coupled constraint. There is a coupling relationship between variables x and y.*

Definition 2 (Self-coupling). *In Definition 1, if $c_x = c_y$ and the range of c_x is equal to the range of $c(x, y)$, then $x = y$. x and y can be considered as the same variable acting on different objects, so we can combine then with a single vector variable $\boldsymbol{x}$. The specific definition is as follows: given a function $f(x_n)$ and constraints c_{x_n} and $c(\boldsymbol{x})$, $\boldsymbol{x} = [x_1, \cdots, x_N]$, where N is the number of variables. If the range of $c_{x_n}, \forall n \in \{1, \cdots, N\}$ is equal to the range of $c(\boldsymbol{x})$. Then, $\boldsymbol{x}$ is a self-coupling variable.*

Thus, the CCMDP is defined as $<\mathcal{S}, \mathcal{A}, \mathcal{P}, \mathcal{R}, \mathcal{B}, \gamma, \mathcal{C}, \mathcal{T}>$, where $\boldsymbol{a}_k = [a_{k_1}, \cdots, a_{k_N}] \in \mathcal{A}$ is an N-dimensional vector action that guides N objects, where $k \in \mathcal{K} = \{1, \cdots, K\}$ and K represents the dimensional size of action space. We denote the actions in CCMDP uniformly as bold $\boldsymbol{a}_k$. This is because the actions in most CCMDP problems is a vector actions that affect multiple objects. $\mathcal{B} = \{\boldsymbol{b}_k\}$ represents the balancing terms related to self-coupled actions. Specifically, if action k is a self-coupled action, then $\boldsymbol{b}_k = b_k$, otherwise $\boldsymbol{b}_k = \boldsymbol{0}$. $\mathcal{C} = \{c_k\}$ is the set of constraints. c_k represents the set of all constraints associated with the sub-action k. Thus, the action space affected by constraints is redefined as

$$<\mathcal{A}, \mathcal{C}> = \{<\boldsymbol{a}_k, c_k> \mid \boldsymbol{a}_k \in c_k\}. \tag{1}$$

We denote the action constraint pair as $<\boldsymbol{a}_k, c_k>$. Therefore, at time epoch t, the sub-action taken by the agent follows the policy $\boldsymbol{a}_{k,t} \sim \pi_k(s_t, c_k)$. It is worth noting that a particular constraint in c_k maybe also exist in $c_{\bar{k}}$, $k \in \mathcal{K}$ and $\bar{k} \neq k$. Indeed, this is quite common, as illustrated by action y and f in Fig. 1, both of which are associated with the delay constraint. Such constraints are referred to as coupled constraints. Having defined policy and value/action-value functions for the discounted CCMDP, the objective of the agent is to explore a policy that maximizes the expected task reward subject to coupled constraints.

$$P1 : \max_{\pi \in \mathcal{C}} \sum_{t=0}^{T} \mathcal{J}(\pi), \tag{2}$$

where $\pi \in \mathcal{C}$ denotes that the policy π is affected by the constraint set $\mathcal{C}$. Many existing works employ classic Lagrangian multiplier methods or RL-assisted primal-dual methods. However, both methods require the imposition of additional assumption conditions. In this paper, we first attempt to incorporate an InIt framework into the policy gradient method to guide the agent in searching for policies under the coupled constraints.

2.2   Policy Gradient

Most of the early works in RL are centered around the value-function approach, in which all function approximation effort goes into estimating a value function. The limitation is that the value-function method lies in its inability to capture discontinuous changes in actions and overlooks the fact that the optimal policy is often stochastic. Subsequently, the action-value method, known as the policy gradient, has been proposed, which has the advantage of being able to select actions without consulting a value function. Admittedly, a value function may still be used to learn the policy parameter, but is not required for action selection.

Although the policy gradient has worked well in many applications [4,13], it is not compatible with the coupled action space. This is because policy gradient can only perceive the boundaries of constraints through interaction with the environment as constraint information cannot be directly incorporated as inputs to the neural network. In high-dimensional coupled action space, classic policy gradients such as Multinomial distribution or Gaussian distribution cannot model the coupling among actions, resulting in the inherent trade-off between maximizing task reward and satisfying constraints. Implicit policies derived by action value functions, often adopted in value-based methods, also fail due to intractable maximization over high-dimensional coupled actions. For instance, in a 1-dimensional action space with simple constraints, action-value methods can implicitly understand constraints through interaction with the environment. However, for high-dimensional action space with non-convex coupled constraints, traditional action-value methods also struggle to learn it.

According to Theorem 1, $P1$ with hybrid action space is a non-convex problem. Therefore, we first leverage a relaxation approach to transform discrete actions into continuous actions, thereby converting the corresponding discrete non-convex constraints into continuous convex constraints, such as $a_k = \{0,1\} \rightarrow [0,1]$. Then, the CCMDP with high-dimensional hybrid action space is decoupled into multiple sub-CCMDPs, each of which has a single continuous action space. Further, programming problems with complex coupled constraints can be simplified for analysis. As Lemma 1 demonstrates, each sub-CCMDP problem is convex and thus easier to solve.

Theorem 1. *The CCMDP problem with non-zero hybrid action space is non-convex.*

Proof. Since the action space is hybrid, CCMDP problems have both discrete actions and continuous actions. Constraints related to discrete actions are non-convex, so CCMDP problems are not differentiable. Therefore, the CCMDP problem with hybrid action space is non-convex. ◻

Lemma 1. *By decoupling a CCMDP problem with non-zero high-dimensional continuous action space into multiple CCMDP problems with a single continuous action space, each sub-CCDMP problem is convex.*

Proof. For CCMDP with a single action $k, k \in \mathcal{K}$, $\boldsymbol{a}_k$ is bound by c_k that is a continuous convex constraint and each non-zero $\boldsymbol{a}_k$ is independent of each other. Thus, any sub-CCMDP problem with a single continuous action space is convex after decoupling the CCMDP. ◻

Additionally, by decoupling high-dimensional coupled action space, we can ensure the Bellman optimality equation holds in each sub-CCMDP with a single action as much as possible. In other words, if the constraints within a sub-CCMDP are not self-coupled, then the Bellman optimality equation holds in the sub-CCMDP; otherwise, it does not hold. For the detailed proof, please refer to Lemma 2.

Lemma 2. *For any CCMDP with a single non-zero action space, if the action is a self-coupled variable, the Bellman optimality equation does not hold in this CCMDP, otherwise the Bellman optimality equation holds.*

Proof. There are generally two Bellman optimal equations, namely $v^*(s) = \max_a \sum_{s'} \sum_r p(s', r|s, \boldsymbol{a})[r + \gamma v^*(s')]$ and $q^*(s, \boldsymbol{a}) = \sum_{s'} \sum_r p(s', r|s, \boldsymbol{a})[r + \gamma \max_{a'} q^*(s', \boldsymbol{a}')]$, both of which are equivalent. Thus, we prove the first equation here, and the second follows naturally.

Taking a non-zero action space k as an example, if $\boldsymbol{a}_k$ is a self-coupling variable, we have $\boldsymbol{a}_k = [a_{k_1}, \cdots, a_{k_N}]$, and constraints $a_{k_n} \in c_k, n \in \{1, \cdots, N\}$ and $\boldsymbol{a}_k \in c_k$. We assume that $\boldsymbol{a}_k^*$ is the optimal action to obtain $v^*(s) = \max_a \sum_{s'} \sum_r p(s', r|s, \boldsymbol{a}_k^*)[r + \gamma v^*(s')]$, where $\boldsymbol{a}_k^* = [a_{k_1}, \cdots, a_{k_N}]^*$. This means that the combined action is optimal and $\boldsymbol{a}_k^* \in c_k$. However, a_{k_n} is not necessarily constrained. For example, given a function $f(\boldsymbol{x}) = \sum_1^3 x_n$, and constraints $0 < x_n \le 1$ and $0 < \sum_1^3 x_n \le 1$. The objective is maximize $f(\boldsymbol{x})$. It is not difficult to observe that $x_1 = 1, x_2 = 0$ and $x_3 = 0$ are an optimal solution and it satisfy the constraint $0 < \sum_1^3 x_n \le 1$. Nonetheless, $[x_1, x_2, x_3]^* = [1, 0, 0]$ does not satisfy the constraint $0 < x_n \le 1$.

Further, if $\boldsymbol{a}_k$ is not a self-coupling variable, we have $\boldsymbol{a}_k = [a_{k_1}, \cdots, a_{k_N}]$ and constraint $a_{k_n} \in c_k$. Furthermore, we assume the optimal state value $v^*(s) = \sum_{s'} \sum_r p(s', r|s, \boldsymbol{a}_k^*)[r + \gamma v^*(s')]$, where $\boldsymbol{a}_k^* = [a_{k_1}, \cdots, a_{k_N}]^*$. Each a_{k_n} is the optimal action made under the premise of satisfying the constraint $a_{k_n} \in c_{k_n}$. Accordingly, in a CCMDP with a single non-zero action space, if the action is a self-coupling variable, the Bellman optimality equation does not hold in this CCMDP, otherwise the Bellman optimality equation holds.   □

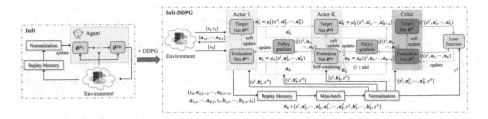

Fig. 2. The left figure depicts the overview of InIt, while the right figure illustrates the overall workflow of InIt incorporating DDPG (InIt-DDPG), including the balancing loss for self-coupled constraints.

3   InIt for Decoupling High-Dimensional Coupled Action Space

3.1   InIt Design

The policy-making for high-dimensional coupled action space should take the coupling between actions into account. Moreover, we need to account for the self-coupling of actions. To this end, we propose an InIt framework to decouple high-dimensional coupled action space. The intuition behind InIt is that actions in high-dimensional action space are coupled to each other, implying that we cannot make decisions on individual actions without considering the changes caused by altering other actions. In principle, our design is agnostic to specific algorithms and serves as an extension to existing policy gradient methods. In this paper, we demonstrate our approach by using the model-free RL algorithm, Deep Deterministic Policy Gradient (DDPG) that is easy to visualize [15].

The overview of InIt is illustrated on the left-hand side of Fig. 2. For the sake of clarity, we provide a detailed explanation by incorporating the easily visualizable DDPG. The overall workflow of InIt-DDPG is depicted on the right-hand side of Fig. 2. Initially, we decouple the coupled action space into K sub-action spaces. During the execution phase, the agent observes the environmental state s_0 and then keeps the action $a_{\bar{k},t}$, $k = 1$ unchanged while deciding the action $a_{1,t}$ iteratively according to $a_{1,t} = \mu_1(s_0, a_{2,t-1}, \cdots, a_{K,t-1})$, following the same procedure until $a_{K,t}$ is selected. Subsequently, the agent interacts with the environment based on the action $\mathcal{A}_t = \{a_{1,t}, \cdots, a_{K,t}\}$ to obtain the next state s_{t+1}, the reward r_t, and the balancing item $b_{k,t}$, generating a transition $\{s_0, a_{2,t-1}, \cdots, a_{K,t-1}, a_{1,t}, \cdots, a_{K,t}, r_t, b_{1,t}, \cdots, b_{K,t}, s_{t+1}\}$ stored in the replay buffer. The above process is iterated repeatedly until the solution converges. Unlike traditional policy gradient methods, the agent always starts from state s_0. This is because we are only concerned with input and solution in programming. Additionally, most programming problems are unsolvable [3], which means we cannot fit an optimal path. Moreover, due to the presence of the coupled constraints, it is also impossible to transition from one state to another with a single action while keeping other actions invariant.

As a policy gradient algorithm, the main idea of InIt-DDPG is to obtain an optimal policy π^* and learn the state-action function $Q_{\pi^*}(s_0, A^*)$, $A^* = (a_{1,t}, \cdots, a_{K,t})^*$ corresponding to π^*, which is carried out by adjusting the parameters of the evaluation and target networks for the actor and the critic until convergence. As shown on the right side of Fig. 2, the evaluation networks' parameters, θ^{μ_k} and θ^Q, are updated in real time. Specifically, a mini-batch of transitions with size N_b are randomly sampled from the replay buffer and inputted into the agent one by one. According to each inputted transition, the actor and the critic then update the parameters of the evaluation networks during the training stage.

Taking the i-th transition, $\{s^i, a_2^i, \cdots, a_K^i, a_1^{i\prime}, \cdots, a_K^{i\prime}, r^i, b_1^i, \cdots, b_K^i, s^{i\prime}\}$, as an example, Actor 1 formulates the decision $a_1 = \mu_1(s^i, a_{2,t-1}, \cdots, a_{K,t-1})$ based on s^i and the preceding actions of other actors. Subsequently, Actor 2 formulates the decision $a_2 = \mu_2(s^i, a_{1,t}, a_{3,t-1}, \cdots, a_{K,t-1})$ based on s^i, $a_{1,t}$ and the preceding actions of other actors. This process is repeated until action $a_K = \mu_2(s^i, a_{1,t}, \cdots, a_{K-1,t})$ is determined. Then, the critic evaluates all selected actions and update the evaluation network parameters θ^{μ_1} and θ^{μ_2}. Regarding the evaluation network parameters θ^Q in the critic, the agent first feeds the state and actions from the i-th transition separately into the state-action function of both the evaluation and target networks, obtaining $Q(s^i, a_1^i, \cdots, a_K^i)$ and $Q'(s^{i\prime}, a_1^{i\prime}, \cdots, a_K^{i\prime})$. Finally, the critic adjusts the evaluation network's parameters by minimizing the following loss.

$$L(\theta^Q) = \mathbb{E}_\pi[(Q(s^i, a_1^i, \cdots, a_K^i) - (r^i + Q'(s^{i\prime}, a_1^{i\prime}, \cdots, a_K^{i\prime})))^2]. \qquad (3)$$

Therefore, we can adjust the parameters θ^Q by computing the gradient of the loss function using Eq. (3). Furthermore, as the actor network makes action decisions based on each observation and the agent aims to maximize the cumulative reward, the evaluation network's parameters of the actor are updated by maximizing the policy objective function.

$$J(\theta^{\mu_k}) = \mathbb{E}_\pi[(Q(s^i, a_1^i, \cdots, a_K^i)|a_k = \mu_k(s^i, a_1, \cdots, a_{k-1}, a_{k+1}^i, \cdots, a_K^i)]. \quad (4)$$

where $\mu_k(\cdot)$ is the evaluation network's parameters of Actor k, which represents the deterministic policy $\pi_k : \mathcal{S} \times \mathcal{A} \mapsto \mathcal{A}$. Before decoupling high-dimensional coupled action space, we firstly relax the discrete actions. As each selection decision variable is relaxed to $[0, 1]$, the action space of the agent, $\mathcal{A}_k$, is continuous, and thus $\mu_k(\cdot)$ is a continuous function. Under this condition, it can be concluded that $J(\theta_k^\mu)$ is continuously differentiable according to [15], such that θ_k^μ can be adjusted in the direction of $\nabla_{\theta_k^\mu} J(\theta_k^\mu)$. However, for self-coupled actions, Eq. (4) is evidently insufficient to ensure that the network updates consistently move towards the global optimum. As demonstrated in Lemma 2, the Bellman optimality equation does not hold in the CCMDP with the self-coupled action space. This is because the policy may violate the objective constraint while adhering to the self-coupling constraint. To address this issue, we additionally propose a balancing loss function that enables partial network updates based on the feedback obtained from the interaction between self-coupled actions and the environment.

$$L_b(\boldsymbol{\theta}^{\mu_k}) = \frac{1}{N} \sum_{n=1}^{N} (|a_{k_n} - b_{k_n}|)^2, \tag{5}$$

where b_{k_n} is the balancing feedback obtained from interacting with the environment. Specifically, if a_{k_n} satisfies the constraints, then $(a_{k_n} - b_{k_n}) > 0$ denotes the surplus portion. On the other hand, if a_{k_n} fails to satisfy the constraint, then $(a_{k_n} - b_{k_n}) < 0$ represents the deficit portion. For instance, given a function $f(a_n) = \frac{\tilde{c}_n}{a_n}, n \in \{1, 2\}$, where $\tilde{c}_1 = 1$, $\tilde{c}_2 = 2$ and the objective of minimizing $\sum_n f(a_n)$. The corresponding constraints are $0 < a_n \leq 1.5$, $0 < \sum_n a_n \leq 1.5$ and $f(a_n) \leq 2$. At a certain time epoch, the agent makes a decision of $a_1 = 1$ and $a_2 = 0.5$. According to the constraint $f(a_n) \leq 2$, we have $a_1 - b_1 = 1 - 1/2 = 0.5$ and $a_2 - b_2 = 0.5 - 2/2 = -0.5$. Intuitively, new a_1 and new a_2 can be updated by $a_1 - 0.5$ and $a_2 - (-0.5)$. Therefore, through the design of the balancing loss, the agent can not only achieve higher reward returns but also avoid violation of the constraint. Specifically, the self-coupled action network updates the evaluation network's parameters by minimizing the following loss function.

$$L_{self} = \alpha L_b - (1 - \alpha)J(\boldsymbol{\theta}^{\mu_k}), \tag{6}$$

where α is the weight coefficient used to measure the importance of the balancing loss in the loss function, and correspondingly, $(1 - \alpha)$ represents the importance of the Q-network. With the real-time updated $\boldsymbol{\theta}^{\mu_k}$ and $\boldsymbol{\theta}^Q$, the parameters of the target networks, $\boldsymbol{\theta}^{\mu'_k}$ and $\boldsymbol{\theta}^{Q'}$, then can be softly updated as follows,

$$\boldsymbol{\theta}^{\mu'_k} = \xi^{\mu_k}\boldsymbol{\theta}^{\mu_k} + (1 - \xi^{\mu_k})\boldsymbol{\theta}^{\mu'_k}, \tag{7}$$

$$\boldsymbol{\theta}^{Q'} = \xi^Q\boldsymbol{\theta}^Q + (1 - \xi^Q)\boldsymbol{\theta}^{Q'}, \tag{8}$$

where $\xi^{\mu_k} \ll 1$ and $\xi^Q \ll 1$.

3.2   InIt-Based Solution

The overall workflow of InIt with DDPG is introduced in the previous section. According to the above discussion and Fig. 2, the proposed InIt-DDPG-based decoupling scheme for high-dimensional coupled action space can be summarized in Algorithm 1. We first randomly initialize the evaluation network and replay buffer, and then use the evaluation network to initialize the target network. Continuous time slots are grouped into different episodes, each consisting of T time epochs. To better describe the convergence performance, we use r_{ep} to represent the total reward obtained in each episode. For an episode starting from time epoch t_0, we have $r_{ep} = \sum_{t=t_0}^{T+t_0} r_t$.

Algorithm 1. InIt-DDPG

1: Initialize the actor's and the critic's evaluation networks with random weights θ^{μ_k}, for all $k \in \mathcal{K}$ and θ^Q.
2: Initialize target networks with weights $\theta^{\mu'_k} \leftarrow \theta^{\mu_k}$, for all $k \in \mathcal{K}$ and $\theta^{Q'} \leftarrow \theta^Q$.
3: Initialize replay buffer $\mathcal{R}$.
4: **for** $episode \leftarrow 1, \cdots, \infty$ **do**
5:     Receive initial observation state s_0, and set $r_{ep} = 0$.
6:     **for** $t \leftarrow 1, \cdots, T$ **do**
7:         Make decision $a_{k,t} = \mu_k(s_0, a_{1,t}, \cdots, a_{k-1,t}, a_{k+1,t-1}, \cdots, a_{K,t-1}) + \mathcal{N}_k(t)$ according to the policy π_k and exploration noise sequentially from $k = 1$ to $k = K$.
8:         Execute $\mathcal{A}_t = (a_{1,t}, \cdots, a_{K,t})$, receive reward r_t and observe balancing item $\mathcal{B}_t = (b_{1,t}, \cdots, b_{K,t})$ and new state s'_t.
9:         Store transition $\{s_0, a_{2,t-1}, \cdots, a_{K,t-1}, a_{1,t}, \cdots, a_{K,t}, r_t, b_{1,t}, \cdots, b_{K,t}, s'_t\}$ in $\mathcal{R}$.
10:        Sample a random minibatch of N_b transitions $\{s^i, a_2^i, \cdots, a_K^i, a_1^{i'}, \cdots, a_K^{i'}, r^i, b_1^i, \cdots, b_K^i, s^{i'}\}$
11:        Set $target = (r^i + Q'(s^{i'}, a_1^{i'}, \cdots, a_K^{i'}))$
12:        Update critic by minimizing the loss:
             $L(\theta^Q) = \frac{1}{N_b} \sum_i (Q(s^i, a_1^i, \cdots, a_K^i) - target)^2$
13:        Update policy π_k using the sampled policy gradient (and balancing loss):
             $\nabla_{\theta^{\mu_k}} J \approx \frac{1}{N_b} \sum_i \nabla_{a_k} Q(s^i, a_1, \cdots, a_K) \times \nabla_{\theta^{\mu_k}} \mu_k(s^i, a_1, \cdots, a_{k-1},$
             $a_{k+1}^i, \cdots, a_K^i)$ ▷ if a_k is not self-coupled
        or
             $\nabla_{\theta^{\mu_k}} J \approx \frac{1}{N_b} \sum_i \nabla_{a_k} (Q(s^i, a_1, \cdots, a_K) + L_b(\theta^{\mu_k})) \times \nabla_{\theta^{\mu_k}} \mu_k(s^i, a_1,$
             $\cdots, a_{k-1}, a_{k+1}^i, \cdots, a_K^i)$ ▷ if a_k is self-coupled
14:        Update actor's and critic's target networks based on Eq. (7) and Eq. (8).
15:        $r_{ep} = r_{ep} + r_t$
16:     **end for**
17: **end for**

3.3 Extension to High-Dimensional Action Space

High-dimensional action space is decoupled into multiple sub-action spaces to reduce the dimensionality of action space and the coupling of constraints, and then sequentially decides the optimal sub-policies to derive the optimal hybrid policy. However, the InIt framework can introduces redundancy to the network when the dimensionality is high. The most straightforward approach to address this issue is to combine multiple action into a single policy network for output, which makes it difficult to guarantee high reward when it does not satisfy the Bellman optimality equation. To tackle this issue, we further conceive a notion, that is the coupling compactibility, to guide action merging by utilizing the coupling relationship between actions. Specifically, actions with low coupling compactibility can be concatenated together, while actions with high coupling compactibility should be decided separately.

To measure the degree of coupling correlation between actions, we first define the coupling compactibility as follows.

Definition 3 (Coupling Compactibility).   *According to Definition 1, the value ranges of variables x and y ($y \neq x$) are constrained by the new constraints c'_x and c'_y due to coupling constraint $c(x, y)$, where $c'_x \subseteq c_x$ and $c'_y \subseteq c_y$. In this case, we refer to it as the present constraint. If variable x is increased by Δ_x, c'_y is increased or decreased by $\Delta_{c'_y}$ accordingly. In particular, $\Delta_{c'_y}$ represents the variation of the present constraint on y, which is given by*

$$\Delta_{c'_y} = \frac{||c''_y| - |c'_y||}{|c_y|}, \tag{9}$$

where $c''_y \in c_y$ denotes the new constraint of y after the variable x is increased by Δ_x. $|\cdot|$ indicates the interval size of the constraint. Specifically, if y is continuous, then $|\cdot|$ is the interval length; if y is discrete, then $|\cdot|$ is the number of the constraint set. Therefore, the varaible coupling between variables x and y is denoted as

$$\cos\theta = \frac{\Delta_x \Delta_{c'_y} + \Delta_y \Delta_{c'_x}}{\Delta_x \Delta_y}, \tag{10}$$

where θ represents teh coupling angle between x and y.

According to Definition 3, it can be inferred that $\cos\theta \geq 0$. If two variables are not coupled to each other, then $\cos\theta = 0$ because $\Delta_{c'_x} = 0$ and $\Delta_{c'_y} = 0$; conversely, $\cos\theta > 0$ because $\Delta_{c'_x} > 0$ or $\Delta_{c'_y} > 0$. Specifically, in real-world scenarios, the constraint range $\Delta_{c'_x}$ of variable x is sometimes insensitive to small variations Δ_{c_y} in its coupled variables. Consequently, we can appropriately amplify the value of Δ_{c_y} to clearly observe $\Delta_{c'_x}$. For a single variable with multiple coupling constraints (i.e., multiple objectives), we can take the maximum change value. Therefore, in CCMDP with high-dimensional action space, we can merge actions with low $\cos\theta$ together and separate actions with high $\cos\theta$ based on the degree of coupling correlation between actions. This metric allows us to significantly reduce model redundancy with minimal reward loss.

4   Experiments

We select two cases from the CCMDP scenarios to evaluate InIt and compare it against existing representative algorithms. Subsequently, we conduct detailed ablation experiments to verify the contribution of each component in InIt.

4.1   Experiment Setups

We consider a class of programming problems characterized by high coupling relationship among actions, which is the task offloading in edge computing. Based on most existing research [7,12,26], we design two task offloading scenarios. Scenario (1) represents a simple dual-action CCMDP, while Scenario (2) considers a more complex cooperative offloading scenario. Due to space limitations, we refrain from extensively introducing the functions related to computation and transmission in task offloading. These repetitive works can be found in most task offloading literature [7,12,18], with references [12,18] being important references for this paper. The details of the two scenarios are described as follows:

Scenario (1): We consider a square area of 3km × 3km, where 6 edge servers and 40 edge devices are uniformly deployed within the area [12,18]. The servers' computing resources are 5GHz. Each device has a computational task assigned to it. The input data size, required computing resources, and maximum task processing delay are uniformly distributed in the range of $U[10, 640]$ KB, $U[0.2, 2]$ GHz, and $U[0.5, 2]$ s, respectively. In this scenario, similar to Fig. 1, we only consider two actions: selection decision and computation resource allocation. The selection decision refers to which server the device's task is offloaded to, thus being a binary variable, while the computation resource allocation is a self-coupled action. Their constraints are $\{1, 2, 3, 4, 5, 6\}$ and $[0, 5]$ GHz, respectively. Moreover, both variables are subject to the delay constraint, hence they are coupled with each other. The objective is to minimize the task processing delay.

Scenario (2): We consider a more promising collaborative offloading scenario, namely vehicle-assisted edge computing, situated within a square area of approximately 18 km × 18 km located within the Fourth Ring Road in Beijing. Compared to existing UAV-assisted schemes [22,23], vehicle-assisted schemes offer greater cost-effectiveness [12]. As shown in Fig. 3, we utilize the publicly available taxi dataset T-Drive [25] to plot the one-minute trajectory map of February 3rd, 2008, at 9:30 AM. The red dots represent trajectory points, and the blue lines represent trajectories. It can be observed that within one minute, taxis have already covered the entire area within the Fourth Ring Road, further supporting the feasibility of vehicle-assisted edge computing. In this scenario, there are 8 edge servers and 100 mobile devices uniformly deployed within the area. Due to their simple hardware structure, these edge devices often have limited computation and communication capabilities, which are insufficient to access servers and accomplish their own tasks [12]. Therefore, we consider vehicles as relay to assist in task offloading for IoT devices, where vehicles can offload tasks themselves or further upload tasks to the servers [12]. Considering the limited resources of vehicles, it can provide computation resources of 1 GHz.

Compared to Scenario (1), task divisibility and device energy optimization are additionally considered in Scenario (2). Therefore, in the scenario, there are five optimization variables: selection decision (including vehicle offloading decision and server selection decision), computation resource allocation, bandwidth resource allocation, offloading ratio and transmission power of tasks, with respective constraints denoted as $U\{1, 2, 3, 4, 5, 6, 7, 8\}$, $U[0, 5]$ GHz and $U[0, 3]$ MHz, $U[0, 1]$ and $U[0, 37]$ dBm. The input data size, required computation resources, local computation resources, maximum task processing delay, and maximum permissible energy consumption are uniformly distributed as $U[25, 1e3]$ KB, $U[0.4, 3]$ GHz, $U[0.1, 1]$ GHz, $U[0.5, 2.5]$ s, and $U[0.01, 1.8]$ J, respectively. Unlike Scenario (1), the objective is to minimize task processing delay and device energy consumption, making it a multi-objective optimization problem.

Our code is implemented using Python 3.7.9 and Torch 1.11.0. All experiments are conducted on an NVIDIA GeForce GTX 3090 GPU. Each individual training process took approximately 2 to 18 h, depending on the specific algorithm and environment. For the detailed information about the network, please refer to the next section.

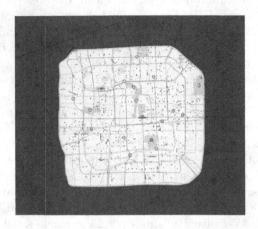

Fig. 3. Waypoint Map of Beijing Taxi for 1 min (09:30–09:31) on February 3, 2008.

Table 1. Network structures for the actor network and the critic network (Q-Network or V-Network), where $(+\mathbb{R}^x)$ in the actor network indicates that the InIt framework is used and $(+\mathbb{R}^x)$ in the critic network represents Q-network, and $|\mathcal{X}_k|$ denotes the size of the action space k.

Layer	Actor Network ($\pi(\cdot)$)	Critic Network ($Q(s, \boldsymbol{a}_1, \cdots, \boldsymbol{a}_K)$ or $(V(s))$
Fully Connected	(state dim $(+\mathbb{R}^{\sum_{\bar{k}} |x_{\bar{k}}|})$, 256)	(state dim $(+\mathbb{R}^{\sum_k |x_k|})$, 256)
Activation	ReLU	ReLU
Fully Connected	(256, 128)	(256, 128)
Activation	ReLU	ReLU
Fully Connected	$(128, \mathbb{R}^{|\mathcal{X}_k|})$	(128, 1)
Activation	Softmax or Tanh	None

4.2 Baselines

The InIt framework is initially proposed to address the issue of action coupling in programming problems that existing policy gradient methods fail to learn. We introduce four popular policy gradient methods, namely DDPG [15], Twin Delayed Deep Deterministic Policy Gradient (TD3) [9], Soft Actor-Critic (SAC)

[10], and Proximal Policy Optimization (PPO) [20], to design experiments. Algorithms utilizing the InIt framework are identified with the prefix "InIt-". The suffix "-b" indicates whether a balancing loss is applied to address self-coupled actions. Since action space in CCMDP is typically hybrid and high-dimensional, to ensure fairness, we decouple the original action space in the base policy gradient algorithms but do not apply InIt further. Furthermore, as the CCMDP can be viewed as the programming, we also introduce two state-of-the-art algorithms as follows:

1) The classical Lagrangian multiplier method [7,12,22], hereafter referred to as Lagr, is characterized by its advantage not requiring training, which takes the input directly and seeks solutions accordingly. However, it suffers from two main issues. Firstly, task reward is limited because the algorithm simultaneously addresses the problem understanding and decision-making, particularly in high-dimensional action space. Secondly, the algorithm's convergence is contingent upon imposing additional conditions.

2) The primal-dual method combined with deep reinforcement learning [2,8,11], hereafter referred to as PD-DRL, shares similarities with the Lagrangian approach but incorporates neural networks to approximate the objective function. However, similar to Lagr, additional conditions are typically imposed to ensure convergence. Furthermore, these methods are predominantly employed for addressing CMDP problems at a macro level, with limited attention given to the coupling constraints within CMDP.

The network architectures used in this study are presented in Table 1. For all algorithms, we employ a two-layer feedforward neural network for both the actor and critic networks, with 256 and 128 hidden units, respectively, and ReLU activation (excluding the output layer). The learning rate for the actor network is set to 2.5e−4, while the learning rate for the critic network is set to 3e-4. The critic represents the Q-network. Specifically, the critic network for PPO also serves as the value network, i.e., V-network. The batch size is set to 16. In policy gradient methods, the maximum number of interactions between the agent and environment is set to 20, and models are trained for 5e5 iterations. The parameter α in the balancing loss is set to 0.7. On the other hand, Lagr and PD-DRL do not engage in interactions with the environment but rather receive inputs for solving. The maximum number of iterations for the Lagr and PD-DRL cycles is set to 1e3 and 1e4, respectively.

4.3   Performance Evaluation

Reward and Loss. In scenario (1), we conduct multiple experiments to evaluate the convergence of reward and loss among different models. As shown in Fig. 4(a), it can be observed that the policy gradient method with the InIt framework consistently outperforms the original policy gradient method. This is because, the InIt framework aims to reduce the dimensionality of action space and the coupling of constraints via decoupling high-dimensional coupled action space,

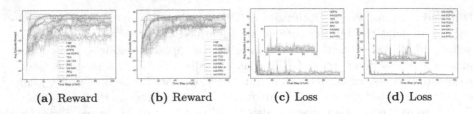

(a) Reward   (b) Reward   (c) Loss   (d) Loss

Fig. 4. Comparisons of algorithms in Scenario (1). The x- and y-axis denote the environment steps and average reward over recent 1e4 episodes in (a) and (b). The x- and y-axis denote the environment steps and average critic loss over recent 1e4 episodes in (c) and (d).

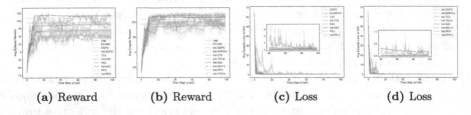

(a) Reward   (b) Reward   (c) Loss   (d) Loss

Fig. 5. Comparisons of algorithms in Scenario (2). The description of the x- and y-axis is the same as that of Fig. 4.

thereby maximizing the task reward over time. Furthermore, approaches based on DDPG and TD3 perform worse compared to those based on PPO and SAC. This is because deterministic policy gradient methods (DDPG and TD3) tend to get trapped in local optimum during global exploration due to their reliance on the optimal action at the next time step. On the other hand, the SAC-based approaches are more stable than the PPO-based approaches, as it utilizes past transitions for exploration while PPO only utilizes data from the previous episode. Additionally, similar to InIt-PPO and InIt-SAC, Lagr and PD-DRL achieve higher reward due to their low-dimensional action spaces. Similarly, it can be observed in Fig. 4(b) that even with the application of the balancing loss to self-coupled actions, the approaches do not show significant improvement. This is because the design aims to allocate resources more unbiased to find a single policy that achieves high task reward without violating the objective constraints. A small number of users allows the server to have sufficient resources for allocation, while a low dimensionality of action space implies a relatively simpler solving space.

In Fig. 4(c) and Fig. 4(d), it can be observed that the loss consistently decreases rapidly. There are primarily two reasons for this phenomenon. Firstly, unlike sequence learning and feature learning, both of which are supervised learning tasks with fixed labels. The loss can be regarded as the distance between predicted values and ground truth. On the other hand, RL is an unsupervised learning paradigm where the next optimal state-action value is used to guide the progression of the current state and provide appropriate reward.

Table 2. Comparisons of algorithms with different α in balancing loss. Each experiment is repeated three times in Scenario (2), and the highest value is taken to calculate the mean and error.

	$\alpha = 0.0$	$\alpha = 0.4$	$\alpha = 0.6$	$\alpha = 0.7$	$\alpha = 1.0$
InIt-DDPG-b	102.77 ± 1.95	108.91 ± 1.21	114.41 ± 8.67	$\mathbf{116.04 \pm 4.69}$	103.49 ± 7.78
InIt-TD3-b	94.92 ± 1.19	99.17 ± 1.94	$\mathbf{109.17 \pm 4.47}$	105.95 ± 1.42	90.41 ± 5.04
InIt-SAC-b	118.19 ± 0.74	122.25 ± 3.41	119.49 ± 5.64	$\mathbf{123.65 \pm 1.07}$	115.68 ± 1.43
InIt-PPO-b	128.33 ± 6.43	129.15 ± 4.39	133.11 ± 4.86	$\mathbf{133.52 \pm 5.85}$	123.39 ± 4.15

However, the next state-action value is also estimated by the network and the reward are fixed, which may cause the network to converge towards the vicinity of the reward even if it performs poorly. Taking DDPG as an example, $criticLoss = \sum_i q - (r + \gamma * q\_ * done)$, where q and $q\_$ represent the current and the next state-action value, respectively. $done$ represents game progress. It is evident from this equation that γ is a constant, and r represents the fixed reward obtained from the environment. The major factor affecting $criticLoss$ is $(q - q\_)$ when $done$ is a non-zero value. Furthermore, in contrast to controlling scenarios, optimization problems have a well-defined objective function that is often unsolvable [3]. We can only violate constraints to set an unattainable upper bound and further approximate done as $[0, 1]$ to evaluate the progress of the environment reaching the terminal state. This explains why the model consistently achieves high reward and low loss rapidly.

Accordingly, the same experiments are conducted in Scenario (2). However, there are notable differences in the performance between the PPO and SAC methods with the InIt framework compared to Lagr and PD-DRL. Specifically, the rewards for Lagr and PD-DRL are 93.6 and 108.5, respectively. The average maximum rewards (averaged over three experiments) for InIt-PPO and InIt-SAC are 128.33 and 118.19, respectively, representing improvements of 18.27% and 8.93% compared to PD-DRL. In Fig. 5(b), the average maximum rewards for InIt-PPO-P and InIt-SAC-P are 133.52 and 123.65, respectively, indicating improvements of 23.0% and 13.96% compared to PD-DRL. This is because as the dimensionality of the coupled action space increases, the gap arising from the need for additional conditions to ensure model convergence also grows larger. Similar to Fig. 4, Fig. 5(c) and Fig. 5(d) show that the loss for algorithms based on PPO and DDPG is relatively higher and more unstable than DDPG and TD3, which is reflected in the reward experiments as well. DDPG is prone to converge to local optimum due to its deterministic policy gradient, while PPO consistently explores new policies, resulting in poorer stability.

Balancing Loss. Next, extensive ablation experiments are conducted in Scenario (2) to evaluate the impact of balancing loss (as shown in Eq. 6). Firstly, the reward can be observed to assess the influence of different α values on model performance, as presented in Table 2. It can be summarized that initially, as α increases, the reward obtained by all approaches also increases. When α is around 0.7, the achieved reward reaches their maximum value and then starts to decline. This is because the Bellman optimality equation does not hold in the CCMDP with the self-coupled action space, so updating actions solely based on the state-action value function is insufficient. By incorporating a balancing loss, the agent can be guided to find a policy that achieves high task reward while adhering to the constraints by taking into consideration the real objective constraints. However, if we only utilize the balancing loss to learn action selection ($\alpha = 1$), the evaluation of the state-action value becomes blurred because the balancing feedback then becomes random noise. In conclusion, the balancing loss is designed to guide the agent to pursue high task reward without violating the objective constraints.

Furthermore, to carefully examine whether the self-coupled action f is effectively balanced when applying balancing loss, server load balancing experiments are conducted in two scenarios. As illustrated in Fig. 6 higher server utilization rate indicates a heavier load on the server, while a variance can be used to evaluate load balancing [7]. Lower variance indicates a more balanced load allocation among servers. In Fig. 6(a), the variances obtained by each approach are 0.017, 0.014, 0.038, 0.008, 0.016, and 0.004, respectively. It can be observed that InIt-DDPG achieves the lowest utilization variance since Server 0 carries an excessive load while Server 1 remains idle. With the employing of the balancing loss, the servers achieve a fairer load allocation.

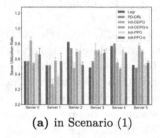 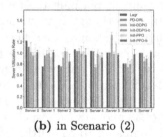

(a) in Scenario (1)   (b) in Scenario (2)

Fig. 6. Comparisons of load balancing under six algorithms. The x-axis denotes servers, and the y-axis denotes the server utilization rate.

Similarly, in Fig. 6(b), the variances obtained by each approach are 0.024, 0.013, 0.032, 0.005, 0.014, and 0.003, respectively. The deterministic policy InIt-DDPG still exhibits the most unfair load allocation. Furthermore, in the experiments depicted in Fig. 4 and Fig. 5, InIt-DDPG achieves also the lowest reward among the six approaches. This is because deterministic policy gradient methods tend to get trapped in local optimum due to the lack of exploration in optimization problems. Moreover, the task reward is determined by hybrid actions rather than a single action, which leads to InIt-DDPG-b having lower reward even after employing the balancing loss compared to Lagr and PD-DRL. In conclusion, the design of the balancing loss guides the agent to improve load allocation, resulting in finding a policy that achieves high task reward without violating the constraints.

5   Conclusion and Future Work

In this paper, we propose an InIt framework to address the inability of policy gradient to handle CCMDPs. The intuition behind InIt is that policy gradient are unable to incorporate constraint information into the network and thus rely on the interactions with the environment to perceive the boundaries of constraints. Consequently, high-dimensional coupled action space disable the agent to find the policy that achieves high task reward without violating the constraints. InIt reduces the space dimensionality and the constraint complexity by decoupling high-dimensional coupled action space into multiple independent sub-action spaces. It then utilizes the strong convergence properties of policy gradient to iteratively solve for sub-actions and derive optimal hybrid actions to avoid the trade-off. Additionally, we propose a balancing loss for self-coupled actions to balance self-coupled actions, which pursues high task reward while adhering to the constraints through feedback from interactions with the environment.

In future research, we will focus on the notion of coupling compactibility, which can be used to guide the significant reduction of model redundancy with minimal reward loss.

Acknowledgements. This work was supported by the National Natural Science Foundation of China (61836016).

References

1. Altman, E.: Constrained Markov Decision Processes, vol. 7. CRC Press, Boca Raton (1999)
2. Bai, Q., Bedi, A.S., Agarwal, M., Koppel, A., Aggarwal, V.: Achieving zero constraint violation for constrained reinforcement learning via primal-dual approach. In: Proceedings of the AAAI Conference on Artificial Intelligence, vol. 36, pp. 3682–3689 (2022)
3. Bazaraa, M.S., Sherali, H.D., Shetty, C.M.: Nonlinear Programming: Theory and Algorithms. Wiley, Hoboken (2013)
4. Berner, C., et al.: Dota 2 with large scale deep reinforcement learning. arXiv preprint arXiv:1912.06680 (2019)
5. Boyd, S.P., Vandenberghe, L.: Convex Optimization. Cambridge University Press, Cambridge (2004)
6. Cheng, Z., Liao, B.: QoS-aware hybrid beamforming and DOA estimation in multi-carrier dual-function radar-communication systems. IEEE J. Sel. Areas Commun. **40**(6), 1890–1905 (2022)
7. Dai, Y., Xu, D., Maharjan, S., Zhang, Y.: Joint load balancing and offloading in vehicular edge computing and networks. IEEE Internet Things J. **6**(3), 4377–4387 (2018)

8. Ding, D., Zhang, K., Basar, T., Jovanovic, M.: Natural policy gradient primal-dual method for constrained Markov decision processes. Adv. Neural. Inf. Process. Syst. **33**, 8378–8390 (2020)
9. Fujimoto, S., Hoof, H., Meger, D.: Addressing function approximation error in actor-critic methods. In: International Conference on Machine Learning, pp. 1587–1596. PMLR (2018)
10. Haarnoja, T., Zhou, A., Abbeel, P., Levine, S.: Soft actor-critic: off-policy maximum entropy deep reinforcement learning with a stochastic actor. In: International Conference on Machine Learning, pp. 1861–1870. PMLR (2018)
11. Huang, S., et al.: A constrained multi-objective reinforcement learning framework. In: Conference on Robot Learning, pp. 883–893. PMLR (2022)
12. Huang, W., Zeng, Z., Xiong, N.N., Mumtaz, S.: JOET: sustainable vehicle-assisted edge computing for IoT devices. J. Syst. Architect. **131**, 102686 (2022)
13. Kober, J., Bagnell, J.A., Peters, J.: Reinforcement learning in robotics: a survey. Int. J. Robot. Res. **32**(11), 1238–1274 (2013)
14. Lazic, N., et al.: Data center cooling using model-predictive control. Adv. Neural. Inf. Process. Syst. **31** (2018)
15. Lillicrap, T.P., et al.: Continuous control with deep reinforcement learning. arXiv preprint arXiv:1509.02971 (2015)
16. Ma, H., Huang, P., Zhou, Z., Zhang, X., Chen, X.: GreenEdge: joint green energy scheduling and dynamic task offloading in multi-tier edge computing systems. IEEE Trans. Veh. Technol. **71**(4), 4322–4335 (2022)
17. Miryoosefi, S., Jin, C.: A simple reward-free approach to constrained reinforcement learning. In: International Conference on Machine Learning, pp. 15666–15698. PMLR (2022)
18. Ndikumana, A., et al.: Joint communication, computation, caching, and control in big data multi-access edge computing. IEEE Trans. Mob. Comput. **19**(6), 1359–1374 (2019)
19. Sadamoto, T.: On equivalence of data informativity for identification and data-driven control of partially observable systems. IEEE Trans. Autom. Control (2022)
20. Schulman, J., Wolski, F., Dhariwal, P., Radford, A., Klimov, O.: Proximal policy optimization algorithms. arXiv preprint arXiv:1707.06347 (2017)
21. Sutton, R.S., Barto, A.G.: Reinforcement Learning: An Introduction. MIT Press, Cambridge (2018)
22. Tan, T., Zhao, M., Zeng, Z.: Joint offloading and resource allocation based on UAV-assisted mobile edge computing. ACM Trans. Sens. Netw. (TOSN) **18**(3), 1–21 (2022)
23. Wang, L., Wang, K., Pan, C., Xu, W., Aslam, N., Nallanathan, A.: Deep reinforcement learning based dynamic trajectory control for UAV-assisted mobile edge computing. IEEE Trans. Mob. Comput. **21**(10), 3536–3550 (2021)
24. Wu, R., Zhang, Y., Yang, Z., Wang, Z.: Offline constrained multi-objective reinforcement learning via pessimistic dual value iteration. Adv. Neural. Inf. Process. Syst. **34**, 25439–25451 (2021)
25. Yuan, J., Zheng, Y., Xie, X., Sun, G.: Driving with knowledge from the physical world. In: Proceedings of the 17th ACM SIGKDD International Conference on Knowledge Discovery and Data Mining, pp. 316–324 (2011)
26. Zhang, T., Xu, Y., Loo, J., Yang, D., Xiao, L.: Joint computation and communication design for UAV-assisted mobile edge computing in IoT. IEEE Trans. Industr. Inf. **16**(8), 5505–5516 (2019)

A Path Planning and Obstacle Avoidance Method for USV Based on Dynamic-Target APF Algorithm in Edge

Di Wang[1], Haiming Chen[1,2](✉) [ID], and Cangchen Wu[1]

[1] Faculty of Electrical Engineering and Computer Science, Ningbo University, Ningbo 315211, Zhejiang, China
chenhaiming@nbu.edu.cn
[2] Zhejiang Key Laboratory of Mobile Network Application Technology, Ningbo 315211, Zhejiang, China

Abstract. Unmanned Surface Vessel (USV) has been widely used in various fields due to its autonomous advantages, and path planning is a crucial technology for autonomy. However, using global path planning alone cannot avoid moving obstacles, while using local path planning alone may lead to falling into local minima and fail to reach the target. Therefore, this paper proposed the Dynamic Target Artificial Potential Field (DTAPF) algorithm which use a dynamic point that follows the global path generated by the A* algorithm as the target point of the Artificial Potential Field (APF). In addition, in order to improve the safety of USV navigation and response time of the traditional centralized path planning methods, we proposed an edge computing architecture for global path planning and an Offset Guidance method to avoid moving obstacles while confirming to the Collision Regulation (CORLEGs) for navigation safety. The experimental results show that, using the method proposed in this paper, USV can reach the target in an environment with moving obstacles with high probability (about 99.4%), and compared to traditional APF algorithm, our method can reduce collision probability by 71% with almost no increase in average path length and average navigation time. Besides, our architecture has much lower computing delay than local computing, and also lower than cloud computing.

Keywords: USV · path planning · edge computing · COLREGs

1 Introduction

In the past decade, research on Unmanned Ground Vehicle (UGV), Unmanned Surface Vessel (USV), and Unmanned Aerial Vehicle (UAV) has been steadily increasing [1]. USV is defined as unmanned, autonomous or remotely controlled vehicle that navigate on the surface of the water. With the advancement of artificial intelligence and sensor technologies, the applications of USV have been increasingly widespread, including scientific research, ocean exploration, marine rescue, and military operations [2]. These missions require technologies such as target recognition, path planning, control, localization and navigation, and communication. Given the limited energy carried by USV [3] and the complex requirements in terms of navigation paths and arrival times, path planning and collision avoidance are crucial for USV, which require intelligent navigation and collision avoidance capabilities.

Path planning is generally divided into global path planning and local path planning based on known environmental information. Most of global path planning algorithms such as Dijkstra [4] and A* algorithm [5], which use electronic maps or satellite maps to obtain global environmental information and complete path planning before the USV departs. However, such algorithms exhibit significant computing delays, resulting in low real-time performance. Additionally, they typically only account for static environmental factors, failing to consider dynamic obstacles that may appear and result in collisions between USV and obstacles. Therefore, using global path planning algorithms alone may not always guarantee safe arrival of USV at the target point in environments with dynamic obstacles. Another type is local path planning algorithms, such as Artificial Potential Field (APF) [6] and Dynamic Window Approach (DWA) [7]. These algorithms utilize sensors to gather information on the immediate surroundings of the USV and accordingly plan its path. Despite their advantages, including high real-time performance and short computing delays, they only consider the local environment. Consequently, these algorithms may fail to obtain the optimal path and encounter issues such as local minima and unreachable target. To overcome these limitations, it is crucial to combine global path planning with local path planning, especially in environments with dynamic obstacles, to ensure the safe and efficient navigation of USV.

Currently, most path planning problems are solved using centralized methods [8], which involve either local computing or processing all the path planning data and requests in the cloud. However, USV have limited computing performance, and performing large-scale global path planning tasks can result in high computing delays. If cloud computing is used, there can be high transmission delays between the USV and the cloud, and the cloud can also experience high waiting delays when processing large amounts of computing tasks. Edge computing is a new computing paradigm that offloads computing and storage resources from the cloud to the network edge closer to the end user, and its computing performance is not constrained by power consumption. Therefore, edge computing is better suited for time-sensitive tasks such as path planning. Yan et al. [9] proposed a cloud-edge collaborative path planning system and a grid-based caching path planning algorithm. By deploying the traditionally centralized path planning system on a collaborative architecture between the cloud and the edge, it can reduce transmission delays and alleviate the load on cloud servers. The grid-based caching path planning algorithm is applied on edge servers to further reduce the processing time for path planning. However, the caching-based path planning algorithm proposed in that article may have limitations in environments with dynamic obstacles, despite its good performance in static scenes.

In addition, it is worth noting that during the navigation and collision avoidance of USV, the majority of encountered dynamic obstacles are other vessels. The article [10] shows that many maritime collision accidents occur due to ship operators' failure to comply with the rules of Collision Regulation (CORLEGs). Therefore, it is necessary to incorporate the rules of COLREGs into the path planning and collision avoidance system for USV, for minimizing the probability of maritime accidents.

Thus, an edge-device collaborative architecture for path planning and obstacle avoidance of USV is proposed in this paper, where the global path planning tasks are offloaded to the edge and local path planning tasks are remained on the local. Furthermore, a Dynamic Target Artificial Potential Field (DTAPF) method that combines A* and APF algorithms and complies with the COLREGs rules is proposed to achieve path planning and safe obstacle avoidance for USV. The purpose of adopting dynamic targets is to integrate global and local path planning, enable the USV to avoid moving obstacles while following the global path, and guides the USV by adjusting the position of the dynamic targets, while ensuring compliance with COLREGs rules. The main contributions of this article are:

- To address the limitations of traditional centralized path planning algorithms, we proposed an edge-end collaborative architecture for USV path planning and obstacle avoidance, which was designed to meet the real-time requirement better of USV path planning.
- We proposed a DTAPF method to address the issue of USV path planning and obstacle avoidance in dynamic environments. By integrating the advantages of the A* and APF algorithm and considering the COLREGs rules, our method solved the inherent problems of traditional A* and APF algorithm in the presence of dynamic obstacles and improved navigation safety.
- We conducted experiments with the system architecture and algorithms proposed in this paper. The results indicate that DTAPF has improvements in both response speed and navigation safety.

The rest of the paper is organized as follows. Section 2 covers the related works of path planning. Section 3 introduces the system architecture and environmental modeling. Section 4 presents the algorithms including improved A*, improved APF, DTAPF and Offset Guidance. In Sect. 5, we present simulation experiments to verify the algorithm and architecture proposed in this paper. The conclusions are presented in Sect. 6.

2   Related Works

Path planning aims to calculate the shortest collision-free path from the starting point to the destination point, considering multiple constraints such as energy consumption, travel time, and safety, while avoiding both static and dynamic obstacles. Path planning is usually divided into global path planning and local path planning based on the type of known environmental information. Global path planning requires obtaining all environmental information from the starting point to the destination point before the USV departs and then performing path planning. One of the most commonly used methods for global path planning is the A* algorithm, which was first proposed by Hart et al. [5]. This algorithm is a heuristic-based graph search algorithm used to find the shortest path between two points in a graph. In the context of USV path planning, many studies have been conducted to improve the performance of the A* algorithm, such as adding heuristic functions or considering environmental constraints. Song et al. [11] proposed an improved A* algorithm that uses three path smoothers to optimize the path, resulting in a smoother and more continuous path. Wang et al. [12] proposed an improved A* algorithm for optimizing navigation cost based on electronic navigational charts. They utilized electronic navigational charts to establish an eight-directional grid environmental model, taking into account safety weights and the number of nodes, which improved navigational safety, reduced planning time, and increased path smoothness. Yang et al. [13] proposed a Finite Angle A* algorithm (FFA*), which took into account the size and turning capability of the USV, and introduced some constraints to improve the quality of the path planning algorithm and make the navigation path smoother. This algorithm aims to calculate a relatively short but safer path, rather than the shortest distance path without collision.

In local path planning, only a portion of the environment information is known, which is collected by sensors carried by the USV. The artificial potential field (APF) algorithm is a path planning method based on the potential field concept, first proposed by Khatib [6]. It used artificial potential fields to guide the USV towards the target point and avoids obstacles. The APF algorithm established an attractive force towards the target point and a repulsive force away from obstacles. By balancing these two forces, the USV can avoid obstacles and reach the target point successfully. Compared with other path planning algorithms, the APF algorithm has high real-time performance and computational efficiency, and therefore is widely used in USV navigation and obstacle avoidance. Xie et al. [14] proposed a modified potential field method to address the issues of local minima and target unreachable in the traditional APF. They introduced an adjusting factor to control the linear decrease of the attractive force and exponential decrease of the repulsive force when the USV approaches the target, thus preventing the motion path from diverging. Li et al. [15] proposed a swerving force method (SFM) based on the APF algorithm by introducing a perpendicular swerving force to the repulsive force direction in the traditional APF, which can prevent the USV from falling in local minima.

According to the characteristics of global and local path planning, global path planning is usually used to compute the approximate global path of USV, while local path planning is usually used to avoid obstacles in unknown environments. In recent years, many studies have proposed integrated planning methods that complement the advantages of global and local path planning, and many achievements have been made. Chen et al. [16] proposed an improved ant colony optimization- artificial potential field (ACO-APF) algorithm for USV path planning and obstacle avoidance in unknown environments. The algorithm utilized an improved ACO algorithm to search for the global optimal path in a grid environment, followed by an improved APF algorithm to avoid unknown obstacles during USV navigation. In addition, an equipotential line outer tangent circle and redefine potential functions was proposed to address the problem of local minima. Sang et al. [17] proposed a multiple sub-target artificial potential field (MTAPF) method based on an improved APF. The MTAPF algorithm divides the global optimal path generated by the improved A* algorithm into a sequence of sub-target points. The improved APF algorithm is used between these sub-target points and avoid local minima by switching target. Chen et al. [18] proposed a hybrid algorithm that combines A* and DWA. The global path is generated by A* algorithm, and the DWA is used to track the local target point which is the intersection point of the global and local maps, to avoid dynamic obstacles and follow the global path. Considering the sea state, a weight coefficient is added to the objective function of DWA, which ensures the safety of USV by decreasing the velocity weight and increasing the distance weight when the sea state level is high.

However, none of the aforementioned studies has incorporated COLREGs into their path planning and obstacle avoidance systems, which may lead to potential safety risks for USV encountering other vessels in complex waterway environments. Lazarowska [19] proposed a Discrete Artificial Potential Field (DAPF) algorithm that enforce compliance with CORLEGs by decreasing the potential field on the right side cells from the line segment connecting the start and target cells. However, forcing right side navigation may not be suitable for certain encounter situations.

3   System Design and Environment Modeling

3.1   System Overview

The USV path planning system based on edge computing is shown in Fig. 1. The system consists of edge server, USV, and roadside units (RSU). The edge server is responsible for storing the global map of the current area and providing services for the USV's global path planning requests. The GPS sensor carried by the USV provides location services to determine the absolute position of the USV, and the LIDAR is used to locate nearby obstacles to determine their relative position. The embedded device carried by the USV combines the global path provided by the edge server with the data provided by the sensors to perform local path planning to avoid unknown obstacles.

The flowchart of this system is illustrated in Fig. 2. When the USV generates a demand for a target point, the USV first sends the position of the target point and its own location to the edge server. The edge server constructs a global path planning map based on the electronic map and performs global path planning for the USV based on its information. The planned global path is then simplified and

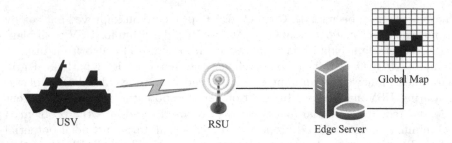

Fig. 1. System overview.

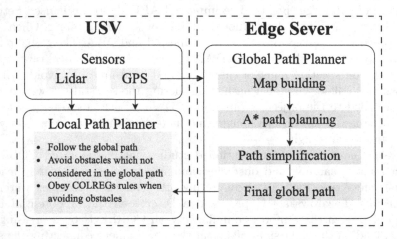

Fig. 2. System flowchart

returned to the USV. In this paper, we make the USV obtain the global path and avoid moving obstacles not considered in the global path while following the global path, ensuring compliance with COLREGs rules until it reaches the desired target point.

3.2 Environment Modeling

In this paper, we assume that the USV operates and moves in a general two-dimensional space, and is treated as a point mass. In the path planning process, we use a grid map to construct the static environment. The method of constructing the grid map will be mentioned in the following subsection. The black areas in the grid represent obstacles, such as islands, rocks, and floating objects, while the blank areas represent navigable areas, with each grid length set to a distance unit.

Global Path Planning Model. For global path planning, the current position of the USV is not taken into consideration, only the start point p_{g_start} and the

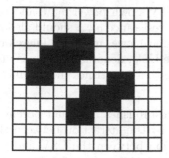

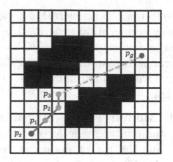

Fig. 3. Grid binary map.   **Fig. 4.** A* algorithm diagram.

destination point p_{g_goal} are considered. In this paper, a grid map as shown in Fig. 3 is used as the environment model for global path planning, and the USV is treated as a point that can move in eight adjacent grid directions from its current grid.

Map Expansion: In order to improve redundancy and navigation safety, we expand the obstacle areas $pxl_b = 1$ in the grid binary map $M_b (pxl_b)$ by k_m units, where k_m is determined by the current map size and the length of a single grid. The specific approach is to take all obstacle areas in the original binary map as the center and fill all navigable areas $pxl_b = 0$ within a radius of k_m units as 1, to obtain the expanded grid binary map $M_{bk} (pxl_b)$.

A* Model: The traditional A* algorithm searches 8 adjacent nodes of the current node at a time. Its heuristic function is defined as $F(n) = G(n) + H(n)$, where n is the current node being searched, $F(n)$ is the estimated cost from the start point to the target node t passing through node n, $G(n)$ is the actual cost from the start point to the current node n, which is the length of the path taken from the initial node through the selected node sequence to reach the current node, as shown by the blue solid line in the Fig. 4. $H(n)$ is the heuristic function that estimates the cost from node n to the target node t, as shown by the yellow dashed line in the Fig. 4. In this paper, we use the Manhattan distance to estimate $H(n)$, i.e., $H(n) = |x_g - x_p| + |y_g - y_p|$, where (x_g, y_g) is the coordinate of the target node and (x_p, y_p) is the coordinate of the current node.

Local Path Planning Model. In local path planning, the start point is the current position of the USV, denoted as p_{usv} , and the target point is denoted as p_{l_goal}. In addition to considering the positions of known static obstacles, the local path planning also needs to take into account the real-time positions of n_{mo} mobile obstacles, denoted as $P_{mo} = \{p_{mo}(1), p_{mo}(2), \ldots, p_{mo}(n_{mo})\}$. During local path planning, it is also necessary to consider the velocity of the USV, and limit it to not exceed the maximum velocity of v_{max}. While considering the velocity, the acceleration generated by the USV's own power a_{usv}, and the braking acceleration generated by environmental resistance a_{res}, should also be taken into account. Unlike UGVs and UAV, USV have lower mobility. Therefore, in local path planning, the angular velocity of the USV, denoted as ω_{usv}, is also limited with a maximum value of ω_{max} to ensure that it satisfies the kinematic constraints.

APF Model: The traditional APF algorithm establishes a virtual potential field and analyzes the forces acting on the USV by the attraction between the USV and the target point, as well as the repulsion between the USV and the obstacles. The commonly used function for the attractive potential field is:

$$U_{attr} = \frac{1}{2}\lambda_{attr}d_{ug}^2 \tag{1}$$

The parameter $\lambda_{att} > 0$ is the attraction scale factor, and d_{ug} represents the distance between the current position of the USV and the target point. The commonly used function for repulsive potential field is:

$$U_{repu} = \sum_{i=0}^{n_{mo}} U_{repu}^i \tag{2}$$

$$U_{repu}^i = \begin{cases} \frac{1}{2}\lambda_{repu}\left(\frac{1}{d_{um}(i)} - \frac{1}{d_r}\right)^2, d_{um}(i) < d_r \\ 0 \qquad\qquad\qquad\qquad , d_{um}(i) > d_r \end{cases} \tag{3}$$

U_{repu}^i is the repulsive potential field strength of i-th obstacle, and U_{repu} is the sum of all obstacle repulsive potential field. The parameter $\lambda_{att} > 0$ is the repulsion scale factor, and d_r is the influence radius of the obstacle. $d_{um}(i)$ represents the distance between the current position of the USV and the i-th obstacle. The repulsive potential field is defined as 0 outside the obstacle's influence range, and the closer the distance between the USV and the obstacle inside the influence range, the stronger the repulsive potential field. The total potential field strength $U_{total} = U_{attr} + U_{repu}$ is the sum of the attractive potential field strength and the repulsive potential field strength. The visualization of the total potential field is shown in Fig 5.

4 Algorithm Design

4.1 Improved A* Algorithm

The A* algorithm is a classical graph search algorithm widely used in global path planning. During the search process, the A* algorithm uses a heuristic function to evaluate the value of nodes, thus enabling it to find the optimal solution within a limited search space. However, traditional A* algorithm has the problem of generating excessive path nodes and turns. To address this issue, this paper optimizes the path points obtained from using the A* algorithm to

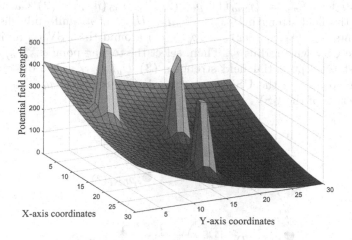

Fig. 5. 3D visualization of artificial potential field.

obtain the global path, by removing unnecessary path points to reduce the length and turns of the global path. The specific steps are as follows: (1) The global path, $P_g = \{p_g(1), p_g(2), \ldots, p_g(n)\}$ consisting of n path points is produced by the A* algorithm. (2) Take $p_g(1)$ as the first key point $p_k(1)$ and check if the line segment between $p_k(1)$ and the next path point $p_g(2)$ passes through any obstacles. If the line segment does not intersect with any obstacles, then $p_g(2)$ is a redundant point and will not be added to the key point sequence P_k. (3) The operation in step (2) is repeated between the key point $p_k(1)$ and $p_g(3)$, $p_g(4)$, etc., by checking line segments if they intersect with obstacles. This process continues until the line segment between $p_k(1)$ and $p_g(i+1)$ intersects with an obstacle. Then, $p_g(i)$ becomes the second key point $p_k(2)$. This process is repeated iteratively using $p_k(1)$ as the endpoint and $p_g(i+1)$, $p_g(i+2)$, etc., as the other endpoint until all target points are covered. Finally, the last path point $p_g(n)$ is added to the key point sequence P_k. Now, we have obtained a key point sequence P_k with the minimum number of turns.

4.2   Improved APF Algorithm

In this paper, the discretization idea of A* algorithm is incorporated into the APF local path planning algorithm. Unlike traditional APF algorithms, this paper calculates the total potential field of n_t test points within a distance of r_t around the current USV position p_{usv} (as shown in the Fig. 6, taking $n_t = 8$ as an example), and selects the direction of the point with the minimum total potential field strength as the propulsion force direction for the USV's next step. The specific steps are as follows: (1) The current position of USV is p_{usv}, and the target position of local path planning is p_{l_goal}. The n_t test points uniformly distributed on a circle centered at the USV with a radius of r_t are denoted as $P_{test} = \{p_{test}(1), p_{test}(2), \ldots, p_{test}(n_t)\}$. There are n_{mo} moving obstacles near

USV, denoted as $P_{mo} = \{p_{mo}(1), p_{mo}(2), \ldots, p_{mo}(n_{mo})\}$. **(2)** Calculate the total potential field strength $U_{total} = U_{attr} + U_{repu}$ of n_t uniformly distributed testing points $p_{test}(1), p_{test}(2), \ldots, p_{test}(n_t)$ around the USV on a circle centered at the USV with radius r_t. Then, select the testing point $p_{min\_of\_test}$ with the smallest total potential field strength. **(3)** D_{usv_next} is the direction from the current position of the USV, p_{usv}, towards the test point $p_{min\_of\_test}$ is the target direction of the USV.

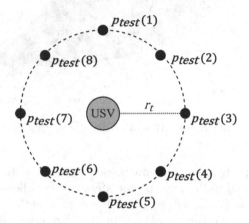

Fig. 6. Test point distribution diagram.

4.3   DTAPF Method

In general, global path planning algorithms can provide a shortest global path but take a long computing time and are unable to avoid dynamic obstacles, lacking real-time performance. On the other hand, local path planning algorithms can avoid dynamic obstacles in a local area in a short time but are prone to losing direction and falling into local minima. Furthermore, the path obtained from a local path planning algorithm alone may not be the optimal path. By combining A* algorithm and APF algorithm, USV can benefit from the effective path planning capability of A* algorithm and the ability of APF algorithm to handle unknown dynamic environments. A* algorithm is used to find the optimal path in a static environment, while APF algorithm is used to handle moving obstacles encountered in unknown environment during navigation. This paper proposes a Dynamic Target Artificial Potential Field (DTAPF) method that combines the advantages of A* algorithm and APF algorithm to solve the problem of USV collision avoidance in dynamic environment. The main idea of the algorithm is to introduce a dynamic point that follow the global path and set it as the target of the APF algorithm, thus achieving the avoidance of moving obstacles while following the global path. The specific steps are as follows: **(1)** First, a global path P_k consisting of n_k key path points is obtained by improving

---

Algorithm 1: Improved APF Algorithm

Input: USV current position p_{usv}, local path planning target p_{l_goal}, position of n_{mo} moving obstacles $P_{mo} = \{p_{mo}(1), p_{mo}(2), \ldots, p_{mo}(n_{mo})\}$.

Output: Target direction of USV D_{usv_next}.

1 Initialize: Attraction scale factor λ_{attr}, repulsion scale factor λ_{repu}, influence radius of an obstacle d_r, number of test points n_t.

2 **for** $i = 1 : n_t$ **do**

3 $\quad U(i)_{attr} = \frac{1}{2}\lambda_{attr}d_{ug}^2$

4 $\quad$ **for** $j = 1 : n_{mo}$ **do**

5 $\quad\quad$ **if** $d_{um}(j) < d_r$ **then**

6 $\quad\quad\quad U(i)_{repu}^j = \frac{1}{2}\lambda_{repu}\left(\frac{1}{d_{um}(j)} - \frac{1}{d_r}\right)^2$

7 $\quad\quad$ **else**

8 $\quad\quad\quad U(i)_{repu}^j = 0$

9 $\quad\quad$ **end**

10 $\quad\quad U(i)_{repu} = U(i)_{repu} + U(i)_{repu}^j$

11 $\quad$ **end**

12 $\quad U(i)_{total} = U(i)_{attr} + U(i)_{repu}$

13 **end**

14 $p_{min\_of\_test}$ is the point of p_{test} which U_{total} is the minimum, and D_{usv_next} is the direction from the current position of the USV p_{usv}, towards the test point $p_{min\_of\_test}$

15 **return** D_{usv_next}

---

the A* algorithm. **(2)** Taking the current position of the USV p_{usv}, as the star point for local path planning and the dynamic target point p_{d_goal} as the local target (p_{l_goal}), use the improved APF algorithm to calculate the next navigation direction. The USV moves in this direction. **(3)** The USV performs step (2) every time it moves. It calculate the distance between its current position p_{usv} and the dynamic target point p_{d_goal} in each round. If the distance is not greater than r_l, the dynamic target point moves along the sequence of key points in P_k for a distance of d_{dp}. Otherwise, it remains in its current position. When the USV moves near the global path planning target, the path planning ends.

4.4   Offset Guidance

When planning paths and avoiding obstacles, USV must also consider the COL-REGs rules to ensure safe navigation. COLREGs is a set of international rules that govern the behavior of ships at sea to avoid collisions. It covers various aspects of maritime navigation, including the use of lights, shapes, sound signals, and other means of communication to ensure the safety and efficient movement of ships. Therefore, when planning paths and avoiding obstacles, it is necessary to consider COLREGs to ensure that navigation is in compliance with regulations and does not pose a threat to other vessels in the area.

Traditional APF algorithms can easily plan paths to avoid static obstacles without considering the rules of COLREGs. However, when avoiding dynamic

---

Algorithm 2: Dynamic-Target APF Method

Input: USV current position p_{usv}, global path key points P_k, position of n_{mo}
moving obstacles P_{mo}.

1 Initialize: Initial position of dynamic target point $p_{d\_goal} = p_k(1)$
2 **while** *USV did not reach global target* **do**
3 | **Algorithm 1** Input: $(p_{usv}, p_{d\_goal}, P_{mo})$, Output:$D_{usv\_next}$
4 | **if** $\|p_{usv} - p_{d\_goal}\| < r_l$ **then**
5 | | p_{d_goal} moves along the sequence of key points in P_k for a distance of d_{dp}
6 | **end**
7 | USV navigates in the D_{usv_next} direction
8 **end**

---

obstacles, the path planned by traditional APF algorithms often fails to ensure compliance with the rules of COLREGs. Therefore, based on the DTAPF method, we propose an algorithm to comply with the rules of COLREGs and improve USV's security. This algorithm mainly utilizes the characteristics of APF algorithms, which always have an attractive force pointing towards the target, and guide the USV to follow a rule-compliant path to avoid moving obstacles by modifying the trajectory of the dynamic target point. The specific steps are as follows: **(1)** As shown in the Fig. 7, when the USV moves to the boundary of obstacle range, the direction of the moving obstacle's movement and the angle θ between the line connecting the USV and the moving obstacle are determined. The direction of the moving obstacle's movement and the angle θ between the line connecting the USV and the moving obstacle are determined. When the angle θ is in the range of $(-90°, 90°)$, offset guidance is required, and the direction of the offset guidance is the direction in which the USV is moving relative to the moving obstacle. **(2)** The trajectory of the dynamic target point, which should have been moved according to the global path, is translated in the direction of the offset guidance by a distance of d_{co}, generating a lateral force to guide the USV to avoid the moving obstacle along a compliant path. **(3)** When the angle θ is not within the range of $(-90°, 90°)$, the dynamic target point returns to the original global path. The offset guidance algorithm ends after the USV completely leaves the influence range of the moving obstacle.

5   Evaluation

5.1   Simulation Environment Configurations

In this section, simulation verifications were conducted for algorithm design and system architecture design. The verification for algorithm design is carried out in MATLAB simulation environment, and executed on a PC with an i5-10500 six-core CPU and 16 GB memory. As for the verification of system architecture design, considering the compatibility on different platforms, the MATLAB program of global path planning algorithm is ported to Python program. The local

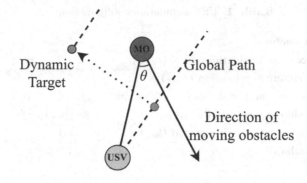

Fig. 7. Offset guidance diagram.

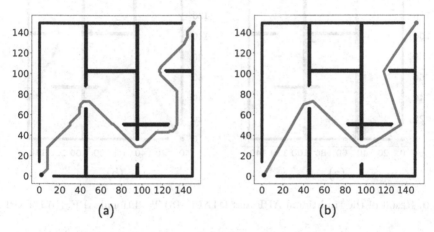

Fig. 8. Result of the traditional A* algorithm and the improved A* algorithm. (a) Traditional A*. (b) Improved A*.

device is a Raspberry Pi 4B, the edge server is equipped with an Intel i5-10500 six-core CPU and 16GB memory, and the cloud server is a Tencent cloud server.

In the experiment result figures, the static obstacle area is represented in black, and the navigable area is represented in white. Black point is used to mark the global start point, and red point is used to mark the global target point. In the global path planning algorithm, the generated global path is represented by a blue line. In the USV simulation experiment, the USV's path is represented by a blue-green gradient line, where the color closer to green represents the velocity closer to 0, and the color closer to blue represents the velocity closer to the maximum sailing velocity of the USV. The parameters of USV are shown in Table 1.

Table 1. USV simulation configuration

Parameter	Value	Unit
Maximum navigation speed (v_{max})	1.2	m/s
Maximum acceleration (a_{usv})	2	m/s^2
Environmental resistance deceleration (a_{res})	−0.4	m/s^2
Maximum yaw rate (ω_{max})	0.625	π/s
Simulation sampling period (t_{int})	0.1	s
Simulation yaw resolution	1/16	π

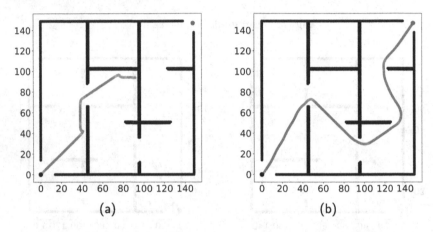

(a)   (b)

Fig. 9. Result of the traditional APF and DTAPF. (a) Traditional APF. (b) DTAPF.

5.2   Performance of Algorithms

Improved A*. In MATLAB, the traditional A* algorithm and the improved A* algorithm are used to calculate the global path in the same static environment. We set the start point in the bottom left corner (0, 0) and the goal point in the top right corner (150, 150). The global path generated by the traditional A* algorithm is shown in Fig. 8(a) with a length of 327.8 and 38 turns, while the global path generated by the improved A* algorithm is shown in Fig. 8(b) with a length of 310.7 and 6 turns. It can be observed that the improved A* algorithm greatly reduces the number of turns and optimizes the path length compared to the traditional A* algorithm.

DTAPF. To validate the effectiveness of the proposed algorithm for path planning and obstacle avoidance of moving obstacles, MATLAB was used to compare the traditional APF and the DTAPF method. The simulation parameters were set as follows: $\lambda_{attr} = 1$, $\lambda_{repu} = 100$, test point distribution radius $r_t = 0.1$, moving obstacle influence range radius $R_{mo} = 2$, maximum velocity of USV $v_{max} = 1.2$, acceleration of USV $a_{usv} = 2$, and environmental resistance braking acceleration of $a_{res} = -0.4$. For traditional APF, the target point was kept consistent with the global target point; for the DTAPF method, the maximum distance between the dynamic target point and the USV was set to $r_l = 1.5$. For the motion of the USV, a time interval of $t_{int} = 0.1$ was used to update the scene. Due to the short time interval, the motion of the USV can be simplified as uniform accelerated linear motion, and the next moment's motion state is updated according to the following equation.

$$v_{usv\_next} = v_{usv} + t_{int} * a_{usv} + t_{int} * a_{res}$$
$$p_{usv\_next} = p_{usv} + \frac{1}{2} (v_{usv} + v_{usv\_next}) * t_{int} \tag{4}$$

In the situation without moving obstacles, the simulation path using the traditional APF is shown in Fig. 9(a), the USV falls into local minima, and cannot reach the global target. The simulation path using the DTAPF method is shown in Fig. 9(b), where the global path guides the USV to easily reach the global goal point.

Offset Guidance. To validate the proposed offset guidance algorithm, we compared the dynamic target approach with and without the safety avoidance algorithm using MATLAB. We chose a common situation where path crossings are prone to occur during obstacle avoidance, as shown in Fig. 10. The start point of the USV is located at (0, 0) marked as black point, and the target point is located at (40, 70) marked as red point. A series of red circles marks the trajectory of the moving obstacle, which moves from top to bottom. As shown in Fig. 10(a), although the DTAPF method without the offset guidance algorithm can avoid the moving obstacle, it uses the path in front of the moving obstacle, which may lead collision risk. The DTAPF with the offset guidance algorithm, as shown in Fig. 10(b), uses the path behind the moving obstacle, which is safer than the path in front.

To verify the superiority of the proposed algorithm in terms of safety, this paper conducted a random obstacle avoidance experiment. The USV's starting and target points are the same as those shown in the Figure. The start point coordinate is (0, 0), and the target point coordinate is (40, 70). In this scenario, 1000 random obstacle avoidance experiments are conducted. For each experiment, the start point and direction of the random obstacle are randomly generated. The velocity of the obstacle is in the range of [0.2, 1]. In each randomly generated scenario, the USV moved from the start point to the target point and completed obstacle avoidance using four different methods: stop-wait-go (SG), traditional APF (APF), Dynamic-Target APF (DTAPF), and DTAPF with the Offset Guidance (DOG). The collision count, average travel time, and actual travel path length were recorded. As shown in Fig. 11, the DTAPF with offset guidance can achieve greater safety improvement with relatively small costs in terms of time and path length. Compared with the traditional APF method, the collision count was reduced by 71.4%, while the average travel time and average path length increased by only 1.8% and 6.3%.

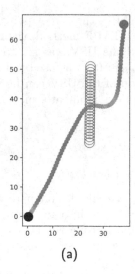 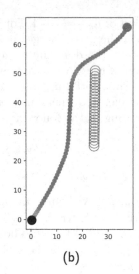

(a)   (b)

Fig. 10. Obstacle avoidance path of DTAPF and DTAPF with offset guidance, moving obstacle moves from top to bottom. (a) DTAPF. (b) DTAPF with offset guidance.

Comprehensive Simulation. Then, we added two moving obstacles to the same scenario as in Fig. 8. The obstacle on the left moves from top to bottom, while the one in the middle moves from bottom to top. The USV starts from the bottom left corner at (0,0) and the goal is the top right corner at (150,150). Figure 12 shows the result of this experiment, and the pink dotted line is the global path. It can be seen from the figure that the USV can successfully reach the global goal while avoiding the moving obstacles, and it complies with the rules of CORLEGs.

5.3   Verification of System Architecture

To compare the proposed architecture with traditional local computing and cloud computing architectures, we evaluated it in simulation using the global path generation time T_{Gp} as a metric. In order to compare with the traditional computing architecture, we considered the time it takes to generate a global path $T_{Gp} = T_c + T_{tr}$, where T_c is the delay during global path computing and T_{tr} is the transmission delay. In addition, we tested the algorithm's generalizability in different scenarios by using three different map sizes (60*60, 150*150, and 300*300) as inputs. The results is shown in Fig. 13. It can be seen that for each size of maps, the delay of obtaining the global path using edge computing is much better than using local computing, and it is also lower than the delay of traditional cloud computing.

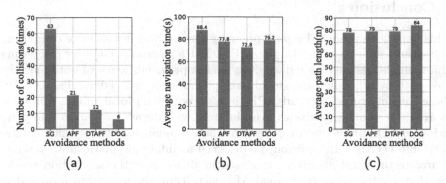

Fig. 11. Result of random obstacle avoidance experiment. (a) Number of collisions. (b) Average time. (c) Average path length.

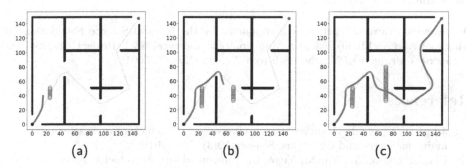

Fig. 12. Results of USV in a scene with 2 moving obstacles. The left one moves from top to bottom, while the middle one moves from bottom to top. (a) USV meets the first moving obstacle. (b) USV meets the second moving obstacle. (c) Complete path of USV.

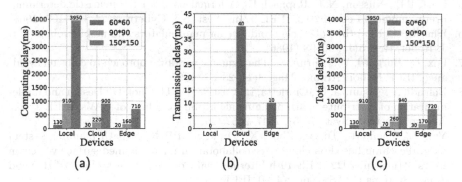

Fig. 13. Delay of generating global path. (a) Computing delay. (b) Transmission delay. (c) Total delay.

6   Conclusion

In this paper, we proposed a path planning system includes an edge computing architecture and a Dynamic Target APF (DTAPF) method. The edge computing architecture is designed to reduce delay and improve efficiency of obtaining the global path by offloading the global path planning tasks to edges. The DTAPF is based on improved A* and APF, it is to solve the local minima problem and avoid the moving obstacles. And in order to improve navigation safety, we considered CORLEGs and proposed an Offset Guidance method which is integrated into DTAPF. In addition, we designed a simulation platform to validate the architecture and algorithm proposed in this paper, the simulation results show that, compared to traditional APF algorithm, the algorithm proposed in this article can reduce collision probability by 71% with almost no increase in average path length and average navigation time. And the architecture proposed in this paper has much lower computing delay than local computing, and also lower than cloud computing.

Acknowledgments. This work is supported by the Natural Science Foundation of Ningbo City (2021J090), Ningbo Municipal Commonweal S&T Project (2022S005), Major S&T Projects of Ningbo High-tech Zone (2022BCX05001).

References

1. Jorge, V.A., et al.: A survey on unmanned surface vehicles for disaster robotics: main challenges and directions. Sensors **19**(3), 702 (2019)
2. Liu, Z., Zhang, Y., Yu, X., Yuan, C.: Unmanned surface vehicles: an overview of developments and challenges. Annu. Rev. Control. **41**, 71–93 (2016)
3. Ding, F., Zhang, Z., Fu, M., Wang, Y., Wang, C.: Energy-efficient path planning and control approach of USV based on particle swarm optimization. In: OCEANS 2018 MTS/IEEE Charleston, pp. 1–6. IEEE (2018)
4. Dijkstra, E.: A note on two problems in connexion with graphs. Numer. Math. **1**, 269–271 (1959)
5. Hart, P.E., Nilsson, N.J., Raphael, B.: A formal basis for the heuristic determination of minimum cost paths. IEEE Trans. Syst. Sci. Cybern. **4**(2), 100–107 (1968)
6. Khatib, O.: Real-time obstacle avoidance for manipulators and mobile robots. Int. J. Robot. Res. **5**(1), 90–98 (1986)
7. Fox, D., Burgard, W., Thrun, S.: The dynamic window approach to collision avoidance. IEEE Robot. Autom. Mag. **4**(1), 23–33 (1997)
8. Belkadi, A., Abaunza, H., Ciarletta, L., Castillo, P., Theilliol, D.: Design and implementation of distributed path planning algorithm for a fleet of UAVs. IEEE Trans. Aerosp. Electron. Syst. **55**(6), 2647–2657 (2019)
9. Yan, L., Chen, H., Tu, Y., Zhou, X., Drew, S.: PPGC: a path planning system by grid caching based on cloud-edge collaboration for unmanned surface vehicle in IoT systems. In: 2022 IEEE 19th International Conference on Mobile Ad Hoc and Smart Systems (MASS), pp. 74–80. IEEE (2022)
10. Uğurlu, Ö., Köse, E., Yıldırım, U., Yüksekyıldız, E.: Marine accident analysis for collision and grounding in oil tanker using FTA method. Maritime Policy Manage. **42**(2), 163–185 (2015)

11. Song, R., Liu, Y., Bucknall, R.: Smoothed A* algorithm for practical unmanned surface vehicle path planning. Appl. Ocean Res. **83**, 9–20 (2019)

12. Wang, Y., Liang, X., Li, B., Yu, X.: Research and implementation of global path planning for unmanned surface vehicle based on electronic chart. In: Qiao, F., Patnaik, S., Wang, J. (eds.) ICMIR 2017. AISC, vol. 690, pp. 534–539. Springer, Cham (2018). https://doi.org/10.1007/978-3-319-65978-7_80

13. Yang, J.M., Tseng, C.M., Tseng, P.: Path planning on satellite images for unmanned surface vehicles. Int. J. Naval Archit. Ocean Eng. **7**(1), 87–99 (2015)

14. Xie, S., et al.: The obstacle avoidance planning of USV based on improved artificial potential field. In: 2014 IEEE International Conference on Information and Automation (ICIA), pp. 746–751. IEEE (2014)

15. Li, X., Song, H., Han, Z., Zhang, D., Peng, Y.: An improved artificial potential field algorithm with swerving force for USV path planning. In: 2021 IEEE International Conference on Unmanned Systems (ICUS), pp. 1019–1024. IEEE (2021)

16. Chen, Y., Bai, G., Zhan, Y., Hu, X., Liu, J.: Path planning and obstacle avoiding of the USV based on improved ACO-APF hybrid algorithm with adaptive early-warning. IEEE Access **9**, 40728–40742 (2021)

17. Sang, H., You, Y., Sun, X., Zhou, Y., Liu, F.: The hybrid path planning algorithm based on improved A* and artificial potential field for unmanned surface vehicle formations. Ocean Eng. **223**, 108709 (2021)

18. Chen, Z., Zhang, Y., Zhang, Y., Nie, Y., Tang, J., Zhu, S.: A hybrid path planning algorithm for unmanned surface vehicles in complex environment with dynamic obstacles. IEEE Access **7**, 126439–126449 (2019)

19. Lazarowska, A.: Comparison of discrete artificial potential field algorithm and wave-front algorithm for autonomous ship trajectory planning. IEEE Access **8**, 221013–221026 (2020)

Node-Disjoint Paths in Balanced Hypercubes with Application to Fault-Tolerant Routing

Shuai Liu, Yan Wang$^{(\boxtimes)}$, Jianxi Fan, and Baolei Cheng

School of Computer Science and Technology, Soochow University, Suzhou 215006,
China
wangyanme@suda.edu.cn

Abstract. The study of interconnection networks plays an essential role
in the design of parallel computing systems because their topological
properties make a great impact on the performance and reliability of
the systems. The balanced hypercube, designed for fault tolerance, is a
variant of the hypercube with desirable properties of strong connectivity,
regularity, and symmetry. Over the past decade, the node-disjoint paths
problem has received much attention. The existence of these parallel
paths can improve reliability, fault tolerance, message throughput, and
information security. In this paper, we propose algorithms to construct a
maximal number of node-disjoint paths between any two distinct nodes
of an n-dimensional balanced hypercube in $O(n^2)$ time. The lengths of
these parallel paths exceed the internode distance by no more than six.
In addition, we conduct simulation experiments to evaluate the performance of the fault-tolerant routing using multiple node-disjoint paths as
transmission channels.

Keywords: Interconnection networks · Balanced hypercubes ·
Node-disjoint paths · Fault tolerance · Algorithm

1 Introduction

Parallel computing systems are the solution to large-scale computing problems
such as engineering simulation, weather forecasting, genetic analysis, and oil
exploration. The well-known hypercube is one of the most popular interconnection networks for parallel computing systems and has practical applications
in real systems. The balanced hypercube, as a variant of the hypercube, was
designed to tolerate processor failure by Huang and Wu [18]. The balanced hypercube retains many attractive properties, such as recursiveness, strong connectivity, node symmetry [18], and edge symmetry [21]. Furthermore, the diameter
of the odd-dimensional balanced hypercube is smaller than that of the hypercube with the same order. Particularly, each processor in an n-dimensional balanced hypercube, denoted as BH_n, has a backup processor that shares $2n$ same

neighboring nodes, which means that tasks running on a faulty processor can be transferred to its backup processor to make system reconfiguration efficient. With so desired performances, the balanced hypercube has been extensively investigated, such as Hamiltonian cycle embedding [7], super connectivity [19], restricted connectivity [1], reliability analysis [17], structure fault tolerance [10], and so on.

The concept of disjoint paths arose naturally from the study of routing in parallel and distributed systems. As one of the disjoint path problems, node-disjoint paths are those paths that do not share any common node except the source and destination nodes. The existence of such parallel paths can improve reliability, fault tolerance, message throughput, and information security. For example, we consider one-to-one communication, where one node sends a message to another. We assume that there exist k node-disjoint paths between any two nodes in the network. As long as the number of faulty nodes does not exceed $k - 1$, there is always at least one fault-free path to transmit the message. Furthermore, to increase transmission rate, the message can be divided into k segments, which are sent simultaneously on each path. In addition, by applying different encryption information to the segment transmitted in each path, the secure transmission of the message can be ensured. Recently, based on the property of node-disjoint paths, Pai et al. configured secure-protection routing schemes for Möbius cubes and crossed cubes [13, 14]. To sum up, constructing such parallel paths is of great practical significance.

The famous Menger's Theorem [3] reveals the relationship between graph connectivity and node-disjoint paths in a graph G. It proves that the necessary and sufficient conditions for the graph G to be k-connected is that any two distinct nodes in G connect at least k node-disjoint paths. Furthermore, the maximum flow algorithm can be applied to obtain node-disjoint paths for an arbitrary graph, but the time complexity is considerable. To improve fault tolerance and reduce transmission latency, the construction of node-disjoint paths in networks is expected to make their maximal length and the time complexity as small as possible. At present, with regard to research on disjoint paths of networks, many results have been obtained. Lai [8] studied the optimal construction of all shortest node-disjoint paths in the hypercube. Guo et al. [6] gave an $O(k^2)$ algorithm of finding the $k + 1$ node-disjoint paths between any two nodes in a k-dimensional data center network BCube, and they have proved that the longest length of these disjoint paths is $k + 1$. Zhou et al. [22] characterized a maximal number of node-disjoint paths between any two nodes of alternating group networks and demonstrated that these parallel paths are optimal or near-optimal, in the sense of their lengths exceeding the internode distance by no more than four. Please refer to [4, 5, 9, 15, 16] for more research on other networks.

So far, many scholars have studied the problem of disjoint paths problem on BH_n. There has been some research to deal with the disjoint path cover (DPC for short) problem in BH_n [2, 11, 12]. The DPC problem is also one of the disjoint path problems, which requires that disjoint paths cover all nodes in the network. Considering that paths in the DPC problem need to cover all nodes in the network, the length of some paths may be considerable for communication. Furthermore, these study were focused on the proof of the existence of disjoint paths, and no specific algorithms were given. In addition, Yang et al. [20] constructed two completely independent spanning trees (CISTs for short) in 2019. For any two distinct nodes in two CISTs, there exist two node-disjoint paths.

In this paper, we propose algorithms for constructing a maximal number (i.e., $2n$) of node-disjoint paths between any two distinct nodes in BH_n. The construction procedure has worst case time complexity $O(n^2)$, and the lengths of these parallel paths exceed the internode distance by no more than six. Furthermore, taking multiple node-disjoint paths as transmission channels, we simulate fault-tolerant data transmission in balanced hypercubes under multiple node failures.

The rest of this paper is organized as follows. In Sect. 2, the structure of balanced hypercubes is introduced, and some preliminary knowledge is given. Section 3 presents the construction of $2n$ node-disjoint paths between any two distinct nodes in BH_n. In Sect. 4, the results of simulation experiments are presented to evaluate the performance of the fault-tolerant routing. The conclusions of this paper and future work are given in Sect. 5.

2   Preliminaries

A network can generally be represented by a simple graph $G = (V(G), E(G))$, where $V(G)$ and $E(G)$ represent the node set and the edge set of G, respectively. Two nodes, u and v, in G are adjacent if $(u, v) \in E(G)$. A neighbor of a node u in G is any node adjacent to u. A path $\pi = \langle v_0, v_1, \cdots, v_k \rangle$ is a sequence of distinct nodes in which any two consecutive nodes are adjacent. For brevity, the path π can also be denoted by $v_0 \rightarrow v_1 \rightarrow \cdots \rightarrow v_k$ or $v_0 \sim v_k$. The length of a path π, denoted by $|\pi|$, is the number of edges in π. We use π_t to denote the t-th node in π, where $0 \leq t \leq k$. For any two nodes $u, v \in V(G)$, $d(u, v)$ denotes the distance between u and v, that is, the length of the shortest path connecting u and v. The diameter of G, denoted by $d(G)$, is defined by $d(G) = \max\{d(u, v) \mid u, v \in V(G)\}$. The set of integers $\{m, m + 1, \cdots, n\}$ can be denoted by $[m, n]$, where $m < n$.

Now, we introduce balanced hypercubes. There is one definition for an n-dimensional balanced hypercube BH_n.

Definition 1. *[18] An n-dimensional balanced hypercube consists of 4^n nodes* $(a_0, a_1, \cdots, a_{i-1}, a_i, a_{i+1}, \cdots, a_{n-1})$, *where a_0 and $a_i \in \{0, 1, 2, 3\}(1 \leq i \leq n - 1)$. An arbitrary node $v = (a_0, a_1, \cdots, a_{i-1}, a_i, a_{i+1}, \cdots, a_{n-1})$ in BH_n has the following $2n$ neighbors:*

1. $((a_0 \pm 1) \bmod 4, a_1, \cdots, a_{i-1}, a_i, a_{i+1}, \cdots, a_{n-1})$, and
2. $((a_0 \pm 1) \bmod 4, a_1, \cdots, a_{i-1}, (a_i + (-1)^{a_0}) \bmod 4, a_{i+1}, \cdots, a_{n-1})$.

The first element a_0 in the n-tuple of node v is called the *inner index* or 0-index, and the element a_i $(1 \leq i \leq n - 1)$ is called the i-*index*. By Definition 1, for $j \in [0, n-1]$, each node v of BH_n has two j-dimensional neighbors (say v'

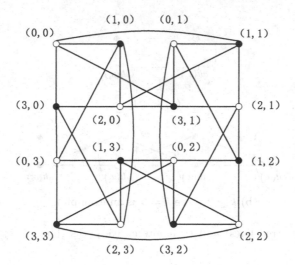

Fig. 1. BH_2.

and v^c). We use $N^j(v) \in \{v', v^c\}$ to denote an arbitrary j-dimensional neighbor of v. If the inner index of v is odd (resp. even), we call $v \xrightarrow{j} N^j(v)$ the *odd (resp. even) conversion*.

BH_2 is depicted in Fig. 1, and we have the following proposition.

Proposition 1. *[18] The balanced hypercube BH_n has the following properties.*

1. *BH_n is bipartite, node-symmetric, and $2n$-connected;*
2. *The diameter of BH_n is $2n$ when n is even or $n = 1$, and is $2n - 1$ when n is odd other than 1;*
3. *Nodes $u = (a_0, a_1, \cdots, a_{n-1})$ and $v = ((a_0 + 2) \bmod 4, a_1, \cdots, a_{n-1})$ in BH_n have the same neighbors.*

By Property 3, it implies that $u = (a_0, a_1, \cdots, a_{n-1})$ and $v = ((a_0 + 2) \bmod 4, a_1, \cdots, a_{n-1})$ are pairwise backup nodes in BH_n. We use $alt(u)$ to denote v, which indicates that v is the alternative node of u. Furthermore, if there are no pairwise backup nodes in a node set $V' \in V(BH_n)$, we call it a *non-pair set*. That is to say, if $u \in V'$, then $alt(u) \notin V'$. Now, we can extend to the representation of alternative paths in BH_n. Let $\pi = \langle v_0, v_1 \cdots, v_k \rangle$ be a path. If $k = 1$, we use $alt(\pi)$ to denote the path $\langle v_0, alt(v_1), alt(v_0), v_1 \rangle$. If $k \geq 2$ and $V(\pi)$ is a non-pair set, we use $alt(\pi)$ to denote the path $\langle v_0, alt(v_1), alt(v_2), \cdots, alt(v_{k-1}), v_k \rangle$. Obviously, π and $alt(\pi)$ are node-disjoint. Figure 2 illustrates such two kinds of paths and their alternative paths.

The following lemma will be used in the main result of this paper.

Lemma 1. *Assume π and π' are two node-disjoint paths from u to v, where u and v are two distinct nodes in BH_n. Then, π, π', $alt(\pi)$, and $alt(\pi')$ are node-disjoint if $V(\pi) \cup V(\pi')$ is a non-pair set.*

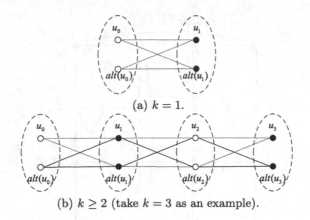

(a) $k = 1$.

(b) $k \geq 2$ (take $k = 3$ as an example).

Fig. 2. Two paths (red lines) and their alternative paths (blue lines). (Color figure online)

Proof. It is clear that the lemma holds when the length of an arbitrary path is 1. Suppose $\pi = \langle u, u_1, \cdots, u_i, \cdots, u_k, v \rangle$ and $\pi' = \langle u, u'_1, \cdots, u'_j, \cdots, u'_l, v \rangle$, where $1 \leq i \leq k$, $1 \leq j \leq l$, and $u_i \neq u'_j$. When $V(\pi) \cup V(\pi')$ is a non-pair set, we have $alt(u_i) \neq u'_j, u_i \neq alt(u'_j)$, and $alt(u_i) \neq alt(u'_j)$. Thus, π, π', $alt(\pi)$, and $alt(\pi')$ are node-disjoint.   □

We now give the definition of relative address between two distinct nodes in BH_n, which will be used in the rest of this paper.

Definition 2. *The relative address r of nodes $u = (a_0, a_1, \cdots, a_{n-1})$ and $v = (b_0, b_1, \cdots, b_{n-1})$ in BH_n is denoted by $r = v - u = (r_0 r_1 \cdots r_{n-1})$, where $r_0 = (b_0 \bmod 2 - a_0 \bmod 2 - \sum_{j=1}^{n-1}(b_j - a_j)) \bmod 4$ and $r_i = (b_i - a_i) \bmod 4$ ($1 \leq i \leq n - 1$).*

We use r^k ($k = 0, 1, 2, 3$) to denote a set of bit positions in r, the value of which is k. For example, for two nodes u and v in BH_6, (0,2,1,2,1,3) and (0,2,0,3,1,1), the relative address r of two nodes is (203102), and $r^0 = \{1, 4\}$.

3   Node-Disjoint Paths

Suppose that u and v are two distinct nodes of an n-dimensional balanced hypercube. According to Menger's theorem [3], there are at most $2n$ node-disjoint paths between u and v. In this section, $2n$ node-disjoint paths from u to v are constructed. Since an n-dimensional balanced hypercube is node-symmetric, we assume

$$u = (\overbrace{0, 0, \cdots, 0}^{n}) = 0^n$$

i.e., the source node, without losing generality. In the rest of this section, we first design the mixed sequence, which can be used to simplify the path representation. Then the construction of node-disjoint paths is shown, and the correctness of the algorithms is proved.

3.1 The Mixed Sequence

In this paper, the construction of a path between two distinct nodes in BH_n is actually the process by which several indexes are corrected. For example, a path $\pi^c : u = (0,0,0,0,0,0) \xrightarrow{2} (1,0,1,0,0,0) \xrightarrow{3} (2,0,1,3,0,0) \xrightarrow{4} (1,0,1,3,1,0) \xrightarrow{1} (0,3,1,3,\ 1,0) \xrightarrow{5} (3,3,1,3,1,1) \xrightarrow{1} (0,2,1,3,1,1) = v$. During the even (resp. odd) conversions in the construction process, the 2-index, 4-index, and 5-index (resp. the 3-index and 1-index) are sequentially corrected. By adjusting the correction order of several indexes, we can construct different paths from u to v.

To indicate the order of correction, we design a data structure named *mixed sequence*. The mixed sequence ξ is denoted by $\eta(S; T)$, where S and T are two sequences. The two sequences, S and T, record the bit positions of several indexes. Multiple mixed sequences can be merged into a mixed sequence, i.e., $\langle \xi_0, \xi_1, \cdots, \xi_{k-1} \rangle \rightarrow \eta(\langle S_0, S_1, \cdots, S_{k-1} \rangle; \langle T_0, T_1, \cdots, T_{k-1} \rangle) = \eta(S; T)$, where $\xi_i = \eta(S_i; T_i)(0 \le i \le k - 1)$. For example, for $k = 2$, $\xi_0 = \eta(2; 3)$, $\xi_1 = \eta(4, 5; 1)$, and $\langle \xi_0, \xi_1 \rangle \rightarrow \eta(\langle \langle 2 \rangle, \langle 4, 5 \rangle \rangle; \langle \langle 3 \rangle, \langle 1 \rangle \rangle) = \eta(2, 4, 5; 3, 1)$.

Now, we give the function $\text{Mseq}(u, v)$ to generate a mixed sequence ξ, which will contribute to the construction of node-disjoint paths between u and v. We briefly describe its process here. We first compute the relative address $r = v - u = (r_0 r_1 \cdots r_{n-1})$ and set ξ to $\langle \rangle$. Then bit positions are selected into ξ based on the value of $r_i (0 \le i \le n - 1)$. With the help of r^1, r^2 and r^3, it takes $O(n)$ time to generate a mixed sequence. The formal description of Function $\text{Mseq}(u, v)$ is given below.

Algorithm $\text{Mseq}(u, v)$

1: **function** $\text{MSEQ}(u, v)$
2:　　$r \leftarrow v - u$, $\Omega \leftarrow r^1 \cup r^2 \cup r^3$, $\xi \leftarrow \langle \rangle$;
3:　　**while** $\exists a, b \in \Omega \ (r_a = 1 \wedge r_b = 3)$ **do**
4:　　　　$\xi \leftarrow \langle \xi, \eta(a; b) \rangle$, $\Omega \leftarrow \Omega \setminus \{a, b\}$;
5:　　**end while**
6:　　**while** $\exists a, b, c \in \Omega \ (r_a = r_b = 1 \wedge r_c = 2 \ (\text{or } r_a = 2 \wedge r_b = r_c = 3))$ **do**
7:　　　　$\xi \leftarrow \langle \xi, \eta(a, b; c) \rangle$ (or $\xi \leftarrow \langle \xi, \eta(a; b, c) \rangle$), $\Omega \leftarrow \Omega \setminus \{a, b, c\}$;
8:　　**end while**
9:　　**while** $\exists a, b \in \Omega \ (r_a = 2 \wedge r_b = 2)$ **do**
10:　　　$\xi \leftarrow \langle \xi, \eta(a; b) \rangle$, $\Omega \leftarrow \Omega \setminus \{a, b\}$;
11:　　**end while**
12:　　**while** $\exists a, b, c, d \in \Omega \ (r_a = r_b = r_c = r_d = 1 \ (\text{or } r_a = r_b = r_c = r_d = 3))$ **do**
13:　　　$\xi \leftarrow \langle \xi, \eta(a, b, c; d) \rangle$ (or $\xi \leftarrow \langle \xi, \eta(a; b, c, d) \rangle$), $\Omega \leftarrow \Omega \setminus \{a, b, c, d\}$;
14:　　**end while**
15:　　/* If the inner index of v is even, $|\Omega| = 0$. */
16:　　**if** $|\Omega| = 1$ **then** $\xi \leftarrow \langle \xi, \eta(a; \langle \rangle) \rangle$ $(a \in \Omega, r_a = 1)$;
17:　　**else if** $|\Omega| = 2$ **then** $\xi \leftarrow \langle \xi, \eta(a; b) \rangle$ $(a, b \in \Omega, r_a = 2 \wedge r_b = 3)$;
18:　　**else if** $|\Omega| = 3$ **then** $\xi \leftarrow \langle \xi, \eta(a; b, c) \rangle$ $(a, b, c \in \Omega \wedge r_a = r_b = r_c = 3)$;
19:　　**end if**
20:　　**return** ξ; /* Obviouly, we can also retain the form of $\langle \xi_0, \xi_1, \cdots, \xi_{k-1} \rangle$. */
21: **end function**

In the following, we demonstrate how to derive the corresponding path π with the help of a mixed sequence $\xi = \eta(S; T)$. As described above, the mixed sequence is designed to indicate the correction order of several indexes during the construction of the path. The arrangement of the elements in S (resp. T) is used to imply the correction order of several indexes during the even (resp. odd) conversions. Since the correction order is determined, the corresponding path is determined. Based on the method, the function Rpath(s, d, ξ) is proposed. With inputs s, d, ξ, Function Rpath(s, d, ξ) can generate a path from s to d in $O(n)$ time, where s and d are two distinct nodes in BH_n. For brevity, a call to Function Rpath(s, d, ξ) can be written as $s \sim d \Leftarrow \xi$. The formal description of Function Rpath(s, d, ξ) is given below.

Algorithm Rpath(s, d, ξ)

1: **function** RPATH(s, d, ξ)
2:    /* $s = (a_0, a_1, \cdots, a_{n-1})$, $d = (b_0, b_1, \cdots, b_{n-1})$, and $\xi = \eta(S, T)$ */
3:    $\pi \leftarrow \langle s \rangle, r \leftarrow d - s$;
4:    let j, k be the first element of S, T, respectively;
5:    **while** $s \neq d$ **do**
6:        **if** the inner index of s is even **then**
7:            **if** $r_j = 0$ **then**
8:                let m be the next element of S, $j \leftarrow m$;
9:            **end if**
10:           $i \leftarrow j, r_j \leftarrow r_j - 1$;
11:       **else**
12:           **if** $r_k = 0$ **then**
13:               let m be the next element of T, $k \leftarrow m$;
14:           **end if**
15:           $i \leftarrow k, r_k \leftarrow (r_k + 1) \bmod 4$;
16:       **end if**
17:       select a_0' randomly from $\{(a_0 \pm 1) \bmod 4\}$ such that $(b_0 - a_0') \bmod 4 \neq 2$;
18:       **if** $i = 0$ **then**
19:           $s \leftarrow (a_0', a_1, \cdots, a_{n-1})$;
20:       **else**
21:           $s \leftarrow (a_0', a_1, \cdots, (a_i + (-1)^{a_0}) \bmod 4, \cdots, a_{n-1})$;
22:       **end if**
23:       $\pi \leftarrow \langle \pi, s \rangle$;
24:   **end while**
25:   **return** π;
26: **end function**

Then, we will see how the functions Mesq() and Rpath() work by going over an example. In BH_6, let $u = (0, 0, 0, 0, 0, 0)$ and $v = (0, 2, 1, 3, 1, 1)$ be the source node and the destination node, respectively. The mixed sequence ξ can be generated by applying Mesq(u, v), and the process is as follows.

(1) $r = u - v = (021311)$, $\Omega = \{1, 2, 3, 4, 5\}$, and $\xi = \langle \rangle$.
(2) Since $2, 3 \in \Omega$, $r_2 = 1$, and $r_3 = 3$, we have $\xi = \eta(2; 3)$ and $\Omega = \{1, 4, 5\}$.
(3) Since $1, 4, 5 \in \Omega$, $r_4 = r_5 = 1$ and $r_1 = 2$, we have $\xi = \eta(2, 4, 5; 3, 1)$ and $\Omega = \emptyset$.

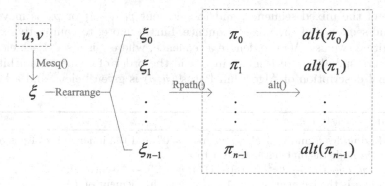

Fig. 3. The construction process of the $2n$ node-disjoint paths between u and v in BH_n.

Next, by applying $Rpath(u, v, \xi)$, a required path π between the source and destination nodes can be generated: $(0,0,0,0,0,0) \xrightarrow{2} (1,0,1,0,0,0) \xrightarrow{3} (2,0,1,3,0,0) \xrightarrow{4} (1,0,1,3,1,0) \xrightarrow{1} (0,3,1,3,1,0) \xrightarrow{5} (3,3,1,3,1,1) \xrightarrow{1} (0,2,1,3, 1,1)$.

Finally, we rearrange the mixed sequence ξ into $\eta(4, 2, 5; 1, 3)$. By applying $Rpath(u, v, \xi)$, the corresponding path π' can be generated: $(0,0,0,0,0,0) \xrightarrow{4} (1,0,0,0,1,0) \xrightarrow{1} (2,3,0,0,1,0) \xrightarrow{2} (1,3,1,0,1,0) \xrightarrow{1} (0,2,1,0,1,0) \xrightarrow{5} (1,2,1,0, 1,1) \xrightarrow{3} (0,2,1,3,1,1)$.

3.2 Construction of Node-Disjoint Paths

In this subsection, we show the construction of $2n$ node-disjoint paths from u to v in BH_n. As described in the previous subsection, a mixed sequence can derive the corresponding path. By adjusting finely the arrangement of the elements in the mixed sequence (the correction order of indexes is changed), we can derive n node-disjoint paths in BH_n. During the construction procedure of the n paths, we only choose one of pairwise backup nodes. Thus, by Lemma 1, the remaining n node-disjoint paths can be constructed by applying $alt()$. Figure 3 illustrates the construction process of the $2n$ node-disjoint paths between u and v in BH_n. Based on the construction strategy above, two algorithms, $Paths0(u, v)$ and $Paths1(u, v)$, are proposed. With inputs, u and v, $Paths0(u, v)$ (resp. $Paths1(u, v)$) can produce $2n$ node-disjoint paths from u to v in BH_n, where the inner index of v is odd (resp. even). The details are given below.

Firstly, we consider that the inner index of v is odd. In Algorithm $Paths0(u, v)$, the n paths derived by mixed sequences are divided into two types. For the first type (corresponds to lines 8 and 16 of the algorithm), several elements of the mixed sequence are in left circular shifts. For the second type (corresponds to lines 11 and 20 of the algorithm), neighbors of u and v are considered as new source and destination nodes, the element q_j is added to the first sequence of the mixed sequence, and the element $p_{|r^1|-c-1}$ or p_w is moved to the second sequence of the mixed sequence. Furthermore, to compactly express the algorithm, we use $\Delta(\tau)$ to denote a sequence, where τ is a set. Each element in $\Delta(\tau)$ corresponds to an element in τ, and the order of elements is arbitrary. The formal description of Algorithm $Paths0(u,v)$ is given below, and a lemma follows.

Algorithm $Paths0(u,v)$

Input: Two nodes u and v in BH_n, where $u = 0^n$ and the inner index of v is odd.
Output: $2n$ node-disjoint paths from u to v.
1: $r \leftarrow v - u$, $c \leftarrow (|r^1| - 2|r^2| - |r^3| - 1)/4$, $\eta(S,T) \leftarrow \text{Mseq}(u,v)$;
2: $\Gamma \leftarrow \{a : a \text{ is the element of } S\}$, $\Lambda \leftarrow \{a : a \text{ is the element of } T\}$;
3: /* The n paths derived by mixed sequences. */
4: **if** $c > 0$ **then**
5:    /* Each p_i and q_j correspond to an element in r^1 and $[0, n-1]\setminus r^1$, respectively.*/
6:    **for** each $p_i \in r^1$ $(0 \leq i \leq |r^1| - 1)$ **do**
7:       $d \leftarrow i + |r^1| - 4c$;
8:       $\pi(p_i) : u \sim v \Leftarrow \eta(p_i, p_{(i+1) \bmod |r^1|}, \cdots, p_{(d+1) \bmod |r^1|}, p_{(d+3) \bmod |r^1|},$
           $p_{(d+4) \bmod |r^1|}, p_{(d+5) \bmod |r^1|}, p_{(d+7) \bmod |r^1|}, \cdots, p_{(i-1) \bmod |r^1|},$
           $\Delta(r^2 \cup r^3), p_{(d+2) \bmod |r^1|}, p_{(d+6) \bmod |r^1|}, \cdots, p_{(i-2) \bmod |r^1|})$;
9:    **end for**
10:   **for** each $q_j \in [0, n-1] \setminus r^1$ $(0 \leq j \leq n - |r^1| - 1)$ **do**
11:      $\pi(q_j) : u \to N^{q_j}(u) \sim N^{q_j}(v) \to v$, where $N^{q_j}(u) \sim N^{q_j}(v) \Leftarrow$
          $\eta(q_j, p_0, p_1, \cdots, p_{|r^1|-c-2}; \Delta(r^2 \cup r^3 \setminus \{q_j\} \cup \{p_{|r^1|-c-1}, p_{|r^1|-c}, \cdots, p_{|r^1|-1}\}))$;
12:   **end for**
13: **else**
14:    /* Each p_i and q_j correspond to an element in Γ and $[0, n-1]\setminus\Gamma$, respectively.*/
15:    **for** each $p_i \in \Gamma$ $(0 \leq i \leq |\Gamma| - 1)$ **do**
16:       $\pi(p_i) : u \sim v \Leftarrow \eta(p_i, p_{(i+1) \bmod |\Gamma|}, \cdots, p_{(i-1) \bmod |\Gamma|}; \Delta(\Lambda))$;
17:    **end for**
18:    randomly select an integer $p_w \in \Gamma$ such that r_{p_w} is the maximum;
19:    **for** each $q_j \in [0, n-1] \setminus \Gamma$ $(0 \leq j \leq n - |\Gamma| - 1)$ **do**
20:       $\pi(q_j) : u \to N^{q_j}(u) \sim N^{q_j}(v) \to v$, where $N^{q_j}(u) \sim N^{q_j}(v) \Leftarrow$
          $\eta(q_j, p_0, p_1, \cdots, p_{w-1}, p_{w+1}, \cdots, p_{|\Gamma|-1}; \Delta(\Lambda \setminus \{q_j\} \cup \{p_w\}))$;
21:    **end for**
22: **end if**
23: /* The remaining n paths. */
24: construct the alternative path of each path constructed above by applying $alt()$;

Lemma 2. *Algorithm $Paths0(u,v)$ can construct $2n$ node-disjoint paths from u to v in BH_n, where the inner index of v is odd, with the maximum path length not exceeding $d(u,v) + 6$.*

Proof. By Lemma 1, we only need to prove the n paths constructed by lines $5 \sim 12$ or lines $14 \sim 21$ of Algorithm $Paths0(u,v)$ are node-disjoint. Since the proofs in both cases are similar, we only prove the n paths constructed by lines $5 \sim 12$ of Algorithm $Paths0(u,v)$ are node-disjoint for brevity.

Suppose $\pi(p_{i_0})$ and $\pi(p_{i_1})$ (resp. $\pi(q_{j_0})$ and $\pi(q_{j_1})$) are two paths from u to v constructed by lines $6 \sim 9$ (resp. $10 \sim 12$) of Algorithm Paths0(u, v). Each node $\pi_t(p_{i_0})$, $1 \leq t \leq |\pi(p_{i_0})| - 1$, has the p_{i_0}-index corrected, but $p_{(i_0-1) \bmod |r^1|}$-index uncorrected. By the rotation property, the correction order of $p_{(i_0-1) \bmod |r^1|}$-index precedes that of p_{i_0}-index in $\pi(p_{i_1})$. Thus, $\pi(p_{i_0})$ and $\pi(p_{i_1})$ are node-disjoint. For paths $\pi(q_{j_0})$ and $\pi(q_{j_1})$, the q_{j_0}-index of each node $\pi_{t_0}(q_{j_0})$ is always different from that of each node $\pi_{t_1}(q_{j_1})$, where $1 \leq t_0 \leq |\pi(q_{j_0})| - 1$ and $1 \leq t_1 \leq |\pi(q_{j_1})| - 1$. Thus, $\pi(q_{j_0})$ and $\pi(q_{j_1})$ are node-disjoint. Similarly, $\pi(p_{i_0})$ and $\pi(q_{j_0})$ are node-disjoint. Therefore, the n paths constructed by lines $5 \sim 12$ of Algorithm Paths0(u, v) are node-disjoint.

At last, we analyze the path lengths. Without adding the number of changes, each $\pi(p_i)$ is the shortest path with the length of $d(u, v)$. Under the worst case of $r_{q_j} = 0$ and $r_{p_{|r^1|-c-1}} = 1$ (or $r_{p_w} = 1$), the length of such a path $\pi(q_j)$ is $d(u, v) + 6$. $\square$

To illustrate the flow of Algorithm Paths0(u, v), two examples are given in Table 1. In both examples, the rearranged mixed sequences are listed. By applying Rpath(s, d, ξ) and $alt()$, a maximal number of node-disjoint paths between two nodes can be obtained.

Table 1. Two examples illustrating the flow of Algorithm Paths0(u, v).

Example 1	Example 2
$u = 0^{11}$ $v = (1, 1, 1, 1, 1, 1, 1, 2, 3, 0)$	$u = 0^9$ $v = (1, 1, 1, 1, 2, 2, 2, 3, 0)$
$\xi = \eta(0, 1, 2, 3, 4, 5, 7; 8, 9, 6) \leftarrow$ Mesq(u, v)	$\xi = \eta(0, 1, 2, 5, 3; 7, 4, 6) \leftarrow$ Mesq(u, v)
$\pi(p_i) : u \sim v \Leftarrow \xi_i$	
$i \in \{0, 1, 2, 3, 4, 5, 6, 7\}$	$i \in \{0, 1, 2, 5, 3\}$
$\xi_0 = \eta(0, 1, 2, 3, 4, 5, 7; \Delta(\{8, 9\}), 6)$	$\xi_0 = \eta(0, 1, 2, 5, 3; \Delta(\{7, 4, 6\}))$
$\xi_1 = \eta(1, 2, 3, 4, 5, 6, 0; \Delta(\{8, 9\}), 7)$	$\xi_1 = \eta(1, 2, 5, 3, 0; \Delta(\{7, 4, 6\}))$
$\xi_2 = \eta(2, 3, 4, 5, 6, 7, 1; \Delta(\{8, 9\}), 0)$	$\xi_2 = \eta(2, 5, 3, 0, 1; \Delta(\{7, 4, 6\}))$
$\cdots$	$\xi_5 = \eta(5, 3, 0, 1, 2; \Delta(\{7, 4, 6\}))$
$\xi_7 = \eta(7, 0, 1, 2, 3, 4, 6; \Delta(\{8, 9\}), 5)$	$\xi_3 = \eta(3, 0, 1, 2, 5; \Delta(\{7, 4, 6\}))$
$\pi(q_j) : u \to N^{q_j}(u) \sim N^{q_j}(v) \to v$, where $N^{q_j}(u) \sim N^{q_j}(v) \Leftarrow \xi_j$	
$j \in \{8, 9, 10\}$	$j \in \{7, 4, 6, 8\}$
$\xi_8 = \eta(8, 0, 1, 2, 3, 4, 5; \Delta(\{9, 6, 7\}))$	$\xi_7 = \eta(7, 0, 1, 2, 3; \Delta(\{4, 6, 5\}))$
$\xi_9 = \eta(9, 0, 1, 2, 3, 4, 5; \Delta(\{8, 6, 7\}))$	$\xi_4 = \eta(4, 0, 1, 2, 3; \Delta(\{7, 6, 5\}))$
$\xi_{10} = \eta(10, 0, 1, 2, 3, 4, 5; \Delta(\{8, 9, 6, 7\}))$	$\xi_6 = \eta(6, 0, 1, 2, 3; \Delta(\{7, 4, 5\}))$
	$\xi_8 = \eta(8, 0, 1, 2, 3; \Delta(\{7, 4, 6, 5\}))$
construct the alternative path of each path constructed above by applying $alt()$	

Now, we consider that the inner index of v is even. By Function Mseq(u, v), there are six types of mixed sequences, which are determined by the relative address r of u and v. We list them $(\xi_{i,0})$ and rearrange them as follows, which will contribute to Algorithm Paths1(u, v).

1. $r_a = 1 \wedge r_b = 3 : \xi_{i,0} = \eta(a; b)$ and $\xi_{i,1} = \eta(b; a)$.
2. $r_a = r_b = 1 \wedge r_c = 2 : \xi_{i,0} = \eta(a, b; c), \xi_{i,1} = \eta(b, c; a)$, and $\xi_{i,2} = \eta(c, a; b)$.
3. $r_a = 2 \wedge r_b = r_c = 3 : \xi_{i,0} = \eta(a; b, c), \xi_{i,1} = \eta(b; c, a)$, and $\xi_{i,2} = \eta(c; a, b)$.
4. $r_a = 2 \wedge r_b = 2 : \xi_{i,0} = \eta(a; b)$ and $\xi_{i,1} = \eta(b; a)$.
5. $r_a = r_b = r_c = r_d = 1 : \xi_{i,0} = \eta(a, b, c; d), \xi_{i,1} = \eta(b, c, d; a), \xi_{i,2} = \eta(c, d, a; b)$, and $\xi_{i,3} = \eta(d, a, b; c)$.
6. $r_a = r_b = r_c = r_d = 3 : \xi_{i,0} = \eta(a; b, c, d), \xi_{i,1} = \eta(b; c, d, a), \xi_{i,2} = \eta(c; d, a, b)$, and $\xi_{i,3} = \eta(d; a, b, c)$.

In Algorithm Paths1(u, v), the n paths derived by mixed sequences are also divided into two types. For the first type (corresponds to line 10 of the algorithm), the neighbor of u is considered as the new source node, the mixed sequences are in left circular shifts, and the last mixed sequence is rearranged. For the second type (corresponds to line 14 of the algorithm), neighbors of u and v are considered as new source and destination nodes, and the mixed sequence does not need to be changed. Furthermore, to compactly express the algorithm, we use $|\xi|$ to denote the number of elements in the mixed sequence. The formal description of Algorithm Paths1(u, v) is given below, and a lemma follows.

Algorithm Paths1(u, v)

Input: Two nodes u and v in BH_n, where $u = 0^n$ and the inner index of v is even.
Output: $2n$ node-disjoint paths from u to v.
1: **if** u and v are pairwise backup nodes **then**
2:    construct the $2n$ paths: $\langle u, u_0, v \rangle, \langle u, u_1, v \rangle, \cdots, \langle u, u_{2n-1}, v \rangle$, where $u_0, u_1, \cdots,$ u_{2n-1} are $2n$ common neighbors of u and v;
3: **else**
4:    /* The n paths derived by mixed sequences. */
5:    $r \leftarrow v - u$, $\langle \xi_0, \xi_1, \cdots, \xi_{k-1} \rangle \leftarrow \text{Mseq}(u, v)$;
6:    **for** each ξ_i $(0 \leq i \leq k - 1)$ **do**
7:       determine its type by the relative addrese r;
8:       **for** each $j = 0, 1, \cdots, |\xi_i| - 1$ **do**
9:          rearrange ξ_i to obtain $\xi_{i,j}$ and let k be the first element in $\xi_{i,j}$;
10:          $\pi(i, j) : u \rightarrow N^k(u) \sim v$, where $N^k(u) \sim v \Leftarrow$ $\langle \xi_{(i+1) \bmod k}, \xi_{(i+2) \bmod k} \cdots, \xi_{(i-1) \bmod k}, \xi_{i,j} \rangle$;
11:       **end for**
12:    **end for**
13:    **for** each $i \in r^0$ **do**
14:       $\pi(i) : u \rightarrow N^i(u) \sim N^i(v) \rightarrow v$, where $N^i(u) \sim N^i(v) \Leftarrow \langle \xi_0, \xi_1, \cdots, \xi_{k-1} \rangle$;
15:    **end for**
16:    /* The remaining n paths. */
17:    construct the alternative path of each path constructed above by applying $alt()$.
18: **end if**

Lemma 3. *Algorithm Paths1(u, v) can construct $2n$ node-disjoint paths from u to v in BH_n, where the inner index of v is even, with the maximum path length not exceeding $d(u, v) + 4$.*

Proof. Obviously, we only need to consider that u and v are not pairwise backup nodes. Furthermore, by Lemma 1, we only need to prove the n paths constructed by lines $6 \sim 15$ of Algorithm Paths1(u, v) are node-disjoint.

Suppose $\pi(i_0, j_0)$ and $\pi(i_1, j_1)$ (resp. $\pi(i_2)$ and $\pi(i_3)$) are two paths from u to v constructed by lines $6 \sim 12$ (resp. $13 \sim 15$) of Algorithm Paths1(u, v). We first prove $\pi(i_0, j_0)$ and $\pi(i_1, j_1)$ are node-disjoint. Obviously, the node $\pi_1(i_0, j_0)$ is different from each node in $\pi(i_1, j_1)$. When $i_0 = i_1$, the different correction order of ξ_{i_0,j_0} and ξ_{i_1,j_1} guarantees that $\pi(i_0, j_0)$ and $\pi(i_1, j_1)$ are node-disjoint. When $i_0 \neq i_1$, the execution of $\xi_{(i_0+1) \bmod k}$ takes precedence over ξ_{i_0,j_0} in $\pi(i_0, j_0)$. However, by the rotation property, the execution of ξ_{i_0} takes precedence over $\xi_{(i_0+1) \bmod k}$ in $\pi(i_1, j_1)$. Thus, $\pi(i_0, j_0)$ and $\pi(i_1, j_1)$ are node-disjoint. For paths $\pi(i_2)$ and $\pi(i_3)$, the i_2-index of each node $\pi_{t_2}(i_2)$ are always different from that of each node $\pi_{t_3}(i_3)$, where $1 \leq t_2 \leq |\pi(i_2)| - 1$ and $1 \leq t_1 \leq |\pi(i_3)| - 1$. Thus, $\pi(i_2)$ and $\pi(i_3)$ are node-disjoint. Similarly, $\pi(i_0, j_0)$ and $\pi(i_2)$ are node-disjoint. Therefore, the n paths constructed by lines $6 \sim 15$ of Algorithm Paths0(u, v) are node-disjoint.

At last, we analyze the path lengths. The rearrangement of ξ_i will at worst increase the length of the path by 4, i.e., $d(u, v) \leq |\pi(i, j)| \leq d(u, v) + 4$. Due to the addition of two changes, the length of each $\pi(i)$ is $d(u, v) + 2$.   □

To illustrate the flow of Algorithm Paths1(u, v), an example is given in Table 2. The rearranged mixed sequences are listed. By applying Rpath(s, d, ξ) and $alt()$, a maximal number of node-disjoint paths between two nodes can be obtained.

Table 2. An example illustrating the flow of Algorithm Paths1(u, v).

$u = 0^8$ $v = (0, 1, 1, 1, 2, 2, 2, 3)$	
$\xi = \langle \eta(1; 7), \eta(2, 3; 4), \eta(5; 6) \rangle = \eta(1, 2, 3, 5; 7, 4, 6) \leftarrow \text{Mesq}(u, v)$	
$\pi(i, j) : u \rightarrow N^k(u) \sim v$, where $N^k(u) \sim v \Leftarrow \xi'$	
$i = 0, j = k = 1$	$\xi' = \langle \eta(2, 3; 4), \eta(5; 6), \eta(1; 7) \rangle = \eta(2, 3, 5, 1; 4, 6, 7)$
$i = 0, j = k = 7$	$\xi' = \langle \eta(2, 3; 4), \eta(5; 6), \eta(7; 1) \rangle = \eta(2, 3, 5, 7; 4, 6, 1)$
$i = 1, j = k = 2$	$\xi' = \langle \eta(5; 6), \eta(1; 7), \eta(2, 3; 4) \rangle = \eta(5, 1, 2, 3; 6, 7, 4)$
$i = 1, j = k = 3$	$\xi' = \langle \eta(5; 6), \eta(1; 7), \eta(3, 4; 2) \rangle = \eta(5, 1, 3, 4; 6, 7, 2)$
$i = 1, j = k = 4$	$\xi' = \langle \eta(5; 6), \eta(1; 7), \eta(4, 2; 3) \rangle = \eta(5, 1, 4, 2; 6, 7, 3)$
$i = 2, j = k = 5$	$\xi' = \langle \eta(1; 7), \eta(2, 3; 4), \eta(5; 6) \rangle = \eta(1, 2, 3, 5; 7, 4, 6)$
$i = 2, j = k = 6$	$\xi' = \langle \eta(1; 7), \eta(2, 3; 4), \eta(6; 5) \rangle = \eta(1, 2, 3, 6; 7, 4, 5)$
$\pi(i) : u \rightarrow N^i(u) \sim N^i(v) \rightarrow v$, where $N^i(u) \sim N^i(v) \Leftarrow \xi''$	
$i = 0$	$\xi'' = \langle \eta(1; 7), \eta(2, 3; 4), \eta(5; 6) \rangle = \eta(1, 2, 3, 5; 7, 4, 6)$
construct the alternative path of each path constructed above by applying $alt()$	

It is clear that each path can be constructed in $O(n)$ time, and the number of node-disjoint paths is $2n$. By Lemmas 2 and 3, we have the following theorem.

Theorem 1. *The $2n$ node-disjoint paths between any two distinct nodes of BH_n can be constructed in $O(n^2)$ time, with the maximum path length not exceeding $d(u,v) + 6$.*

4   Simulations and Performance Evaluation

In this section, we simulate the data transmission in balanced hypercubes under multiple node failures. Simulation results will be used to assess the performance of fault-tolerant routing using multiple node-disjoint paths as transmission channels. Since BH_n is a symmetric graph, Algorithms Paths0(u,v) and Paths1(u,v) can be applied to the construction of node-disjoint paths between any two distinct nodes. All algorithms for constructing node-disjoint paths and the required routing algorithms are implemented by using the programming language C++. We carry out the simulation on a Lenovo computer with a 2.90 GHz Intel®Core™ i7-10700 CPU, 16 GB RAM, and a Windows 10 operating system.

Considering the network size, we simulate on balanced hypercubes $BH_n (3 \leq n \leq 6)$ under different scenarios of node failures. For each scenario, we randomly generate 100,000 instances, each consisting of a node pair (s,d) with $s \neq d$ and a set of faulty nodes not containing the source node s and the destination node d. We are interested in computing the transmission failure rate (TFR for short) to reflect the capability of fault tolerance, which is the percentage of the number of failed transmissions over generated instances. The transmission is successful if the message can be received by the destination node through at least one transmitting path. A transmitting path is called *a regular route* provided that it guarantees the successful transmission of the message. Otherwise, it is *a failed route*. Also, the path lengths of all regular routes for different scenarios are measured to evaluate the corresponding performance of fault-tolerant routing. The number of faulty nodes is initialized to 0 and then added one at a time until the TFR exceeds the threshold, which we set up 1‰.

For each instance of transmission from s to d in BH_n, let P_i be the transmitting path for $i = 0, 1, 2, \cdots, 2n-1$. Under 100,000 times simulation, we count the number of failed routes of P_i and P_{all} (denoted by $\#P_i$ and $\#P_{all}$, respectively), where P_{all} indicates that each P_i for $i = 0, 1, 2, \cdots, 2n-1$ is a failed route. We compute $\text{TFR}_{all} = \#P_{all}/10^5 \times 100\%$ and $\text{TFR}_{avg} = \sum_0^{2n-1}(\#P_i)/(10^5 \times 2n) \times 100\%$. Figure 4 (a)~(d) shows the corresponding trend of TFR_{all} and TFR_{avg} with respect to the number of faulty nodes for $BH_n (3 \leq n \leq 6)$. From Fig. 4 (a)~(d), we are conscious of the following three phenomenons:

1. The fault-tolerant routing using multiple node-disjoint paths can guarantee a successful transmission if the number of faulty nodes is no more than $2n-1$ in BH_n. Furthermore, we can observe that a low number of node failures over the generated 100,000 instances have almost no effect on the data transmission, i.e., the TFR_{all} is close to or equal to 0.

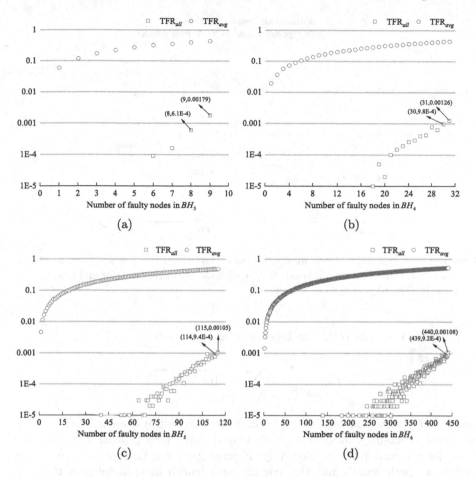

Fig. 4. (a) ~ (d): The corresponding trend of TFR_{all} and TFR_{avg} with respect to the number of faulty nodes for $BH_n (3 \leq n \leq 6)$.

2. As we expect, the TFR_{all} and TFR_{avg} tend to increase as the number of faulty nodes increases. The reasons responsible for this phenomenon are obvious. As a consequence, for $BH_n (3 \leq n \leq 6)$ under TFR no more than 1‰, the fault-tolerant routing can tolerate 8, 30, 114, and 439 faulty nodes, respectively.
3. The adoption of fault-tolerant routing will lead to a significant drop in TFR. For instance, if we consider the situation that there are 439 faulty nodes in BH_6, more than half of the generated 1,200,000 paths are failed routes, but TFR_{all} is only 0.092%.

The network diameter and the average distance are two of the criteria used to evaluate interconnection networks. As we have mentioned before, Huang and Wu [18] showed that the diameter of BH_n is $2n$ when n is even or $n = 1$, and is $2n - 1$ when n is odd other than 1. Moreover, the average distance between any

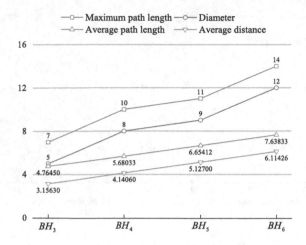

Fig. 5. The comparsion of the maximum path length and the average path length in fault-tolerant routings under no node failure with the network diameter and the average distance in $BH_n(3 \leq n \leq 6)$.

two distinct nodes in BH_n can be defined as

$$\overline{d}(BH_n) = \frac{1}{4^n} \sum_{v \in V(BH_n)} d((\overbrace{0,0,\cdots,0}^{n}),v)$$

since BH_n is a symmetric graph. In order to reduce transmission latency and cost, node-disjoint paths are desired to have their maximal length and total length minimized, respectively. Figure 5 provides the comparison of the maximum path length and the average path length in fault-tolerant routings under no node failure with the network diameter and the average distance in $BH_n(3 \leq n \leq 6)$. It can be found that the maximum path length is close to the diameter of BH_n, and the average path length is close to the average distance of BH_n. In fact, the maximum path length is only $d(BH_n)+2$, and the average path length is approximately equal to $\overline{d}(BH_n)+1.5$.

In addition, to facilitate the evaluation of the corresponding performance of fault-tolerant routing, Fig. 6 shows the average path lengths of the three scenarios, i.e., no node failure, $2n-1$ faulty nodes, and under TFR no more than 1‰ in the respective networks $BH_n(3 \leq n \leq 6)$. The ratio of the average path length under node failures to the average path length under no node failure is calculated. Longer paths are more likely to fail, resulting in a ratio of less than 1. When considering the situation of $2n-1$ faulty nodes, no matter what the scale of BH_n, there always exists at least one fault-free path to transmit messages. According to the comparison, we find that the ratio is getting closer to 1, and thus we conclude that the failure seemingly has no effect on the transmission performance of the fault-tolerant routing with respect to path length.

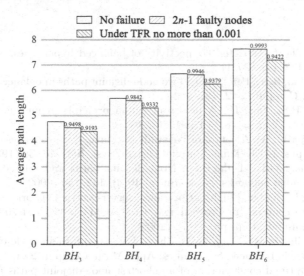

Fig. 6. The comparison of average path lengths in fault-tolerant routings under situations with no failure, $2n - 1$ nodes, and under TFR no more than 1‰ in $BH_n(3 \leq n \leq 6)$.

5   Conclusion

The node-disjoint paths problem is to find k paths between two arbitrary nodes, u and v, in a k-connected graph. The existence of node-disjoint paths reflects the robustness of the network, which can avoid congestion and improve fault tolerance. In this paper, we construct $2n$ node-disjoint paths between any two distinct nodes in BH_n by designing algorithms. The number of node-disjoint paths is the best possible, the time complexity of algorithms is $O(n^2)$, and the lengths of these parallel paths exceed the internode distance by no more than six. In addition, we conduct simulation experiments to evaluate the performance of the fault-tolerant routing using multiple node-disjoint paths as transmission channels. The simulation results in terms of transmission failure rate and path length imply that the fault-tolerant routing can ensure stable and low-latency transmissions. A useful variation on the node-disjoint paths problem is the node-to-set disjoint paths problem, which is worth studying. Furthermore, we are interested in whether the maximal length or the total length of node-disjoint paths could be smaller.

Acknowledgement. This work was supported by the National Natural Science Foundation of China (Nos. 62172291, 62272333) and A Project Funded by the Priority Academic Program Development of Jiangsu Higher Education Institutions (PAPD).

References

1. Cheng, D.: The h-restricted connectivity of balanced hypercubes. Discret. Appl. Math. **305**, 133–141 (2021)
2. Cheng, D., Hao, R.X., Feng, Y.Q.: Two node-disjoint paths in balanced hypercubes. Appl. Math. Comput. **242**, 127–142 (2014)
3. Choudhury, B.S., Das, K.: A new contraction principle in menger spaces. Acta Math. Sin. Engl. Ser. **24**(8), 1379–1386 (2008)
4. Day, K., Tripathi, A.: A comparative study of topological properties of hypercubes and star graphs. IEEE Trans. Parallel Distrib. Syst. **5**(1), 31–38 (1994)
5. Fu, J.S., Chen, G.H., Duh, D.R.: Node-disjoint paths and related problems on hierarchical cubic networks. Netw. Int. J. **40**(3), 142–154 (2002)
6. Guo, C., et al.: Bcube: a high performance, server-centric network architecture for modular data centers. In: Proceedings of the ACM SIGCOMM 2009 Conference on Data Communication, pp. 63–74 (2009)
7. Hao, R.X., Zhang, R., Feng, Y.Q., Zhou, J.X.: Hamiltonian cycle embedding for fault tolerance in balanced hypercubes. Appl. Math. Comput. **244**, 447–456 (2014)
8. Lai, C.N.: Optimal construction of all shortest node-disjoint paths in hypercubes with applications. IEEE Trans. Parallel Distrib. Syst. **23**(6), 1129–1134 (2011)
9. Latifi, S.: On the fault-diameter of the star graph. Inf. Process. Lett. **46**(3), 143–150 (1993)
10. Liu, H., Cheng, D.: Structure fault tolerance of balanced hypercubes. Theor. Comput. Sci. **845**, 198–207 (2020)
11. Lü, H.: Paired many-to-many two-disjoint path cover of balanced hypercubes with faulty edges. J. Supercomput. **75**, 400–424 (2019)
12. Lü, H., Wu, T.: Unpaired many-to-many disjoint path cover of balanced hypercubest. Int. J. Found. Comput. Sci. **32**(08), 943–956 (2021)
13. Pai, K.J., Chang, R.S., Chang, J.M.: A protection routing with secure mechanism in möbius cubes. J. Parallel Distrib. Comput. **140**, 1–12 (2020)
14. Pai, K.J., Chang, R.S., Wu, R.Y., Chang, J.M.: Three completely independent spanning trees of crossed cubes with application to secure-protection routing. Inf. Sci. **541**, 516–530 (2020)
15. Wang, G., Lin, C.K., Fan, J., Cheng, B., Jia, X.: A novel low cost interconnection architecture based on the generalized hypercube. IEEE Trans. Parallel Distrib. Syst. **31**(3), 647–662 (2019)
16. Wang, X., Fan, J., Lin, C.K., Jia, X.: Vertex-disjoint paths in DCell networks. J. Parallel Distrib. Comput. **96**, 38–44 (2016)
17. Wei, C., Hao, R.X., Chang, J.M.: The reliability analysis based on the generalized connectivity in balanced hypercubes. Discret. Appl. Math. **292**, 19–32 (2021)
18. Wu, J., Huang, K.: The balanced hypercube: a cube-based system for fault-tolerant applications. IEEE Trans. Comput. **46**(4), 484–490 (1997)
19. Yang, M.C.: Super connectivity of balanced hypercubes. Appl. Math. Comput. **219**(3), 970–975 (2012)
20. Yang, Y.X., Pai, K.J., Chang, R.S., Chang, J.M.: Constructing two completely independent spanning trees in balanced hypercubes. IEICE Trans. Inf. Syst. **102**(12), 2409–2412 (2019)
21. Zhou, J.X., Wu, Z.L., Yang, S.C., Yuan, K.W.: Symmetric property and reliability of balanced hypercube. IEEE Trans. Comput. **64**(3), 876–881 (2014)
22. Zhou, S., Xiao, W., Parhami, B.: Construction of vertex-disjoint paths in alternating group networks. J. Supercomput. **54**, 206–228 (2010)

Performance Evaluation of Spark, Ray and MPI: A Case Study on Long Read Alignment Algorithm

Kun Ran[1], Yingbo Cui[2(✉)], Zihang Wang[1], and Shaoliang Peng[1(✉)]

[1] College of Computer Science and Electronic Engineering, Hunan University, Changsha, Hunan 410082, China
{rankunkk,wangzihang,slpeng}@hnu.edu.cn
[2] School of Computer, National University of Defense Technology, Changsha, Hunan 410073, China
yingbocui@nudt.edu.cn

Abstract. The utilization of large-scale datasets in various fields is increasing due to the advancement of big data technology. Due to limited computing resources, traditional serial frameworks are no longer efficient in processing such massive data. Furthermore, as Moore's Law gradually loses its effect, improving program performance from the hardware level becomes increasingly challenging. Consequently, numerous parallel frameworks with distinct features and architectures have emerged, and selecting an appropriate one can enhance researchers' performance across various tasks. This paper evaluates three prominent parallel frameworks-Spark, Ray, and MPI-and employs minimap2, a third-generation CPU-based sequence alignment tool, as the benchmark program. The experimental results are discussed comprehensively. To evaluate the three frameworks, we devised a parallel algorithm for minimap2 and implemented its parallel versions using Ray and MPI, respectively. Furthermore, we selected IMOS as the Spark version of minimap2. The experiments involved six real datasets and one simulated dataset to evaluate and compare speedup, efficiency, throughput, scalability, peak memory, latency, and load balance. The findings demonstrate that MPI outperforms Apache Spark and Ray in terms of achieving a maximum speedup of 104.019, 81.3% efficiency, 33.510 MB/s throughput, the lowest latency, and better load balance. However, MPI exhibits poor fault tolerance. Apache Spark demonstrated the second-best performance, with a speedup of 88.937, efficiency of 69.5%, throughput of 29.546 MB/s, low latency, and the best load balance. Furthermore, it exhibited good fault tolerance and benefited from a mature ecosystem. Ray achieves a speedup of 76.828, efficiency of 60.0%, and throughput of 25.009 MB/s. However, it experiences high latency fluctuations, possesses less load balance compared to the previous two frameworks, and maintains good fault tolerance. The source code and a comprehensive user manual for these parallel programs are available at https://github.com/Geehome/minimapR and https://github.com/Geehome/minimapM, respectively.

Z. Tari et al. (Eds.): ICA3PP 2023, LNCS 14489, pp. 57–76, 2024.
https://doi.org/10.1007/978-981-97-0798-0_4

Keywords: parallel processing · Ray · Apache Spark · MPI · sequence alignment

1   Introduction

The rapid development of Internet technology has led to an unprecedented growth of data, propelling us into the era of big data. Consequently, large-scale data is increasingly utilized across diverse fields. The data size has evolved from tens of MB to tens of GB, and even reaching TB. Given the enormous volume of data, the limited computing resources of personal computers and traditional serial programs are inadequate to meet the demands of data processing and analysis [1]. Ongoing advancements in Internet technologies have made cloud computing, parallel computing, and distributed computing indispensable tools for handling big data analysis tasks. Since users typically do not complete the entire computation lifecycle on a single machine, the significance of parallel or distributed frameworks becomes evident. Moreover, at the hardware level, the progression of the semiconductor industry, driven by the miniaturization of transistors, has propelled computer performance. However, as miniaturization nears its limits and Moore's Law begins to wane, enhancing performance at the hardware level becomes progressively challenging and labor-intensive. Hence, alternative avenues for boosting parallel computing power at the software or algorithmic level must be explored [2]. Addressing this challenge necessitates the development of high-performance or distributed clusters, efficient scalable algorithms, and the proposal of diverse parallel frameworks to enhance program performance.

Numerous research teams are currently engaged in parallel framework development. AMP lab proposes Apache Spark [3] as the dominant distributed framework. Resilient Distributed Datasets(RDDs) are the fundamental data types in Spark. They are a fault-tolerant, distributed, and immutable memory abstraction. RDDs group a number of components that can be subjected to several operations to generate other RDDs (transformations) or return data (actions). Spark boasts a diverse range of features and a mature ecosystem, making it a popular choice for parallel optimization in numerous programs. RISELab introduces Ray [4], a high-performance distributed framework that adopts a distinct architecture and abstraction approach compared to traditional distributed computing systems. It supports the training, simulation, or interaction of models with the environment and the servicing of models. Ray can solve a wide range of tasks, from lightweight, stateless computational tasks (e.g. simulation) to long-running, stateful computational tasks (e.g. training). Ray also provides Lineage-based fault tolerance for tasks and executors, as well as replication-based fault tolerance for the metadata store. High-Performance Computing (HPC) technologies, such as the Message Passing Interface (MPI) [5], enhance performance and scalability but lack reliable fault tolerance. MPI is a message passing application programming interface that includes protocols and semantic descriptions that specify how it should be described in various implementations. High performance, scalability, and portability are the objectives of MPI, which is still the dominant model in use today. In the past, integrating MPI and Spark has been a major research direction [6]. MPI implementations often outperform Apache Spark significantly in numerous analytic operations, yet they should strive to leverage Spark's strengths in terms of availability, productivity, and fault tolerance.

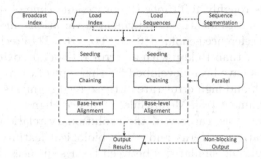

Fig. 1. Workflow of Parallel minimap2. Once the master node has successfully acquired the index, it needs to broadcast the index to all the remaining nodes. All nodes then perform a sequence assignment and load the sequence. Next, all nodes perform a parallel minimap2 alignment process. When the nodes have completed their tasks, the alignment results are output asynchronously.

As the number of parallel frameworks continues to increase, conducting comparisons among them has become crucial in the field of parallel computing. These comparisons aid researchers and developers in making informed decisions regarding the selection of a parallel framework, ensuring optimal performance and reliability for specific application scenarios. Through parallel framework comparisons, one can ascertain the strengths and weaknesses of each framework, determining which is better suited for processing large-scale data, real-time data, or specific application scenarios. Additionally, the comparison facilitates evaluating the performance of each framework on clusters of varying sizes, aiding in the selection of the most suitable framework for a specific hardware environment. Moreover, by comparing parallel frameworks, it helps researchers and developers understand the architecture and implementation of each framework, contributing to a better understanding of how the framework works and how it is used. Such understanding expands the range of options available to researchers and developers, fostering innovation and advancement in the realm of parallel computing.

Reyes-Ortiz et al. [7] conducted the initial comparison between Apache Spark and MPI/OpenMP in their study. In a separate study, Damian et al. [8] assessed the performance of MPI using Unified Parallel C (UPC) and OpenMP on multi-core architectures. However, their evaluation was limited to general scenarios for parallel frameworks. The findings from these evaluations offer limited assistance in specific domains, such as bioinformatics. Consequently, researchers would benefit more from a domain-specific evaluation of parallel frameworks.

This study focuses on the field of bioinformatics to determine the most suitable parallel framework for the data characteristics in this domain. Sequence alignment is a fundamental step in the analysis of genomic data, generating input data for many downstream analysis tools. With the development of DNA sequencing technology, third-generation long sequence sequencing (TGS) offers the opportunity to obtain high-quality genomic data. Third-generation sequencing techniques take roughly 100 times the processing time of second-generation sequencing techniques to identify novel genomic variations and close significant gaps in the human reference genome [10–13]. For now, TGS technology produces read segments longer than 10 kbp with an error rate of approximately 15%. Several alignment tools have been developed for this type of data [14–17]. However, the excessive length and high error rate of these read segments severely increase the cost of the alignment [13]. Meanwhile, sequence alignment, a key technique in genomic data analysis, can reveal the evolutionary relationships, functional relationships, mutation patterns and other biological features of two or more organisms by calculating similarities between their sequences. In genomic studies, sequence alignment is a key step in discovering homologous genes, studying gene family evolution and the structure of the entire genome. Hence, the research on sequence alignment necessitates algorithm optimization to address practical requirements, yet there has been limited exploration of algorithm parallelism. Optimal selection of an appropriate parallel framework can expedite algorithm execution time, enhance execution efficiency, and significantly improve algorithm application.

Therefore, we selected sequence alignment, a key technique for data processing in bioinformatics, as a benchmark and devised a parallel algorithm for the sequence alignment tool minimap2 [9] for a comprehensive evaluation of Spark, Ray and MPI. Dr. Heng Li created the flexible sequence alignment tool Minimap2. It enhances the capacity of base-level alignment using a seed-chain-expansion heuristic paradigm similar to that of minimap [18]. Minimap2 allows for the alignment of spliced read portions, which makes the minimap even more useful. Minimap2 is more than 30 times faster than other long read segment or cDNA alignment tools, and it is 3–4 times faster than popular short read segment alignment tools, according to experimental results [9]. Nonetheless, the current version of minimap2 is confined to utilizing multi-threaded acceleration. This limitation propelled us to create a parallel version of minimap2, serving as a benchmark program to fulfill our framework evaluation and enhance the alignment performance of minimap2 to aid interested researchers.

In summary, this paper aims to address the gap in cross-sectional evaluation of multiple frameworks and provide assistance to researchers in comprehending the performance of each framework within specific contexts, aiding in the selection of suitable frameworks for different tasks. The present study employs minimap2, a third-generation sequence alignment tool, as the benchmark program to develop parallel versions of Spark, Ray, and MPI for the purpose of evaluation. The study performs a thorough assessment of Ray, Apache Spark, and MPI based on criteria such as performance, scalability, ease of use, peak memory, latency, and load balance. The parallel outcomes of the three frameworks are extensively discussed to elucidate their characteristics within specific application scenarios, facilitating the selection and effective implementation of parallel computing frameworks in relevant contexts, thus benefiting related research endeavors.

The contributions of this paper are summarized as follows:

- It conducts the first comprehensive evaluation of Spark, Ray, and MPI to assess their performance in specific domains, enabling researchers to make informed decisions when selecting a parallel framework for particular tasks.
- An in-depth discussion of the three frameworks-Spark, Ray, and MPI-is provided along with an analysis of the variations in experimental results. The results reveal that each framework possesses distinctive advantages. When considering performance alone, MPI outperforms the others, whereas Spark and Ray excel in terms of latency, load balancing, ease of use, ecosystem support, and fault tolerance.
- To the best of our knowledge, this is the first study to propose a parallel algorithm with index sharing, sequence segmentation, parallel alignment, and asynchronous output in order to accelerate a long read segment alignment algorithm with MPI. We also implemented the minimapMPI tool in order to achieve leading performance.

The remainder of the paper is structured as follows: Sect. 2 introduces the parallel algorithm for minimap2, the benchmark program used in our evaluation of the three parallel frameworks, along with the specific implementation of each parallel version. Section 3 presents the experimental environment, setup, and results. In Sect. 4, we discuss the work related to the evaluation of parallel frameworks. Section 5 provides a comprehensive analysis of the experimental results. Finally, in Sect. 6, we summarize our conclusions and outline future research directions.

2   Methods

A seed-chain-extension heuristic model serves as the foundation for the minimap2 utility. Index sharing, sequence segmentation, parallel alignment, and asynchronous output are the four components of the parallel alignment algorithm based on minimap2. In this section, the parallel algorithm of minimap2 and the parallelization of each step for the Spark, Ray, and MPI versions are described in detail. The parallel version of Spark is based on a modified version of IMOS [19], the Ray version refers to our previous work minimapR [20] on accelerating sequence alignment via Ray, and the MPI version is based on the proposed parallel optimization process for minimap2 is shown in Fig. 1. The workflow of minimapMPI is shown in Fig. 2.

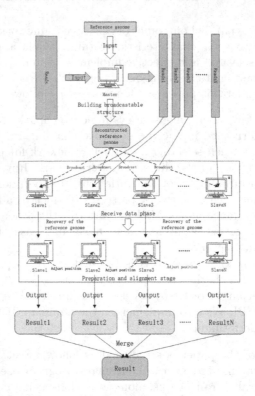

Fig. 2. Workflow of minimapMPI. The master node first reads the reference genome, then constructs the reference genome index and broadcasts the index to the rest of the nodes. The master node then performs sequence segmentation, distributing sequences evenly to the slave nodes. After obtaining the reference genome index, the slave node reads the allocated sequences to be aligned. It then communicates with the neighbouring nodes to adjust the start and end positions of the sequences. Then, the parallel alignment process begins. Finally, the alignment results are output asynchronously and merged by the master node.

2.1   Parallel Algorithm Design

To parallelize minimap2, we propose a general framework-independent parallelism algorithm. As shown in Algorithm 1. The overall idea of the algorithm is designed using a master-slave model. First, one node among all nodes is selected as the master node, and the rest of the nodes are all slave nodes. The master node reads the reference genome and constructs the reference genome index. Then, depending on the parallel framework used, either shared memory or message passing is chosen to enable the slave nodes to obtain a copy of the reference genome index. Based on the number of nodes, the master node initially divides the genome read segments to be processed and distributes them to the slave nodes. The slave nodes only need to read part of the genome read segment data. Next, the slave nodes fine-tune the read segment data based on the divided

---

Algorithm 1. Parallel minimap2

Input: $reference$ reference genome,$reads$ reads to be aligned,n number of nodes, $node$ node type, num node number, $nodes$ node collection

Output: alignment results

1: **function** PARALLELMINIMAP2($reference, reads, n, node$)
2:     $result \leftarrow null$
3:     **if** $node$ is master **then**
4:         $ref, length \leftarrow$ MINIMAP2($reference$),$reads/n$
5:         **for** $i = 0 \rightarrow n$ **do**
6:             SEND($ref, nodes[i]$)
7:             SEND($length, nodes[i]$)
8:         **end for**
9:     **else**
10:         $ref, length \leftarrow$ RECEIVE(ref),RECEIVE($length$)
11:     **end if**
12:     $start, end \leftarrow num * length, (num + 1) * length$
13:     $start, end \leftarrow$ POSITIONCALIBRATION($num, start, end$)
14:     $read, result\_num \leftarrow$ READ($reads, start, end$), MINIMAP2($ref, read$)
15:     **if** $node$ is slave **then** SEND($1, master$)
16:     **else**
17:         $count \leftarrow 0$
18:         **while** $count < n$ **do**
19:             $tmp, count \leftarrow$ REVEIVE(1), $count + tmp$
20:         **end while**
21:         **for** $i = 0 \rightarrow n$ **do**
22:             $result \leftarrow result + result\_i$
23:         **end for**
24:     **end if**
25:     **return** $result$
26: **end function**

---

read segments. Depending on the format, different fine-tuning methods are performed. For fasta format data, it is necessary to identify the start character of the sequence and then communicate with the neighboring nodes to adjust the start-end position. For fastq format data, every four rows are used as a division unit. Once the nodes are ready, the serial alignment process of minimap2 can be started. The results of each process are output in a separate result file. When a process finishes alignment, it reports to the master node and blocks until all the other nodes have finished alignment. When all the nodes are finished, the master node is responsible for merging the results of all the nodes and outputting them.

2.2 Index Sharing

First, after the master node has constructed the reference genome into a reference genome index, it needs to be passed on to each node so that the remaining nodes can get a copy of the reference genome index.

MinimapMPI uses a master-slave mode to read the index. The process that reads the index into memory is identified as the master process. Then, it uses the function MPI_Bcast to broadcast the index to all other processes so that each process gets a copy of the index. The advantage of this method is that only one process needs to access the index file, reducing the I/O overhead of the file. However, while the main process is reading the index, the other processes are inactive, resulting in some resource waste. This is a good strategy since the master process can read the index file rapidly and finish the broadcast in $O(\log p)$ execution time, where p is the number of minimapMPI processes. The data type of the index is a complex nested structure. The parameters of the function MPI_Bcast need to specify the MPI data type and the number of broadcast objects, so the structure of the index cannot be broadcast directly.

To be able to successfully broadcast the reference genome index, we propose a broadcastable reconfigured genome index structure based on the composition structure of the reference genome index. This broadcastable index structure is made up of a number of distinct broadcastable data structures rather than just one array or data structure. We extract and store each part of useful values separately from the original reference genome index, and then put them into the broadcastable data type. This process was adapted to store the values of the structure with arrays and then broadcast those arrays. The complexity of array dimensions was required to be balanced with the number of arrays. The uncertain array length is needed to obtain information through the value variables of the index. The length of arrays was first put into an array and broadcast so that each slave processor could request space for receiving the arrays. These slave processors create arrays, and then uniformly execute broadcast functions. Finally, the received arrays are stored in their index structure, so that each processor has a copy of the index. Only non-zero data was broadcast which can greatly reduce the amount of communication data.

Ray separates the execution logic of task scheduling and the state information of task scheduling. The Global Control State(GCS) stores and manages all kinds of task control and state information, including task topology information, data and task production relationship information, function (task) invocation relationship topology information, and so on. After stripping out this state information for unified management, the scheduler itself becomes a stateless service and therefore has the ability to realize task migration, scaling, and information sharing as mentioned above. Ray uses shared memory to implement an object store on each node, allowing jobs executing on the same node to exchange zero-copy data. For internal scheduling, each node has a local scheduler, and the local scheduler manages the information flow between each node's workers. Each node may include numerous workers, and they work together to store local data, or object-store, using the memory data structure *Apache arrow*. The object is saved in the object store and an ID is returned by the procedure *ray.put()*. GCS may be indexed by ID to gain access to any node's object store and obtain the item. This allows all other nodes to access the reference genome index after it has been produced by the master node.

Spark utilizes HDFS to share memory among several nodes (Hadoop Distributed File System). After the reference genome index is uploaded to HDFS, each node can access the reference genome.

2.3   Sequence Segmentation

Then, the query sequences need to be evenly distributed to each processor. Parallel computing is advantageous because it can complete multiple tasks at the same time. This means that a task can be divided and assigned to several processors to reduce the processing time. The key to dividing tasks is to reasonably divide the data.

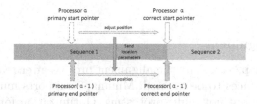

Fig. 3. After dividing the file, the pointer bit required adjusting by identifying the sequence start identifiers such as ">" or "@" Set.

The minimal influence on the original code must be taken into account by minimapMPI. While this is happening, it is important to guarantee load balance as much as possible while separating the tasks of the query sequence. To begin with, minimapMPI measures the size of the query sequence file using the function *stat*. The query sequence file size that each process requires for alignment is then N/p, which can be calculated by dividing the number of processes by the input value. The number of processors is indicated by p, and the size of the query sequence file is indicated by N. This method ensures that the workloads are distributed evenly among the processors with minimum impact on the original code. After dividing the query sequence, the file pointer of each process moved to its corresponding starting position to start reading the query sequence. When the pointer moves to the end position, the reading is stopped. It should be noted that the file position pointer obtained using this division method was probably not at the beginning of a complete sequence. As a result, a sequence between every two processors will be divided into two segments and read into different processors. Therefore, as shown in Fig. 3, after dividing the file, the pointer bit required adjusting by identifying the sequence start identifiers such as ">" or "@" Set. Each non-0 processor moves from its starting position to the next character. If the ">" or "@" identifier is recognized, this position will be set as the new start reading position. This position will be sent to its previous processor so that the former processor can modify its end position. In this way, each process will synchronize the new sequence, and read in the start and end positions to ensure each of the sequence columns can be read correctly.

MinimapR uses a similar approach to minimapMPI, but it needs to be combined with the format of the input file. MinimapR reads the sequence file using the Python function *open*. Every four lines of the file are a sequence, as specified by the fastq file format. For every four lines in the sequence file, MinimapR stores a sequence in the array. Unlike minimapMPI which reads the file directly, minimapR first reads the content of the sequence file when dividing the sequence and divides it into different sub-sequences depending on the content of the sequence file, so it does not need to adjust the position of each processor to align the sequence.

Through HDFS, IMOS distributes memory among numerous nodes, but this is unsuitable for allocating query sequences. Therefore, the entire query sequence needs to be read at each node and then assigned via RDD for subsequent alignment.

2.4   Parallel Alignment

The original minimap2 can start the alignment process after reading the genome index and the sequences to be queried. Minimap2 supports multi-threaded execution of programs, but not multi-processing. Optimization for multi-threading is done in the I/O phase of the file. For example, the overlapping sequence alignment and file I/O times significantly reduce the overall execution time. However, the capacity of a single compute node is always limited, requiring days or even weeks of alignment time when the query sequence is too large.

MinimapMPI implements multi-level parallel alignment, which can accelerate alignment that could take several hours to several weeks and fully utilize the powerful computing resources of large-scale computer clusters. Thus, when multi-process optimization for the computing cluster, a pthread is used based on the multi-thread alignment within the nodes, and multi-process parallelization is conducted based on MPI between nodes.

Ray can define an actor through the function *remote* to handle stateful calculations. *Actor* is actually a class of *remote*. After it is created, the global scheduler will assign *Actor* to the local scheduler and execute the constructor. When the driver calls the function *Actor*, the task will be directly assigned to the local scheduler that creates the constructor for execution without going through the global scheduler. After processing the sequence file, minimapR creates the function *Actor* to run minimap2. Then, all the classes defined by Ray with *@remote* become an *Actor* object. When defining *Actor*, it will be assigned to the node. When using class functions, it will initialize *Actors* and execute functions, and these *Actors* will execute in parallel. It is stateful to compute with the same *Actor*. Then use the method *subprocess* to create the process of minimap2.

2.5   Asynchronous Output

The original minimap2 requires a shell command to redirect the alignment results to a file when the program is executed. Otherwise, the alignment results will be output to the screen, which requires more execution time. When multiple processes output data to the screen at the same time, it causes process blocking to ensure correct output, which can reduce the efficiency of program execution. IMOS, minimapR, and minimapMPI output directly to a file, avoiding process blocking and improving the efficiency of program operation.

Since multiple processes of a parallel program cannot share file pointers, variables, and memory, when multiple processes write a file at the same time, the processes write to the file serially to ensure correct results. While one process writes, the other processes are forced to wait, which can result in a huge waste of resources. To avoid this problem, IMOS, minimapR, and mminimapMPI use a separate output file for each process. Each process creates its output file based on the process ID number. After the alignment of all sequences, all the result files can be immediately combined into a general file using the straightforward shell command *cat* because there is no correlation between the query sequences.

3   Results

3.1   Experimental Setup

Table 1 displays the computing nodes' configuration for the system. In our research, we used the "-ax map-ont" option of minimap2, which employs parameters tailored for Oxford nanopore read segments, to align the dataset to a reference genome. Each experiment was carried out three times, with the average findings being compared. For reasons of space, only some of the graphs of the results are shown in the following experiments.

Table 1. Hardware configurations

CPU model		Intel(R) Xeon(R) CPU E7-8890 v3	
Architecture	x86_64	**Base Frequency**	2.50 GHz
# Cores/nodes	144	**Device Memory Size**	2015 GB
Threads/cores	2	**Compiler**	gcc 7.5.0

3.2   Input Data

Real and simulated datasets were used in the experiments. The human reference genome, GRCh38/hg38, serves as the reference genome. Pbsim [21] was used to produce the simulated reads from chromosome 1 of the human reference genome hg38. We used six real datasets that are samples from WGS of intact genomic DNA on promethION platform with accession numbers SRX14742398, SRX15792102, SRX13323058, SRX14498310, SRX14498314 and SRX12255900 published in NCBI Sequence Read Archive [22]. They are used for measuring speed on a real dataset. Table 2 presents the characteristics of seven datasets used to examine the performance. These datasets can be accessed at https://www.ncbi.nlm.nih.gov/sra.

Table 2. Datasets Characteristic

Dataset	Organism	Platform	Size on disk
SRX14742398	Homo sapiens	Oxford Nanopore	216MB
SRX15792102	Homo sapiens	Oxford Nanopore	1.3GB
SRX13323058	Homo sapiens	Oxford Nanopore	9.9GB
SRX14498310	Homo sapiens	Oxford Nanopore	35GB
SRX14498314	Homo sapiens	Oxford Nanopore	61GB
SRX12255900	Homo sapiens	Oxford Nanopore	88.9GB
Pbsim200000	Homo sapiens	Simulation sequence	1.12GB/cores

3.3 Results Consistency

Since the output of the programs cannot be modified after parallel optimization, the correctness of the parallel programs was evaluated. We used minimap2, minimapR, and minimapMPI to process the datasets SRX14742398, SRX15792102, SRX13323058, SRX14498310, and SRX14498314. IMOS, minimapR, and minimapMPI run under 2, 4, 8, 16, 32, 64, and 128 processors with 1 thread, respectively. Minimap2 runs under 1 processor with 1 thread. To make sure that all sequences were in the same order, we sorted the result files. The output files from minimap2 and the parallel programs were then compared. Our tests revealed that IMOS, minimapR, and minimapMPI had no bearing on the program's outcome.

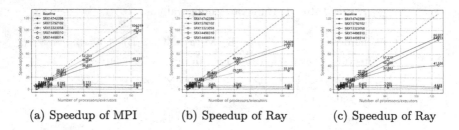

| (a) Speedup of MPI | (b) Speedup of Ray | (c) Speedup of Ray |

Fig. 4. Speedup of MPI, Ray, and Spark on different number of processors. The experiments were conducted at parallelism levels of 2, 4, 8, 16, 32, 64, and 128 respectively. When the dataset is small, the speedup of increases slowly and remains relatively low as the parallelism increases. When the dataset is large enough, the speedup is noticeable. MPI achieves the best performance.

3.4 Strong Scalability

Strong scalability refers to the ability of a program to effectively utilize an increasing number of processors while keeping the problem size constant. The objective is to achieve faster calculation completion. Ideally, the runtime of the program should approach $1/p$, where p represents the number of processors, in comparison to a serial program.

We compared the strong scalability of IMOS, minimapR, and minimapMPI. The parallelization results for the data sets SRX14742398, SRX15792102, SRX13323058, SRX14498310, and SRX14498314 in the three frameworks are shown in Fig. 4. When the data size is relatively small, the effect of parallelization is not obvious due to the serial overhead of the program itself. As the data size increases, the speedup of the three frameworks increases significantly. MinimapMPI has the highest speedup and is close to the linear speedup. MinimapR has the lowest speedup ratio of the three frameworks. It is worth noting that at a moderate experiment size, Ray's speedup ratio does not increase significantly as the number of nodes increases, implying that Ray and scalability are not as good as MPI and Spark at this size.

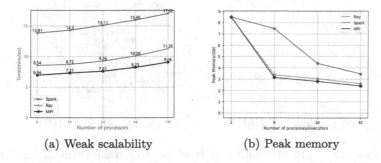

(a) Weak scalability   (b) Peak memory

Fig. 5. Wall-time with increasing numbers of processors and data size and comparison of peak memory in GB per process.

3.5   Weak Scalability and Peak RAM Memory

Weak scalability is to keep the computational size of a single node constant and to keep the parallel efficiency constant as the number of processors increases. We want to be able to process more data in the same amount of time. The size of the problem that can be addressed grows p times the initial problem size in the ideal situation.

The same work was assigned to each processor in the weak scalability test. To create datasets of simulation sequences, we employ pbsim. Each processor processes 200000 sequences from the Pbsim200000 dataset. The average read length of the two datasets is 3000bp. The alignment results are shown in Fig. 5.a. The completion time of IMOS, minimapR and minimapMPI is relatively stable.

Figure 5.b compares the peak memory in GB used by each process for IMOS, minimapR, and minimapMPI. The difference in memory use when employing 8 cores is obvious from the image, as can be deduced from it. Additionally, minimapR and minimapMPI always use less RAM than IMOS. This is mostly due to the fact that just a portion of the query sequence needs to be read by each minimapMPI and minimapR process.

3.6 Efficiency

In addition, the efficiency of parallelism is one of the important metrics to evaluate the performance of parallelism. The ideal speedup for a parallel program with p processors is p. However, since each processor cannot devote 100% of its computing power to computation, the effective utilization time metric for each processor becomes efficient. We measured the efficiency of minimapR, IMOS, and minimapMPI at five different data sizes, 216 MB, and 61G, and the results are shown in Fig. 6.

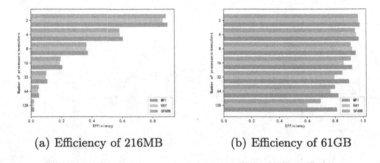

(a) Efficiency of 216MB   (b) Efficiency of 61GB

Fig. 6. Efficiency of 216 MB, and 61 GB on different frameworks. In the 216 MB size experiment, MPI performed the best, outperforming the remaining two parallel frameworks. In the 61 GB size experiment, Ray, Spark and MPI were all at a high level of efficiency. However, at lower parallelism levels, the difference in efficiency between the three is not significant, but as parallelism increases, MPI has a clear advantage.

Where, the efficiency of minimapMPI is the most efficient among all experiments, and is more stable as the number of processors increases when the data size is large enough. The efficiency of IMOS and minimapR decreases significantly more than minimapMPI as the number of processors increases for all data sizes. This shows that our proposed minimapMPI tool itself has a smaller overhead and is able to devote more computational resources to efficient parallel alignment. The difference in efficiency between Ray and Spark is small, but Ray's efficiency drops off more quickly as more processors are added.

3.7 Throughput

Throughput is another important measure of a parallel program, reflecting how many tasks the parallel program can handle per unit of time. Therefore, we conducted experiments on two real datasets(SRX14742398, and SRX14498314) to measure the throughput of minimapR, IMOS, and minimapMPI, and the results are shown in Fig. 7.

The experiments on the small dataset show that the overall throughput of the program is not high due to the overhead of minimap2 itself and the parallel framework, and the increase in throughput is not significant as the number of processors increases. However, as the data size grows, the throughput increases rapidly for minimapR, IMOS, and minimapMPI. However, the throughput of minimapR, IMOS and minimapMPI does not increase all the time and reaches its maximum. As the size of the dataset continues to increase from 9.9 GB to 61 GB, the throughput even decreases to a certain extent. MinimapMPI achieves the best results, being consistently above average.

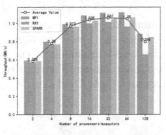
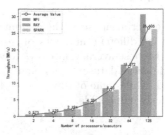

(a) Throughput of 216MB   (b) Throughput of 61GB

Fig. 7. Throughput of 216 MB, and 61 GB on different frameworks. When the size of the dataset is small, MPI's throughput is slightly higher than Spark and Ray's. The difference in throughput between Spark and Ray is not significant. The throughput of the three frameworks increases rapidly as the parallelism increases. MPI has the highest throughput and is always in the lead. Ray and Spark are not very different in the low parallelism experiments, but Spark performs better at high parallelism. In experiments with a dataset size of 61 GB, the throughput of the three frameworks did not differ much when the parallelism was low. As parallelism increases, MPI outperforms Ray and Spark. Spark performs slightly better than Ray.

3.8   Latency

Latency is used to measure the ability of a node to complete a task within a specified time. The latency of completing a task varies from node to node due to the different division of tasks between nodes. As shown in Fig. 8.a, the minimum, average and maximum latency for the three frameworks are shown for a node count of 128. MPI has the lowest average and maximum latency, while Ray has the highest average and maximum latency, but the lowest node has less latency than MPI. The latency of Spark is close to MPI, but the maximum latency is higher than MPI, and it is worth noting that the minimum latency of Ray is higher than the other two frameworks when the dataset is small.

Figure 8.b shows a scatter plot of node latencies for experiments conducted on dataset SRX14498314. As shown, MPI and Spark have a more concentrated latency distribution, while Ray nodes have a wider latency distribution, while having the lowest minimum latency and highest maximum latency.

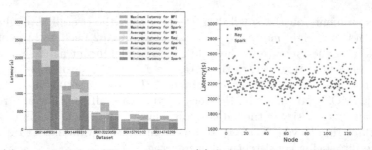

(a) Latency of 61GB on different frameworks.

(b) Scatter plot of 61GB with different number of nodes.

Fig. 8. The experiments show the minimum, maximum and average latency of the three frameworks for different numbers of nodes. MPI has the lowest maximum latency. Spark has the best average latency. Ray has the lowest minimum latency with the large experimental dataset. MPI and Spark and node latency distribution is more concentrated, while some of Ray's nodes have very low latency and can complete their tasks as quickly as possible.

3.9 Load Balance

A parallel framework should be able to balance the workload and ensure that tasks are distributed evenly across all nodes. Load balancing demonstrates the ability of parallel frameworks to balance workloads. As shown in Fig. 9, MPI and Spark overall load balancing are more focused and better than Ray, while Spark still has excellent load balancing as the dataset size decreases, while MPI shows some degradation. However, when the dataset size is further reduced, MPI performs better in load balancing, with little difference between Ray and Spark.

4 Related Work

Although Spark, Ray, and MPI have been around for some time, and Spark and MPI in particular have been widely used in several domains, performance research work on these frameworks is still limited. Currently, only a fraction of researchers has begun to look at the performance evaluation of the frameworks themselves.

Rajkumar et al. [24] compared the performance of multi-threaded fine-grained and coarse-grained computational problems by using a hybrid MPI and MPI+OpenMP approach and evaluated the applicability of multi-core clusters based on the nature of the problem. The results show that the hybrid programming approach yields better performance results in most cases. For fine-grained or coarse-grained problems and computationally intensive or data-intensive problems, multi-core processor clusters are most suitable for CGCI-type problems and least suitable for FGDI-type problems.

The Apache Hadoop and high-performance computing paradigms were described by Jha et al. [23] as solutions to a number of issues related to processing massive volumes of data that call for handling large-scale data distribution,

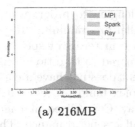
(a) 216MB

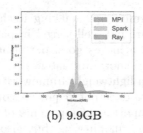
(b) 9.9GB

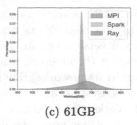
(c) 61GB

Fig. 9. Load balance of 216MB, 9.9GB, 61GB on different frameworks. Experiments have found that MPI has better load balancing, with the majority of access points concentrating around 2.5MB of load. And in experiments of 216MB, Ray and Spark performed similarly in terms of load balancing. In the 9.9GB size experiment, Spark performs far better than MPI and Ray in terms of load balancing, with more even node loads. Ray, on the other hand, has the worst load balancing. In the 61GB size experiment, Both MPI and Spark achieved good load balancing performance, with node workloads concentrated around 670MB. the load balancing of Ray was poor.

co-location, and scheduling. Their micro-benchmark demonstrates that MPI performs better than the Hadoop-based implementation and also made note of the fact that Spark, the Hadoop framework, has significantly improved performance by incorporating HPC techniques while maintaining a very high and accessible level of abstraction and coupling to resources managed in HDFS or RDD chunk size. The use of communication primitives in HPC applications improves performance since communication operations and application-specific files don't have a common runtime framework for processing data objects efficiently.

Due to the limitations of technology development, these research efforts have mainly focused on the performance of MPI itself in clusters, or two-by-two evaluation of Hadoop, Spark, and MPI, while few researchers have shifted their attention to a comprehensive evaluation of more frameworks. In particular, no research has been conducted on some of the later proposed frameworks (e.g. Ray).

5 Discussion

This paper comprehensively compares the performance of three frameworks: Spark, Ray, and MPI. The benchmark program used for evaluation is the sequence alignment tool minimap2. We propose a parallel algorithm for minimap2 and parallelize it with Spark, Ray, and MPI respectively. The results demonstrate that the parallelized version of minimap2 using MPI achieves superior performance in terms of speedup, efficiency, throughput, peak memory, and latency. MPI, a general-purpose programming framework for parallel computing applications, does not use shared memory but communicates between processes via message passing. The efficiency of communication is strongly influenced by the size of the data being communicated and the communication network. Additionally, the implementation of the MPI version involves substantial modifications to the source code of minimap2, making it challenging to program and modify for researchers who are not well-versed in parallel programming. Ray's parallel version of minimap2 achieves good performance in terms of speedup, efficiency, throughput, peak memory and latency. Compared to traditional distributed frameworks, Ray is a lightweight framework that does not have a complete ecosystem and therefore does not require a complex build process. GCS and Workers are the most important components of Ray. Centralized management provides good scheduling monitoring, but also affects performance. The parallel version of minimap2 implemented with Spark demonstrates excellent performance in terms of speedup, efficiency, throughput, peak memory, and load balance. Spark is a mature distributed framework, but it still has some limitations. Spark is designed to be fault-tolerant, so the data flow needs to be disk-based, which ensures data consistency, but also affects the data processing time. In addition, Spark is written in Scala, and the JVM-based management brings problems such as GC, more memory usage, and excessive objects created. However, Spark offers a comprehensive ecosystem and, although moderately challenging to program, it remains a favorable choice for distributed computing.

One advantage of Apache Spark is that it relaxes the constraints of the programming model and increases the flexibility of the system, but the core programming model itself is more difficult to use and implement, requiring developers to consider more stability and performance during development. In contrast, MPI provides higher flexibility and is compatible with a wide range of algorithms, albeit demanding more programming effort. Ray integrates certain advantages from both Spark and MPI. As a general-purpose parallel framework, Ray prioritizes minimalism in its API design. If we compare only from the perspective of the core framework, we can see that Ray's most prominent advantage is its ability to dynamically build task topology logic diagrams. Hence, Ray is particularly well-suited for scenarios characterized by complex task flows that require dynamic adjustments.

6   Conclusion

This paper presents a comprehensive evaluation of Apache Spark, Ray, and MPI regarding scalability, efficiency, throughput, peak memory, latency, and load balance. We utilize the sequence alignment tool minimap2 as a benchmark program to aid researchers in selecting appropriate parallel frameworks. Furthermore, we introduce a parallelization algorithm for minimap2 and develop minimapMPI, a parallelization tool that outperforms Spark and Ray in most performance comparisons. This tool serves as a valuable reference for sequence alignment research. The results of our experiments demonstrate that MPI outperforms in terms of speedup, efficiency, throughput, and latency. The performance of Spark-accelerated minimap2 is also commendable, displaying the optimal load balance. Although Ray performs slightly below the other two frameworks, it excels in achieving the lowest minimum latency.

This paper focuses on evaluating performance within the specific domain of bioinformatics, which is of particular relevance to certain researchers. Additionally, we do not compare the fault tolerance of these three frameworks, but according to some research, Apache Spark differs from the other two in terms of fault-tolerant recovery and data replication [3]. Spark deals with these issues effectively at the cost of execution speed. While MPI provides a solution geared towards High Performance Computing, it is vulnerable to failures. In the past, the integration of Apache Spark and MPI has been a mainstream research direction. According to the work of Philipp Moritz et al. [4], Ray has excellent performance and fault-tolerance support, and it offers a new option for parallel computing. Next, we will continue our comparison of more parallel frameworks from a more comprehensive perspective.

Acknowledgements. This work was supported by NSFC Grants U19A2067 and NSFC Grant 62102427; National Key R&D Program of China 2022YFC3400400; Top 10 Technical Key Project in Hunan Province 2023GK1010, Key Technologies R&D Program of Guangdong Province (2023B1111030004 to FFH). The Funds of State Key Laboratory of Chemo/Biosensing and Chemometrics, the National Supercomputing Center in Changsha (http://nscc.hnu.edu.cn/), and Peng Cheng Lab.

References

1. Abuín, J.M., Lopes, N., Ferreira, L., et al.: Big data in metagenomics: apache spark vs MPI. PLOS ONE **15** (2020)
2. Leiserson, C.E., et al.: There's plenty of room at the top: what will drive computer performance after Moore's law? Science (New York, N.Y.) **368**, 6495 (2020). https://doi.org/10.1126/science.aam9744
3. Zaharia, M.A. et al.: Spark: cluster computing with working sets. In: USENIX Workshop on Hot Topics in Cloud Computing (2010)
4. Moritz, P., Nishihara, R., Wang, S., et al.: Ray: a distributed framework for emerging AI applications (2017)
5. The MPI Forum, Corporate. MPI: a message passing interface. Supercomputing'93 (1993)
6. Kumar, D.S., Rahman, M.A.: Performance evaluation of apache spark vs MPI: a practical case study on twitter sentiment analysis. J. Comput. Sci. **13**(12), 781–794 (2017)
7. Reyes-Ortiz, J.L., Oneto, L., Anguita, D.: Big data analytics in the cloud: spark on hadoop vs MPI/OpenMP on Beowulf. Procdia Comput. Sci. **53**, 121–130 (2015)
8. Mallón, D.A., et al.: Performance Evaluation of MPI, UPC and OpenMP on Multicore Architectures. PVM/MPI (2009)
9. Li, H.: Minimap2: pairwise alignment for nucleotide sequences. Bioinformatics **34**(18), 3094–3100 (2018)
10. Lee, H., et al.: Third-generation sequencing and the future of genomics. BioRxiv, p. 048603 (2016)
11. Rhoads, A., Au, K.F.: PacBio sequencing and its applications. Genomics Proteomics Bioinform. **13**(5), 278–289 (2015)
12. Jain, M., Olsen, H.E., et al.: The Oxford nanopore MinION: delivery of nanopore sequencing to the genomics community. Genome Biol. **17**(1), 239 (2016)

13. Alser, M., Rotman, J., et al.: Technology dictates algorithms: recent developments in read alignment (2020). arXiv preprint arXiv:2003.00110

14. Li, H.: Aligning sequence reads, clone sequences and assembly contigs with BWA-MEM (2013). arXiv, 1303.3997

15. Lin, H.-N., Hsu, W.-L.: Kart: a divide-and-conquer algorithm for NGS read alignment. Bioinformatics **33**, 2281–2287 (2017)

16. Sedlazeck, F.J., et al.: Accurate detection of complex structural variations using single-molecule sequencing. Nat. Methods (2018). https://doi.org/10.1038/s41592-018-0001-7

17. Sović, I., et al.: Fast and sensitive mapping of nanopore sequencing reads with GraphMap. Nat. Commun. **7**, 11307 (2016)

18. Li, H.: Minimap and miniasm: fast mapping and de novo assembly for noisy long sequences. Bioinformatics **32**(14), 2103 (2015)

19. Yousefi, M.H.N., Goudarzi, M., et al.: IMOS: improved meta-aligner and Minimap2 on spark. BMC Bioinform. **20**(1), 51 (2019)

20. Wang, Z., et al.: MinimapR: a parallel alignment tool for the analysis of large-scale third-generation sequencing data. Comput. Biol. Chem. **99** (2022)

21. Ono, Y., et al.: PBSIM: pacbio reads simulator-toward accurate genome assembly. Bioinformatics (Oxford, England) **29**(1) (2013). https://doi.org/10.1093/bioinformatics/bts649

22. Ncbi Sequence Read Archive (SRA). www.ncbi.nlm.nih.gov/sra. Accessed 2018

23. Jha, S., Qiu, J., Luckow, A., Mantha, P., Fox, G.C.: A tale of two data-intensive paradigms: applications, abstractions and architectures. In: Proceedings of the IEEE International Congress on Big Data, June 27–2 July, IEEE Xplore Press, Anchorage, AK, USA, pp. 645–652 (2014)

24. Sharma, R., Kanungo, P.: Performance evaluation of MPI and hybrid MPI+OpenMP programming paradigms on multi-core processors cluster. In: 2011 International Conference on Recent Trends in Information Systems, pp. 137–140 (2011)

Fairness Analysis and Optimization of BBR Congestion Control Algorithm

Bo Zhang(✉), Ying Wang, and Xiya Yang

Communication University of China, Beijing 100000, China
{zhangbo2015,yingwang,yangxiya}@cuc.edu.cn

Abstract. In distributed computing, data transmission is a crucial component, and the efficiency and fairness of transmission are key factors influencing system performance. Bottleneck bandwidth and round-trip time (BBR) is a new congestion control algorithm, it can improve the efficiency and stability of data transmission, thereby enhancing the overall performance of distributed computing. It aims to improve the performance of traditional Transmission Control Protocol (TCP) by measuring the bottleneck bandwidth and round-trip propagation time. However, in many cases, the BBR algorithm suffers from serious fairness problems, mainly in terms of its internal round-trip time (RTT) fairness. We construct a simple network topology based on NS-3 to evaluate the fairness problem. To optimize the RTT fairness problem of BBR, we propose the BBR-Optimization (BBR-O) algorithm, its pacing_gain is related to the size of the RTT. The experimental results show that the BBR algorithm prefers long RTT flows. In contrast, the BBR-O algorithm can effectively reduce the goodput difference between flows with different RTT sizes, increasing the values of inflight and sendrate for short RTT flows. The BBR-O algorithm balances the sendrate between different flows by setting the pacing_gain concerning the RTT size, which effectively alleviates the RTT fairness problem of BBR.

Keywords: BBR · Congestion control algorithm · Fairness · RTT

1 Introduction

A new TCP congestion control algorithm called bottleneck bandwidth and round-trip time was proposed by Cardwell, Jacobson [1], et al. in 2016. The algorithm explores a high throughput and low latency workspace by estimating the bottleneck bandwidth and round-trip propagation delay of the network path. The goal of BBR is to ensure that packets arrive at a rate equal to the bottleneck bandwidth while filling the pipe so that the number of packet in the path that is in transmission equals the bandwidth-delay product (BDP). The bandwidth-delay product is the product of maximizing throughput and minimizing latency.

© The Author(s), under exclusive license to Springer Nature Singapore Pte Ltd. 2024
Z. Tari et al. (Eds.): ICA3PP 2023, LNCS 14489, pp. 77–88, 2024.
https://doi.org/10.1007/978-981-97-0798-0_5

However, BBR has drawbacks and challenges, such as potential buffer over-flow and unfair competition with other congestion control algorithms that follow AIMD principle. Some literature has pointed out that although BBR can significantly improve throughput and reduce latency, there are fairness issues between flows with different Round-Trip Time (RTT). This means that, in the presence of multiple BBR flows sharing a bottleneck link, flows with longer RTT receive more bandwidth than those with shorter RTT, resulting in unfairness. This unfairness affects the quality and efficiency of the network, and highlighting the importance of studying and improving the fairness issue of the BBR congestion control algorithm to enhance network performance and efficiency.

In this paper, we aim to investigate and study the RTT fairness of the BBR algorithm, the effect of the number of flows in the link on it. We propose a new optimization algorithm, BBR-O, for the RTT fairness problem.

The rest of this paper is organized as follows: In the next section, the performance of the fairness problem of the BBR algorithm and the reason for it is analyzed. The third section proposes a new optimization algorithm, BBR-O, for the RTT fairness problem of BBR. The fourth section tests the fairness issues of BBR and the evaluation of the performance of BBR-O. The last section summarizes the main findings of this paper and gives a brief overview of future work.

2   Related Work

Fairness is an important metric in network congestion control algorithms. If an algorithm improves the utilization of network resources while not ensuring fair access for all users, it will affect the overall quality of the network and the user's experience. Therefore, the study of the fairness of the BBR algorithm can provide better direction and guidance for the development of the network congestion control algorithm, so as to better balance the relationship between network resource utilization and fairness.

In 2017, Mario Hock [2] et al. evaluated the performance of BBR in Linux Kernel 4.9, and found that it performed well for a single flow at the bottleneck. However, the observed behavior of multiple flows did not align with BBR's original goals.

In 2018, Atxutegi [3] et al. compared New Reno [4], CUBIC [5], and BBR in a 4G network using the MONROE simulation environment. They concluded that TCP BBR outperformed TCP New Reno and TCP CUBIC in most cases in cellular network environments. Vivek Jain [6] et al. designed and implemented an evaluation of BBR in NS-3, validating the proposed model by executing different simulation sets to ensure that the model demonstrated key features of TCP BBR and performed tests on the network simulator.

Zhang [7] et al. used NS-3 to establish different network communication models. Through analysis of the simulation data, they found that BBR performed better than the BIC algorithm on high latency, high bandwidth networks, and there were fairness issues between BBR and other versions of TCP protocols. They also confirmed the presence of RTT fairness issues in BBR flows.

In 2022, Daniela M. Casas-Velasco [8] et al. analyzed the performance of TCP CUBIC and TCP BBR protocols in the presence of background traffic. The analysis was performed using TCP simulations and considered high-capacity end-to-end data connections and different connection duration. The results showed that the BBR protocol had higher throughput and fairness than the CUBIC protocol.

In 2023, Syed Z. Ahmad [9] et al. proposed an incast recovery BBR algorithm that introduced additional controls such as incast shaping to handle queue buildup during TCP incast. Compared to standard BBR and CUBIC implementations, the modifications in the BBR implementation were studied in terms of performance parameters such as flow completion time, throughput, RTT variation, and fairness to other competing flows.

The current optimization algorithm mainly focuses on reducing the throughput of long RTT streams, so that the throughput of short RTT is slightly improved, thereby alleviating the unfairness problem between the two, and ignoring the throughput improvement of short RTT streams. Moreover, most algorithms conservatively set the cut-off point of buffer overflow to 1.25BDP, which affects the improvement of link utilization. In view of the above problems, the algorithm design goal is to improve the throughput of short RTT streams, consider the overall situation of the link, and improve the link utilization.

3 Problem Analysis

3.1 RTT Fairness in BBR Algorithm

In traditional TCP congestion control algorithms, both loss-based and delay-based algorithm, shorter RTT flows have higher performance. However, in BBR, long RTT flows dominate over short RTT flows, in contrast to the preference for short RTT flows in traditional TCP congestion control algorithms. The phenomenon is that long RTT flows occupy a large amount of bandwidth, while short RTT flows suffer from significant throughput degradation. The latest version of BBR, BBRv2 [10], proposed by Google, still does not optimize the RTT fairness issue.

In terms of RTT fairness analysis, the disparity between long RTT flows and short RTT flows arises due to queue buildup, leading to substantial imbalances in their sending rates. Consequently, our emphasis is on scrutinizing the correlation between sending rates and RTT.

When multiple BBR flows share a common bottleneck link, the application of Little's Law enables us to calculate the approximate throughput of flow i at time t (i.e., the sending rate of BBR flow i at time t):

$$V_i(t) = \frac{B_i(t)}{T_i + D_i(t)} \tag{1}$$

where $B_i(t)$ represents the packets sent by flow i at time t, T_i denotes the minimum RTT of flow i, and $D_i(t)$ is the queuing delay of flow i at time t. In steady-state, the value of parameter $B_i(t)$ may be limited by pacing or CWND, depending on the accuracy of bandwidth share estimation.

When the queuing delay $D_i(t)$ is small, whether due to inaccurate bandwidth estimation or insufficient utilization, inflight packets are constrained by pacing. Thus, the average value of $B_i(t)$ is $C_i(t - \Delta t)T_i$, where $C_i(t - \Delta t)T_i$ is the previous bandwidth estimate and reaches the maximum value of $1.25 * C_i(t - \Delta t)T_i$ during the first probing stage.

$$B_i \max(t) = 1.25 * C_i(t - \Delta t)T_i \tag{2}$$

From the formula $V_i(t) = B_i(t)/RTT$,

$$V_i \max(t) = \frac{B_i \max(t)}{RTT} = \frac{1.25 * C_i(t - \Delta t)T_i}{T_i + D_i(t)} \tag{3}$$

The probing cycle of BBR flow is 8 $RTprop$, i.e., $\Delta t = 8T_i$. Thus, we have:

$$V_i \max(t) = \frac{1.25 * C_i(t - 8T_i) * T_i}{T_i + D_i(t)} \tag{4}$$

Rearranging this equation, we obtain:

$$V_i \max(t) = \frac{1.25 * T_i}{T_i + D_i(t)} * C_i(t - 8T_i) \tag{5}$$

Since all flows share the same bottleneck buffer, and packets from any flow are uniformly distributed in the queue, we can effectively assume that the queuing delay $D_i(t)$ observed by each flow is the same at any time t, which we denote as $D(t)$. Therefore, the effective probing gain

$$pacing\_gain = \frac{1.25 * T_i}{T_i + D(t)} < 1.25 \tag{6}$$

This means that once a queue is formed, the actual bandwidth cannot increase at the rate of pacing_gain $= 1.25$, but rather at a rate lower than 1.25 due to the presence of queuing delay $D(t)$.

The effective probing gain $(1.25 * T_i)/(T_i + D(t))$ for bandwidth measurement shows that T_i and $D(t)$ have an impact on the bandwidth probing of flows with different RTT lengths, revealing the RTT unfairness of BBR. That is, the larger the RTT, the greater the effective probing gain of BBR flow, and the stronger the bandwidth preemption ability.

Firstly, during the ProbeBW state, when exploring more bandwidth with pacing_gain $= 1.25$, [11] too many persistent queues are created at the bottleneck, resulting in greater queuing delay. Compared with short RTT flows, long RTT flows generate larger surplus queue backlog and delivery rate measurements. Short RTT flows measure a lower delivery rate and slow down to match the measurement result, making them more disadvantaged in the competition. Secondly, the BDP of the data to be transmitted is calculated by multiplying the measured BtlBw and RTprop. Due to the higher BDP of long RTT flows, they transmit more excess data than short RTT flows, resulting in more buffer backlog. In contrast, short RTT flows do not use enough bandwidth or bottleneck buffers, and the difference in buffer occupancy causes fairness issues in throughput between the final hosts. This bias of BBR towards long RTT flows can be easily exploited by malicious users to increase network latency and obtain more bandwidth.

4   Algorithm Optimization

Based on the RTT fairness model analysis, we made relevant optimizations and proposed the BBR-O algorithm. The BBR-O algorithm was designed to adjust the pacing_gain between long and short RTT flows and make it adaptive to the RTT size, so that the sending rate is correlated with the RTT size. A parameter α was introduced, which is defined as the ratio of the current delay to the maximum delay and can reflect the link utilization well. Its definition is shown in the following equation:

$$\alpha = \frac{T_n}{T_{\max}} \qquad \alpha \in (0, 1] \tag{7}$$

Here, T_n represents the delay obtained from the last ACK of the ith flow, $T_{\max}$ represents the maximum delay for ACK updates, and the value range of α is kept between $(0, 1]$.

I_i represents the number of inflight packet, and $Q_i = I_i - RTT_{\min} * V_i(t)$ represents the remaining number of packet in the bottleneck link buffer. The BBR algorithm adds the rs structure, which can directly obtain the value of rtt_us, that is, the value of T_n. Then, rtt_us is used to calculate the values of $T_{\min}$ and $T_{\max}$.

Algorithm 1 describes the logical implementation of the BBR-O algorithm, which incorporates the adjustment factor β, γ in relation to the parameter α to enable the adaptive modification of pacing_gain. Firstly, the update of the α-related coefficient is implemented, and the current delay T_n and the maximum delay $T_{\max}$ after ACK update are recorded. The BBR-O algorithm is implemented in the ProbeBW state. If there is too much packets transmission, BBR will work at a point to the right of the original Kleinrock working point. If there are two BBR flows with different RTTs, both flows will transmit an additional $1 \sim 1.5BDP$ of inflight data, so 1.5 BDP is used as a boundary point for judgment.

When the number of inflight data flows in the link is less than the BDP size or there are no packet in the bottleneck link buffer, it means that the bottleneck link pipeline has not yet been occupied or filled, allowing BBR flows with different RTTs to send more packets to occupy the idle bandwidth. At this time, the upper limit of *pacing_gain* for upward probing can be slightly increased from

---

Algorithm 1: BBR-O

---

    input : $T_{\min}$ $T_{\max}$
    output: *pacing_gain*

1 **for** *every ACK* **do**
2   **if** $T_n > T_{\max}$ **then**
3         $T_{\max} \leftarrow T_n$;
4   **end**
5   **if** $T_n < T_{\min}$ **then**
6         $T_{\max} \leftarrow T_n$;
7   **end**
8 **end**
9 **if** $BBR \rightarrow \text{mode} = \text{Pr}\,obeBW$ **then**
10   **if** $I_i < BDP$ *or* $Q_i = 0$ **then**
11   *pacing_gain* = *pacing_gain* + β;
12   **end**
13   **if** $BDP \leq I_i \leq 1.5BDP$ **then**
14   *pacing_gain* = 1;
15   **end**
16   **if** $I_i > 1.5BDP$ *or* $loss \geq 2\%$ **then**
17   *pacing_gain* = *pacing_gain* − γ;
18   **end**
19 **end**

1.25 to 1.5, and the range of values for *pacing_gain* under this state can be set to [1.25, 1.5) by adjusting the factor β.

When the number of inflight data flows in the link exceeds the BDP size but is less than 1.5 times the BDP size, it means that the bottleneck link has been filled, and the number of data packets in the bottleneck buffer is slowly increasing, gradually forming a queue. At this time, the bottleneck link no longer has enough space to transmit more data packets. Therefore, *pacing_gain* only needs to be kept at 1 for steady-state probing of bandwidth.

When the number of inflight data flows in the link exceeds 1.5 BDP size or the packet loss rate of more than 2% occurs, it means that the bottleneck link buffer has been filled or even packet loss has occurred. At this time, the *pacing_gain* used in the original BBR for emptying the queue downward is 0.75, but in order to empty the excess queue more quickly, *pacing_gain* can be slightly adaptively reduced to 0.5 by adjusting the factor γ, and the range of values for *pacing_gain* under this state can be set to (0.5, 0.75].

In order to maintain the adaptive change of pacing_gain within the previously mentioned range, it is essential to limit the variation of the adjustment factor β to the interval [0, 0.25) and limit the variation of the adjustment factor γ to the interval (0, 0.25]. Moreover, it is crucial to establish an inverse proportionality between β and α and establish a proportionality between γ and α. By introducing the adjustment factor β and γ, the BBR-O algorithm achieves adaptive changes to the pacing_gain parameter, enabling efficient data transmission and optimizing network performance in various conditions. The relationship between these parameters is illustrated by the fitting curve shown in Fig. 1.

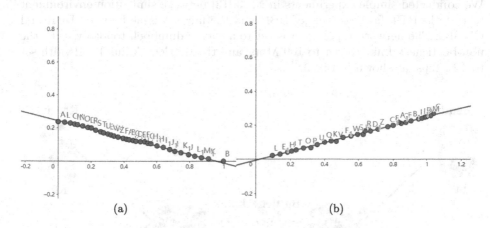

(a)   (b)

Fig. 1. The curve fitting graph of β and γ

Where the abscissa is α, and the ordinates of the two figures are β and γ. According to the fitted curve obtained in Geogebra [12], the expression of β can be derived as:

$$\beta = -\frac{\alpha}{4} + \frac{1}{4} \qquad \beta \in (0, 0.25] \tag{8}$$

$$\gamma = \frac{\alpha}{4} \qquad \gamma \in (0, 0.25] \tag{9}$$

For long flows, their T_n values are larger than those of short flows, which means that their corresponding α values are relatively large, resulting in smaller β values. It effectively prevents long flows from occupying too much bandwidth when competing with short flows. For short RTT flows, their corresponding β values are relatively large, leading to an increase in their sending rate and thus improving their competitiveness in the competition.

For long RTT flows, their T_n are larger than those of short streams, so the corresponding α value is relatively large, so the corresponding γ will be larger, and the emptying speed is faster, effectively avoiding the long flows from occupying too much bandwidth when competing with the short flows; For short RTT flows, the corresponding γ value is relatively small, and the sending rate is reduced less than that of long RTT flows, effectively improving the advantage of short flows in the competition.

5   Experiment

5.1   Experimental Setup

We conducted simple experiments in NS-3 [13] network simulation environment to test the RTT fairness issue of BBR and the fairness issue between BBR and CUBIC. The network topology was set to a classic dumbbell topology, with the non-bottleneck link size set to 100 Mbps and the bottleneck link bandwidth set to 12 Mbps, as shown in Fig. 2.

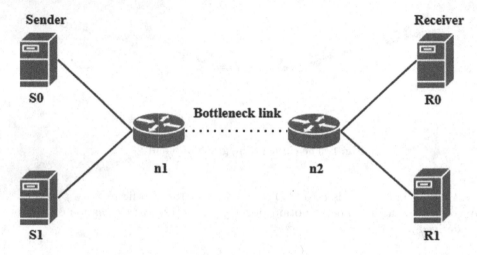

Fig. 2. Network Topology

5.2   Results on RTT Fairness

– **BBR**

When testing RTT fairness, two flows were sent by the sender side. The two flows were transmitted from S0 and S1, and arrived at R0 and R1, respectively. Each test lasted for 200 s. The test results are shown in Fig. 3. Moreover, we increased the number of BBR flows competing in the same path to four to observed the impact of the number of flows on RTT fairness. The test results are shown in Fig. 4. The setting of RTT size is shown in Table 1.

Table 1. RTT size in Fig. 4 and Fig. 5

	BBR1	BBR2	BBR3	BBR4
Figure 3	220 ms	60 ms		
Figure 4	220 ms	220 ms	60 ms	60 ms

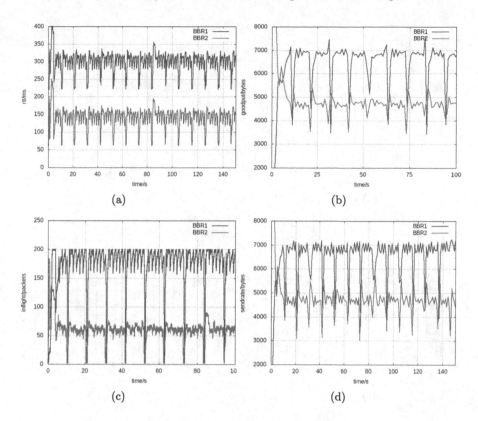

Fig. 3. Competition between two BBR flows with different RTT sizes

As shown in Fig. 3(a), there is a significant difference ratio in the RTT of the two BBR flows. In this scenario, the goodput, inflight, and sendrate of the two flows competing for the same bottleneck link bandwidth over time are shown in Fig. 3(b), (c), and (d), respectively. The performance indicators of the BBR flow with a longer RTT, such as goodput, inflight, and sendrate, are all higher than those of the BBR flow with a shorter RTT, indicating that the BBR algorithm has significant RTT unfairness.

By comparing Fig. 3 and 4, it can be seen that as the number of flows increases, the goodput, inflight, and sendrate of the two short RTT flows have all decreased, and the long RTT flow significantly suppresses the performance of the short RTT flow. Therefore, we can draw the conclusion that regardless of the number of flows in the link, the RTT fairness of BBR still exists, meaning that the issue of RTT fairness is unrelated to the number of flows in the link.

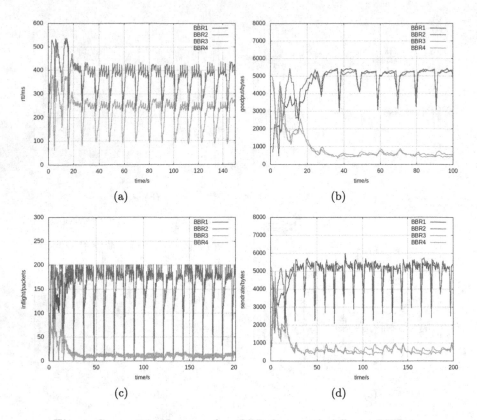

Fig. 4. Competition between four BBR flows with different RTT sizes

- **BBR-O**

To ensure the accuracy of the experiments, we selected four BBR flows for comparison in the following tests of the BBR-O algorithm. The test results are shown below.

By comparing Fig. 4 and Fig. 5, we can observe that the fairness of good-put, inflight, and sendrate between long and short BBR-O flows has been improved, and long RTT flows can maintain relative fairness with short RTT flows. In the original BBR, the goodput of short flows was suppressed by long flows, maintaining at below 1000 bytes, while in BBR-O, the values of goodput for both long and short flows can be maintained at around 3000 bytes. Similarly, the inflight and sendrate of short RTT flows have also been improved, indicating that the BBR-O algorithm effectively mitigates the RTT fairness problem.

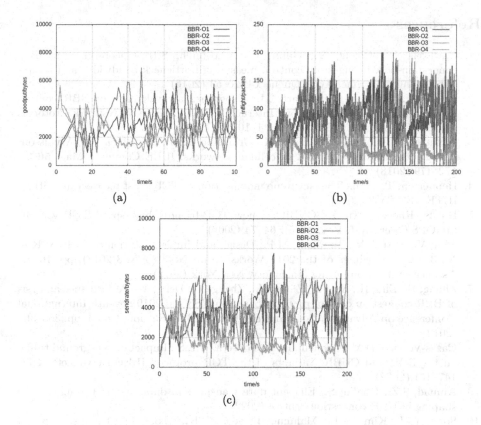

Fig. 5. Competition of four BBR-O flows with different RTT sizes

6 Conclusion

In this study, we evaluated the RTT fairness of BBR in different network scenarios and proposed a corresponding optimization algorithm, BBR-O, to address the RTT fairness issue. Our experimental results showed that BBR exhibited significant RTT unfairness, with long RTT flows occupying most of the bandwidth and severely suppressing the performance of short RTT flows. This problem persisted regardless of the number of competing flows in the bottleneck link, but the design of the BBR-O algorithm was able to effectively alleviate the RTT fairness issue of the original BBR algorithm.

Our research results indicate that although BBR is a promising congestion control algorithm, its RTT fairness issue and fairness issue with other congestion control algorithms need to be addressed. Our further research is needed to explore optimization methods for the RTT fairness issue of BBR and its fairness with other congestion control algorithms in different network scenarios.

Acknowledgements. This research was supported by the Fundamental Research Funds for the Central Universities.

References

1. Cardwell, N., Cheng, Y., Gunn, C.S., Yeganeh, S.H., Jacobson, V.: BBR: congestion-based congestion control: measuring bottleneck bandwidth and round-trip propagation time. ACM Queue **14**, 20–53 (2016)
2. Hock, M., Bless, R., Zitterbart, M.: Experimental evaluation of BBR congestion control. In: 2017 IEEE 25th International Conference on Network Protocols (ICNP), Toronto, ON, Canada, pp. 1–10 (2017)
3. Atxutegi, E., Liberal, F., Haile, H.K., Grinnemo, K.-J., Brunstrom, A., Arvidsson, A.: On the use of TCP BBR in cellular networks. IEEE Commun. Mag. **56**(3), 172–179 (2018)
4. Henderson, T., et al.: The NewReno modification to TCP's fast recovery algorithm. IETF RFC 6582 (2012)
5. Ha, S., Rhee, I., Xu, L.: CUBIC: a new TCP-friendly high-speed TCP variant. SIGOPS Oper. Syst. Rev. **42**(5), 64–74 (2008)
6. Jain, V., Mittal, V., Tahiliani, M.P.: Design and implementation of TCP BBR in NS-3. In: Proceedings of the 2018 Workshop on NS-3 (WNS3 2018), pp. 16–22. Association for Computing Machinery, New York (2018)
7. Zhang, H., Zhu, H., Xia, Y., Zhang, L., Zhang, Y., Deng, Y.: Performance analysis of BBR congestion control protocol based on NS3. In: 2019 Seventh International Conference on Advanced Cloud and Big Data (CBD), Suzhou, China, pp. 363–368 (2019)
8. Casas-Velasco, D.M., Granelli, F., da Fonseca, N.L.S.: Impact of background traffic on the BBR and CUBIC variants of the TCP protocol. IEEE Netw. Lett. **4**(3), 147–151 (2022)
9. Ahmad, S.Z., Khalid, S.: Efficient data transport in data centers through traffic shaping in BBR congestion control (2022)
10. Song, Y.-J., Kim, G.-H., Mahmud, I., Seo, W.-K., Cho, Y.-Z.: Understanding of BBRv2: evaluation and comparison with BBRv1 congestion control algorithm. IEEE Access **9**, 37131–37145 (2021)
11. Kim, G.-H., Cho, Y.-Z.: Delay-aware BBR congestion control algorithm for RTT fairness improvement. IEEE Access **8**, 4099–4109 (2020)
12. https://www.geogebra.org/
13. https://www.nsnam.org/

Accelerating QUIC with AF_XDP

Tianyi Huang[(⊠)] and Shizhen Zhao[(⊠)]

School of Electronic Information and Electrical Engineering,
Shanghai Jiao Tong University, Shanghai 200240, China
{hty690,shizhenzhao}@sjtu.edu.cn

Abstract. QUIC is a high-performance and secure transport layer protocol that has been applied in many scenarios. Considering that kernel
bypass techniques can avoid the overhead of the kernel's network stack
and therefore improve the performance greatly, we choose AF_XDP,
a compatible kernel bypass technique to improve the performance of
QUIC. We present XSKConn, a Golang package that implements UDP
connection based on AF_XDP and we integrate it into quic-go. The
experiments show that XSKConn improves the RPS of quic-go by 5% to
40%, while the CPU usage also decreases by 5% to 50%.

Keywords: network system · eBPF · QUIC · kernel bypass

1 Introduction

QUIC is a UDP-based transport layer protocol, which is designed to solve problems of TCP and HTTP like protocol entrenchment, handshake delay, Head-of-
Line block delay, etc. [23]. Considering its great performance and security, QUIC
is getting applied in many scenarios. HTTP/3 chooses QUIC as the standard
protocol [16]. In serverless computing, QUIC is also widely used. QFaas [24]
leverages QUIC to secure the connections between functions and the gateway.

Therefore, it's attractive for researchers and developers to improve the performance of QUIC. Wang et al. [30] implements a NanoBPF model to offload the
QUIC's encryption module and improve the throughput. MPQUIC [20] improves
the performance of QUIC in lossy scenarios.

Though QUIC is implemented in the user space, it still relies on the kernel to
handle packets and suffers from the overhead of switch between kernel space and
user space. Moreover, many researches have shown that the kernel network stack
introduces heavy overheads, especially in high-speed networks [17,18]. Therefore, reducing the cost of the kernel's network stack is also effective in optimizing QUIC's performance. Some kernel optimizations like GRO/GSO [5,6] and
packet steering [11] reduce the overhead of the kernel's network stack. Some
frameworks like netmap [27] and DPDK [3] leverage the kernel-bypass technique
to completely avoid the processing of the kernel's network stack.

We leverage AF_XDP to optimize the performance of QUIC. AF_XDP is
also a kernel bypass technique proposed by kernel developers [22]. Compared
with DPDK, AF_XDP doesn't need to occupy several CPU cores or NICs for

Z. Tari et al. (Eds.): ICA3PP 2023, LNCS 14489, pp. 89–100, 2024.
https://doi.org/10.1007/978-981-97-0798-0_6

polling, while the performance is close to DPDK [3]. Some previous works leverage AF_XDP for performance improvement or migration to userspace application [14,29,31].

In this paper, we first depict different data paths of AF_XDP and Linux kernel, and we analyze how AF_XDP can achieve better performance. We choose quic-go [9], a famous open-source QUIC implementation to optimize. We implement an AF_XDP package in Golang named XSKConn, which implements socket interfaces in Golang standard library. We import the package and change only about 10 lines of code in quic-go so that quic-go is able to avoid most processings of the kernel's network stack. Moreover, we provide high-performance network policy support which allows users to enforce access control or rate control.

We evaluate the performance of vanilla quic-go and the revised quic-go with XSKConn. The results show that XSKConn improve the RPS of quic-go by 5%–40%, while CPU usage decrease by 5%-50%. We also implement support for simple network policies in XSKConn and evaluate their effectiveness and performance. The results show that our XSKConn helps reduce the overhead of the kernel's network stack and improve the performance of QUIC significantly.

The paper is organized as follows: Sect. 2 introduces the backgrounds of this paper and explains why AF_XDP is effective by analyzing its data path and CPU usage. Section 3 describes the design of XSKConn and explains how we implement the package. Section 4 evaluates the performance of XSKConn in different situations. Section 5 discusses some related topics of the paper and future work. We conclude the paper in Sect. 6.

2 Background

2.1 QUIC Acceleration

QUIC is a transport layer protocol which is designed to solve problems of TCP and HTTP like protocol entrenchment, handshake delay, Head-of-Line block delay, etc. [23]. UDP is selected as the bottom layer of QUIC since UDP is simple enough to implement different features.

QUIC is becoming more and more widely used in recent years. HTTP/3 chooses QUIC as the standard protocol [16]. There are also many open-source projects which implement QUIC following the IETF standard [9,10]. Therefore, improving the performance of QUIC also draws the attention of researchers. eQUIC [25] uses XDP to implement a gateway to accelerate QUIC. Wang et al. [30] implement a NanoBPF model to offload the QUIC's encryption module and improve the throughput. DCQUIC [28] is designed for data center network. MPQUIC [20] provides a Multi-path QUIC that improves the performance in lossy scenarios.

2.2 Network Stack Optimization

Improving the performance of the network stack is always an appealing issue for developers and researchers. Kernel developers have made many patches for it. GRO/GSO [5,6] focuses on offloading specific jobs to hardware aggregating small packets into large ones. Zero-copy support in Linux kernel [13] is also an attempt to improve network performance.

Frameworks like netmap and DPDK leverage the idea of kernel bypass and build network stack in userspace [19]. It is believed that kernel bypass is able to achieve much higher performance against kernel network stack. However, compatibility is the biggest downside of kernel bypass. DPDK occupies several CPU cores and NICs for busy polling. It reduces the overheads of context switch while incurring high CPU utilization. Also, since general protocol processing is bypassed, developers have to thoroughly rewrite the source codes of applications if they wish to leverage the advantages of DPDK.

2.3   eBPF and AF_XDP

eBPF. The extended Berkeley Packet Filter (eBPF) [4] allows users to inject codes into kernel safely. Users can get information about the kernel or change specific behaviors of the kernel with eBPF. Considering the flexibility of eBPF, it has been widely applied especially in networking. OVS uses eBPF to extend its data-path [29]. eBPF-iptables [15] improves the performance and security of iptables. eBPF is the fundamental technique of Cilium [2], which is the most famous CNI plugin in Kubernetes. SPRIGHT [26] uses eBPF in serverless to avoid overheads of sidecar proxy.

AF_XDP. AF_XDP is a new type of socket handling raw frames [22]. Developers can create this type of socket in their applications, and bind one NIC queue to one AF_XDP socket (XSK). Moreover, developers have to attach an XDP program to the NIC, which is responsible for filtering the packets and redirecting the packets to the corresponding XSK. The kernel network stack is also bypassed in AF_XDP's processing. Therefore developers have to handle the raw frames in their userspace applications. Though AF_XDP is a little worse than DPDK in performance, it doesn't need to occupy CPU cores and it will not interfere with other applications on one host machine compared with DPDK.

Since AF_XDP is performant and more compatible than DPDK, developers leverage AF_XDP to achieve higher performance in their applications. Generally, AF_XDP is used to accelerate packets' processing or fastly forward packets to userspace [14,21,29].

2.4   AF_XDP Datapath

To understand why AF_XDP can achieve better performance, we have to understand how it works and what the difference is between AF_XDP and the kernel network stack. Figure 1(a) shows different datapaths on RX. Suppose the application creates two sockets, one is a normal socket and the other is XDP socket (XSK). Then the user attaches an XDP program to the NIC so that the packets needed can be forwarded to XSK.

When a packet arrives at the NIC, it will be copied to the ring buffer in memory by DMA. Soft interrupt will be called to handle the packet. The packet

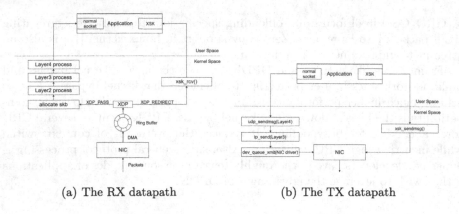

(a) The RX datapath   (b) The TX datapath

Fig. 1. The datapath of AF_XDP

will then be filtered by the XDP program. As for normal sockets, the packet will be passed to the kernel network stack. The kernel will allocate sk_buff to process the packet, which incurs another data copy. Then the packet will be processed per layer. Different functions will be called depending on different type of the packet. Finally, the packet will be copied into user space and transmitted to the socket's buffer.

Compared with the kernel network stack, processing of XDP socket is far more simple. The packet will be redirected if XDP program manages to find the corresponding XSK's file descriptor in the eBPF map. The kernel will copy the packet into XSK's ring buffer. The application is responsible for parsing this raw frame.

Figure 1(b) shows the two different TX datapaths. In the TX datapath, AF_XDP is also much more compact. Suppose the application is to send a UDP packet. In the kernel network stack, the packet will be encapsulated per layer. Modules like netfilter are still involved in TX datapath, which will also influence the performance. After that, the packet will be copied to NIC's buffer and wait to be sent. If we use AF_XDP, we can build a complete packet. Then the kernel call xsk_sendmsg() to handle the packet. Therefore, AF_XDP can avoid complicated processings in both RX and TX data-path.

3   Design and Implementation

In this section, we will introduce how we design and implement the Golang package which is able to replace UDP with AF_XDP. We will illustrate how the packet is redirected to the userspace socket and how we handle the raw frame. We will also introduce how we solve some problems and improve the performance of our implementation.

3.1 Overview

As introduced in Sect. 2.1, UDP is the bottom layer of QUIC. Therefore it's obvious that optimizing UDP will also improve the performance of QUIC. Since protocol processing of UDP is comparatively simple, implementing the UDP processing stack in userspace will be concise and performant.

We choose to implement a Golang package that supports UDP above the AF_XDP socket. The reason is that Golang supports interface. In many cases, the UDP socket in Golang can be replaced by other user-defined sockets if this socket has implemented the necessary interface methods. The other reason is that Golang is widely used in cloud native applications and serverless computing. The Golang implementation will be practical in these scenarios.

3.2 Architecture

Pre-configuration. Before the application uses the AF_XDP package, the developers has to do some pre-configurations for XDP sockets. First, the developer have to create two eBPF maps in userspace and pin them to the file system. One of the maps is the port map, which records the desired port numbers. The other map is the socket map, which combines the queue number of the NIC with the XDP socket fd.

The developer has to write an XDP program for filtering packets and a userspace program to attach the XDP program. The XDP program will search the maps and redirect desired packets to corresponding XDP sockets.

Note that in most cases, the pre-configurations are almost the same. Therefore the developers just have to write these programs once. These configurations can also be set in a bash file.

Initialization. To use XSKConn, some revisions have to be made to the application's source code. First, the package should be added to the application so that XSKConn can be imported. The developers also have to figure out where the application uses UDPConn and then replace it with XSKConn. XSKConn implements the basic interface methods as UDPConn and provides the same semantics as the methods in UDPConn. Therefore the developers only need to change the code where they create UDPConn originally with about 10 lines of code.

XSKConn can be created by NewXSKConn(). The developers have to pass parameters like link information, and configuration of the XDP socket to the function. Then the XDP socket will be created which contains four ring buffers to buffer packets. XSKConn will handle the n-tuple rules and operations with eBPF maps to ensure that the desired packets can be redirected.

Network Policy. In Linux, network policy is usually enforced by iptables and TC Qdisc [8,12]. However, AF_XDP bypasses the two modules in its datapath. Therefore we implement a module to support simple network policies.

Currently, we support simple access control and rate control. We leverage the eBPF map to implement the functions. We use one map named admission map and the other named rate map. It's optional for the users whether to use network policy and the option can be set during the pre-configuration phase. If the option is turned on, the extra module in the XDP program will be compiled. The admission map acts as a blacklist. If the source IP address of the packet is recorded in the map, the packet will be dropped by the XDP program. The users can add or delete rules by using APIs provided by XSKConn.

The rate map acts as the bucket in the token bucket algorithm. To enable rate control in XSKConn, the pre-configuration script will launch a process that will timely add value to the rate map. The rate of adding and the maximum value in the map, which refers to the volume of the bucket in the token bucket algorithm can be assigned by users. We allow users to set the expected rate and will calculate corresponding parameters in the script. The rate users set will limit the RPS of the application. In the XDP program, it will refer to the rate map and fetch the value. If the value is zero, which means the bucket is used up, the packet will be dropped.

3.3   Implementation

Userspace Network Stack. Since AF_XDP bypasses the kernel's network stack, we have to implement the network stack in our Golang package. Considering that only UDP packets over IPV4 will be redirected to the XDP socket, we just have to handle this type of packets. That's part of the reason why userspace network stack performs better than the kernel network stack.

The userspace network stack is implemented in a parsing function and will be called in interface methods like Read()/ReadFrom(). First, it parses the raw frame and gets the information. Then it calculates the IP checksum and UDP checksum, and invalid packets will be dropped. The destination IP and port will be returned to the user if the user calls ReadFrom() and the payload will be copied to the user's buffer which is passed to the interface method.

If Write()/WriteTo() is called, XSKConn will also serve as the network stack. It parses the IP address and port number that users provide, and sets the headers of UDP and IPV4. It will also handle ARP requests, which will be introduced later, to construct the MAC header. Calculation of checksum will also be done in userspace. We also process the ARP queries to get MAC address.

Optimization of Batch Read/Write. The Golang standard package net/ipv4 implements ReadBatch()/WriteBatch() which supports reading/writing a batch of messages by one function call. This function will improve the efficiency of the receiver/sender side since it reduces the number of switches between kernel space and user space. Some Golang frameworks like quic-go support batch operation which can improve the performance [1].

We also implement batch operation in our package since the package aims at improving performance. It's natural to read or write packets in a batch in

AF_XDP since we just fetch the raw frames from the ring. We just need to add a loop to fetch from the ring. But the messages passed to the user contain Out-of-Band (OOB) data other than payload. Therefore we parse the flags the user provides and extract the corresponding information and write it into the message. Now we implement support for IP_PKTINFO and IP_TOS.

4   Evaluation

4.1   Experiment Setup

The experiments are conducted on two symmetric host machines. One serves as the server and the other serves as the client. The CPU is Intel Xeon Silver 4110 @ 2.10 GHZ. The memory is DDR4 @ 2400 MT/s and the size is 128 GB. The system is Ubuntu 20.04.5 LTS and the kernel version is 5.15. The two machines are connected with Mellanox ConnectX-5 with link speed of 40 Gbps. We use h2load [7] to test the performance of the QUIC server. The version of h2load is 1.53.0 compiled with the support of http/3. XDP program is attached to the NIC of the server in native XDP mode.

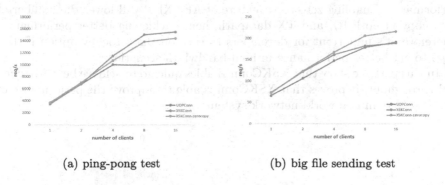

(a) ping-pong test   (b) big file sending test

Fig. 2. The RPS of the QUIC server

4.2   QUIC Performance

We integrate XSKConn into quic-go so that the quic-go server can use AF_XDP for communication.

In this part, UDPConn denotes the vanilla quic-go version which uses UDP-Conn to receive and send packets. XSKConn denotes the version which uses XSKConn for communication and the XDP socket is in copy mode. XSKConn-zerocopy denotes that the zero-copy mode is enabled for XDP sockets, which avoids data copy in processing.

Ping-Pong Test. In this situation, we set the body size of the HTTP request as 0 and the QUIC server will only respond with a 403 Forbidden status code in the HTTP response. The situation primarily tests the latency between the QUIC server and the client.

Figure 2(a) shows the req/s of the QUIC server in ping-pong test. When the client number is small, the performance is close. But when the number of clients is larger than 4, XSKConn shows more than 30% RPS improvement than UDPConn. XSKConn with zero-copy is even a little worse than normal XSKConn. We suppose the reason is that in ping-pong test the overhead of data copy is trivial.

Sending Big Files. We also evaluate the throughput performance of the QUIC server. The response of the server in this situation contains a 1MB picture.

The results in Fig. 2(b) show that using XSKConn with zero-copy achieves about 10% more RPS than using original UDPConn, while using XSKConn without zero-copy doesn't show obvious advantages against UDPConn. When the number of clients is 16, XSKConn without zero-copy is even worse than UDPConn.

In big file sending scenario, the overhead of data copy is the key factor of the performance. Enabling zero-copy feature of AF_XDP will lower the number of data copy in both RX and TX datapath, hence achieving better performance. Therefore, it's important for developers to turn on the zerocopy option if they hope to get better performance in high-bandwidth scenarios.

In this part, we show that XSKConn enables quic-go to achieve better latency and throughput. It proves that XSKConn is able to improve the performance of QUIC servers in real-world network systems.

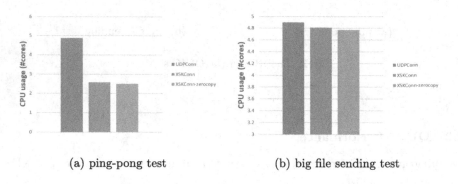

(a) ping-pong test   (b) big file sending test

Fig. 3. The CPU usage of the QUIC server

4.3   CPU Usage

Besides the latency and throughput of the server, the CPU usage of the server process is also an important metric. We still test the QUIC server in the ping-pong test and big file sending test. We set 10 clients in h2load and limit the RPS of each client to the same value. In this way we make sure that the workload of the QUIC server with UDPConn and with XSKConn is almost the same.

Figure 3(a) shows the CPU usage of the QUIC server in ping-pong test. XSKConn lowers the CPU usage by more than 40% in ping-pong test. Whether turn on or turn off zero-copy feature doesn't show big difference in this scenario. The result is not surprising since AF_XDP can avoid many complex processes which may consume large CPU times.

Figure 3(b) shows the CPU usage in big file sending test. The CPU usage of UDPConn and XSKConn is almost the same. XSKConn only lowers about 5% CPU usage. We believe the reason is that data copy consumes most CPU time and the advantage of AF_XDP is not obvious in this scenario.

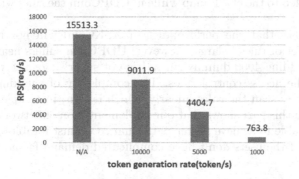

Fig. 4. The server's RPS in different token generation rate

4.4   Network Policy

In this part, we will analyze the effect of our network policy implementation. We will also compare the performance of our XDP solution with the iptables solution. We use XSKConn with zero-copy enabled and the test case is the ping-pong test.

Effectiveness of Rate Control. We test the effect of rate control by changing the token generation rate. N/A means we don't enable the rate control feature. As shown in Fig. 4, the RPS of the server is almost linear to the token generation rate. The result proves that the rate control module of XSKConn works well.

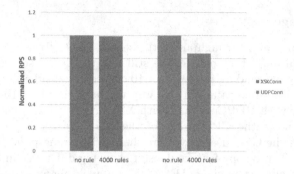

Fig. 5. Normalized RPS of the server

Performance of Access Control. In this part, we evaluate the influence of access control rules. We first test the RPS of the server with no access control rule. Then we add 4000 dummy rules and test the RPS. In XSKConn, the dummy rules will be added to the eBPF map while in UDPConn the rules will be enforced by iptables.

Figure 5 shows that the performance of XSKConn will not be influenced when the number of rules is large. However, UDPConn suffers nearly 20% RPS decrease after adding 4000 dummy rules. It's not surprising since we use a hash map to store the access control rules. The complexity of checking a packet is $O(1)$. But iptables scan the rules linearly, which is $O(n)$ in complexity.

The result in this part shows that our implementation of network policy takes effect, and it is able to achieve better performance against iptables. XSKConn is a better choice if the users don't have complicated demand for network policy.

5 Discussion

Non-intrusive AF_XDP Implementation. XSKConn offers the same interface methods to the users and the users just have to revise about 10 lines of code to use XSKConn. However, the need for revision still hurts the compatibility of XSKConn. Moreover, applications not implemented in Golang can't get improvement. We are working on a non-intrusive XDP socket implementation so that applications in all languages can use the XDP socket and users don't have to change the codes. Using LD_PRELOAD to redirect system calls may be a good solution, which is implemented in Slim [32] to improve the performance of container networking.

Support Multi-queue. Currently, we just implement the support for single queue NIC. If the NIC is multi-queue, we use n-tuple filter to direct the corresponding UDP packets to the specified queue. In the future, we will implement a manager in XSKConn to manage the sockets created for different NIC queues. We will also manage the map of port number and queue number so that a host machine can run different applications which use XSKConn for communication.

6   Conclusion

AF_XDP is a high-performance kernel bypass solution and it is more compatible than other frameworks like DPDK. In this paper, we depict the datapath of AF_XDP and analyze how it can achieve high performance. We propose a Golang-based package named XSKConn, which enables users to leverage AF_XDP for UDP connections. We also support simple network policies in this package. We test UDP throughput and QUIC performance by revising the code of quic-go. The results show that XSKConn can improve the throughput by 5–10% when the payload is large and improve the RPS of the QUIC server by over 30% when the payload is small. The average CPU usage decreases by 5% to 50% when using XSKConn.

We are still making XSKConn more compatible. We will add multi-queue support in XSKConn and we are also exploring the solution to provide non-intrusive AF_XDP support. Improving XSKConn's performance in high-bandwidth situations is also our future work.

References

1. Boosting performance: UDP offloading and GSO. https://github.com/quic-go/quic-go/issues/2877
2. Cilium. https://cilium.io/
3. Dpdk. https://www.dpdk.org/
4. ebpf. https://ebpf.io/
5. Generic receive offload. https://lwn.net/Articles/358910/
6. Gso: Generic segmentation offload. https://lwn.net/Articles/188489/
7. h2load. https://nghttp2.org/documentation/h2load-howto.html
8. iptables project. https://www.netfilter.org/projects/iptables/index.html
9. quic-go. https://github.com/quic-go/quic-go
10. quiche. https://github.com/cloudflare/quiche
11. Receive flow steering. https://lwn.net/Articles/382428/
12. tc(8)—linux manual page. https://man7.org/linux/man-pages/man8/tc.8.htmlQDISCS
13. Zero-copy networking. https://lwn.net/Articles/726917/
14. Abranches, M., Keller, E.: A userspace transport stack doesn't have to mean losing Linux processing. In: 2020 IEEE Conference on Network Function Virtualization and Software Defined Networks (NFV-SDN), pp. 84–90. IEEE (2020)
15. Bertrone, M., Miano, S., Risso, F., Tumolo, M.: Accelerating Linux security with eBPF iptables. In: Proceedings of the ACM SIGCOMM 2018 Conference on Posters and Demos, pp. 108–110 (2018)
16. Bishop, M.: RFC 9114: HTTP/3 (2022)
17. Cai, Q., Chaudhary, S., Vuppalapati, M., Hwang, J., Agarwal, R.: Understanding host network stack overheads. In: Proceedings of the 2021 ACM SIGCOMM 2021 Conference, pp. 65–77 (2021)
18. Cai, Q., Vuppalapati, M., Hwang, J., Kozyrakis, C., Agarwal, R.: Towards μs tail latency and terabit ethernet: disaggregating the host network stack. In: Proceedings of the ACM SIGCOMM 2022 Conference, pp. 767–779 (2022)

19. Chen, R., Sun, G.: A survey of kernel-bypass techniques in network stack. In: Proceedings of the 2018 2nd International Conference on Computer Science and Artificial Intelligence (2018)
20. De Coninck, Q., Bonaventure, O.: Multipath QUIC: design and evaluation. In: Proceedings of the 13th International Conference on Emerging Networking Experiments and Technologies, pp. 160–166 (2017)
21. Enberg, P., Rao, A., Tarkoma, S.: Partition-aware packet steering using XDP and eBPF for improving application-level parallelism. In: Proceedings of the 1st ACM CoNEXT Workshop on Emerging in-Network Computing Paradigms, pp. 27–33 (2019)
22. Karlsson, M., Töpel, B.: The path to DPDK speeds for AF XDP. In: Linux Plumbers Conference (2018)
23. Langley, A., et al.: The QUIC transport protocol: design and internet-scale deployment. In: Proceedings of the Conference of the ACM Special Interest Group on Data Communication, pp. 183–196 (2017)
24. Nguyen, H.T., Usman, M., Buyya, R.: QFaaS: a serverless function-as-a-service framework for quantum computing. arXiv preprint arXiv:2205.14845 (2022)
25. Pantuza, G., Vieira, M.A., Vieira, L.F.: eQUIC gateway: maximizing QUIC throughput using a gateway service based on eBPF+ XDP. In: 2021 IEEE Symposium on Computers and Communications (ISCC), pp. 1–6. IEEE (2021)
26. Qi, S., Monis, L., Zeng, Z., Wang, I.C., Ramakrishnan, K.: SPRIGHT: extracting the server from serverless computing! High-performance eBPF-based event-driven, shared-memory processing. In: Proceedings of the ACM SIGCOMM 2022 Conference, pp. 780–794 (2022)
27. Rizzo, L.: netmap: a novel framework for fast packet I/O. In: 21st USENIX Security Symposium (USENIX Security 2012), pp. 101–112 (2012)
28. Tan, L., Su, W., Liu, Y., Gao, X., Zhang, W.: DCQUIC: flexible and reliable software-defined data center transport. In: IEEE INFOCOM 2021-IEEE Conference on Computer Communications Workshops (INFOCOM WKSHPS), pp. 1–8. IEEE (2021)
29. Tu, W., Wei, Y.H., Antichi, G., Pfaff, B.: Revisiting the open vSwitch dataplane ten years later. In: Proceedings of the 2021 ACM SIGCOMM 2021 Conference, pp. 245–257 (2021)
30. Wang, J., Lv, G., Liu, Z., Yang, X.: QUIC cryption offloading based on NanoBPF. In: 2022 23rd Asia-Pacific Network Operations and Management Symposium (APNOMS), pp. 1–4. IEEE (2022)
31. Zhao, B., Qin, Y., Yang, W., Fan, P., Zhou, X.: SRA: leveraging AF_XDP for programmable network functions with IPv6 segment routing. In: 2022 IEEE 47th Conference on Local Computer Networks (LCN), pp. 455–462. IEEE (2022)
32. Zhuo, D., et al.: Slim: OS kernel support for a low-overhead container overlay network. In: 16th USENIX Symposium on Networked Systems Design and Implementation (NSDI 2019), pp. 331–344 (2019)

Segmenta: Pipelined BFT Consensus with Slicing Broadcast

Xiang Li[1]([⊠]), Liang Cai[1]([⊠]), Weiwei Qiu[2], Fanglei Huang[2], and Zhigang Lei[1]

[1] Zhejiang University, Hangzhou, China
alang172lee@gmail.com, leoncai@zju.edu.cn
[2] Hangzhou Qulian Technology Co., Ltd., Hangzhou, China
{qiuweiwei,fanglei.huang}@hyperchain.cn

Abstract. Bandwidth is an essential resource for blockchain systems since BFT protocols require a series of communication steps to reach a consensus. Reducing available bandwidth will directly hurt system performance, especially in leader-based protocols where the leader carries the main communication overhead and becomes a bottleneck. Current approaches reduce the load on the leader by replacing its original star topology communication model with a tree. However, using trees increases the round latency and makes the consensus view change extremely complicated.

In this paper, we propose Segmenta, a BFT consensus protocol leveraging slicing broadcasts to achieve low latency in the case of restricted bandwidth. We use the erasure code to slice the transaction blocks in the consensus proposals and provide verifiability with the Merkle tree. The origin blocks will be reconstructed with the participation of all replicas, thus reducing the leader's network overhead. To avoid the increase in communication steps, the additional actions brought by the block shards broadcast are integrated into different consensus phases. Meanwhile, we propose a pipelined version of Segmenta, which further optimizes the transaction waiting time and lowers latency. We evaluate our protocol through a simple implementation in the testbed of Aliyun. Compared to Hotstuff, Segmenta shows a smoother and slower trend of latency increase and gets 4.6× higher throughput while requiring only 17% of the latency in limited bandwidth (10 Mbps).

Keywords: Slice · Fault Tolerance · Consensus · Blockchain

1 Introduction

State Machine Replication (SMR) guarantees data consistency in distributed systems. Since blockchain [22] was thrust into the limelight, a series of new Byzantine fault-tolerant (BFT) [15] SMR protocols have been designed to improve the efficiency and scalability of the system.

Bandwidth is a critical resource for blockchain systems because of the frequent communication steps in BFT protocols. On the one hand, the communication overhead increases along with the number of replicas in the system, which

© The Author(s), under exclusive license to Springer Nature Singapore Pte Ltd. 2024
Z. Tari et al. (Eds.): ICA3PP 2023, LNCS 14489, pp. 101–120, 2024.
https://doi.org/10.1007/978-981-97-0798-0_7

will finally become a performance bottleneck since the available bandwidth is limited. On the other hand, in some special application scenarios [21,25], blockchain systems are challenged by poor network conditions and insufficient bandwidth. For example, when applying blockchain in drones, their communication methods are different from typical cases, which usually provide only a minimal amount of bandwidth and are vulnerable to electromagnetic interference. In such cases, it is much easier to reach the bandwidth bottleneck, even though the system is at a tiny replica size.

BFT protocols suffer from a single-node bottleneck, especially for those leader-based BFT protocols [6,28]. In each round, the leader generates a quorum certificate (QC) with a list of $secp256k1$ signatures [26] from other replicas and broadcasts the list with a new block, while the other replicas only need to vote for the leader. Obviously, the leader carries a $O(N)$ complexity communication overhead, which is much higher than other replicas. Therefore, when the system size grows or the available bandwidth decreases, the leader will reach the network bottleneck much earlier than others and become the bottleneck.

Earlier approaches to reduce communication load are either choosing a smaller committee like [8,24] or doing consensus in hierarchical levels like [2, 11,16], which impair resilience and compromise the deterministic finality. Recent research, such as [12–14,23], uses trees to replace the original star-topology communication pattern. These approaches balance the network overhead, but the single-round latency increases since each communication step is separated into H (the tree height) communication steps. The latest research, Kauri [23], tries to solve this problem through the pipeline, but it actually has no effect but reduces the waiting time for transactions.

Some asynchronous BFT protocols [10,20,27] using verifiable information dispersal (VID) [5] to handle the problem of a single-node bottleneck. VID stores a big file among servers by distributing it so that each server holds a small slice of the file. The HoneyBadger BFT [20] uses VID as a reliable broadcast module (RBC) which causes additional communication rounds for the consensus. Recently, DispersedLedger [27] fixed it by separating the block-delivering phase, which breaks the integrity of the consensus.

In this paper, we propose Segmenta, a new semi-synchronous BFT protocol that can improve the performance and scalability of the blockchain system under low bandwidth. We first attempt to replace the broadcast of Hotstuff with the RBC module in the HoneyBadger BFT [20]. However, it does not make a breakthrough performance improvement. Then we prune the RBC protocol and integrate only those key steps into Segmenta to disseminate the block in the proposal without increasing additional communication rounds for the consensus. Slicing the block will not destroy the availability of the proposal since the consensus is stepped forward by QC. The Segmenta protocol fully utilizes the bandwidth of each node, reduces the bandwidth consumption of the leader, and thus improves the system's performance when the bandwidth decreases.

In short, we make the following contributions.

- We propose the Segmenta consensus protocol (Sect. 5) that reduces the leader's communication overhead from $n|m|$ to $\lambda|m|(1 < \lambda < 1.5)$ by slicing the consensus proposal. It adds no additional message communication round compared to those tree-based models.
- We optimize Segmenta through a pipelined model (Sect. 6) to further reduce the transaction waiting time. Chained Segmenta implements an accurate pipeline model with the same efficiency as the Hotstuff pipeline [28].
- We implement Segmenta by C++ and evaluate it on the test bed of Aliyun (Sect. 7). The results show that Segmenta achieves a smoother and slower trend of latency increase compared to Hotstuff. It has 4.6× higher throughput while requiring only 17% of the latency in 10 Mbps bandwidth.

2   Related Work

Permissioned blockchains are expected to scale to hundreds of replicas [3,9, 29]. The communication load can be huge since a block may contain a large number of transactions and the size of the quorum certificate (QC) grows linearly along with the number of replicas. Based on such observations, researchers have proposed several ways to improve the consensus protocol.

The early approaches choose a smaller committee [8,24] or do consensus at hierarchical levels [2,11,16]. Such protocols reduce message complexity and balance bandwidth. However, they impair censorship resilience by reducing the number of committee nodes to a subgroup of $K(K < N)$ replicas. Some solutions compromise the deterministic finality of the system since a block finalizes after multiple subsequent blocks.

The second approach organizes the processes into a tree topology [12–14,23], which places the leader at the root to weaken the bottleneck. In traditional consensus protocols, the leader disseminates and aggregates messages while the other replicas rarely communicate with each other, which is called a star topology communication model. Through a tree, messages are delivered level by level, thus reducing the number of messages the leader sends, receives, and processes. However, using the tree topology increases the latency of each consensus round since each single communication step is separated into H (the tree height) communication steps. Kauri [23] tries to optimize this problem through a pipeline. However, it only reduces the waiting time of transactions rather than fixing the root causes of the high single-round consensus latency. At the same time, using trees makes the system slow to recover from node failures because finding a tree configuration without failed internal nodes is of combinatorial complexity.

Other approaches use the verifiable information dispersing model (VID) [5] under the asynchronous network assumption [19]. Asynchronous BFT protocols like [10,18,20,27] suffer from a single-node bottleneck that a round cannot be completed until the message from the weakest correct node has been delivered. In the HoneyBadger BFT [20], researchers use VID as an efficient broadcasting method, called reliable broadcast (RBC), to accomplish the task of broadcasting blocks. RBC reduces the complexity from $O(n^2|v|)$ to $O(n|v| + \lambda n^2 log n)$

through erasure coding even in the worst case. The total number of messages each node needs to send reduces from $n|v|$ to only $\lambda|v|(1 < \lambda < 2)$. DispersedLedger [27] achieves better system performance by using VID to separate the block-delivering phase from the consensus.

Inspired by the VID protocols, the Segmenta consensus protocol proposed in this paper leverages erasure code to slice the proposal and then provide available verifiability, which is an unprecedented design to our knowledge. The Segmenta protocol achieves a balanced communication overload similar to tree-based systems with no additional communication steps and does not compromise the fault tolerance of the consensus, which is $f \leq \frac{n-1}{3}$.

3   Analysis and Design

Different rounds in Hotstuff have a high degree of similarity, which can be abstracted into two steps: the broadcast step and the voting step.

Broadcast: When the leader has received enough votes, it generates the votes into a quorum certificate QC and broadcasts the QC with a possible transaction block.

Vote: When a replica has received the message from the leader, it verifies the QC in the message. If the QC is valid, then it does some operations according to the round tag and sends a vote back to the leader.

As we have discussed in Sect. 1, the broadcast step consumes much more bandwidth compared to the voting step (similar analyses can be found in [23, 27]). In the HoneyBadger protocol [20], VID [5] has been used as a reliable broadcast module (RBC) to disseminate transaction blocks, which lowers the communication load of the dissemination initiator. To disseminate a block of size $|v|$, using the RBC module in HoneyBadger reduces this cost from $O(N^2|v|)$ to $O(N|v| + \lambda N^2 log N)$ even in the worst case. Thus, we are inspired to use the RBC module in the Hotstuff protocol to reduce the leader's communication overload.

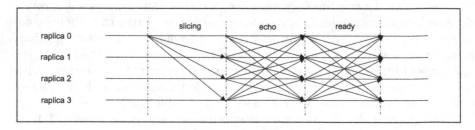

Fig. 1. Communication pattern of the RBC module with 4 replicas (replica 0 is the dissemination initiator).

3.1   Directly Apply RBC Module in Hotstuff

We first directly use the RBC module [20] to replace all the broadcast steps in Hotstuff. The communication pattern of Segmenta is shown in Fig. 1, we briefly describe the three rounds of RBC as follows.

Slicing round: The dissemination initiator uses erasure code to encode the message m and gets n different shards of message m. Then it sends a distinct shard to each replica.

Echo round: When a replica receives a new shard s from the dissemination initiator, it broadcasts s to all replicas.

Ready round: When a replica receives a shard s from the other replica, it stores the shard s. When a replica has already been stored with $n - 2f$ different shards, it broadcasts a ready message. When a replica has already received $f + 1$ ready messages, it uses any $n - 2f$ shards to decode the origin message m.

The RBC module provides two external interfaces: $disseminate(m)$ and $deliver(m)$.

func $disseminate(m)$: Using erasure code to slice message m and send one shard to each replica.

func $deliver(m)$: When there are enough $ready$ message, use erasure code to reconstruct original message m and return m.

When $disseminate(m)$ is invoked, it continues with several shards echoing and $ready$ message broadcast steps. However, these steps are not necessary to know by the consensus protocol. In the consensus module, the leader news an RBC instance for each round and uses the $disseminate(m)$ function to broadcast a message, and for the replicas, when there is a message delivered by $deliver(m)$, it drops the RBC instance and then invokes the voting step (Algorithm 1). The new system will not break the safety and liveness of the Hotstuff protocol, since the timeout clock of the Hotstuff protocol controls when it should drop the RBC instance in advance to stop waiting for the delivery and start a new view.

Algorithm 1. Hotstuff with RBC Module

1: as a leader (broadcast):	9: as a replica (vote):
2: **upon** receive $n - f$ votes:	10: **upon** $deliver(m)$:
3: generate QC	11: verify QC
4: $txs \leftarrow \emptyset$	12: if available QC
5: if $roundTag == PREPARE$:	13: do $roundTag$ operations
6: generate transactions into txs	14: send $vote$ to leader
7: $m \leftarrow Msg < roundTag, txs, QC >$	15:
8: $disseminate(m)$	16:

The use of the RBC module reduces the communication load of the leader and achieves a more balanced system bandwidth consumption. We implemented the above algorithm in the open-source platform Diem [1] in our early study and compared the performance in four experimental boards of Inter celeron J1800

2.41 GHz CPU and 16 GiB RAM. We ran one Diem node in each board. We restricted the available bandwidth of each board from 100 Mbps to 7 Mbps. The result is shown in Fig. 2.

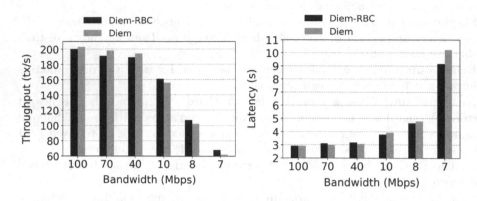

Fig. 2. Throughput and Latency of 4 Diem processes.

We found that only a little optimization of max 1.1× throughput and 0.9× latency had been made by applying the algorithm. The main reason is that the RBC protocol has three rounds of communication. Therefore, when the consensus invokes *desaminate*(m), it actually extends the origin one-round broadcast to three rounds, which impairs the consensus efficiency.

3.2   Design to Reduce the Additional Communication Round

Algorithm 1 slices both transaction blocks and QC, which triggers up to six additional communication rounds. An effective improvement to reduce the additional round is that we can only use RBC to disseminate the proposal in the *PREPARE* round since other rounds only broadcast QC and the QC is tiny compared to the proposal. However, it still triggers two additional communication rounds.

Can we design a protocol with no additional communication but still reduce the leader's communication overload? To make progress, we analyze the RBC module in-depth. Actually, the RBC module can be seen as several VID [5] instances working in parallel. The RBC module reduces the leader's communication overhead by slicing the message m and sends a distinct shard to each replica. The following two communication rounds are to share the shards and make sure there are enough replicas received enough shards to reconstruct m.

We make a further observation: The consensus proposal consists of a transaction block b and a QC. Replicas use the QC to verify the validity of the message. Compared to the QC, block b can be much bigger. Thus, we can use the erasure code to slice only the block b and use the QC in the following round to check there are enough replicas ready for the block reconstruction.

Based on the above thoughts, we designed the Segmenta protocol. Segmenta sends proposals with distinct shards of the transaction block and leverages QC for both consensus verification and shards adequacy check, which has no additional communication rounds but achieves the positive effect of the Algorithm 1.

4   System Model

We consider a system with n replicas, where each replica is indexed by $i \in [n]$ where $[n] = 1, ..., n$. We assume the Byzantine fault mode of up to $f < N/3$ faulty replicas, which can send arbitrary values, delay or omit messages, and even collude with other faulty replicas.

Segmenta makes use of a (k, n)-threshold signature, where a single public key is held by all replicas and each replica holds a distinct private key. Replicas will not change their key during the protocol execution and the faulty replicas do not have the ability to compromise the cryptographic primitives.

Replicas communicate with each other through perfect point-to-point channels, which provides validity and termination properties:

Validity: if a replica i delivers a message m on the channel c_{ij}, m was sent by replica j.

Termination: If both replica i and j are correct, then messages in channel c_{ij} and c_{ji} will eventually delivered.

Facing the FLP impossibility [7], we use the semi-synchronous network assumption to guarantee the liveness, where there is a known bound Δ after an unknown Global Stabilization Time (GST), such that all messages will be delivered within a period Δ.

5   The Segmenta Protocol

The Segmenta protocol is an extension of the Hotstuff protocol [28], and its core idea is to slice the transaction blocks through erasure code and broadcast proposals with distinct shards. Although the Hotstuff protocol is the subject of this paper, this idea can still be used in any other leader-based BFT protocol.

Using the list of signatures as the QC in Segmenta is suggested since the calculation consumption of erasure coding can be pretty high. When computing power is sufficient, multi-signatures, like BLS [4], can be used to further reduce the communication overhead of the leader.

5.1   Segmenta-Specific Data Structures

Algorithm 2. Utilities

1: **function** Msg($tag, node, slice, qc$)
2:     $m.tag \leftarrow tag$
3:     $m.node \leftarrow node$
4:     $m.slice \leftarrow slice$
5:     $m.qc \leftarrow qc$
6:     **return** m
7: **function** Vote($tag, node, qc$)
8:     $m \leftarrow$ Msg($tag, node, slice, qc$)
9:     $m.sig \leftarrow sign\{m.tag, m.view, m.node\}$
10:    **return** m
11: **function** GenerateSlice()
12:     $block \leftarrow$ client's cmds
13:     $\{shards\} \leftarrow$ EEncode($block$)
14:     $root, \{proof\} \leftarrow$ MerkleTree($\{shards\}$)
15:     **for** i-th $slice_i$ in $\{slice\}$:

16:         $slice_i.id \leftarrow root$
17:         $slice_i.shard \leftarrow shard_i$
18:         $slice_i.proof \leftarrow proof_i$
19:     **return** $\{slice\}$
20: **function** QC(M)
21:     $qc.tag \leftarrow m.tag : m \in M$
22:     $qc.view \leftarrow m.view : m \in M$
23:     $qc.node \leftarrow m.node : m \in M$
24:     $qc.sig \leftarrow combine\ m.sig : m \in M$
25:     **return** qc
26: **function** safeNode($node, qc$)
27:     **return**
28:         $node$ extends from $lockedQC.node$
29:         $\lor\ qc.view > lockedQC.view$

Messages. A message m contains the following fields:

- $m.view$ is the view in which m is sent.
- $m.tag \in \{new\text{-}view, prepare, pre\text{-}commit, commit, decide\}$ refers to the different phases of Segmenta.
- $m.node$ contains a leaf node of a proposed branch.
- $m.slice$ contains a distinct shard of the sliced transaction block for each replica.
- $m.justify$ includes a quorum certificate (QC) for different phases.
- $m.sig$ contains a partial signature of the vote sender.

Transaction Block Shard. The leader uses $EEncode(block)$ function to apply $\{n, n-f\}$-erasure encoding towards a transaction block and generates n different $shard$. Then, the leader uses $MerkleTree(\{shard\})$ to build a Merkle tree for these n $shard$ and calculates a Merkle branch for each $shard$ as proof. A $m.slice$ contains several fields:

- $m.slice.id$ uses the Merkle tree root hash as the identity of this block.
- $m.slice.shard$ is a distinct shard of n transaction block shards.
- $m.slice.proof$ is the Merkle branch of $m.slice.shard$.

Quorum Certificates. A quorum certificate (QC) is a threshold signature of a message for the node of a proposed branch. Given a quorum certificate qc for message m:

- $qc.view$ is $m.view$ in which the qc is generated.
- $qc.tag$ is $m.tag$, which refers to different phases.
- $qc.node$ refers to the proposed node $m.node$.
- $qc.sig$ is the combined signature.

Local Variables. A replica uses some local variables to keep some Segmenta protocol state:

- $CurView$ is initialized as 1 and incremented when making a decision or a new-view interruption occurred.
- $prepareQC$, initially $\perp$, stores the highest voted QC for *pre-commit* phase.
- $lockedQC$, initially $\perp$, stores the highest voted QC for *commit* phase.
- $myslice$ is the slice $m.slice$ in the proposal message m from the leader for the current round.
- $sMap$ is a map, which key is $slice.id$ (which is also $m.node.hash$) and the value is a set of shards $slice.shard$, $sMap_r$ refers to the slice set for $slice.id = r$.
- $rblock$ stores the reconstructed block for the current round.

5.2   Segmenta Phases

The Segmenta protocol is shown in Algorithm 3 and functions used in Segmenta can be found in Algorithm 2. Segmenta contains the following four phases.

Prepare Phase: When the leader of the current view has received enough *newview* message $m \in M$, it starts *prepare phase* by generating $highQC$ from $m.justify : m \in M$ with the highest view and creating a new node on the proposed branch. The leader selects a list of client commands as the new block and then gets the set $\{slice\}$ by executing the function GenerateSlice(), in which the block is erasure encoded and a Merkle proof is generated for each shard. $m.node$ records the Merkle tree root hash $slice.hash$ to lock the block with the new proposed node. Then the leader generates n *prepare* messages with a non-repetitive $slice_i$, which will be sent to $replica_i$ according to the identity i.

Upon receiving a *prepare* message m, a replica will vote for the proposal only when both $m.justify$ and $m.slice.proof$ are valid. $sMap$ is a global map of which the key is $m.node.hash$ and the value is a set $sMap_r$ of shards $m.slice.shard$. If the shard $m.slice.shard$ is valid through the check of the Merkle branch $m.slice.proof$, an empty set $sMap_r$ will be created, where $r = m.node.hash$ and the $m.slice.shard$ will be inserted into it. The replica will also store $m.slice$ into its local variable $myslice$ for the future *echo* slice sharing. The set $sMap_r$ corresponding to the r is unique, which means if there exists $sMap_r$, the *prepare* message m containing a slice that $m.slice.id = r$ will be rejected.

Pre-commit Phase: The leader half operates the same as it is in basic Hotstuff.

Upon receiving a *precommit* message m from the leader, a replica will verify the $m.justify$. If the $m.justify$ is valid, it declares that there are at least $n - f$ replicas that have received their distinct shards. Thus, the replica will respond to the leader with *precommitVote* and start a concurrent communication step to share its recorded slice $myslice$ by broadcasting *echo*.

The delivering of *echo* message will be handled by replicas concurrently with consensus phases in *parallel echo*. Whenever an *echo* message is delivered, the replica will verify $echo.proof$. If the shard $echo.shard$ in $echo$ belongs to the tree with root $echo.id$, it will be stored in $sMap_{echo.id}$.

Commit Phase: The *commit phase* for the leader is the same as basic Hotstuff. Upon receiving a *commit* message m, if there exist $n - f$ items in $sMap_r$, where $r = m.justify.node.hash$, and the $m.justify$ is valid, a replica will start the *reconstruct* and use the erasure code to decode shards and get $Rblock$. The $RblkHash$ is calculated towards $Rblock$ to check the validity of the reconstructed block. If the $RblkHash$ matches $blkHash$, a replica will store the $Rblock$ into its local variable $rblock$ and respond with *commitVote*.

Algorithm 3. The Segmenta Protocol

1: **for** $CurView \leftarrow 1, 2, 3...$ **do**
2: **Prepare Phase:**
3: as a leader:
4:   **upon** receive $n - f$ m
5:     **where** $m.tag == new\text{-}view$:
6:     $highQC \leftarrow$ the $m.justify$ in $\{m\}$
7:     **which** has the highest $m.view$
8:     $\{slice\} \leftarrow$ GenerateSlice()
9:     $node.parent \leftarrow highQC.node$
10:     $node.hash \leftarrow slice.id$
11:     send to $replica_i$ in $Replicas$:
12:       Msg($prepare, node, slice_i, highQC$)
13: as a replica:
14:   **upon** receive m from current leader
15:     **where** $m.tag == prepare$:
16:     **if** $m.slice.proof$ is valid **then**
17:       $myslice \leftarrow m.slice.shard$
18:       $sMap_r \leftarrow \emptyset$
19:       $sMap_r \leftarrow$ insert $m.slice$
20:     **if** $m.node.parent == m.justify.node$
21:     and safeNode($m.node, m.justify$)
22:     **then** send to the current leader
23:       Vote($prepare, m.node, \bot$)
24: **Pre-Commit Phase:**
25: as a leader:
26:   **upon** receive $n - f$ votes $m \in M$
27:     **where** $m.tag == prepare$:
28:     $prepareQC \leftarrow$ QC(M)
29:     broadcast
30:       Msg($pre\text{-}commit, \bot, \bot, prepareQC$)
31: as a replica:
32:   **upon** receive m from current leader
33:     **where** $m.tag == pre\text{-}commit$:
34:     $echo.id \leftarrow myslice.id$
35:     $echo.shard \leftarrow myslice.shard$
36:     $echo.proof \leftarrow myslice.proof$
37:     broadcast $echo$
38:     $prepareQC \leftarrow m.justify$
39:     send to the current leader

40:     Vote($pre\text{-}commit, m.justify.node, \bot$)
41: **Commit Phase:**
42: as a leader:
43:   **upon** receive $n - f$ votes $m \in M$
44:     **where** $m.tag == pre\text{-}commit$:
45:     $qc \leftarrow$ QC(M)
46:     broadcast
47:       Msg($commit, \bot, \bot, qc$)
48: as a replica:
49:   **upon** receive m from current leader
50:     **where** $m.tag == commit$:
51:     **if** $|sMap_r| \geq N - f$ **then**
52:     $Rblock, RblkHash$
53:       $\leftarrow$ EDecode($sMap_r$)
53:     **if** $RblkHash = blkHash$ **then**
54:       $lockedQC \leftarrow m.justify$
55:       $rblock \leftarrow Rblock$
56:     send to the current leader
57:     Vote($commit, m.justify.node, \bot$)
58: **Decide Phase:**
59: as a leader:
60:   **upon** receive $n - f$ votes $m \in M$
61:     **where** $m.tag == commit$:
62:     $qc \leftarrow$ QC(M)
63:     broadcast
64:       Msg($decide, \bot, \bot, qc$)
65: as a replica:
66:   **upon** receive m from current leader
67:     **where** $m.tag == decide$:
68:     execute commands in $rblock$
69: **newview interrupt:**
70:   **Whenever** a $timeout$ occurred:
71:     send to the leader for $CurView + 1$
72:     Msg($new\text{-}view, \bot, \bot, prepareQC$)
73: **Parallel Echo:**
74:   **Whenever** receive $echo$:
75:     **if** $echo.proof$ is valid **then**
76:       $sMap_r \leftarrow insertecho$

Theorem 1. *After GST, at least $n - f$ different echo will be delivered before the replica has received commit.*

Proof. Upon receiving valid $m.slice$, correct replicas will broadcast $echo$, which is expected to deliver after a period of half RTT according to the network assumption. The timeout between *pre-commit phase* and *commit phase* is an RTT. Therefore, the *echo* message will be delivered before *commit*. Since there are $n - f$ correct replicas, there will be at least $n - f$ *echo*.

Decide Phase: The *decided phase* is exactly the same as in basic Hotstuff. Upon receiving a *decided* message, a replica knows that there are at least $n - f$ replicas that have received $n - f$ different valid shards and reconstructed the transaction block. Thus, it will execute *cmds* in its *rblock*.

Apart from these phases, a *newview interrupt* will be triggered whenever a timeout occurs. According to the election algorithm, a timeout replica will broadcast a *newview* message with its *prepareQC* to the next leader. Therefore, replicas start new views whenever failures occur and keep the consensus alive. The communication pattern is shown in Fig. 3.

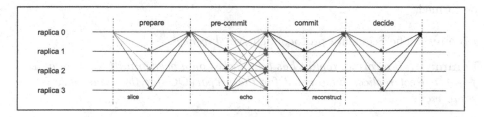

Fig. 3. Communication pattern of Segmenta with 4 replicas (replica 0 is the current leader).

5.3   Safety, Liveness, and Communication Complexity

Before discussing the safty and liveness properties of the Segmenta protocol, we first define two quorum certificates qc_1 and qc_2 are conflicting if $qc_1.node$ and $qc_2.node$ are not on the same branch. Two reconstructed blocks $Rblock_1$ and $Rblock_2$ are conflicting if $Rblock_1.id \neq Rblock_2.id$.

Safety: The Segmenta protocol maintains *Safety*.

Lemma 1. *If there exist two conflicting valid quorum certificates qc_1 and qc_2 with the same round tag, then $qc_1.view \neq qc_2.view$.*

Proof. Assume there are two conflicting valid quorum certificates qc_1 and qc_2 with the same round tag and $qc_1.view = qc_2.view$. Then qc_1 is generated by votes from the replica set $|s_1| = 2f + 1$, and qc_2 is generated by votes from the replica set $|s_2| = 2f + 1$. Let the system replica set be s_0, then $|s_1 \cup s_2 \cap s_0| = f + 1 > f$. Thus, there must be a correct replica who has voted twice for the same phase in a view, which is not allowed by the pseudocode. Therefore, $qc_1.view \neq qc_2.view$.

Lemma 2. *If there exist two conflicting valid reconstructed blocks $Rblock_1$ and $Rblock_2$, then $Rblock_1.view \neq Rblock_2.view$.*

Proof. Assume there are two conflicting valid reconstructed blocks $Rblock_1$ and $Rblock_2$ and $Rblock_1.view = Rblock_2.view$. Then $Rblock_1$ is reconstructed by an existing valid $prepareQC_1$ and an existing valid $commitQC_1$. At the same time, there exists valid $prepareQC_2$ valid $commitQC_2$ for $Rblock_2$. For these quorum certificates, $prepareQC_1 \neq prepareQC_2$ and $commitQC_1 \neq commitQC_2$, which is contrary to Lemma 1. Thus, $Rblock_1.view \neq Rblock_2.view$.

Theorem 2. *If n_1 and n_2 are two conflicting nodes, they cannot be committed, each by a correct replica.*

Proof. Assume both n_1 and n_2 are committed, then two reconstructed blocks $Rblock_1$ and $Rblock_2$ are formed for each node. According to Lemma 2, we can see that $Rblock_1.view(v_1) \neq Rblock_2.view(v_2)$. Assume $v_1 \leq v_2$. As $Rblock_1$ is formed, there are at least $f + 1$ correct nodes (set s_1)that have locked and prepared on n_1. And there are at least $f + 1$ correct nodes (set s_2)that have locked and prepared on n_2. Thus, there is at least one correct first locked at n_1 and then locked on a conflict n_2, which is impossible since $commitQC_2$ cannot be formed.

Liveness: The Segmenta protocol achieves *Liveness*.

Lemma 3. *If a correct replica has been locked on a precommitQC, then at least $f + 1$ correct replicas have voted for the prepareQC for the same node in the same view.*

Proof. Assume replica r has been locked on $precommitQC$ for node n, then $2f + 1$ votes have been generated into the $prepareQC$ in the $prepare$ phase of the same view for n according to the pseudocode. Since there are at most f faulty nodes, at least $f + 1$ votes are from correct replicas.

Theorem 3. *After GST, there exists a bounded period T_f such that all correct replicas reach a decision if they remain in view v during T_f and the leader for view v is correct.*

Proof. To start a new view, the leader selects the highest qc in the *newview* message to extend a new node n and makes a proposal p. Replicas send their $prepareQC$ to the leader in the *newview* message. Assume that a node has been locked on a $precommitQC_h$ that is highest among replicas. According to Lemma 3, at least $f + 1$ correct replicas has voted for $prepareQC_h$ for the same node as $precommitQC_h$ in the same view. Thus, at least one *newview* message containing $prepareQC_h$ will be received by the leader and the leader then chooses this $prepareQC_h$ to propose. Therefore, all correct replicas will vote for this proposal since n is extended from its $lockedQC$ or $prepareQC_h.view$ is higher than $lockedQC.view$. Then the system will follow the normal case in Segmenta and reach a decision. After GST, the duration T_f of finishing a view will be a bounded period.

Communication Complexity: The typical broadcast method carries an $n|m|$ communication overhead for the proposer as well as the overall system. A message size of $|m|$ will grow to $\frac{n}{n-f}|m|$ after encode with $\{n, n-f\}$-*Erasure*. For the leader in Segmenta, it broadcasts sliced proposal messages of the total size $n\frac{\frac{n}{n-f}|m|}{n}$. Each replica broadcasts messages of the total size $n\frac{1}{n-f}|m|$ to echo their shards. Therefore, the overall overhead of the leader is $2\frac{n}{n-f}|m| + \lambda$ and $\frac{n}{n-f}|m| + \lambda$ for a single replica, where the λ stands for the additional field carried by the protocol. Thus, the system's overhead will be $(n+1)\frac{n}{n-f}|m| + \lambda$ which means Segmenta sacrifices a small amount of overall communication overhead to ease the burden on the leader. The communication complexity reduces from $O(n)$ to $O(1)$ for the block-broadcast phase. Since the QC contains at least $n-f$ votes, the overall communication complexity of Segmenta per view is $O(n)$. BLS can be used to further reduce the overall complexity to $O(1)$.

5.4   Echoing Shards Optimistically

In Algorithm 3, a replica echoes its shards when it convinces that there are at least $n-f$ different shards that have been delivered to $n-f$ replicas by verifying the $precommitQC$. In this way, Segmenta avoids meaningless shards echoing when the prepare phase fails due to the unstable network or faulty leader. However, the erasure coding is computationally intensive, which will take a long time and reduce consensus efficiency.

If we can expect that the system will be in perfect condition in the most period, we can use an optimistic echo version of Segmenta (Fig. 4). In this protocol, replicas echo their shards when they receive the proposal in the *prepare* phase and start reconstruction as soon as there are $n-f$ matching shards. Then, when the QC is delivered in the *commit* phase, it can directly check whether the block has successfully reconstructed and vote to the leader according to the result. If a *decidedQC* has been generated, a replica can be sure that there are $n-f$ replicas that have reconstructed the same block. In this way, we give one more RTT for replicas to do the reconstruction, which further improves the consensus efficiency.

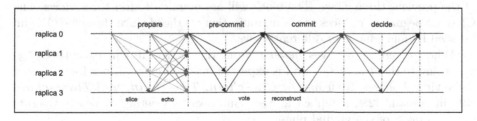

Fig. 4. Communication pattern of optimistic echo Segmenta with 4 replicas (replica 0 is the current leader).

6   Chained Segmenta Protocol

We can abstract Segmenta phases as following two parts:

1) For a leader: collect votes, (propose), broadcast QC
2) For a replica: wait for QC, execute some operation $Opt.$, vote

The Chained Segmenta parallels several views, leveraging the similarity of each phase to achieve a simplified process. The key is to start a new view whenever a replica receives QC from the current leader, which is the same as Chained Hotstuff [28]. A proposal will be broadcast along with QC for each round.

Since each phase has a unique identifier p, we can distinguish $Opt.$ by mark them with p (use $p_1..p_4$ in this chapter), donating $Opt._p$:

$Opt._{p1}$: Validate shard s_i and broadcast $echo$
$Opt._{p2}$: update $genericQC(prepareQC)$
$Opt._{p3}$: reconstruct $block$ and update $lockedQC$
$Opt._{p4}$: execute $block$

The Chained Segmenta (Algorithm 4) processes as below: Upon collecting $n - f$ $newview$ message, the leader of $CurView$ generates $genericQC$ from the votes, processes $propose$ operations, and starts the $prepare$ $phase$ by sending $proposal_i$ together with $genericQC$ to each $replica_i$. Then the $genericQC$ will be relayed to the leader of $CurView + 1$. Upon the leader of $CurView + 1$ receiving $genericQC$, it starts the $prepare$ $phase$ for $CurView + 1$. It virtually carries the $precommit$ $phase$ for $CurView$ since its operations are covered by the $prepare$ $phase$. $CurView + 2$ starts simultaneously as $CurView + 1$, virtually carrying the $prepare$ $phase$ of $CurView + 1$ and the $precommit$ phases of $CurView$. $CurView + 3$ starts simultaneously, virtually carrying the $prepare$ $phase$ of $CurView + 2$, the $precommit$ $phase$ of $CurView + 1$ and $commit$ $phase$ of $CurView$. $CurView + 4$ starts simultaneously, virtually carrying the $prepare$ $phase$ of $CurView + 3$, the $precommit$ $phase$ of $CurView + 2$, $commit$ $phase$ of $CurView + 1$, $decided$ $phase$ of $CurView$.

In short, the leader under $View$ sends $generic\{proposal_i, genericQC\}$ to each $replica_i$, which starts $prepare$ $phase$ for this view and three virtual phases of the previous three views. The block will be committed after three views. The Chained Segmenta reduces the communication overhead since the same QC can be used by four phases of different views.

When a replica receives $generic$ from the leader, it will execute $Opt._{p1}$ according to the $genericQC$ and is expected to execute operations for the other three virtual phases. By using the **One-chain, Two-chain, and Three-chain** rule in Hotstuff [28], a replica in Segmenta decides whether or not to operate $Opt._{p2}..Opt._{p4}$ of the virtual phases.

Algorithm 4. Chained-Segmenta

1: **for** $CurView \leftarrow 1, 2, 3...$ **do**	16: $r = m.slice.id$
2: as a leader:	17: $sMap_r \leftarrow \emptyset$
3: **upon** receive $n - f$ $newview_{CurView-1}$:	18: $sMap_r \leftarrow$ insert $m.slice.shard$
4: $genericQC \leftarrow$ HighestQC($newview$)	19: broadcast $echo$
5: $\{slice\} \leftarrow$ GenerateSlice()	20: **if** $safeNode$ **then**
6: $node.parent \leftarrow genericQC.node$	21: send $vote$ to the leader
7: $node.hash \leftarrow slice.id$	22: **if** $one\text{-}chain \& |sMap_{r'}| \geq N - f$ **then**
8: send to $replica_i$ in $Replicas$:	23: update $genericQC$
9: Msg($generic, node, slice_i, genericQC$)	24: **if** $two\text{-}chain$ **then**
	25: $Rblock'', RblkHash'' \leftarrow$
10: as a replica:	EDecode($sMap_{r''}$)
11: **upon** receive $echo$:	26: update $lockedQC$
12: **if** $echo.proof$ is valid **then**	27: **if** $three\text{-}chain$ &
13: $sMap_{echo.id} \leftarrow insertecho.shard$	$RblkHash''' = blkHash^*$ **then**
14: **upon** receive $generic$ message m:	28: **execute** $Rblock^*$
15: **if** $m.slice.proof$ is valid **then**	

7   Evaluation

We have implemented Segmenta[1] by extending the public *libhotstuff*[2] for about 1000 C++ code lines of addition/adaption. A Reed Solomon Erasure module is implemented with the algorithm leopard GF16 [17], which supports *encode/decode* for up to 65536 shards. We provide *buildMerKleTree* and *proof* functions in the sub-module Merkle Tree using sha256 to generate the hash. In this section, we examine the throughput and latency of Segmenta by comparing it to Hotstuff in large-scale systems under restricted bandwidth.

7.1   Experimental Setup

Our experiments were performed on the testbed of Aliyun using ecs.c6.4xlarge instance. Each instance had 32 GiB RAM and 16vCPUs supported by Intel Xeon (Cascade Lake) Platinum 8269CY processors. We deployed 4 Segmenta nodes on each physical machine and 4 clients for each node. We scaled the system by joining physical machines (from 25 to 100). We evaluated Segmenta under different network scenarios. We restricted the available bandwidth of each physical machine through the Wonder Shaper[3], which is a script using iproute's tc command.

7.2   Prediction Model

We provide a prediction model which shows the trade-off between computation and communication in Segmenta. Since the erasure encode/decode function is

---

[1] Available at https://github.com/1401MIDA/Segmenta.

[2] Available at https://github.com/hot-stuff/libhotstuff.

[3] Available at https://github.com/magnific0/wondershaper.

sensitive for computation, it will cause additional latency t_c for each round in Segmenta. t_c grows with the block size and system size from tens to hundreds of milliseconds. In Hotstuff, the period of the proposal (size $|m|$) delivering is $\frac{|m|}{b}$ when the available bandwidth is b. Though slicing broadcast, the proposal of Segmenta is $\frac{1}{n-f}|m| + \lambda$, where λ refers to the size of the Merkle proof of a slice. Thus, the period of proposal delivering is $\frac{\frac{1}{n-f}|m|+\lambda}{b}$ in Segmenta. Therefore, Segmenta donates an approximate latency optimization of:

$$\Delta l = \frac{(1 - \frac{1}{n-f})|m|}{b} - t_c \tag{1}$$

The additional message size λ caused by the Merkle proof is ignored here since it is tiny compared to the size of block slices.

We can predict a *balancepoint* where the reduction of communication latency is covered by the latency caused by computation. According to Eq. 1, the performance improvement of Segmenta (related to Δl) is affected by block size, system size, as well as available bandwidth.

7.3   Throughput and Latency

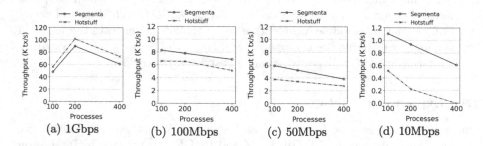

Fig. 5. Throughput of different numbers of processes (block size: 3 Mb).

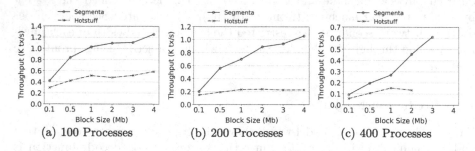

Fig. 6. Throughput of different block sizes (available bandwidth: 10 Mbps).

We measured the throughput and latency in multiple scenarios. For most experiments, we used three system sizes, which are $n = 100, 200, 400$ nodes, and tested them under 10 Mbps, 25 Mbps, 100 Mbps, and 1 Gbps bandwidth. Some of the results are displayed in Fig. 5 to Fig. 8, and more results can be found in our git repository.

We first assess the throughput vs. processes of the failure-free scenarios for both Segmenta and HotStuff. We set the system scale to 100, 200, and 400 processes. We report throughput vs. block size under 10 Mbps network in Fig. 6. As shown in Fig. 5, the throughput of Segmenta is 0.9× to 4.6× of HotStuff.

We further report the latency vs. bandwidth for 4° from 10 Mbps to 1024 Mbps. To make the pictures more visual, the logarithm of the latency is taken in Fig. 7. We also show the result of the latency vs. block size for block sizes from 0.1 Mb to 4 Mb in Fig. 8. We can see that Segmenta achieves a peek latency reduction of 83% lower than Hotstuff.

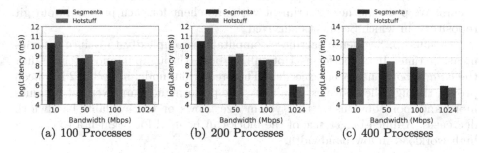

(a) 100 Processes   (b) 200 Processes   (c) 400 Processes

Fig. 7. Latency under different bandwidth (block size: 2 Mb).

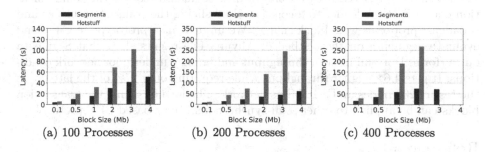

(a) 100 Processes   (b) 200 Processes   (c) 400 Processes

Fig. 8. Latency of different block sizes (available bandwidth: 10 Mbps).

We observed the *balancepoint* in our experiments. Figure 5(a) shows that Segmenta did not exceed the throughput of Hotstuff under 1 Gbps bandwidth. The leader did not arrive at the bottleneck with sufficient bandwidth, so the additional computation overload challenges the system's performance. Then in Fig. 5(b), we observed that Segmenta outperformed Hotstuff. Thus, we can expect that a *balancepoint* will occur between 1 Gbps and 100 Mbps, where the additional computation overload is covered by reduced communication time. Such a phenomenon can also be observed in Fig. 7.

After the *balancepoint*, Segmenta showed better performance in both throughput and latency. Figure 6 shows that Hotstuff keeps a low throughput while it shows a growing trend in Segmenta, as the block size grows. It achieves the best 4.6× higher throughput in Fig. 6(b). Segmenta still kept working in the extremely bad conditions in Fig. 6(c), where Hotstuff had already failed to reach a consensus. Figure 8 demonstrates that the latency of Sementa is much lower than Hotstuff and it shows a smoother growth trend.

Both Segmenta and Hotstuff no longer worked when the block size reached 4M (400 processes, 10 Mbps). According to our configuration, there would be 1600 clients, increasing the deployment difficulty. Since we started clients in the same physical machine with processes, the restricted bandwidth affected the clients. We post another experiment with one client for each process in our git repository, in which Segmenta survived.

To sum up, Segmenta sacrifices computation complexity to reduce communication time. It outperforms Hotstuff in both throughput and latency after the *balancepoint* has been exceeded. The prediction model in Sect. 7.2 has been proven through the above experiments. The performance improvement becomes more and more obvious as the system size grows or the available bandwidth decreases. A much bigger size of the block can be used for Segmenta to carry a high workload in low bandwidth.

8   Conclusion

We present Segmenta, a novel semi-synchronous BFT protocol that achieves low latency under restricted network bandwidth. Segmenta reduces the communication overload of the leader to a const level by slicing the transaction blocks and broadcast proposals with block shards. We implement a prototype in C++ and evaluate Segmenta on the testbed of Aliyun. The results show that Segmenta outperforms Hotstuff in both throughput and latency under poor network conditions. It gets 4.6× higher throughput while requiring only 17% of the latency in the test of 200 nodes and a 10 Mbps network. The idea of slicing the transaction blocks can be applied to other semi-synchronous leader-based BFT protocols.

References

1. Diem. https://www.diem.com
2. Amir, Y., et al.: Steward: scaling byzantine fault-tolerant replication to wide area networks. IEEE Trans. Dependable Secure Comput. **7**(1), 80–93 (2008)
3. Androulaki, E., et al.: Hyperledger fabric: a distributed operating system for permissioned blockchains. In: Proceedings of the Thirteenth EuroSys Conference, pp. 1–15 (2018)

4. Boneh, D., Lynn, B., Shacham, H.: Short signatures from the Weil pairing. In: Boyd, C. (ed.) ASIACRYPT 2001. LNCS, vol. 2248, pp. 514–532. Springer, Heidelberg (2001). https://doi.org/10.1007/3-540-45682-1_30

5. Cachin, C., Tessaro, S.: Asynchronous verifiable information dispersal. In: 24th IEEE Symposium on Reliable Distributed Systems (SRDS 2005), pp. 191–201. IEEE (2005)

6. Castro, M., Liskov, B., et al.: Practical byzantine fault tolerance. In: OsDI 1999, pp. 173–186 (1999)

7. Fischer, M.J., Lynch, N.A., Paterson, M.S.: Impossibility of distributed consensus with one faulty process. J. ACM (JACM) **32**(2), 374–382 (1985)

8. Gilad, Y., Hemo, R., Micali, S., Vlachos, G., Zeldovich, N.: Algorand: scaling byzantine agreements for cryptocurrencies. In: Proceedings of the 26th Symposium on Operating Systems Principles, pp. 51–68 (2017)

9. Gueta, G.G., et al.: SBFT: a scalable and decentralized trust infrastructure. In: 2019 49th Annual IEEE/IFIP International Conference on Dependable Systems and Networks (DSN), pp. 568–580. IEEE (2019)

10. Guo, B., Lu, Z., Tang, Q., Xu, J., Zhang, Z.: Dumbo: faster asynchronous BFT protocols. In: Proceedings of the 2020 ACM SIGSAC Conference on Computer and Communications Security, pp. 803–818 (2020)

11. Gupta, S., Rahnama, S., Hellings, J., Sadoghi, M.: ResilientDB: global scale resilient blockchain fabric. arXiv preprint arXiv:2002.00160 (2020)

12. Kokoris-Kogias, E.: Robust and scalable consensus for sharded distributed ledgers. Cryptology ePrint Archive (2019)

13. Kokoris Kogias, E., Jovanovic, P., Gailly, N., Khoffi, I., Gasser, L., Ford, B.: Enhancing bitcoin security and performance with strong consistency via collective signing. USENIX Association (2016)

14. Kokoris-Kogias, E., Jovanovic, P., Gasser, L., Gailly, N., Syta, E., Ford, B.: OmniLedger: a secure, scale-out, decentralized ledger via sharding. In: 2018 IEEE Symposium on Security and Privacy (SP), pp. 583–598. IEEE (2018)

15. Lamport, L., Shostak, R., Pease, M.: The byzantine generals problem. In: Concurrency: The Works of Leslie Lamport, pp. 203–226 (2019)

16. Li, W., Feng, C., Zhang, L., Xu, H., Cao, B., Imran, M.A.: A scalable multi-layer PBFT consensus for blockchain. IEEE Trans. Parallel Distrib. Syst. **32**(5), 1146–1160 (2020)

17. Lin, S.J., Al-Naffouri, T.Y., Han, Y.S., Chung, W.H.: Novel polynomial basis with fast Fourier transform and its application to Reed-Solomon erasure codes. IEEE Trans. Inf. Theory **62**(11), 6284–6299 (2016)

18. Lu, Y., Lu, Z., Tang, Q., Wang, G.: Dumbo-MVBA: optimal multi-valued validated asynchronous byzantine agreement, revisited. In: Proceedings of the 39th Symposium on Principles of Distributed Computing, pp. 129–138 (2020)

19. Lynch, N.A.: Distributed Algorithms. Elsevier (1996)

20. Miller, A., Xia, Y., Croman, K., Shi, E., Song, D.: The honey badger of BFT protocols. In: Proceedings of the 2016 ACM SIGSAC Conference on Computer and Communications Security, pp. 31–42 (2016)

21. Cheema, M.A., et al.: Blockchain-based secure delivery of medical supplies using drones. Comput. Netw. **204**, 108706 (2022). https://doi.org/10.1016/j.comnet.2021.108706. https://www.sciencedirect.com/science/article/pii/S1389128621005661

22. Nakamoto, S.: Bitcoin: a peer-to-peer electronic cash system. Decentralized business review, p. 21260 (2008)

23. Neiheiser, R., Matos, M., Rodrigues, L.E.T.: Kauri: scalable BFT consensus with pipelined tree-based dissemination and aggregation. In: van Renesse, R., Zeldovich, N. (eds.) SOSP 2021: ACM SIGOPS 28th Symposium on Operating Systems Principles, Virtual Event, Koblenz, Germany, 26-29 October 2021, pp. 35–48. ACM (2021). https://doi.org/10.1145/3477132.3483584

24. Neiheiser, R., Presser, D., Rech, L., Bravo, M., Rodrigues, L., Correia, M.: Fireplug: flexible and robust N-version geo-replication of graph databases. In: 2018 International Conference on Information Networking (ICOIN), pp. 110–115. IEEE (2018)

25. Ossamah, A.: Blockchain as a solution to drone cybersecurity. In: 2020 IEEE 6th World Forum on Internet of Things (WF-IoT), pp. 1–9 (2020). https://doi.org/10.1109/WF-IoT48130.2020.9221466

26. Wuille, P.: libsecp256k1. https://github.com/bitcoin/secp256k1

27. Yang, L., Park, S.J., Alizadeh, M., Kannan, S., Tse, D.: DispersedLedger: high-throughput byzantine consensus on variable bandwidth networks. In: 19th USENIX Symposium on Networked Systems Design and Implementation (NSDI 2022), pp. 493–512 (2022)

28. Yin, M., Malkhi, D., Reiter, M.K., Gueta, G.G., Abraham, I.: HotStuff: BFT consensus with linearity and responsiveness. In: Proceedings of the 2019 ACM Symposium on Principles of Distributed Computing, pp. 347–356 (2019)

29. Zachary, A., Ramnik, A., Shehar, B., Mathieu, B., Sam, B., et al.: The Libra Blockchain (2020). https://developers.diem.com/docs/technical-papers/the-diem-blockchain-paper/

Synthetic Data Generation for Differential Privacy Using Maximum Weight Matching

Miao Zhang[1], Xinxin Ye[1(✉)], and Hai Deng[2]

[1] College of Computer Science and Technology, Nanjing University of Aeronautics and Astronautics, Nanjing 211106, China
xxye@nuaa.edu.cn
[2] Department of Electrical and Computer Engineering, Florida International University, Miami, FL 33174, USA

Abstract. Differential privacy synthetic data is one of the most effective methods for privacy preserving data release. However, the existing schemes still suffer from high computational complexity and inability to directly handle values of large domain size when synthesizing high-dimensional data. To mitigate this gap, we propose synthetic data generation for differential privacy using maximum weight matching (DPMWM), a method for automatically synthesizing tabular data in high-dimensional large domain size via differential privacy. Specifically, DPMWM uses differential privacy maximum weight matching for low-dimensional marginal selection and then automatically synthesizes multiple records based on the filtered marginals. The experimental results show that DPMWM outperforms the state-of-the-art in terms of accuracy for counting queries and classification tasks on datasets with larger domain size.

Keywords: Differential privacy · High-dimensional data · Large domain size · Low-dimensional marginals · Maximum weight matching

1 Introduction

Publishing data with privacy guarantees is a hot issue attracting the attention of a wide range of researchers. Differential privacy (DP) is a common tool for data privacy protection proposed by Dwork et al. [11], which provides a clear definition and effective proof in mathematics. DP can provide strong privacy guarantee and rigorous mathematical theory, and has been widely used in data publishing and analysis [36], data mining [18], machine learning [2], and other fields. Xu et al. [31] uses a histogram approach for the privacy preserving data release and partitions by minimizing the sum of the squared errors of the query set. Inevitably, when the number of query tasks is large, the number of partition bins will be large, resulting in large added noise. In addition, a frequent itemset mining algorithm (FIML) under local differential privacy (LDP) settings for

Z. Tari et al. (Eds.): ICA3PP 2023, LNCS 14489, pp. 121–138, 2024.
https://doi.org/10.1007/978-981-97-0798-0_8

transaction data is proposed [18]. FIML combines frequent oracle filling sampling and interactive query responses satisfying LDP, which can efficiently and accurately identify frequent itemset. However, the method does not effectively handle the publication of generalized high-dimensional data.

The above-mentioned differentially private techniques designed for specific tasks require prior knowledge of specialized technical solutions. The generalization of some specific solutions may have certain limitations. In particular, when there are numerous query tasks, the risk of privacy information leakage increases. A promising solution is to release a differentially private synthetic dataset that is statistically similar to the original dataset. In other words, the same data query or other classification tasks can be performed on the synthetic dataset without compromising individual privacy.

There is a consensus to use differential privacy synthetic data to protect against privacy risks. Currently, these methods can be roughly divided into parametric and non-parametric. Probability graph model is a commonly used technique in parameterization methods. PrivBayes [33, 34] constructs Bayesian networks using mutual information as a metric, and calculates the conditional probability of low dimensional marginals. However, mutual information is highly sensitive and it has a high computational space-time complexity. In addition, McKenna et al. [22] propose a method of applying Markov random fields called PGM. The approach involves constructing an undirected graph and subsequently addresses query tasks via estimation-inference. However, PGM requires manual selection of marginals to construct the graphical model, which entails significant labor costs. One of the most classic non-parametric methods is PrivSyn [35]. This method regards the marginal selection as an optimization problem, and employs a greedy algorithm to select appropriate low-dimensional marginals. Moreover, PrivSyn proposes a gradually update synthesis method (GUM) instead of the random sampling method. Although PrivSyn has improved the accuracy of synthetic datasets, it requires setting different thresholds for different datasets, and selecting the appropriate thresholds still demands extensive experimentation.

For existing methods of differentially private synthetic data, there are still some limitations: (1) High-dimensional data will split too much privacy budget, resulting in lots of noise, which will render the noise distribution ineffective. (2) When an attribute with a large domain size is present, certain methods may not be able to handle it directly. Therefore, it is necessary to specially handle the domain size of attributes to reduce it to an operable range before generating a synthetic dataset. (3) The selection of different thresholds can have a considerable impact on the results. If the threshold is too large, a significant amount of relevant information will be lost. Conversely, if the threshold is too low, it cannot capture a large amount of relevant information, resulting in sparse data.

Motivated by these constraints, this paper makes the following contributions.

- We propose a new scheme for differential privacy synthetic data based on maximum weight matching (DPMWM), which can automatically perform data synthesis with privacy guarantees.

- We select the minimum number of low-dimensional marginals for data synthesis using the maximum weight matching technique to reduce the segmentation of the privacy budget.
- We experimentally evaluated DPMWM against three classical algorithms using three real datasets. Experimental results demonstrate that our approach outperforms state-of-the-art techniques for classification tasks and counting queries on data of large domain size.

Roadmap. In Sect. 2, we briefly introduce related work on current DP synthetic data methods. The preparatory work in the early stage includes problem definition and some theoretical basis required, which will be introduced in Sect. 3. In Sect. 4, we describe the specific content of DPMWM in detail. And then we describe the experimental results in Sect. 5. Finally, we present conclusions in Sect. 6.

2   Related Work

DP synthetic data is one of the effective combination technique to prevent information leakage. It is able to perform tasks such as publishing and statistical analysis instead of the raw dataset, while ensuring that privacy information is not disclosed. According to the diversity of synthetic datasets, we discuss the schemes related to DP synthetic data from three aspects: tabular data, image data, and text data.

Most of the research on the synthetic data of DP focuses on tabular datasets. PrivBayes [33,34] and MODIPS [20] are methods that construct Bayesian networks to approximate the distribution of the original dataset. They both use probabilistic graphical models for data synthesis. Constructing a Bayesian network model consumes a substantial privacy budget, which may result in poor utility of the generated synthetic data. JTree [9] uses the sparse vector technique (SVT) to select marginals and then obtained a Markov network based on the junction tree algorithm. PGM [22] is similar to JTree in that it generates a probabilistic graphical model and then estimates the Markov random field (MRF) based on a given set of marginals. However, it is important to mention that PGM does not provide a method for selecting the marginals. PGM and JTree have a limitation in that they cannot capture more correlation information for data with dense distributions. PrivMRF [6] determines thresholds using θ-useful to construct attribute graphs with high correlations. It further triangulates the attribute graph to prevent the formation of large cliques. However, PrivMRF cannot handle datasets with large domain size directly.

In addition, some methods approximate the distribution of the original dataset by selecting low-dimensional marginals. PriView [25] provides a method for reconstructing contingency tables using views to generate synthetic datasets. However, PriView is more suitable for binary data. Subsequently, PrivSyn [35] is introduced, which greedily selects suitable low-dimensional marginals for generating synthetic data. This method is applicable to high-dimensional datasets and can achieve higher accuracy. After that, HDPView [16] creates views using recursive partitioning, which represents the dataset in a more compact form. Although HDPView reduces space consumption, when the domain size of the dataset is large, it grows exponentially in the number of iterations and thus has a high time complexity.

With the rapid development of machine learning, image datasets are increasingly flooding the Internet. At the same time, they also face the risk of leaking private information. Privacy protection of image datasets needs to address more challenges. Most of the DP synthetic data techniques for image datasets are based on deep learning models. DP-CGAN [28], DP-GAN [30], GS-WGAN [8] etc. utilize the generative adversarial network (GAN) model and apply gradient clipping to control the range of gradients after adding noise, ensuring the quality of the model. There are also methods that utilize variational autoencoders (VAEs) and train parameters using neural networks [26]. Unfortunately, due to the inherent limitations of high-dimensional data, existing DP generative models still cannot guarantee the accuracy of the synthesized data. Chen et al. [7] propose a new perspective, which involves directly synthesizing an optimized set of samples instead of using the overall synthesized data for downstream tasks. Certainly, some deep learning-based methods for DP synthetic data can be applied to both tabular and image datasets [15, 21].

Additionally, different methods have been developed for the DP synthesis of text data containing large amounts of information. Yue et al. [32] propose a simple and practical approach to synthesize text datasets, which involves fine-tuning to a pre-trained generative language model using DP stochastic gradient descent (DP-SGD). Typically, electronic health records collect complete medical information about patients in the form of text, which contains a lot of private information. For this reason, Libbi et al. [19] generate artificial texts with annotations by neural language models, which exhibit sufficient utility for downstream tasks. Also, Torfi et al. [27] utilize a combination of DP on convolutional autoencoders and convolutional generative adversarial networks to generate synthetic data, while preserving the feature correlations of the original data. Currently, DP synthesis methods for textual datasets are not mature enough and are still under development.

3 Preliminary

3.1 Problem Statement

We will consider the following problems. Given a dataset D containing personal sensitive information. Let D has d attributes, denoted as $V = \{V_1, V_2, \ldots, V_d\}$. We expect to generate a synthetic dataset D_s, which is highly approximate to D and does not disclose the user privacy of D. Our goal is to find a set of low-dimensional marginals $\mathcal{M}$ of D to generate a synthetic version using an algorithm satisfying (ε, δ)-differential privacy. The function $\mathrm{dom}\,(m_i) = d_{m_i}{}^1$ is used to calculate the domain size of marginal m_i, where $m_i \in \mathcal{M}$ contains 1-way and 2-way marginals.

3.2 Differential Privacy

DP is a privacy protection model with strict mathematical proof. By adding noise to protect privacy, the task of releasing data can be realized without revealing personal information. DP is defined as follows.

Definition 1. *Neighboring Datasets. Two datasets D and D' are neighboring, if and only if one of them has one more record than the other one.*

Definition 2. *(ε, δ)-Differential Privacy [12]. A random algorithm Γ for neighboring datasets D and D' satisfies (ε, δ)-differential privacy, where $\varepsilon > 0, 0 \leq \delta < 1$, if and only if*

$$\Pr[\Gamma(D) \in O] \leq e^{\varepsilon} \Pr\left[\Gamma\left(D'\right) \in O\right] + \delta \tag{1}$$

where $O \subseteq \mathrm{Output}(\Gamma)$ and $\mathrm{Output}(\Gamma)$ is the set of all possible outputs of the algorithm. ε is the privacy budget that controls the strength of data privacy protection. Without loss of generality, the smaller the epsilon is, the stronger the privacy protection is. δ is the relaxation factor, which indicates the maximum upper bound on the allowed privacy disclosure.

Definition 3. *Global Sensitivity. Define the sensitivity of a function f to any two neighboring datasets D and D' as*

$$\Delta_f = \max_{(D,D)} \|f(D) - f\left(D'\right)\|_2 \tag{2}$$

There are various noise mechanisms that can satisfy DP. In this paper, we use the Gaussian mechanism. Let a function f act on dataset D as $f(D)$, and satisfy the DP requirement by adding random noise to $f(D)$. It is defined as follows.

Definition 4. *Gaussian Mechanism. For the function $f : D \rightarrow \mathbb{R}^n$, add noise to $f(D)$ according to the Gaussian mechanism as*

$$\tilde{f} = f(D) + \mathcal{N}\left(0, \sigma^2\right) \tag{3}$$

where $\mathcal{N}\left(0, \sigma^2\right)$ is the random variable sampled from a normal distribution with a mean of 0 and a standard deviation of σ, and $\sigma = \Delta_f \sqrt{2\ln \frac{1.25}{\delta}} / \varepsilon$.

For the combination of q mechanisms $\mathcal{F}_1, \mathcal{F}_2, \ldots, \mathcal{F}_q$ satisfies $(\varepsilon_i, \delta_i)$-DP for $i = 1, 2, \ldots, q$. According to the combination theorem in [13], the combination mechanism $\mathcal{F} = (\mathcal{F}_1, \mathcal{F}_2, \ldots, \mathcal{F}_q)$ is not a simple linear combination, i.e.

---

1 When m_i is 1-way marginal, $\mathrm{dom}\,(m_i)$ indicates the domain size of a single attribute. When m_i is 2-way marginal, such as $m_i = (V_1, V_2)$, then $\mathrm{dom}\,(m_i) = \mathrm{dom}\,(V_1) \cdot \mathrm{dom}\,(V_2)$.

$(\sum_{i=1}^{q} \varepsilon_i, \sum_{i=1}^{q} \delta_i)$-DP, but the combination mechanism $\mathcal{F}$ satisfies $\left(\varepsilon_i \sqrt{2q \ln (1/\delta')} + q\varepsilon_i (e^{\varepsilon_i} - 1), q\delta_i + \delta'\right)$-DP through complex calculation, where $\delta_i, \delta' > 0$. To simplify the calculation of the combined mechanism privacy budget, the definition of zero-Concentrated Differential Privacy(zCDP for short) is used.

Definition 5. *Zero-Concentrated Differential Privacy [5]. A random mechanism $\mathcal{F}$ for any two neighboring datasets D and D' is satisfied with ρ-zero concentrated difference privacy, $\alpha \in (1, \infty)$, if and only if*

$$\mathcal{D}_\alpha \left(\mathcal{F}(D) \| \mathcal{F}(D')\right) = \frac{1}{\alpha - 1} \log \left(\mathbb{E}\left[e^{(\alpha-1)Z}\right]\right) \leq \rho\alpha \tag{4}$$

where $\mathcal{D}_\alpha \left(\mathcal{F}(D) \| \mathcal{F}(D')\right)$ is the α-Rényi divergence between two distributions $\mathcal{F}(D)$ and $\mathcal{F}(D')$, representing the privacy loss random variable.

Theorem 1. *Combination Theorem [5]. Two random mechanisms $\mathcal{F}_1$ and $\mathcal{F}_2$ satisfy ρ_1-zCDP and ρ_2-zCDP respectively,the combination mechanism $\mathcal{F} = (\mathcal{F}_1, \mathcal{F}_2)$ satisfies $(\rho_1 + \rho_2)$-zCDP.*

4   Our Solution

In this section, we give the detailed implementation steps and basic parameter settings of our scheme (DPMWM) in generating differentially private synthetic datasets.

Algorithm 1. DPMWM

Input: Real dataset D, privacy budget ε
Output: Synthetic dataset D_s
 1: Construct graph $G(V, E, W)$ with real dataset D and $\varepsilon_1 = 0.1 \cdot \varepsilon$;
 2: Select low-dimentional marginals and $\varepsilon_2 = 0.9 \cdot \varepsilon$;
 3: Post-processing of noise distribution;
 4: Generate synthetic dataset D_s;

As shown in Algorithm 1, the execution steps of DPMWM can be divided into four stages. Firstly, we calculate the correlation scores and use them as weights to construct a probabilistic complete graph. Secondly, we apply the maximum weight matching algorithm on the complete graph to automatically select suitable low-dimensional marginals, as described in Sect. 4.2. Thirdly, we adaptively add appropriate noise based on the size of the marginal domain and perform post-processing on the noise distribution. Finally, we generate the synthetic dataset. Among them, the privacy budget ε is divided into two parts, ε_1 and ε_2, where ε_1 is used for graph model construction and ε_2 is used for data synthesis. The main objective of our algorithm is to add less noise to the selected low-dimensional marginals to improve accuracy. For this, we set $\varepsilon_1 = 0.1 \cdot \varepsilon$, $\varepsilon_2 = 0.9 \cdot \varepsilon$. Namely, we strive to allocate more privacy budget to low-dimensional marginals. Furthermore, small changes in the allocation proportion of the privacy budget have little effect on the results.

4.1 Construction of Attribute Graph

Given a dataset D, we construct a weighted complete graph $G(V, E, W)$. The pseudo-code of this process is shown in Algorithm 2. We first let $V = \{V_1, V_2, \ldots, V_d\}$ indicate that there are d attribute nodes in total, and set $E = \{(V_i, V_j) \mid i, j \in \{1, 2, \ldots, d\} \wedge i \neq j\}$ to be the edge set composed of all attribute pairs. For any V_i and V_j, we calculate the weight $w(V_i, V_j)$ of the edge (V_i, V_j) as a correlation score between two attribute nodes:

$$w(V_i, V_j) = \frac{1}{2} \| \Pr[V_i, V_j] - \Pr[V_i] \Pr[V_j] \|_1 \tag{5}$$

where $\Pr[V_i, V_j]$ is the joint probability of V_i and V_j, $\Pr[V_i]$ and $\Pr[V_j]$ respectively are the probability of V_i and V_j. To preserve privacy, we add Gaussian noise $\mathcal{N}(0, \sigma_1{}^2)$ to $w(V_i, V_j)$ to get the noisy version. Then $\tilde{w}(V_i, V_j)$ is inserted into W, so $W = \{\tilde{w}(V_i, V_j) \mid i, j \in \{1, 2, \ldots, d\} \wedge i \neq j\}$ (Lines 4–8).

Lemma 1. *The sensitivity of $w(V_i, V_j)$ is 2, which can be proved in Appendix A.*

Algorithm 2. Construction of Attribute Graph

Input: Real dataset D, noise scale σ_1
Output: A probability complete graph $G(V, E, W)$
 1: Let V be the set of attribute nodes in D;
 2: Let E be the set of all attribute pairs in D;
 3: $W \leftarrow \emptyset$;
 4: **for** each $(V_i, V_j) \in E$ **do**
 5:     $w(V_i, V_j) \leftarrow \frac{1}{2} \| \Pr[V_i, V_j] - \Pr[V_i] \Pr[V_j] \|_1$;
 6:     $\tilde{w}(V_i, V_j) \leftarrow w(V_i, V_j) + \mathcal{N}(0, \sigma_1{}^2)$;
 7:     Insert $\tilde{w}(V_i, V_j)$ into W;
 8: **end for**
 9: **return** Complete graph $G(V, E, W)$

4.2 Marginal Selection

To select the minimum number of low dimensional marginals and capture high correlation between attribute as much as possible, we perform maximum weight matching algorithm [14] on the graph G to obtain a set of disjoint marginals with the maximum global correlation score.

A concrete example of maximum weight matching is presented in Fig. 1, showing a partial process consisting of four rounds. Initially, we select the edge with the maximum weight among all available edges, namely (V_1, V_3). Next, we remove all edges connected to V_1 and V_3. Then, we choose the edge with the maximum weight from the remaining edges. We repeat the process until there are no more edges to select. Finally, we obtain a result for one round as 0.647 from Fig. 1(a). However, this does not guarantee that the obtained total weight is the global maximum. After that, we remove the edge with the highest weight from the first round and continue the aforementioned process. Due to the complexity of the process, only the subsequent three rounds are displayed, as shown in Fig. 1(b), (c) and (d). Upon obtaining all possible results, we ultimately achieve a global maximum weight of 0.686, with the final selected edges being $\{(V_1, V_5), (V_3, V_4)\}$.

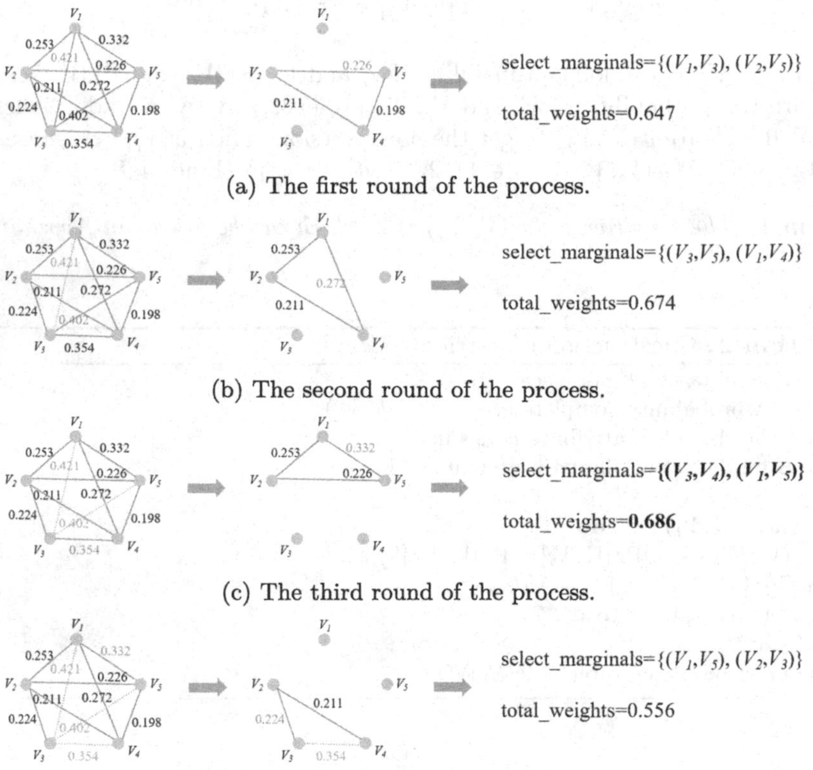

(a) The first round of the process.

(b) The second round of the process.

(c) The third round of the process.

(d) The fourth round of the process.

Fig. 1. The partial process of the example for maximum weight matching.

Algorithm 3 demonstrates marginal selection using maximum weight matching. The probabilistic complete graph G obtained above is taken as the input, and the low-dimensional noise marginal set $\tilde{\mathcal{M}}$ is output. First, we store the set of all marginals with $all\_marginals$ and denote the set of all possible total weights with $all\_total\_weights$, initialised to the empty set. Then, we perform the maximum weight matching algorithm. A set of marginal sets M and the corresponding total weights $total\_weight$ can be obtained by $Maximum\_weight$ representing the process in one round as shown in Fig. 1. We insert them into the set of all marginals $all\_marginals$ and the set of all total weights $all\_total\_weights$, respectively. Repeat this process until all cases have been traversed. Finally, we find the largest total weight value from the set $all\_total\_weights$ of all total weights and get its index, which corresponds to the set $all\_marginals$ of all marginals to get the final result $\mathcal{M}$. Since the selected marginals $\mathcal{M}$ can compromise privacy, we add appropriate noise to the obtained marginals to satisfy DP.

Algorithm 3. Marginal Selection by Maximum Weight Matching

Input: A probability complete graph $G(V, E, W)$, noise scale σ_2
Output: Noise marginal set $\tilde{\mathcal{M}}$
 1: $all\_marginals \leftarrow \emptyset$, $all\_total\_weights \leftarrow \emptyset$;
 2: **while** True **do**
 3:     $M, total\_weight \leftarrow Maximum\_weight\,(G\,(V, E, W))$;
 4:     Insert M into $all\_marginals$ and insert $total\_weight$ into $all\_total\_weights$;
 5:     **if** all cases have been traversed **then**
 6:         **break**
 7:     **end if**
 8: **end while**
 9: $l \leftarrow \arg\max(all\_total\_weights)$;
10: $\mathcal{M} \leftarrow all\_marginals(l)$;
11: **return** $\tilde{\mathcal{M}} = \mathcal{M} + \mathcal{N}\left(0, \sigma_2{}^2\right)$

The process of assigning privacy budgets can be viewed as the optimization problem of minimizing the expected squared difference of the noise scale, which can be expressed as

$$\text{minimize } \frac{d_{m_1}}{p_1^2} + \frac{d_{m_2}}{p_2^2} + \cdots + \frac{d_{m_c}}{p_c^2} \tag{6}$$
$$\text{s.t. } p_1 + p_2 + \cdots + p_c = 1$$

where $d_{m_1}, d_{m_2}, \ldots, d_{m_c}$ denotes the domain size corresponding to the c low-dimensional marginals, and $p_i (i \in \{1, \ldots, c\})$ represents the proportion of privacy budget ε_2 allocated to the i-th marginal. Afterwards, we construct the Lagrangian function $L = \frac{d_{m_1}}{p_1^2} + \frac{d_{m_2}}{p_2^2} + \cdots + \frac{d_{m_c}}{p_c^2} + \lambda\left(p_1 + p_2 + \cdots + p_c - 1\right)$ based on the optimization formula 6. We find the optimal solution of p_i according to the formula $\frac{\partial L}{\partial p_i} = -\frac{2d_{m_i}}{p_i^3} + \lambda = 0$, then $p_i = \left(\frac{2d_{m_i}}{\lambda}\right)^{\frac{1}{3}}$. From formula 6, we can get $\sum_{i=1}^{c} p_i = \sum_{i=1}^{c}\left(\frac{2d_{m_i}}{\lambda}\right)^{\frac{1}{3}} = 1$, then $\lambda = 2 \cdot \left(\sum_{i=1}^{c} d_{m_i}^{\frac{1}{3}}\right)^3$. As a result,

$p_i = \frac{(d_{m_i})^{\frac{1}{3}}}{\sum_{j=1}^{c}(d_{m_i})^{\frac{1}{3}}}$. Therefore, the privacy budget used for the i-th marginal is $p_i \cdot \varepsilon_2$.

4.3 PostProcess

After adding noise, the obtained noise distribution will be negative and the probability sum will not be 1. Therefore, post-processing is required to solve these issues. From the literature [29], we empirically choose a post-processing method that is more applicable to the scheme of this paper.

We implement Norm-Cut [29] for non-negative values and probabilities that do not sum to 1. The core idea of Norm-Cut is to use positive values smaller than the threshold to neutralize the effect of negative values. Specifically, Norm-Cut converts negative values to 0, and for maintaining the noise scale converts positive values below the threshold θ to 0. Typically, the threshold is set to $\theta = 3\sigma$, where σ is the standard deviation of the noise. After that, the normalization process is performed so that the sum is 1.

4.4 Data Synthesis

A commonly used method for generating synthetic datasets is random sampling, which generates synthetic samples based on the approximate distribution obtained. However, this method does not guarantee the accuracy of the synthesis results. Therefore, we use a gradually update method (GUM) [10,35] for synthesis. GUM first initializes a random dataset D_s, and then gradually updates it towards the target value.

The gradual update process of GUM can be divided into two steps. Firstly, the individual attribute values are updated based on the information obtained from the 1-way marginal distributions. We use T_s to denote the target distribution and C_s to denote the current distribution. Our goal is to gradually adjust the value of C_s to T_s. An example is shown in Fig. 2, where the attribute is "education" with five different values {Preschool, Prof-school, Bachelors, Masters, Doctorate}. The first row indicates the distribution for the current D_s, and the second row indicates the target distribution obtained from the approximate distribution. Each node represents an attribute value and its distribution. The update process is mainly to adjust the frequency of 0.1 from Preschool to Bachelors, to move 0.1 from Prof-school to Masters, and to move 0.1 from Doctorate to Masters. The result of the update will approximate the original dataset.

Next, we continue to adjust the 2-way marginal distribution while keeping the one-way marginal distribution unchanged. The target estimation distribution of the 2-way marginals is noted as $T_s(V_i, V_j)$, and the current $C_s(V_i, V_j)$ is adjusted to $T_s(V_i, V_j)$. It is worth noting that the value of only one attribute is updated at a time, i.e., the attribute V_j is guaranteed to be fixed, and then the attribute V_i is changed according to $T_s(V_i, V_j)$. Specifically, for a set of values $(v_i{}^{a_0}, v_j{}^{b_0}) \in (\Omega_i, \Omega_j)$, where Ω_i, Ω_j represent all value sets of V_i and V_j respectively, if $C_s(v_i{}^{a_0}, v_j{}^{b_0}) >$

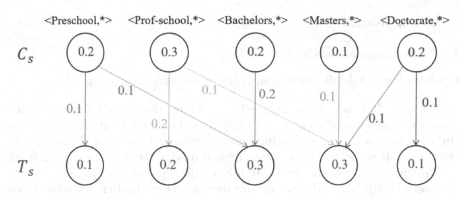

Fig. 2. Example of one-way marginals updating.

$T_s(v_i{}^{a_0}, v_j{}^{b_0})$, there exits $C_s(v_i{}^{a_0}, v_j{}^{b_l}) < T_s(v_i{}^{a_0}, v_j{}^{b_l})$ or $C_s(v_i{}^{a_h}, v_j{}^{b_0}) < T_s(v_i{}^{a_h}, v_j{}^{b_0})$. Here, we assume that $C_s(v_i{}^{a_0}, v_j{}^{b_l}) < T_s(v_i{}^{a_0}, v_j{}^{b_l})$. The minimum value of the quantity to be changed can be calculated as $min_{trans} = \min\left\{(C_s(v_i{}^{a_0}, v_j{}^{b_0}) - T_s(v_i{}^{a_0}, v_j{}^{b_0})), (T_s(v_i{}^{a_0}, v_j{}^{b_l}) - C_s(v_i{}^{a_0}, v_j{}^{b_l}))\right\}$. Then we change the number of records from $(v_i{}^{a_0}, v_j{}^{b_0})$ to $(v_i{}^{a_0}, v_j{}^{b_l})$ based on the value of min_{trans}. The operation is iterated step by step until the result of $C_s(V_i, V_j)$ agrees with the target distribution $T_s(V_i, V_j)$. Figure 3(a) represents the records of Income, Gender attribute pairs in the current unmodified synthetic dataset D_s, and Fig. 3(b) is the statistical distribution table about the 2-way marginals of {Income, Gender}. According to Fig. 3(a) and (b), we adjust the current distribution $C_s(\text{In, Gen})$ to the target distribution $T_s(\text{In, Gen})$ by a combination of duplicating and replacing. For example, by replacing (high, male) with (high, female) and duplicating a record (low, female), we can obtain the target distribution as shown in Fig. 3(c). It is worth noting that the number of records replaced or duplicated is determined by the above min_{trans}.

	Income	Gender
record₁	high	male
record₂	low	male
record₃	high	male
record₄	low	male
record₅	high	female

(a) D_s before updating

	$C_s(\text{In,Gen})$	$T_s(\text{In,Gen})$
<high, male>	0.4	0.2
<high, female>	0.2	0.4
<low, male>	0.4	0.2
<low, female>	0	0.2

(b) Marginal table for {Income,Gender}

	Income	Gender
record₁	high	male
record₂	low	male
record₃	high	female
record₄	low	female
record₅	high	female

(c) D_s after updating

Fig. 3. Example of two-way marginals updating.

5   Experiments

5.1   Experimental Settings

Datasets. We use Adult, Bank, and Nursery datasets for experiments.

- Adult [17]. It is a dataset included on UCI by the Census Bureau, which contains categorical data and numerical data. There are 15 attributes, invalid information such as missing values are removed, and there are 45222 records.
- Bank [23]. It is the user information related to a marketing campaign of Bank of Portugal, and contains attribute information such as age, job, etc.
- Nursery [3,24]. It is a dataset of information about children in custody and their parents, derived from a hierarchical decision model originally developed to rank kindergarten applications, which including attributes such as family structure and financial status.

The details of these datasets are shown in Table 1.

Table 1. Summary of dataset characteristics.

Dataset	Dimension	No. of Point	Domain size	Types
Adult	15	45222	$6.8 \cdot 10^{15}$	Categorical, Numerical
Bank	17	45211	$9.8 \cdot 10^{21}$	Categorical, Numerical
Nursery	7	12960	$6.5 \cdot 10^{4}$	Categorical

Evaluation Metrics. To verify the effectiveness of the synthetic dataset for statistical analysis tasks, we use the following analysis tasks for metrics.

Classification. Classification is a popular method in deep learning, and therefore, classification tasks are often used as a metric to evaluate the utility of synthetic datasets. In this paper, we primarily utilize the widely-used SVM classification method for evaluation purposes.

k-Way Marginals. The metric of k-way marginals is from [4]. We mainly test the total variance distance (TVD) of 1-way, 2-way, and 3-way marginals as a measure of the utility of the synthetic dataset. TVD is the statistical error between the synthetic dataset and the original dataset of marginal M and defined as:

$$\mathrm{TVD} = \frac{1}{n} \left\| T_{D_s,M} - T_{D,M} \right\|_1 \tag{7}$$

Methods. We compare DPMWM with PrivBayes, PGM and PrivSyn. Additionally, to demonstrate the difference more clearly compared to the original dataset, we also include the metrics of the original dataset (Original) for comparison.

Settings. For PrivBayes we implemented the code based on the content of the paper [33]. Since PGM does not provide a method for selecting marginals, we combined PGM with the Adagrid mechanism [1] to obtain synthetic results. All methods generate synthetic records in the same number as the original data. And we define n to denote the number of records in the original dataset, for PrivSyn and our method DPMWM set $\delta = \frac{1}{n^2}$. In addition, for PrivSyn, it is necessary to set a threshold for the domain size of cliques. According to the recommendations in the article, the threshold is typically set to 5000. However, on the Nursery dataset, setting the threshold too high would result in larger clique sizes and excessive noise being added, rendering the synthetic dataset ineffective. Considering the domain size of this dataset, we set the threshold to 200.

The experiments in this paper are run on a machine with an 11th Gen Intel(R) Core(TM) i5-11400F @ 2.60GHz 2.59 GHz processor, 16GB of RAM on board, NVIDIA GeForce GT 730 hardware, 384 CUDA cores, and a driver type and version of DCH 472.98 on the machine. The software used is PyCharm Community Edition 2021.3.3, running in python 3.7 environment.

5.2   Experimental Results

Figure 4 presents the performance of different methods on SVM classification results. We calculate the misclassification rate of the synthetic dataset on the SVM model. We do not showcase the results of PrivBayes on the Bank. This is because the Bank dataset has a large domain size, and computing mutual information leads to memory failures, making it impossible to generate synthetic data.

Regarding the classification results, we can observe that on the datasets of Adult and Bank, the results obtained by the DPMWM method are closer to the original datasets. However, for the Nursery dataset, due to its small domain size, the advantage of DPMWM is not prominent, but it can still ensure results close to other methods. Sometimes, PrivSyn results are better than the Original, probably due to improper setting of the threshold.

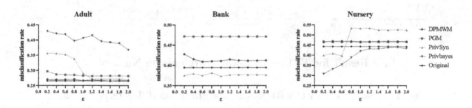

Fig. 4. Results for SVM on different datasets.

Figure 5 demonstrates the TVD of k-way marginals for different methods of generating synthetic datasets on different datasets. The results of PrivBayes are worse, while the results of other methods are closer. For the datasets Adult and Nursery, the domain sizes are relatively small. Therefore, the TVD between the calculated 1-way and 2-way marginals for these two datasets is very small. For example, in the experiment on the Adult, the 1-way marginal TVD of the DPMWM synthesis result is 0.0078, while the 1-way marginal TVD of the PGM synthesis result is 0.0064, and the difference between the two values is very small. Hence, they appear to be quite similar. However, for the Bank dataset, which has a large domain size, the TVD results for k-way marginals are more pronounced. As shown in Fig. 5(b), when the privacy budget ε ranges from 0.2 to 2.0, the TVD for the 3-way marginals consistently decreases and stays between 0.35 and 0.27, outperforming other methods.

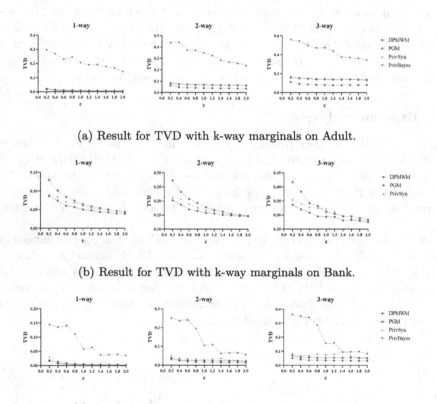

(a) Result for TVD with k-way marginals on Adult.

(b) Result for TVD with k-way marginals on Bank.

(c) Result for TVD with k-way marginals on Nursery.

Fig. 5. Results for TVD with k-way marginals on different datasets.

6   Conclusion

In this paper, we propose a new method for differential privacy data synthesis using maximum weight matching (DPMWM). DPMWM automatically selects the minimum number of low-dimensional marginals to generate synthetic datasets by a maximum weight matching algorithm and consumes less privacy budget. Experimental results on real datasets show that DPMWM can consistently outperform the state-of-the-art methods in terms of accuracy for counting queries and classification tasks on generated synthetic data on large domain size.

A   Proof of Lemma

Proof. We assume a dataset D contains n records, and consider two attributes a and b. Denote the frequency of different values of attribute a is $\{a_1, a_2, \ldots\}$ and the frequency of b is $\{b_1, b_2, \ldots\}$. For 2-way marginal of (a, b), denote its frequency of joint distribution is $\{c_{11}, c_{12}, \ldots\}$.

The metric $w(a, b)$ is

$$w(a,b) = \frac{1}{2} \sum_{ij} \left| \frac{a_i b_j}{n} - c_{ij} \right|$$

If we add a user with value u for a and v for b, then

$$
\begin{aligned}
w'(a,b) = {} & \frac{1}{2} \sum_{i \neq u, j \neq v} \left| \frac{a_i b_j}{n+1} - c_{ij} \right| \\
& + \frac{1}{2} \sum_{i \neq u} \left| \frac{a_i (b_v + 1)}{n+1} - c_{iv} \right| \\
& + \frac{1}{2} \sum_{j \neq v} \left| \frac{(a_u + 1) b_j}{n+1} - c_{uj} \right| \\
& + \frac{1}{2} \left| \frac{(a_u + 1)(b_v + 1)}{n+1} - (c_{uv} + 1) \right|
\end{aligned}
$$

The sensitivity is given by

$$
\begin{aligned}
\Delta_w = {} & \left| w(a,b) - w'(a,b) \right| \\
\leq {} & \frac{1}{2} \sum_{i \neq u, j \neq v} \left| \frac{a_i b_j}{n(n+1)} \right| + \frac{1}{2} \sum_{i \neq u} \left| \frac{a_i b_v}{n(n+1)} - \frac{a_i}{n+1} \right| \\
& + \frac{1}{2} \sum_{j \neq v} \left| \frac{a_u b_j}{n(n+1)} - \frac{b_j}{n+1} \right| + \frac{1}{2} \left| \frac{(n+1)a_u b_v - n(a_u+1)(b_v+1) + n(n+1)}{n(n+1)} \right| \\
= {} & \frac{\frac{1}{2} \sum_{i \neq u, j \neq v} a_i b_j - \frac{1}{2} \sum_{i \neq u} (a_i b_v - n a_i) - \frac{1}{2} \sum_{j \neq v} (a_u b_j - n b_j)}{n(n+1)} \\
& + \frac{(n+1)a_u b_v - n(a_u+1)(b_v+1) + n(n+1)}{2n(n+1)} \\
= {} & \frac{(n - a_u)(n - b_v) - (n - a_u)(b_v - n) - (a_u - n)(n - b_v) + (n - a_u)(n - b_v)}{2n(n+1)} \\
= {} & \frac{4(n - a_u) \cdot (n - b_v)}{2n(n+1)} \\
= {} & \frac{2(n - a_u)(n - b_v)}{n(n+1)} \leq 2
\end{aligned}
$$

(8)

For the above formula, some details are $\sum_{i \neq u} a_i = n - a_u$, $\sum_{j \neq v} b_j = n - b_v$ and $\sum_{i \neq u, j \neq v} a_i b_j = (n - a_u)(n - b_v)$.

References

1. NIST. 2021 differential privacy synthetic data challenge. https://github.com/ryan112358/nist-synthetic-data-2021
2. Abadi, M., et al.: Deep learning with differential privacy. In: Proceedings of the 2016 ACM SIGSAC Conference on Computer and Communications Security, pp. 308–318 (2016)
3. Asuncion, A., Newman, D., Bache, K., Lichman, M.: UCI machine learning repository. Meta 2003 (2003)
4. Barak, B., Chaudhuri, K., Dwork, C., Kale, S., McSherry, F., Talwar, K.: Privacy, accuracy, and consistency too: a holistic solution to contingency table release. In: Proceedings of the Twenty-Sixth ACM SIGMOD-SIGACT-SIGART Symposium on Principles of Database Systems, pp. 273–282 (2007)
5. Bun, M., Steinke, T.: Concentrated differential privacy: simplifications, extensions, and lower bounds. In: Hirt, M., Smith, A. (eds.) TCC 2016. LNCS, vol. 9985, pp. 635–658. Springer, Heidelberg (2016). https://doi.org/10.1007/978-3-662-53641-4_24
6. Cai, K., Lei, X., Wei, J., Xiao, X.: Data synthesis via differentially private Markov random fields. Proc. VLDB Endow. **14**(11), 2190–2202 (2021)
7. Chen, D., Kerkouche, R., Fritz, M.: Private set generation with discriminative information. arXiv preprint arXiv:2211.04446 (2022)
8. Chen, D., Orekondy, T., Fritz, M.: GS-WGAN: a gradient-sanitized approach for learning differentially private generators. In: 34th Conference on Neural Information Processing Systems, pp. 12673–12684. Curran Associates, Inc. (2020)
9. Chen, R., Xiao, Q., Zhang, Y., Xu, J.: Differentially private high-dimensional data publication via sampling-based inference. In: Proceedings of the ACM SIGKDD International Conference on Knowledge Discovery and Data Mining, vol. 2015, p. 129 (2015)
10. Chen, X., Wang, C., Yang, Q., et al.: Locally differentially private high-dimensional data synthesis (2023)
11. Dwork, C.: Differential privacy. In: Bugliesi, M., Preneel, B., Sassone, V., Wegener, I. (eds.) ICALP 2006. LNCS, vol. 4052, pp. 1–12. Springer, Heidelberg (2006). https://doi.org/10.1007/11787006_1
12. Dwork, C., McSherry, F., Nissim, K., Smith, A.: Calibrating noise to sensitivity in private data analysis. In: Halevi, S., Rabin, T. (eds.) TCC 2006. LNCS, vol. 3876, pp. 265–284. Springer, Heidelberg (2006). https://doi.org/10.1007/11681878_14
13. Dwork, C., Rothblum, G.N., Vadhan, S.: Boosting and differential privacy. In: Proceedings of the 2010 IEEE 51st Annual Symposium on Foundations of Computer Science, pp. 51–60 (2010)
14. Yu, W., Iranmanesh, S., Haldar, A., Zhang, M., Ferhatosmanoglu, H.: An axiomatic role similarity measure based on graph topology. In: Qin, L., et al. (eds.) SFDI LSGDA 2020. CCIS, vol. 1281, pp. 33–48. Springer, Cham (2020). https://doi.org/10.1007/978-3-030-61133-0_3
15. Harder, F., Adamczewski, K., Park, M.: DP-MERF: differentially private mean embeddings with random features for practical privacy-preserving data generation. In: Proceedings of the 24th International Conference on Artificial Intelligence and Statistics (AISTATS 2021), vol. 130, pp. 1819–1827. PMLR (2021)

16. Kato, F., Takahashi, T., Takagi, S., Cao, Y., Liew, S.P., Yoshikawa, M.: HDPView: differentially private materialized view for exploring high dimensional relational data. arXiv preprint arXiv:2203.06791 (2022)
17. Kohavi, R.: Scaling up the accuracy of Naive-Bayes classifiers: a decision-tree hybrid. In: Second International Conference on Knowledge Discovery and Data Mining, pp. 202–207 (1996)
18. Li, J., Gan, W., Gui, Y., Wu, Y., Yu, P.S.: Frequent itemset mining with local differential privacy. In: Proceedings of the 31st ACM International Conference on Information & Knowledge Management, pp. 1146–1155 (2022)
19. Libbi, C.A., Trienes, J., Trieschnigg, D., Seifert, C.: Generating synthetic training data for supervised de-identification of electronic health records. Future Internet **13**(5), 136 (2021)
20. Liu, F.: Model-based differentially private data synthesis and statistical inference in multiply synthetic differentially private data. arXiv e-prints, pp. arXiv-1606 (2016)
21. Long, Y., et al.: G-pate: scalable differentially private data generator via private aggregation of teacher discriminators. In: 35th Conference on Neural Information Processing Systems, NeurIPS 2021, pp. 2965–2977. Neural Information Processing Systems Foundation (2021)
22. McKenna, R., Sheldon, D., Miklau, G.: Graphical-model based estimation and inference for differential privacy. In: International Conference on Machine Learning, pp. 4435–4444. PMLR (2019)
23. Moro, S., Cortez, P., Rita, P.: A data-driven approach to predict the success of bank telemarketing. Decis. Support Syst. **62**, 22–31 (2014)
24. Olave, M., Rajkovic, V., Bohanec, M.: An application for admission in public school systems. Expert Syst. Public Adm. **1**, 145–160 (1989)
25. Qardaji, W., Yang, W., Li, N.: Priview: practical differentially private release of marginal contingency tables. In: Proceedings of the 2014 ACM SIGMOD International Conference on Management of Data, pp. 1435–1446 (2014)
26. Takagi, S., Takahashi, T., Cao, Y., Yoshikawa, M.: P3GM: private high-dimensional data release via privacy preserving phased generative model. In: 2021 IEEE 37th International Conference on Data Engineering (ICDE), pp. 169–180. IEEE Computer Society (2021)
27. Torfi, A., Fox, E.A., Reddy, C.K.: Differentially private synthetic medical data generation using convolutional GANs. Inf. Sci. **586**, 485–500 (2022)
28. Torkzadehmahani, R., Kairouz, P., Paten, B.: DP-CGAN: differentially private synthetic data and label generation. In: 2019 IEEE/CVF Conference on Computer Vision and Pattern Recognition Workshops (CVPRW), pp. 98–104. IEEE (2019)
29. Wang, T., Lopuhaa-Zwakenberg, M., Li, Z., Skoric, B., Li, N.: Locally differentially private frequency estimation with consistency. In: NDSS 2020: Proceedings of the NDSS Symposium (2020)
30. Xie, L., Lin, K., Wang, S., Wang, F., Zhou, J.: Differentially private generative adversarial network. arXiv preprint arXiv:1802.06739 (2018)
31. Xu, J., Zhang, Z., Xiao, X., Yang, Y., Yu, G., Winslett, M.: Differentially private histogram publication. VLDB J. **22**, 797–822 (2013)
32. Yue, X., et al.: Synthetic text generation with differential privacy: a simple and practical recipe. arXiv preprint arXiv:2210.14348 (2022)
33. Zhang, J., Cormode, G., Procopiuc, C.M., Srivastava, D., Xiao, X.: Privbayes: private data release via Bayesian networks. In: Proceedings of the 2014 ACM SIGMOD International Conference on Management of Data, pp. 1423–1434 (2014)

34. Zhang, J., Cormode, G., Procopiuc, C.M., Srivastava, D., Xiao, X.: Privbayes: private data release via Bayesian networks. ACM Trans. Database Syst. (TODS) **42**(4), 1–41 (2017)
35. Zhang, Z., et al.: PrivSyn: differentially private data synthesis. In: Proceedings of the 30th USENIX Security Symposium (2021)
36. Zhu, T., Li, G., Zhou, W., Yu, P.S.: Differentially private data publishing and analysis: a survey. IEEE Trans. Knowl. Data Eng. **29**(8), 1619–1638 (2017)

Approximate Multicast Coflow Scheduling in Reconfigurable Data Center Networks

Yuhang Wu⬤, Quan Chen(✉)⬤, Jianglong Liu⬤, Fulong Li⬤,
and Lianglun Cheng

School of Computer Science and Technology, Guangdong University of Technology,
Guangzhou, China
{2112105283,2112205082,2112205020}@mail2.gdut.edu.cn,
{quan.c,llcheng}@gdut.edu.cn

Abstract. The emerging optical circuit technology, capable of establishing circuit connections among switches, has been proposed as a promising paradigm for data center networks. This paper investigates the problem of minimizing the completion time of multicast coflow in optical circuit switches (OCS)-based data center networks. The existing works either only focused on multicast coflow scheduling or focused solely on circuit scheduling in OCS-based networks, which greatly limits their performance. Hence, in this paper, we study how to reduce the completion time of multicast flows by considering circuit scheduling and coflow scheduling simultaneously. Firstly, We formulate the problem of multicast coflow scheduling, and prove it to be NP-hard. We propose a delay-efficient multicast coflow scheduling algorithm by integrating coflow scheduling with circuit scheduling. The proposed algorithm is proved to have an approximate ratio of at most $2\sqrt{n}$, where n represents the number of optical circuit switches. Through extensive simulations, it is shown that the proposed algorithm can achieve high performance compared to state-of-the-art methods.

Keywords: Data center networks · Optical circuit switches · Multicast coflow scheduling

1 Introduction

Recently, optical circuit switches (OCS) have gained significant attention in data center networks due to their robustness, high link capacity, and low power consumption [11–14,20,31,32]. For example, the optical modules with 100 Gbps

This research was supported by China University Industry University Research Innovation Fund under Grant No. 2021FNA02010, the NSFC under Grant No. U20A6003, U2001201, 62372118, the Guangdong Basic and Applied Basic Research Foundation under Grant No. 2022A1515011032, the Guangzhou Science and Technology Plan under Grant 2023A04J1701, and the Guangdong Provincial Key Laboratory of Cyber-Physical System under Grant 2020B1212060069.

Z. Tari et al. (Eds.): ICA3PP 2023, LNCS 14489, pp. 139–154, 2024.
https://doi.org/10.1007/978-981-97-0798-0_9

or faster interfaces are now available in the market, whereas link capacities of packet-switched networks range from 1–10 Gbps. Note that, in the OCS-based networks, establishing a new circuit from an input port to multiple output ports incurs a certain circuit reconfiguration delay, such as 1 millisecond. On the other hand, with the explosive growth of data-centric applications [4,8], the data-parallel computation frameworks, such as MapReduce [1] and Spark [24], have experienced a notable surge in popularity. In data-parallel clusters, a set of flows transferring data for a given application among various clusters of machines, called coflow [25], has emerged as the dominant workload. In addition, data multicast is a common communication pattern in data center networks, where data is disseminated from a source node to multiple destination nodes [2,3,33].

To reduce the multicast coflow completion time (MCCT) in data center networks, some works only consider the multicast coflow scheduling without scheduling the circuit [15,16,23,29,34]. Sun et al. [23] employ the segmented transmission to divide the flow into multiple time slots to avoid small traffic being blocked. While the works in [15] and [16], try to split the multicast flows while the data can be delivered to just a subset of receivers when the circuit capacity is insufficient. Some other works consider the problem of utilizing circuit scheduling to further reduce the latency [5–7,21,26,32]. The authors in [5,21] used the conventional BvN matrix decomposition approach to schedule the circuits. Sunflow [32] studies circuit scheduling using a not-all-stop concept, however its scheduling is on a single optical circuit switch rather than the circuit scheduling of an optical network. Furthermore, studies like c-Through [26] investigated the overall network circuit scheduling with the goal of maximizing the amount of traffic offloaded to the optical network, which may result in a long coflow completion time. However, the problem of joint circuit scheduling and multicast coflow scheduling is not well addressed.

Therefore, in this paper, we investigate the first work to minimize the completion time of multicast flows by concurrently considering circuit scheduling and coflow scheduling. Firstly, we formulate the problem of multicast coflow scheduling, and prove it to be NP-hard. We propose a delay-efficient multicast coflow scheduling algorithm by integrating coflow scheduling with circuit scheduling. The proposed algorithm is proved to have an approximate ratio of at most $2\sqrt{n}$, where n represents the number of optical circuit switches. The major contributions of this paper are listed as follows.

1. We investigate the first multicast coflow scheduling problem in reconfigurable data center networks. The problem of minimizing the multicast coflow completion time is formulated and proved to be NP-hard.
2. We introduce a delay-efficient multicast coflow scheduling algorithm designed to integrate coflow scheduling with circuit scheduling.
3. The proposed algorithm is proved to have an approximate ratio of at most $2\sqrt{n}$, where n represents the number of optical circuit switches.
4. Through extensive simulations, it is shown that the proposed algorithm can achieve high performance compared to state-of-the-art methods.

The rest of this work is organized as follows. Section 2 furnishes a motivating example for our research. Section 3 presents the network model and problem definition. Section 4 presents the detailed design and theoretical analysis of the proposed approximate algorithm. Section 5 presents the simulation results. The survey of related works and the conclusion of this paper find their place in Sects. 6 and 7, respectively.

2 Problem Formulation

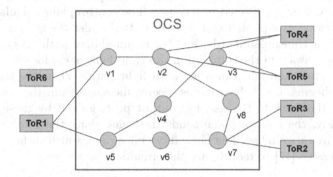

Fig. 1. A data center network model

2.1 System Model

As shown in Fig. 1. A data center network consists of two parts: a set of optical circuit switches $V = \{v_1, v_2, ..., v_n\}$ with $n = |V|$, and a set of ToR switches $R_{set} = \{ToR_1, ToR_2, ..., ToR_m\}$ with $m = |R_{set}|$. Each ToR switch is connected to one or more optical circuit switches. The network can be represented as $G = (R_{set} \cup V, E)$, where E is the set of links, and each link has a bandwidth of b. In terms of circuitry, optical circuit switches use beam steering to guide the light from input ports to output ports with embedded control processors. For the transmission of multicast, optical circuit switches operate directly on the beams of light without decoding packets or buffering them [17]. Since optical circuit switches lack buffering abilities, all data buffering occurs at the network's edge, specifically the ToR switches. It is important to note that we assume the buffering capacity on the ToR switches is sufficient for multicast coflow. This assumption is based on two considerations. Firstly, multicast coflow represents a group of multicast that are dedicated to transmitting data for a specific application, and these multicast can be buffered on the ToR switches [26]. Secondly, various mechanisms have been employed to enhance the buffering capacity of ToR switches. For example, [26] utilizes an application-transparent mechanism to increase the buffering capacity on ToR switches without requiring any modifications to the switches.

To configure optical paths across all optical circuit switches, we assume that the optical network operates in a centralized manner. In [27,30], for example, they use Software Defined Network (SDN) mechanisms in a centralized manner. SDN has been widely applied in data center networks and offers numerous advantages compared to traditional networks. Additionally, we assume that the centralized scheduler has access to information about multicast coflow. Such access can be achieved through upper-layer applications [25] or by employing advanced prediction techniques such as [19].

The system operates as follows: In the control plane, traffic information of the multicast coflow is collected in advance. The control plane calculates the no conflicts multicast groups that can be transmitted under the current link state, along with their transmission order and the required light paths. As a result, the control plane is aware of the completion time and light paths for each multicast, and sets the start and end times for each light path. When the transmission of multicast begins, the ToR switches receive messages from the control plane and forward the flows to the selected output ports guided by these messages. Simultaneously, the control plane sends messages along the routing paths to the optical circuit switches. The optical circuit switches utilize their embedded control processors [17] to reconfigure the circuits.

2.2   Circuit Scheduling Model

In optical circuit switches networks, circuits can be perceived as one-to-many port pairs on the switches. After a multicast transmission is completed, a new circuit needs to be established to transmit another multicast, which is called circuit reconfiguration. For optical circuit switching technology, circuit reconfiguration typically incurs a fixed delay δ (e.g., 1 millisecond). During the reconfiguration process, the affected ports are temporarily inactive while other ports can continue operating.

A routing path, also referred to as a light path, can transmit data only when all circuits along the path are established. For any given light path p, there are two time points, ST and FT, where ST represents the starting time of the light path and FT represents the ending time. Therefore, the task of circuit scheduling is to determine the establishment time and duration for each circuit in order to minimize the completion time of multicast coflow.

Multicast coflow is formed by a set of multicast, denoted as $F_{set} = \{f_1, f_2, ..., f_a\}$. Each multicast sends data from one ToR to multiple ToRs. In general, the number of candidate light paths for each multicast flow can be exponential. To reduce the space complexity, we maintain a set of minimum spanning trees and a set of busy nodes. The set of minimum spanning trees consists of the destination nodes for each multicast group, while the set of busy nodes consists of the currently occupied OCS nodes. During each scheduling iteration, we select the no conflicts multicast transmissions that the current link can accommodate. When a multicast transmission is completed, we update the set of minimum spanning trees and the set of busy nodes. If the destination nodes of a multicast cannot form a minimum spanning tree, that multicast is not considered in the current scheduling round. Once the desired multicast transmission is selected, we determine the required optical path, as well as the start and end times of the circuit.

2.3   Problem Formulation

Given an optical network G and a multicast coflow F_{set}, our objective is to minimize the multicast coflow completion time through both multicast coflow scheduling and optical circuit scheduling. Specifically, we need to arrange the scheduling order of multicast coflow and select the appropriate optical path for each multicast, as well as use the optical network to complete multicast flow transmission as soon as possible.

The integrated multicast coflow and circuit scheduling problem for optical networks to minimize the MCCT can be proved to be NP-hard, which can be reduction from the multi-processor scheduling problem. Considering a group of multicast coflow transmissions that need to be scheduled as a set of tasks, where OCS nodes are regarded as the shared resources required for the multicast coflow transmissions. Each task needs to utilize one or more of these shared resources, and the duration of resource usage is fixed for each task. At any given time, a shared resource can only serve one task. Once a shared resource is allocated to a task, it cannot be assigned to another task until the completion of current task. In this case, the integrated multicast coflow and circuit scheduling problem is equivalent to the multi-processor scheduling problem. Since the multi-processor scheduling problem is NP-hard [18], thus, we have the following theorem.

Theorem 1. *The integrated multicast coflow and circuit scheduling problem is NP-hard.*

3   The Delay-Efficient Multicast Coflow Scheduling Algorithm

In this section, we propose a delay-efficient multicast coflow scheduling algorithm adept at integrating coflow scheduling with circuit scheduling. Unlike the loosely-coupled algorithm proposed in [28], our algorithm integrates multicast coflow scheduling and circuit scheduling within the same framework, ensuring the consistency between multicast coflow scheduling and circuit scheduling. The algorithm follows a greedy approach to find the no conflicts multicast transmissions that can be scheduled at each iteration.

To minimize multicast coflow completion time, we try to integrate coflow scheduling with circuit scheduling. The integrated algorithm consists of two steps: multicast coflow scheduling and circuit scheduling. Considering the mutual influence and dependency between different scheduling, we aim to achieve a more comprehensive system optimization by considering both multicast coflow

---

Algorithm 1. A delay-efficient multicast coflow scheduling algorithm

---

1: **procedure** JOINT SCHEDULING(F_{set})
2:     $Circuit\_information = \emptyset; duration = \emptyset; F_{schedul} = \emptyset; busynode = \emptyset; t = 0;$
3:     Initializing network configuration and circuit setup;
4:     $Minimum\_Spanning\_Tree(F_{set}, busynode);$
5:     **while** $!F_{set}.isEmpty()$ **do**
6:         $F_{schedul} \leftarrow selectflow(F_{set});$
7:         $Circuit\_information, duration \leftarrow selectpath(F_{schedul});$
8:         $t = t + createcircuit(t, Circuit\_information, duration);$ ▷ Start time of the next circuit creation
9:         Update busy link nodes to $busynode;$
10:        $Minimum\_Spanning\_Tree(F_{set}, busynode);$
11:    **end while**
12: **end procedure**

---

scheduling and circuit scheduling factors. Therefore, we propose a joint optimization algorithm.

As shown in Algorithm 1. Let F_{set} represents a group of multicast coflow that needs to be scheduled, where $F_{set}=\{f_1, f_2, ..., f_a\}$. Each multicast has a single sending node and multiple receiving nodes. At the beginning of the algorithm, we introduce the following variables, such as $circuit\_information$, $duration, F_{schedul}, busynode$, and t. $circuit\_information$ represents the transmission light path of multicast that needs to be scheduled. $duration$ represents the duration of the light path. $F_{schedul}$ represents the scheduling queue. $busynode$ represents a set of nodes occupied. t represents the time when the light path was created.

Next, we need to select the flow for transmission. Firstly, initialize all variables and circuits. Construct a minimum spanning tree for the destination node based on $busynode$ and F_{set}. Then, by using the $selectflow$ function, we can obtain the no conflicts multicast transmission group and transmission order. Calculate the required light path and duration of the optical path using the $selectpath$ function for the obtained multicast.

Finally, we will transmit the multicast light path information that needs to be scheduled to the $createcircuit$ function. Through this function, we can create the required light path, set the duration of the light path, and return the earliest end time of the light path as the time when the next circuit starts.

3.1   Multicast Coflow Scheduling

Multicast coflow scheduling is a crucial step in minimizing multicast coflow completion time. Because it can not only determine which multicast to schedule but also determine the sending order and path of these multicast. As shown in Algorithm 2. Firstly, before selecting multicast, we use a minimum spanning tree function to reduce search time as shown in the first function. Since multicast involves sending data from a single source node to multiple destination nodes,

---

Algorithm 2. Multicast Coflow Scheduling

1: **procedure** MINIMUM_ SPANNING_ TREE(F_{set}, $busynode$)
2:    **for** $f \in F_{set}$ **do**
3:        **if** $check(f, busynode)$ **then** ▷ Whether the minimum spanning tree can be constructed
4:            Generate the minimum spanning tree with Prim algorithm for f;
5:        **end if**
6:    **end for**
7: **end procedure**
8: **procedure** SELECTFLOW(F_{set})
9:    $F_{max} = \emptyset$;
10:   **for** $f \in F_{set}$ **do**
11:       **if** $check(f)$ **then**   ▷ Check if the destination node of each multicast f can form a minimum spanning tree
12:           **if** $Conflict\_checking(f, F_{max})$ **then**   ▷ Check if there is a conflict between the flow in F_{max} and f
13:               Add f to F_{max};
14:           **end if**
15:       **end if**
16:   **end for**
17:   **for** $f \in F_{max}$ **do**
18:       Sort F_{max} according to the rules we provide;
19:   **end for**
20:   **return** F_{max};
21: **end procedure**
22: **procedure** SELECTPATH($F_{schedul}$)
23:   **for** $f \in F_{schedul}$ **do**
24:       Find the shortest path through Dijkstra algorithm;
25:       Add light path of f to $Circuit\_information$;
26:       Add transmission time of f to $duration$;
27:   **end for**
28:   **return** $Circuit\_information, duration$;
29: **end procedure**

---

in order to minimize path searching, we abstract the minimum spanning tree formed by the destination nodes into a single virtual destination node. This allows us to avoid traversing multiple destination nodes.

As shown in the *selectflow* function. When traversing the set F_{set}, we need to determine whether the destination nodes of a multicast can form a minimum spanning tree, check if there is a path from the source node to this minimum spanning tree, and identify any conflicts with the flows in F_{max}. If all these conditions are met, we add the multicast to F_{max}. By traversing the F_{set}, we obtain a number of non-conflicting multicasts. Then, we need to determine the transmission order for these multicasts. Therefore, we establish priority setting rules. Since each ToR is connected to a different number of OCSs, we give priority to multicasts from ToRs with fewer connections.

Algorithm 3. Circuit Scheduling

1: **procedure** CREATECIRCUIT(t, $Circuit\_information$, $duration$)
2:     $FT_{set} = \emptyset$
3:     **for** $p \in Circuit\_information$ **do**
4:         **if** all ports along light path p is free at time t **then**
5:             $ST = t + \delta$;   ▷ δ is reconstruction delay
6:             $FT = ST + duration$;
7:             **for** each port along the light path p **do**
8:                 Establish a connection from one input port to one or more output ports;
9:             **end for**
10:            Set p is busy during ST to FT;
11:            Add FT to FT_{set};
12:        **end if**
13:     **end for**
14:     $t_{release} = min\{FT_{set}\}$;
15:     **return** $t_{release}$;
16: **end procedure**

After we determine the multicast that needs to be scheduled, we pass it to the *selectpath* function. By using the Dijktstra algorithm, we store the obtained path into the *circuit_information* and store the transmission time of each flow into the *duration*.

3.2   Circuit Scheduling

The circuit scheduling algorithm effectively coordinates the transmission of multicast flows, achieving efficient data transmission. It can adapt to networks of different scales and complexities, dynamically adjusting based on real-time conditions to provide optimal performance and resource utilization. As shown in Algorithm 3, the *createcircuit* function by receiving information from the multicast coflow scheduling. This information includes the list of paths to be built and their corresponding duration. Upon receiving this information, the circuit scheduling proceeds to construct the paths by checking the availability of each OCS node along the path. If any of the nodes are not available, the circuit establishment fails. On the other hand, if the circuit establishment is successful, the optical circuit switch will reconfigure the circuits with its embedded control processor. After creating the circuit, the start and end times are set for the circuit and multicast sends the ToR port to the specified output port. The end time of each circuit is added to the set FT_{set}. Once all the paths are created, the minimum end time among the paths will be returned.

3.3   Approximation Ratio Analysis

In this section, we first analyze the lower bounds of the multicast coflow scheduling problems with n optical circuit switch nodes. Assuming there is a multicast coflow $F_{set} = \{f_1, f_2, ..., f_a\}$, where each request requires establishing a dedicated light path for transmission. If each multicast chooses the optical path that occupies the most OCS nodes, we can infer that the multicast coflow has the longest completion time at this time. Because this choice of optical path maximizes the probability of conflicts between multicast. The lower bound of the algorithm can be determined by considering all multicast requests transmitted using the longest paths. Therefore, we can define the problem as follows: Given a multicast coflow $F_{set} = \{f_1, f_2, ..., f_a\}$, where each multicast f has a link path p_f and transmission time t_f. Each multicast optical path is transmitted through the longest path and $|p_f| \leq n$. Our goal is to minimize the completion time of multicast coflow under a given optical path.

First, we divide the scheduling instance F_{set} into two sub-instances. It is

$$Q_1 = \{f \in F_{set} : |p_f| \geq k\}, Q_2 = \{f \in F_{set} : |p_f| < k\}. \tag{1}$$

where the k $(1 \leq k \leq n)$ is named as partition parameter.

For these two sub-instances Q_1 and Q_2, we adopt the corresponding scheduling algorithms to schedule them respectively, then the time span is

$$T_{min}(F_{set}) = T_{min}(Q_1) + T_{min}(Q_2). \tag{2}$$

which has

$$\max\{Opt(Q_1), Opt(Q_2)\} \leq Opt(F_{set}) \leq Opt(Q_1) + Opt(Q_2). \tag{3}$$

For each optical circuit switch $i (1 \leq i \leq n)$, its duration is

$$L_i(F_{set}) = \sum_{f \in F_{set} \wedge i \in p_f} t_f. \tag{4}$$

Optical circuit switch average completion time is

$$L'(F_{set}) = \frac{1}{n} \sum_{f \in F_{set}} t_f |p_f|. \tag{5}$$

So $Opt(F_{set}) \geq max(L_i(F_{set})(1 \leq i \leq n), L'(F_{set}))$.

Lemma 1. *For any sub-instance Q_1, regardless of how it is scheduled, its time span is*

$$T_{min}(Q_1) \leq (\frac{n}{k})Opt(F_{set}). \tag{6}$$

Proof. For any sub-instance Q_1, no matter how it is scheduled, the time span $T_{min}(Q_1) \leq \sum_{f \in Q_1} t_f$ is always satisfied. When $f \in Q_1$, $|p_f| \geq k$, so we have:

$$T_{min}(Q_1) \leq \sum_{f \in Q_1} t_f \leq \frac{1}{k} \sum_{f \in Q_1} t_f |p_f| = \frac{n}{k}(\frac{1}{n} \sum_{f \in Q_1} t_f |p_f|) \leq (\frac{n}{k})Opt(F_{set}). \tag{7}$$

Lemma 2. *For any sub-instance Q_2, adopting the first fit algorithm, its time span is*

$$T_{min}(Q_2) \leq kOpt(F_{set}). \tag{8}$$

Proof. For any sub-instance Q_2, let us assume that there is a set of tasks with the longest time span $(f_1, f_2, ..., f_j)$ that can be completed continuously. In other word, at least one of the required nodes of task f_j is not idle at any given time. Therefore, the scheduling time span is $T_{min}(Q_2)$ and the average working time L' of this group of nodes satisfies

$$T_{min}(Q_2) = \sum_{i=1}^{j} t_{f_i}, L' = \frac{1}{|p_{f_j}|} \sum_{i=1}^{j} t_{f_i}. \tag{9}$$

Due to $|p_{f_j}| < k$ and $Opt(F_{set}) \geq max(L_i(F_{set}))$, we can establish:

$$T_{min}(Q_2) \leq kL' \leq kmax(L_i(F_{set})) \leq kOpt(F_{set}). \tag{10}$$

According to Lemma 1 and Lemma 2. So $T_{min}(F_{set}) = T_{min}(Q_1) + T_{min}(Q_2) \leq (n/k + k)Opt(F_{set})$ holds true. When $k = \sqrt{n}$, this approximation ratio is the smallest. Hence we have shown that $T_{min}(F_{set}) \leq 2\sqrt{n}Opt(F_{set})$.

4   Numerical Evaluation

In this section, we will evaluate the performance of our proposed algorithms by conducting sufficient simulations.

4.1   Simulation Settings

In our simulation, the proposed algorithms are specifically designed for optical networks where the switches are assumed to be optical circuit switches. As a commonly adopted parameter in prior works, we set link bandwidth b to 10 Gbps. The average traffic amount of each flow in a multicast coflow is set to 100 Mb, which is same as the work in the context of data center networks. We also have considered two different distributions for flow traffic in the experiment setting: 1) Eighty-Twenty rule: Under this distribution, 80 percent of the flows are considered as mice flows, while the remaining 20 percent are classified as elephant flows. 2) Gaussian distribution: This follows a Gaussian distribution, where the expected flow size is 100 Mb.

For performance comparison, we consider using MCCT as the algorithm performance indicator. As shown below, we will introduce the factors that affect MCCT separately.

- *The number of ToR switches*: In our simulation, we changed the number of ToR switches from 9 to 54. The default number of ToR switches is set to 9.
- *The number of flows per ToR switch pair*: For each ToR switch, we changed the number of multicast sent by each ToR switch from 40 to 70 and set it to 40 by default.

- *The reconfiguration delay*: In optical networks, the circuit reconfiguration latency when constructing a new circuit connect, which is always set up as a fixed duration, ranges from 0 to 10 ms. In our simulation, we consider the default reconfiguration latency as 1 ms, which is widely applied in current optical switches.

We compared and simulated two different algorithms, one is the first fit algorithm(FFA), and the other is the shortest path algorithm(SPA). Our algorithm utilizes both the minimum spanning tree algorithm (MSTA) and the shortest path algorithm (SPA). We conduct at least 100 runs for every experiment setting over each network and report the average results.

4.2   Performance Evaluation

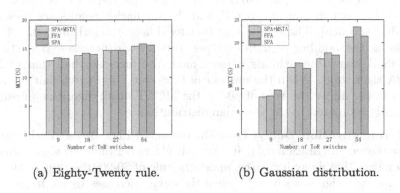

(a) Eighty-Twenty rule.   (b) Gaussian distribution.

Fig. 2. The relationship between the MCCT and the number of ToR switches.

The Number of ToR Switches: It is evident for us to observe the impact of the number of ToR switches on the MCCT, as demonstrated in the Fig. 2(a) and 2(b). For example, when the number of ToR switches changes from 9 to 54, under two different distribution rules, SPA+MSTA algorithm demonstrates more efficient reduction in the MCCT compared to the FFA and SPA. When the number of ToR switches increases, the MCCT of all algorithms tends to increase. In Fig. 2b, we could observe the FFA exhibits the highest MCCT, while our proposed algorithm consistently maintains a lower MCCT compared to others. To explain above phenomenon, we can conclude that as the number of ToR switches increases, there are more ToR switch pairs in the network. This results in a higher number of flows in a coflow. The surge in the number of flows leads to a rise in the MCCT for all scenarios. However, it's worth highlighting that our algorithm can efficiently reduce the MCCT even when the number of ToR increases.

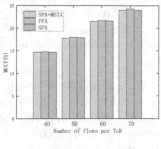 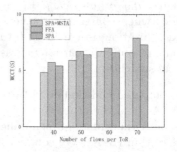

(a) Eighty-Twenty rule.   (b) Gaussian distribution.

Fig. 3. The relationship between the MCCT and the number of flows per ToR.

The Number of Flows per ToR: We investigate the impact of the number of flows per ToR switch pair on MCCT, and the simulation results are presented in Fig. 3(a) and 3(b). The result show that for different algorithms, the MCCT increases as the number of multicast per ToR switch increases. Importantly, SPA+MSTA algorithm exhibits a more moderate increase than that of FFA and SPA algorithms. When the number of flows per ToR switch pair increases from 40 to 60, under the rule of 80-20, the MCCT of all algorithms upsurges significantly compared to the Gaussian distribution.

The Reconfiguration Delay: From the presented Figs. 4(a) and 4(b) we can find how the reconfiguration delay impacts MCCT under different algorithms. As the reconfiguration delay increases, under the rule of 80-20, the time of all algorithms upsurges significantly only when the delay increases to 10. In contrast, under the Gaussian distribution, both the FFA and SPA have a significantly higher delay increase compared to our proposed algorithm. This is because the Baseline requires more reconfiguration time compared to the SPA+MSTA algorithm, our algorithm performs effectively reduction in the MCCT especially as the reconfiguration delay increases.

5   Related Work

There are two important approaches to reduce MCCT: multicast coflow scheduling and circuit scheduling. In the following, we will introduce the related works of these two methods, respectively.

A lot of research has been conducted on coflow scheduling in data center networks. The concept of coflow was introduced and the traffic patterns were summarized by the authors in [25]. Subsequently, studies like [9] started applying the concept of coflow in optimization. However, all of these works primarily focused on determining when to forward traffic and overlooked the issue of routing selection. Only the authors in [35] addressed the routing problem of coflows in data center networks. Nevertheless, it is important to note that classical networks

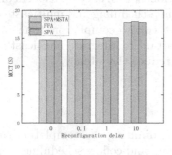

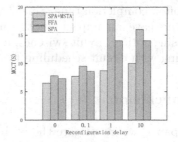

(a) Eighty-Twenty rule.   (b) Gaussian distribution.

Fig. 4. The relationship between the MCCT and Reconfiguration delay.

are based on packet switching, while Optical Circuit Switched networks operate on circuit switching [32]. The coflow scheduling methods designed for packet-switched networks are not applicable to Optical Circuit Switched networks. This incompatibility arises from the exclusive nature of optical switches and the limited availability of optical paths. Furthermore, existing coflow scheduling approaches [28] primarily focus on unicast flows and lack specific scheduling mechanisms for multicast flows. [12–14]are different from the problem studied in this paper. [10]optimize by predicting the job with the shortest remaining processing time (SRPT). [31] optimized coflow to date. But neither [10,31] considered circuit scheduling.

Considering circuit scheduling, since the optical circuit switches need to set up or tear down circuits, a natural topic is scheduling the circuits to minimize the multicast completion time. Steven *et al.* [23] proposed Creek to improve the transmission efficiency of data center network. Creek introduced two methods, namely multi-hop and segmented transmission. Multi-hop increases the reachability of transmission by establishing a tree search form. Segmented transmission divides flows into multiple time slots for transmission to avoid small flows being blocked. Luo *et al.* [15,16] found that conflicts between destination nodes in multicast lead to circuit resources being underutilized in each transmission time slot. Therefore, they proposed Splitcast, which follows a priority strategy with the shortest remaining processing time first and improves the utilization of link resources through splitting multicast transmission. To maximize the utilization of optical circuit resources, the authors in [29] proposed the Blast, which scores each flow and prioritizes the transmission of flows with the highest scores. [22]considered optimizing a multi-stage job, but it abstracted the data center network into a huge non-blocking switch, which conflicts with the model we studied. These studies primarily consider optimizing multicast coflow completion time in scenarios involving a single switch. In contrast, in typical optical networks, data transmission requires traversing multiple switches along the path.

However, the aforementioned studies only focused on specific aspects of multicast coflow completion time. Since both circuit scheduling and multicast coflow scheduling have an impact on the multicast coflow completion time in networks using optical circuit switches, it is important to integrate multicast coflow scheduling with circuit scheduling to further reduce the latency.

6   Conclusion

In this paper, we investigate the problem of minimizing the completion time of multicast flows by considering circuit scheduling and coflow scheduling simultaneously. We commence by formulating the problem of multicast coflow scheduling, and proving that it is NP-hard. We propose a delay-efficient multicast coflow scheduling algorithm adept at integrating coflow scheduling with circuit scheduling. The proposed algorithm is proved to have an approximate ratio of at most $2\sqrt{n}$, where n represents the number of optical circuit switches. Through extensive simulations, it is shown that the proposed algorithm can achieve significant improvements compared to state-of-the-art methods.

References

1. Ajibade, L.S., Bakar, K.A., Aliyu, A., Danish, T.: Straggler mitigation in hadoop mapreduce framework: a review. Int. J. Adv. Comput. Sci. Appl. **13**(8) (2022)
2. Chen, Q., Cai, Z., Cheng, L., Gao, H.: Low latency broadcast scheduling for battery-free wireless networks without predetermined structures. In: 2020 IEEE 40th International Conference on Distributed Computing Systems (ICDCS), pp. 245–255 (2020). https://doi.org/10.1109/ICDCS47774.2020.00052
3. Chen, Q., Gao, H., Cheng, S., Fang, X., Cai, Z., Li, J.: Centralized and distributed delay-bounded scheduling algorithms for multicast in duty-cycled wireless sensor networks. IEEE/ACM Trans. Netw. **25**(6), 3573–3586 (2017)
4. Chen, Q., et al.: Latency-optimal pyramid-based joint communication and computation scheduling for distributed edge computing. In: IEEE Conference on Computer Communications, IEEE INFOCOM 2023, pp. 1–10 (2023). https://doi.org/10.1109/INFOCOM53939.2023.10228964
5. Farrington, N., Porter, G., Fainman, Y., Papen, G., Vahdat, A.: Hunting mice with microsecond circuit switches. In: Proceedings of the 11th ACM Workshop on Hot Topics in Networks, pp. 115–120 (2012)
6. Jia, S., Jin, X., Ghasemiesfeh, G., Ding, J., Gao, J.: Competitive analysis for online scheduling in software-defined optical wan. In: IEEE Conference on Computer Communications, IEEE INFOCOM 2017, pp. 1–9. IEEE (2017)
7. Jin, X., et al.: Optimizing bulk transfers with software-defined optical WAN. In: Proceedings of the 2016 ACM SIGCOMM Conference, pp. 87–100 (2016)
8. Li, J., et al.: Digital twin-assisted, SFC-enabled service provisioning in mobile edge computing. IEEE Trans. Mob. Comput. 1–16 (2022). https://doi.org/10.1109/TMC.2022.3227248
9. Li, W., Chen, S., Li, K., Qi, H., Xu, R., Zhang, S.: Efficient online scheduling for coflow-aware machine learning clusters, vol. 10, pp. 2564–2579. IEEE (2020)

10. Li, W., Chen, S., Li, K., Qi, H., Xu, R., Zhang, S.: Efficient online scheduling for coflow-aware machine learning clusters. IEEE Trans. Cloud Comput. **10**(4), 2564–2579 (2020)
11. Li, W., et al.: CoMan: managing bandwidth across computing frameworks in multiplexed datacenters. IEEE Trans. Parallel Distrib. Syst. **29**(5), 1013–1029 (2017)
12. Li, W., Yuan, X., Li, K., Qi, H., Zhou, X.: Leveraging endpoint flexibility when scheduling coflows across geo-distributed datacenters. In: IEEE Conference on Computer Communications, IEEE INFOCOM 2018, pp. 873–881. IEEE (2018)
13. Li, W., Yuan, X., Li, K., Qi, H., Zhou, X., Xu, R.: Endpoint-flexible coflow scheduling across geo-distributed datacenters. IEEE Trans. Parallel Distrib. Syst. **31**(10), 2466–2481 (2020)
14. Li, W., et al.: Efficient coflow transmission for distributed stream processing. In: IEEE Conference on Computer Communications, IEEE INFOCOM 2020, pp. 1319–1328. IEEE (2020)
15. Luo, L., Foerster, K.T., Schmid, S., Yu, H.: Splitcast: optimizing multicast flows in reconfigurable datacenter networks. In: IEEE Conference on Computer Communications, IEEE INFOCOM 2020, pp. 2559–2568. IEEE (2020)
16. Luo, L., Foerster, K.T., Schmid, S., Yu, H.: Optimizing multicast flows in high-bandwidth reconfigurable datacenter networks. J. Netw. Comput. Appl. **203**, 103399 (2022)
17. Mitsuya, T., Ochiai, T., Kuno, T., Mori, Y., Hasegawa, H., Sato, K.: Highly reliable and large-scale optical circuit switch for intra-datacentre networks. In: 2022 European Conference on Optical Communication (ECOC), pp. 1–4. IEEE (2022)
18. Mosheiov, G.: Multi-machine scheduling with linear deterioration. INFOR: Inf. Syst. Oper. Res. **36**(4), 205–214 (1998)
19. Peng, Y., Chen, K., Wang, G., Bai, W., Ma, Z., Gu, L.: HadoopWatch: a first step towards comprehensive traffic forecasting in cloud computing. In: IEEE Conference on Computer Communications, IEEE INFOCOM 2014, pp. 19–27. IEEE (2014)
20. Perelló, J., et al.: All-optical packet/circuit switching-based data center network for enhanced scalability, latency, and throughput. IEEE Netw. **27**(6), 14–22 (2013)
21. Porter, G., et al.: Integrating microsecond circuit switching into the data center. ACM SIGCOMM Comput. Commun. Rev. **43**(4), 447–458 (2013)
22. Shafiee, M., Ghaderi, J.: Scheduling coflows with dependency graph. IEEE/ACM Trans. Netw. **30**(1), 450–463 (2021)
23. Sun, X.S., Ng, T.E.: When creek meets river: exploiting high-bandwidth circuit switch in scheduling multicast data. In: 2017 IEEE 25th International Conference on Network Protocols (ICNP), pp. 1–6. IEEE (2017)
24. Tang, S., He, B., Yu, C., Li, Y., Li, K.: A survey on spark ecosystem: big data processing infrastructure, machine learning, and applications, vol. 34, pp. 71–91 (2022). https://doi.org/10.1109/TKDE.2020.2975652
25. Tang, Y., Yuan, T., Liu, B., Xiao, C.: Effective*-flow schedule for optical circuit switching based data center networks: a comprehensive survey, vol. 197, p. 108321. Elsevier (2021)
26. Wang, G., et al.: C-through: part-time optics in data centers. In: Proceedings of the ACM SIGCOMM 2010 Conference, pp. 327–338 (2010)
27. Wang, H., Xu, H., Huang, L., Wang, J., Yang, X.: Load-balancing routing in software defined networks with multiple controllers. Comput. Netw. **141**, 82–91 (2018)
28. Wang, H., Yu, X., Xu, H., Fan, J., Qiao, C., Huang, L.: Integrating coflow and circuit scheduling for optical networks. IEEE Trans. Parallel Distrib. Syst. **30**(6), 1346–1358 (2019). https://doi.org/10.1109/TPDS.2018.2889251

29. Xia, Y., Ng, T.E., Sun, X.S.: Blast: accelerating high-performance data analytics applications by optical multicast. In: 2015 IEEE Conference on Computer Communications (INFOCOM), pp. 1930–1938. IEEE (2015)
30. Xu, H., Li, X.Y., Huang, L., Deng, H., Huang, H., Wang, H.: Incremental deployment and throughput maximization routing for a hybrid SDN. IEEE/ACM Trans. Netw. **25**(3), 1861–1875 (2017)
31. Xu, R., Li, W., Li, K., Zhou, X., Qi, H.: Scheduling mix-coflows in datacenter networks. IEEE Trans. Netw. Serv. Manag. **18**(2), 2002–2015 (2020)
32. Yang, H., Zhu, Z.: Topology configuration scheme for accelerating coflows in a hyper-flex-lion, vol. 14, pp. 805–814. Optica Publishing Group (2022)
33. Yao, B., Gao, H., Chen, Q., Li, J.: Energy-adaptive and bottleneck-aware many-to-many communication scheduling for battery-free WSNs. IEEE Internet Things J. **8**(10), 8514–8529 (2021). https://doi.org/10.1109/JIOT.2020.3045979
34. Zeng, Y., Ye, B., Tang, B., Guo, S., Qu, Z.: Scheduling coflows of multi-stage jobs under network resource constraints. Comput. Netw. **184**, 107686 (2021)
35. Zhao, Y., et al.: Rapier: integrating routing and scheduling for coflow-aware data center networks. In: 2015 IEEE Conference on Computer Communications (INFOCOM), pp. 424–432. IEEE (2015)

DAS: A DRAM-Based Annealing System for Solving Large-Scale Combinatorial Optimization Problems

Wenya Deng[1,2], Zhi Wang[1,2(✉)], Yang Guo[1,2], Jian Zhang[1,2], Zhenyu Wu[1,2], and Yaohua Wang[1,2]

[1] National University of Defense Technology, Changsha, China
{dengwenya,guoyang,JianZhang,yhwang}@nudt.edu.cn
[2] Key Laboratory of Advanced Microprocessor Chips and Systems, Changsha, China
zhiwang@nudt.edu.cn

Abstract. As Moore's law approaches its inevitable end, the performance improvement of traditional Von Neumann has encountered challenge. Some dedicated computing architecture for specific domains is seen as one way to meet this challenge, and Ising architecture is one of them, which is mainly used to solve combinatorial optimization problems efficiently. We propose a DRAM-based annealing system (DAS) to realize Ising architecture based on DRAM. The Ising coefficients are transposed and stored in the DRAM cells, and annealing calculations are performed using in-DRAM bulk bitwise operations until the solution to the problem is found. DAS can perform parallel annealing in DRAM, reducing data movement and solution time, making it appropriate for large-scale spin Ising system. We evaluated DAS by segmenting multiple image from the HRSOD dataset and showed that DAS has similar segmentation capabilities to the conventional Onecut method, but improve an average solution time acceleration by 10.2× and an average energy consumption of just 0.4349% compared to the conventional method. Furthermore, the design area we added to DAS accounted for only 7% of its total area.

Keywords: Ising Architecture · In-Memory Computing · DRAM · Combinatorial Optimization Problem

1 Introduction

As Moore's law approaches its inevitable end, the performance improvement of traditional Von Neumann microprocessors becomes more and more challenging [1,2]. In order to further improve data processing performance, some domain-specific dedicated computing architecture has been proposed to fulfil the rising demand for computing. In special computing fields like combinatorial optimization problems, the performance of Von Neumann microprocessors is difficult

Supported by NSF of Hunan Province No.2022JJ10066 and NSFC No.62272477.

to meet the computational demands [3]. In recent years, the Ising model has attracted fresh interest as a promising solution to combinatorial optimization problems.

The Ising architecture is a new computing architecture developed from the Ising model, which can be executed in parallel and is very suitable to solve complex combinatorial optimization problems. The combinatorial optimization problem is firstly mapped to the Ising model and all spin groups are updated simultaneously by annealing to find the ground state, which corresponds to the optimal solution of the problem [4,5].

In 2015, Hitachi Research Center realized a 20k spin Ising chip prototype using CMOS circuits [4], and showed advantages in solving combinatorial optimization problems. The ground state search of the Ising architecture is mainly implemented by spins update, and spin update calculation including local search and probability flipping, local search circuits are generally implemented by adder trees. Local search is the main non-storage overhead of the Ising chip, and for wide bit-width Ising architecture computing systems, the adder trees take up the main overhead of the Ising chip. Waseda University has reduced the bit-widths of coefficients by adding auxiliary spins in 2020 [6], but adder trees are still the dominant computational implementation. Therefore, it is difficult to scale up Ising architecture computing system. Although the Hitachi Research Center has implemented the expansion of the Ising chip by designing a low-latency chip interface [7], it only developed 2 chip expansions and is still not practical for large-scale problems. As far as we know, a majority of CMOS Ising architecture computing systems have been built in SRAM since 2015 [3,4,7–9], which is costly to construct large-scale Ising system. As a result, the scale of the problems that can be solved is often small and the Ising architecture computing system is challenging to apply.

We propose a DRAM-based Annealing System (DAS) to realize Ising architecture. The spins and interaction coefficients are stored in the DRAM memory cells, the annealing computation is carried out by bulk bitwise computation inside DRAM cell arrays. The proposed architecture executes the annealing computation in the memory, which reduces the cost of data movement from memory to calculation unit. And the annealing computation can be executed in parallel in the bitlines of the DRAM, which can improve the speed of solving combinatorial optimization problems. What's more, DRAM has a higher level of integration and lower cost than SRAM, which also makes DAS easier to implement and scale up for large-scale annealing systems.

We performed image segmentation on image from the HRSOD dataset at DAS and compared it to the conventional method OneCut. DAS demonstrated similar capabilities to the OneCut method, but with a 10.2× acceleration and only 0.4349% of the average energy consumption of the OneCut method. In addition, the design we added only represents 7% of the total area of DAS.

This paper is organized as follows. Section 2 introduces the overview of the Ising model, DRAM, and in-DRAM bulk bitwise operations. Section 3 describes each part of DAS in detailed, including spin propagation, calculation of local search term based on in-DRAM bulk bitwise operations and probability flipping. In Sect. 4, we describe evaluation methods and the components of area. And DAS is evaluated by several different image segmentation applications and the results are compared to conventional methods. Finally, in Sect. 5 we conclude the paper.

2 Background

2.1 Ising Model

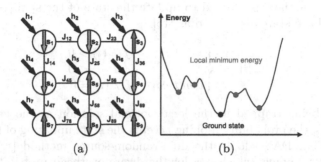

Fig. 1. (a) Ising model. (b) Energy profile of Ising model.

The Ising model was originally used to describe the behavior of magnetic spin [10]. Figure 1(a) shows the Ising model. Each spin is connected to the adjacent spins. Where s_i is the spin state ($+1$ or -1), J_{ij} is an interaction coefficient between s_i and s_j, and h_i denotes the external magnetic field coefficient.

In CMOS Ising system, spins are stored in the memory cells. The value of the memory cell is "0" indicates that the spin state is upward, while the value of the memory cell is "1" indicates that the spin state is downward. The CMOS Ising system contains multiple spin nodes, and there are data paths between the connected spin nodes that transfer spin values to each other. The memory cells in each spin node store its spin value s_i, the interaction coefficient J_{ij} between spin and connected spin nodes, and the external magnetic field coefficient h_i. The spin node also involves an update circuit to calculate the relationship between the spins and determine the next state of the spin. The interaction coefficient J_{ij} and the external magnetic field coefficient h_i are determined by the problem when solving combinatorial optimization problems.

The Ising model's energy process is comprised of the energy generated by spin interactions between spins and the energy that each spin derives from the external magnetic field [11]. And the energy formula is expressed as follows [4]:

$$E_{\text{Ising}} = -\sum_{i<j} J_{ij} s_i s_j - \sum_{i=1}^{N} h_i s_i \qquad (1)$$

Figure 1(b) shows the energy profile of the Ising model, with both local minimum energy and ground state. In order to search the ground state, a series of state search procedures are run in parallel at each spin. The ground state is the spin state when the energy of the system is lowest. The solution to the problem can be found by acquiring the final spin state. State search operations consist of the local search and the probably flipping. The formula of local search term is expressed as follows:

$$L(i) = \sum_{x \in \{L,R,U,D\}} J_{ix} \cdot s_{ix}[t] + h_i \tag{2}$$

Metropolis method is adopted to update the state of the spin [12], the next spin state of local search is defined as [15]:

$$s_i[t]' = \begin{cases} +1 & if\ L(i) > 0 \\ -1 & if\ L(i) < 0 \\ don't\ care & if\ L(i) = 0 \end{cases} \tag{3}$$

To avoid being trapped in the local optimal solution at the ground state search process, the probability flipping accepts the state updating of local search with probability. DAS adopts the dual-random-source method [15] to generate probability flipping, which is a low hardware overhead probability flipping method. The probability flipping term R(i) is defined as follows:

$$R_i(t) = \delta(rn_0 - step(t))\,rn_1 \tag{4}$$

When $x > 0$, $\delta(x)$ is 1, otherwise it is 0. Besides, rn_0 and rn_1 are random numbers, their range is $rn_0 \in (0, step_{final})$, $rn_1 \in (-\alpha\Delta E_{max}(i)/2, \alpha\Delta E_{max}(i)/2)$. $\alpha E_{max}(i)/2$ is the maximum energy difference, α is constant, $step(t)$ and $step_{final}$ are the number of steps for the spin update and the total number of steps for the iterative update. The next spin state is defined as:

$$s_i[t+1] = \begin{cases} +1 & if\ L(i) + R(i) > 0 \\ -1 & if\ L(i) + R(i) \leq 0 \end{cases} \tag{5}$$

Therefore, the next state of the spin can be determined by the sum of the local search term $L_i(t)$ and the probability flip term $R(i)$. The next state of the spin is "+1" when the sum is positive, otherwise it is "−1".

2.2 The Overview of DRAM

As shown in Fig. 2, a modern DRAM chip consists of multiple DRAM banks, and a DRAM bank contains a global decoder, a global row buffer group, and multiple subarrays [14]. A subarray contains DRAM cells connected to sense amplifiers. A set of amplifiers of the subarray constitutes the local row buffer [16]. As the area of the sense amplifier is 100 times than that of a memory cell, a row of sense amplifier is shared by two subarrays in modern DRAM design

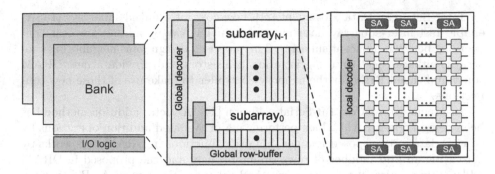

Fig. 2. Bank and subarray organization in a DRAM chip.

to reduce the area of sense amplifier. Therefore, each subarray has two rows of sense amplifiers, one above and one below the subarray [21–23].

There are three steps to access data from a subarray [13]:

1) Activate. A row of memory cells is activated and the values of them are stored in the local row buffer.
2) Read/Write. Read or write data locked in the local row buffer.
3) Precharge. Free the local row buffer and the bitline is restored to VDD/2.

2.3   In-DRAM Bulk Bitwise Operations

Fast Parallel Mode (FPM) [17]. FPM enable copy data between 2 rows within the same subarray. FPM first activates the source row, and then activities the destination row until the data is stable. Precharge is executed in preparation for the next operation. This process is called AAP (ACTIVATE-ACTIVATE-PRECHARGE).

Dual-Contact Cell (DCC) [18]. DCC is a DRAM cell with two transistor and connects to a sense amplifier. It can transmit data from the $\overline{bitline}$ to a cell and can be connected to a bitline at the same time. DCC enable perform a bitwise NOT of row A and store the result in row R.

Bitwise AND and OR [19]. Fast bulk bitwise AND and OR in DRAM are designed using the majority fact. When three DRAM cells are connected to a bitline at the same time, the variance in the bitline voltage after charge sharing tends to be closer to the average value of the three cells. The logical relationship of the three cells is satisfied function: $C(A+B)+\overline{C}(AB)$, so the logical operations of A and B can be determined by controlling the value of C.

Bitwise XOR [18]. Using the above design, a new subarray structure and a small B-group row decoder are designed to completed bitwise XOR. Only 5 AAP operations and 2 AP operations are required for a one-bit bulk bitwise XOR.

Multitude of Activated Subarrays (MASA) [20]. MASA allows multiple subarrays to be activated and read/write within a bank in parallel, which achieve subarray-level parallelism.

Inter-Subarray Data Movement [21]. Low-cost inter-linked subarrays (LISA) enable fast inter-subarray data movement in DRAM by adding low-cost connections between adjacent subarray. The proposed design joins neighboring subarrays' bitlines together using isolation transistors and provide a new DRAM command called RBM for the memory controller to make use of these new connections.

In-Memory Low-Cost Bit-Serial Addition [24]. A vector addition method has been designed in DRAM which performs majority-based addition operations by storing data in transposed manner. Majority functions in DRAM cells works by activating an odd number of rows simultaneously and the proposed in-DRAM adder only requires 9 compute rows. And calculate the sum of A+B according to the following two formulas:

$$C_{out} = Majority\,(A, B, C_{in}) \tag{6}$$

$$S = Majority\,(A, B, C_{in}, \overline{C_{out}}, \overline{C_{out}}) \tag{7}$$

3   DAS: Detailed Design

3.1   An Overview of DAS

Figure 3 shows the overall structure of DAS. The DAS implements a DRAM-based Ising architecture computing system, where the spins and interaction coefficients are stored in memory array and the annealing computational units are distributed in the bottom few rows of each subarray. Without changing the IO logic, logic circuits including shift logic circuits and random logic circuits are added mainly around the subarrays to perform spin update calculations.

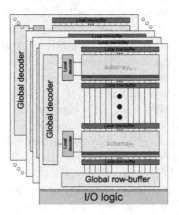

Fig. 3. Overall structure of DAS.

Figure 4 shows a subarray of the proposed DAS. The memory cells in a subarray is divided into 7 parts: 1) Spin rows. 2) Reserved rows. 3) Compute rows.

4) Control group (C-group). 5) Random row. 6) SHF row. 7) Bitwise group (B-group). Part 1–3 constitute data array which stores data for solving combinatorial optimization problems.

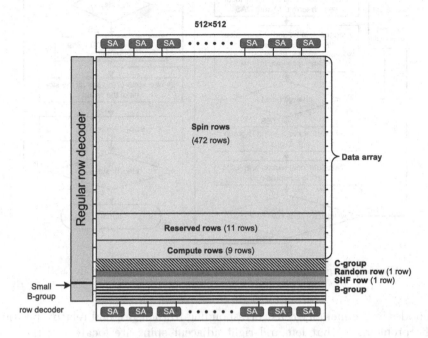

Fig. 4. Row grouping in a subarray.

The spin rows store information about the spin nodes and one column is able to contain 10 spins in a 512×512 subarray when the adjacent coefficient width is 8 bits. In the process of calculating the local search term L(i), it is necessary to read the spin bit and change bit of the adjacent spins to calculate $J_{ij} * s_j$, so it needs additional cells to store these values. And reserved rows are used to save intermediate results that need to be reused during the calculation process. The compute rows contains 9 rows for bulk bitwise addition. The order in which the data is arranged in the compute rows is controlled by the memory controller.

The random row occupies one row and is used to store the generated random numbers. The SHF row also occupies one row and is used to temporarily store values for left and right spin values. B-group and C-group are the structures proposed by Ambit [18], which are the basic structures for bulk bitwise operations of XOR, AND, and OR. C0 and C1 (C-group) store 0 and 1 respectively, which are used to control whether two selected rows perform AND or OR operations [19]. The bitwise group (B-group) consists of the specified rows required to perform bulk bitwise XOR, AND, and OR operations.

The first step in solving a combinatorial optimization problem is to map the problem to the Ising model. When mapping the problem, DAS stores the

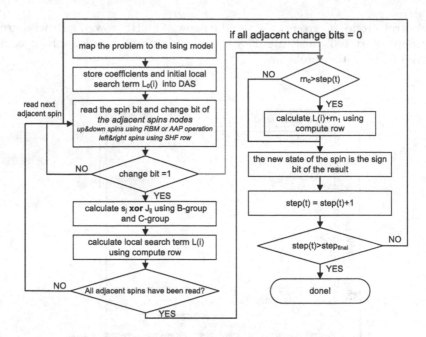

Fig. 5. The process of spin update when solving problem in DAS.

spin nodes in the memory cells corresponding to their original relative positions in the problem, so that left and right adjacent spins are located on the same row of memory cells. Spin updates depend on the spin values of adjacent spins, so interconnected spins cannot update their states simultaneously in the Ising architecture computing system. Since adjacent spins cannot be updated simultaneously, we propose a split-time update for odd and even columns. The design matches the buffer capacity of modern DRAM [21–23], which means that only half a row at a time is updated for a spin update of a subarray.

As shown in Fig. 5, the spin node updates its state in DAS by the following steps:

1) Read the spin bit and change bit of the adjacent spins nodes. Calculate the local search term L(i) based on them. The change bit marks whether the spin node is involved in the spin update calculation, and if the change bit is "1", the local search term in this direction needs to be recalculated.
2) If random number $rn_0 > step(t)$, the other random number rn_1 will be added with L(i). Otherwise, go to step3.
3) The new state of the spin is determined by the sign bit of the result.

In a finite number of steps, the Ising architecture computing system iterates continuously according to the above steps and finally reads the final spin state as the solution to the problem.

3.2   Spin Propagation Between Spin Nodes

According to Formula(2), it is necessary to calculate $s_j * J_{ij}$ of all the adjacent spins for the spin state update computation. In order to calculate $s_j * J_{ij}$, the adjacent spin values need to be read before spins update calculation, and then the adjacent spin values s_j are multiplied by the corresponding interaction coefficients J_{ij} and added one by one to obtain the local search term L(i).

As shown in Fig. 6(a), a spin node contains 45 bits and can be divided into three parts:

1) **Spin state.** s_i is the spin state of the spin and c_i is the change bit to indicate whether this state has changed from the previous spin state. A value of "0" for s_i represents a spin state of "+1", while a value of "1" for s_i represents a spin state of "-1".
2) **Adjacent interaction coefficients.** J_{ij}_L, J_{ij}_R, J_{ij}_D and J_{ij}_U are the interaction coefficient between the spin with 4 adjacent spins in different directions. It is determined by the problem and does not change during the calculation. Each interaction coefficient takes up 8 bits in DAS.
3) **Local search term.** In the initial state, each spin node calculates the respective local search term and stores it in local search cells. The interaction coefficient is stored in transposed way so that different bits of it share the same bitline, which is convenient for computation. According to Formula(2), the sum of five 8-bit signed binary numbers has a maximum value of 11 bits, so it contains 11 bits.

The spin reserved rows contain 11 rows shown in Fig. 6(b) and is used to temporarily store the spin state s_j and change bit c_j of adjacent spins, and the result of $s_j \oplus J_{ij}$.

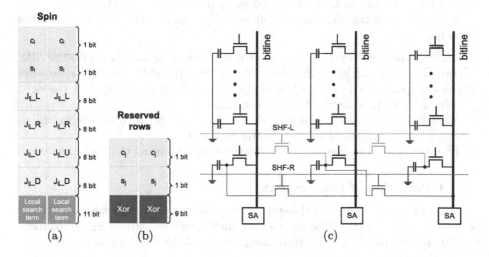

Fig. 6. (a) Row allocation of spin rows. (b) Row allocation of reserved rows. (c) Shift circuit for reading left and right spins.

A subarray only stores a small number of spins and the spins in the system are stored in different subarray, so delivering spin values between different subarray is required. And the spins in the first row of the subarray need to read the last row of the subarray above when reading their up spins. Similarly, the spins in the last row of the subarrays need to link adjacent subarrays to read their down spins. Low-cost inter-linked subarrays (LISA) [21] provides the capability for fast inter-subarray data movement in DRAM, which we have applied in DAS. The Isolation transistor is the transistor connecting the buffers of two adjacent subarrays above and below. When reading the values of adjacent spins at different subarray, the following operations will be performed:

1) AP operation. Read the required value to its subarray buffer.
2) Turn on the Isolation transistors.
3) RBM command. Moves the data from the "source" buffer row (half a row of data) to the "target" buffer row in the same bank.
4) The corresponding values are then read from the buffer to the reserved cells.

For spins whose up spins and down spins are both stored in the same subarray, simply use AAP operation to read out the corresponding values.

To read the left and right spin values, it is necessary to establish a path between two adjacent bitlines to transmit the spin state. We have designed an SHF row in the subarray to read and temporarily store the information of the adjacent bitline shown in Fig. 6(c) [25]. SHF transistors connect the memory cell of the source cell to the bitline of the target cell, with SHF-L controlling movement to the left and SHF-R controlling movement to the right. Bitlines will stabilise at VDD/2 after precharge, the SHF transistors will be turned on when receiving the SHF command and then the value of the source cells will be delivered to the target bitline. Target bitlines will rise or fall depending on the values until reaching at VDD or 0, then the target cells wordlines will be turned on and the value of the bitlines will be written to the target SHF cells.

The process of transferring the left and right spin states is divided into the following steps:

1) Perform the AAP operation to read a half row of spin values into the SHF row. If reading the left spin value, turn on the SHF-R transistor and connect the left spin value to the bitline of the right cell. Similarly, reading the right spin needs to turn on the SHF-L transistor.
2) After the bitlines voltage has reached a steady state, turn on the wordlines and write the values of bitlines to the SHF cells.
3) The spin values are then stored in the reserved cells using the AAP operation.

3.3   Calculation of Local Search Term

Formula(2) is a calculation of the spin energy, where the possible value of $J_{ij}*s_j[t]$ is $+1 * J_{ij}$ or $-1 * J_{ij}$. The interaction coefficients are stored in complement format, so Formula(2) can be translated into the following formula:

$$L_i(t) = \sum_{<i,j>} \left(J_{ij} \oplus \{s_j(t)\}_n + s_j(t) \right) + h_i \qquad (8)$$

$\{s_j(t)\}_n$ is the representation of expanding 1 bit $s_j(t)$ to n bits, and n is the bit width of interaction coefficient J_{ij}. Therefore, the calculation of spin update mainly involves XOR and ADD calculation after the transformation.

Before calculating the local search term, the spin states and change bits of adjacent spins are read and stored in the relevant computing reserved cells, which share the same bitline with the current spin node. According to Formula(8), the steps of calculating the initial local search term are :

1) 0 Compute $J_{ij} \oplus s_j$ bit by bit.
2) Store the computing results to the reserved cells Xor as shown in Fig. 6(b).
3) s_j is written to the carry-bit cell in compute rows as C_{in} for the lowest bit addition.
4) Calculate $J_{ij} \oplus \{s_j(t)\}_n + s_j(t) + h_i$.
5) The result of (4) is written to the local search term cells in spin rows.

After the above five steps, we get $J_{ij} * s_j[t] + h_i$ and then add it to the $J_{ij} * s_j[t]$ of another direction by repeating step1 to step5 until getting the initial local search term. The value of $J_{ij} \oplus s_j$ is computed by bitwise XOR in DRAM [18], and the addition of $J_{ij} * s_j + h_i$ is computed by in-DRAM addition [24].

In DAS, XOR is calculated by controlling the B-group and C-group. A one-bit XOR consumes only 5 AAP operations and 2 AP operations. ADD is implemented by in-memory bulk bitwise addition. First, the interaction coefficient J_{ij} needs to perform XOR operations with the corresponding spin value s_j as shown in Formula(8). Figure 7 shows the command sequences for one-bit XOR where B and C represent B-group and C-group operations [18]. And n indicates the nth bit.

$Xor_n = J_{ij\_n}$ xor s_j

1.AAP(J_{ij_n},	B8);
2.AAP(s_j,	B9);
3.AAP(C0,	B10);
4.AP(B14);	
5.AP(B15);	
6.AAP(C1,	B2);
7.AAP(B12,	Xor_n);

Fig. 7. Command sequences for XOR.

Figure 8 shows an example of calculating the lowest bit of local search term using in-DRAM bulk bitwise addition. The results of $\{s_j(t)\}_n \oplus J_{ij}$ are stored in the appropriate reserved cells before the local search term is calculated. The rows marked in red in Fig. 8 are firstly activated for copy or multi-row activation operations, and then the sense amplifier is enabled. Then, the rows marked in blue are activated to perform the corresponding operation.

Next, we introduce the step-by-step calculation of the local search term for the 0th bit (LS_0) as shown in Fig. 8:

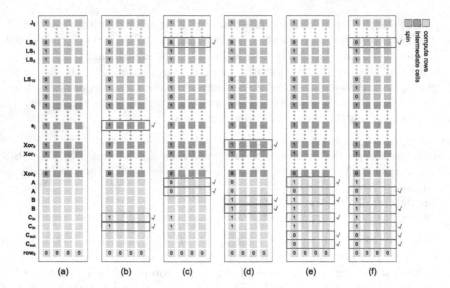

Fig. 8. Local search calculation for the lowest bit using in-DRAM bulk bitwise addition. (a) Initial State. (b) Copy s_j to C_{in}.(c) Copy LS_0 to A.(d) Copy Xor_0 to B. (e) Calculate C_{out} and $\overline{C_{out}}$. (f) Calculate and store sum result to LS_0.

1) According to Formula(8), start by copying s_j to the carry position C_{in} on the calculated row to participate in the calculation of the 0th bit.
2) Then copy the LS_0 at the lowest bit of the local search term to the first two rows in the compute rows (label A).
3) Similarly, copy the 0th bit of $J_{ij} \oplus s_j$ (Xor_0) to the additive rows in the compute row (label B).
4) Thereafter, C_{out} and $\overline{C_{out}}$ are obtained by simultaneously activating A, B and C_{in} of the corresponding row based on the Formula(6).
5) Finally, the sum result LS_0 is calculated by activating the five rows marked in red at the same time according to Formula(7), and then the 0th bit of local search term (LS_0) is stored after the sense amplifier is enabled.

The next-bit addition of local search term will perform the same operation, using the previously calculated C_{out} as C_{in} as shown in Fig. 9.

3.4 Design of Probability Flipping

Formula(5) shows that the calculation of updating the spin state includes two steps: local search and probability flipping. Probability flipping is carried out after local search, which is to avoid the state of the system from being trapped in the local optimum. Probability flipping accepts the state update of local search with a certain probability, which may cause the local energy increase. The next state of the spin after probability flipping does not necessarily reduce the local energy, thus enabling the system to escape from the local minimum energy.

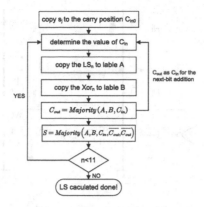

Fig. 9. Computational flow of local search.

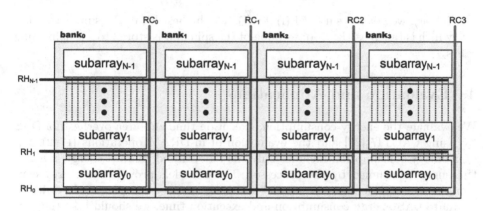

Fig. 10. Dual-random-source approach in DAS.

We adopt dual-random-source approach [15] as the probability flipping approach in the Ising architecture computing system. After getting the local search term, a probability flipping term $R(i)$ is added to it and the next spin state is decided by the sum result. To save the hardware overhead, all subarrays of the same bank share the same column random pulse and all subarrays of the same row between banks share the same random pulse as shown in Fig. 10. Row-column random pulse generators are based on linear feedback shift register implementation method [27].

To get the next state of spin, we need to calculate the addition of $L(i)$ and $R(i)$. $L(i)+R(i)$ is realized by bitwise addition in the bitline of the DRAM. $R(i)$ is produced by the random source, which contains row random pulse generator RH and column random pulse generator RC. The random logic circuit is shown in Fig. 11. $R(i)$ contains 8 bits, the upper 4 bits are determined by the row random pulse RC and the lower 4 bits are determined by the column random pulse RH. After executing n-bit bitwise addition(n is the bit width of the interaction

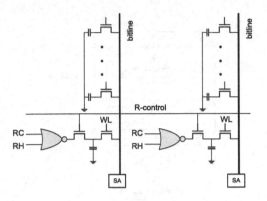

Fig. 11. Random logic circuit.

coefficient), we get the sum of L(i) +R(i). The highest bit of the sum is the sign bit, which is taken as the update value of the spin and is stored to the spin value cell s_i.

4   Evaluation and Applications

We evaluate the energy consumption, execution time and area overhead of DAS by using CACTI7 [26] and the evaluation of in-DRAM operations [17–21,24]. Our energy consumption includes only the DRAM and channel energy, but not the energy consumed by the processor. These models provide the energy consumption of each memory command and each memory component. In order to evaluate DAS energy consumption and execution time, we should take the four parts into account: 1) The added shift logic. 2) The added random logic. 3) The added in-DRAM operations. 4) The basic operation of DRAM. Table 1 shows the main parameters in our evaluations.

Table 1. The configuration of DAS.

Configuration		Configuration	
DRAM Architecture	DDR3	DRAM Tech	22 nm
Number of Banks	16	Channel Size	1
Memory Capacity	8 Gb	Subarray Size	512 × 512

4.1   Area Overhead

The total area of DAS is 42.77476 mm^2, and the design we added to the original DRAM only accounts for 7% of the total area. Table 2 shows the area composition of the proposed design. The in-DRAM bulk bitwise operation used in this paper has been shown to largely exploit the structure and operation of existing DRAM designs. And the added shift logic and random logic design consists of only two consecutive addresses in a subarray, so changes to the decoder logic are easily taken.

Table 2. Composition of the total area.

Component	Area (mm^2)
Dram core	39.7199
The added logic for bitwise operation	0.74786
SHF row	0.767
Random row	1.54
Total area	42.77476

4.2   Application and Analysis

In order to evaluate the performance, a typical combinatorial optimization problem, image segmentation is solved by our design. The image segmentation problem can be mapped to a locally interconnected Ising architecture computing system for accelerating [3].

Image segmentation is a label allocation problem [28,29], which assigns foreground and background labels to all pixels in the image. In essence, image segmentation is also a combinatorial optimization problem. Different pixel tag combinations will result in different image segmentation results. When the image segmentation problem is mapped to the Ising model, a pixel in the image can be represented by a spin, and the spin value can be taken as "+1" and "−1", which just corresponds to two different labels of the pixel. When the pixel belongs to the foreground, the spin value is "+1", and when it belongs to the background, the spin value is "−1". For images containing m×n pixels, it is necessary to map and solve the Ising architecture calculation system with local interconnections with m×n spins.

We conducted experiments on multiple images based on the HRSOD dataset [30], and Fig. 12 shows the evolution of image segmentation in DAS as the number of steps rises. It can be seen that as the number of steps increases, the segmentation results of the images gradually become clearer.

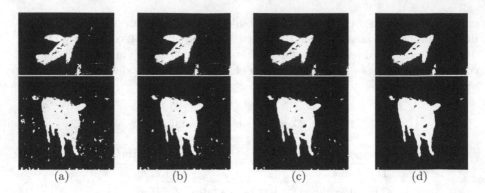

Fig. 12. Update transition process of spin-states when solving combinatorial optimization problems. (a) 25%. (b) 50%. (c) 75%. (d) 100%.

To solve the image segmentation problem by using DAS, we first need to map the image to the Ising model according to the above mapping method, then load the model coefficients into the system. Thereafter, start the ground state search of the Ising model to find the ground state or neighboring ground state, and finally obtains the final image segmentation results according to the spin state of the system. Figure 13 shows the segmentation results of this design system compared to conventional OneCut methods [31].

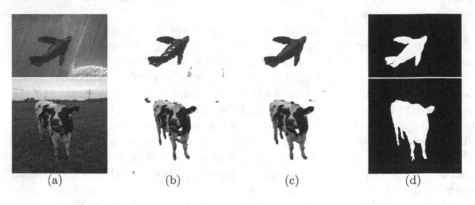

Fig. 13. Examples of image segmentation. (a) Input image. (b) Segmentation results of DAS. (c) Segmentation results of OneCut method. (d) Standard segmentation results.

The basic operation time of the designed system is shown in Table 3. An image with a resolution of 1600×1124 contains 452 subarrays, and a limited number of steps based on previous experience are taken to complete the image segmentation. The parallelism of the subarrays is set to 512 [16,20], and it can perform spin update calculations in 512 subarrays simultaneously to complete the image segmentation. Compared with the conventional method, DAS has similar segmentation capability, but can obtain average 10.2× acceleration shown in Fig. 14. The system's segmentation average energy is only 0.4349% of the conventional OneCut method.

Table 3. The number of needed operations for basic command in DAS.

Command	Word size (bit)	No. of needed AP or AAP operations
Read left/right spin and change bits	2	4AAP
Read up/down spin and change bits	2	2AAP
XOR	9	9×(5AAP+2AP)
ADD	11	45AAP

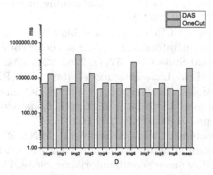

Fig. 14. The time of 10 images segmentation using DAS and OneCut.

5   Conclusion

We propose an Ising architecture computing system based on in-DRAM bulk bitwise operations, which can reduce data movement and latency to perform spin energy calculations quickly. Additionally, the system leverages subarray-level parallelism to further improve performance. The DAS adds logic circuit without changing IO and has a low area cost (only 7% of the area). In image segmentation application, DAS consumes only 0.4349% of the average energy consumption of conventional method, and achieves 10.2× acceleration. The proposed method provides a hardware basis for CMOS Ising machines to solve large-scale combinatorial optimization problems, and can significantly improve performance and efficiency compared to traditional architecture.

References

1. Hennessy, J.L., Patterson, D.A.: A new golden age for computer architecture. Commun. ACM **62**(2), 48–60 (2019)

2. Hill, M.D., Marty, M.R.: Amdahl's law in the multicore era. Computer **41**(7), 33–38 (2008)
3. Zhang, J., Chen, S., Wang, Y.: Advancing CMOS-type Ising arithmetic unit into the domain of real-world applications. IEEE Trans. Comput. **67**(5), 604–616 (2017)
4. Yamaoka, M., Yoshimura, C., Hayashi, M., et al.: A 20k-spin Ising chip to solve combinatorial optimization problems with CMOS annealing. IEEE J. Solid-State Circ. **51**(1), 303–309 (2015)
5. Mohseni, N., McMahon, P.L., Byrnes, T.: Ising machines as hardware solvers of combinatorial optimization problems. Nat. Rev. Phys. **4**(6), 363–379 (2022)
6. Oku, D., Tawada, M., Tanaka, S., et al.: How to reduce the bit-width of an Ising model by adding auxiliary spins. IEEE Trans. Comput. **71**(1), 223–234 (2020)
7. Takemoto, T., Hayashi, M., Yoshimura, C., et al.: A 2×30k-spin multi-chip scalable CMOS annealing processor based on a processing-in-memory approach for solving large-scale combinatorial optimization problems. IEEE J. Solid-State Circuits **55**(1), 145–156 (2019)
8. Wang, Z., Hu, X., Zhang, J., et al.: AIM: annealing in memory for vision applications. Symmetry **12**(3), 480 (2020)
9. Su, Y., Kim, H., Kim, B.: CIM-spin: a scalable CMOS annealing processor with digital in-memory spin operators and register spins for combinatorial optimization problems. IEEE J. Solid-State Circ. **57**(7), 2263–2273 (2022)
10. Brush, S.G.: History of the Lenz-Ising model. Rev. Mod. Phys. **39**(4), 883 (1967)
11. Yoshimura, T., Shirai, T., Tawada, M., et al.: QUBO matrix distorting method for consumer applications. In: 2022 IEEE International Conference on Consumer Electronics (ICCE), pp. 01–06. (IEEE) (2022)
12. Metropolis, N., Rosenbluth, A.W., Rosenbluth, M.N., et al.: Equation of state calculations by fast computing machines. J. Chem. Phys. **21**(6), 1087–1092 (1953)
13. Karp, R.M.: Reducibility among combinatorial problems. Complexity of Computer Computations, pp. 85–103. Springer, Boston (1972)
14. Lee, D., Kim, Y., Pekhimenko, G., et al.: Adaptive-latency DRAM: optimizing DRAM timing for the common-case. In: 2015 IEEE 21st International Symposium on High Performance Computer Architecture (HPCA), pp. 489–501. IEEE (2015)
15. Zhang, J., Chen, S., Yang, C., et al.: Double random sources: low-cost method to enhance local optima escaping ability in CMOS-type Ising chips. Electron. Lett. **52**(21), 1792–1793 (2016)
16. Ferreira, J.D., Falcao, G., Gómez-Luna, J., et al.: PLUTo: enabling massively parallel computation in DRAM via lookup tables. In: 2022 55th IEEE/ACM International Symposium on Microarchitecture (MICRO), pp. 900–919. IEEE (2022)
17. Seshadri, V., Kim, Y., Fallin, C., et al.: RowClone: fast and energy-efficient in-DRAM bulk da-ta copy and initialization. In: Proceedings of the 46th Annual IEEE/ACM International Symposium on Microarchitecture, pp. 185–197 (2013)
18. Seshadri, V., Lee, D., Mullins, T., et al.: Ambit: in-memory accelerator for bulk bitwise operations using commodity DRAM technology. In: 2017 50th Annual IEEE/ACM International Symposium on Microarchitecture (MICRO), pp. 273–287. IEEE (2017)
19. Seshadri, V., Hsieh, K., Boroum, A., et al.: Fast bulk bitwise AND and OR in DRAM. IEEE Comput. Archit. Lett. **14**(2), 127–131 (2015)
20. Kim, Y., Seshadri, V., Lee, D., et al.: A case for exploiting subarray-level parallelism (SALP) in DRAM. In: 2012 39th Annual International Symposium on Computer Architecture (ISCA), pp. 368–379. IEEE (2012)

21. Chang, K.K., Nair, P.J., Lee, D., et al.: Low-cost inter-linked subarrays (LISA): enabling fast inter-subarray data movement in DRAM. In: 2016 IEEE International Symposium on High Performance Computer Architecture (HPCA), pp. 568–580. IEEE (2016)
22. Lim, K.N., Jang, W.J., Won, H.S., et al.: A 1.2 V 23nm 6F 2 4Gb DDR3 SDRAM with local-bitline sense amplifier, hybrid LIO sense amplifier and dummy-less array architecture. In: 2012 IEEE International Solid-State Circuits Conference, pp. 42–44. IEEE (2012)
23. Takahashi, T., Sekiguchi, T., Takemura, R., et al.: A multigigabit DRAM technology with 6F/sup 2/open-bitline cell, distributed overdriven sensing, and stacked-flash fuse. IEEE J. Solid-State Circ. 36(11), 1721–1727 (2001)
24. Ali, M.F., Jaiswal, A., Roy, K.: In-memory low-cost bit-serial addition using commodity DRAM technology. IEEE Trans. Circuits Syst. I Regul. Pap. 67(1), 155–165 (2019)
25. Deng, Q., Jiang, L., Zhang, Y., et al.: DrAcc: a DRAM based accelerator for accurate CNN inference. In: Proceedings of the 55th Annual Design Automation Conference, pp. 1–6 (2018)
26. Balasubramonian, R., Kahng, A.B., Muralimanohar, N., et al.: CACTI 7: new tools for inter-connect exploration in innovative off-chip memories. ACM Trans. Archit. Code Optim. (TACO) 14(2), 1–25 (2017)
27. Tkacik, T.E.: A hardware random number generator. In: International Work-shop on Cryptographic Hardware and Embedded Systems, pp. 450–453 (2002)
28. Boykov, Y., Veksler, O., Zabih, R.: Fast approximate energy minimization via graph cuts. IEEE Trans. Pattern Anal. Mach. Intell. 23(11), 1222–1239 (2001)
29. Kolmogorov, V., Zabin, R.: What energy functions can be minimized via graph-cuts? IEEE Trans. Pattern Anal. Mach. Intell. 26(2), 147–159 (2004)
30. Zeng, Y., Zhang, P., Zhang, J., et al.: Towards high-resolution salient object detection. In: Proceedings of the IEEE International Conference on Computer Vision, pp. 7234–7243 (2019)
31. Tang, M., Gorelick, L., Veksler, O., et al.: Grabcut in one cut. In: Proceedings of the IEEE International Conference on Computer Vision, pp. 1769–1776 (2013)

Graph Neural Network for Critical Class Identification in Software System

Meng-Yi Zhang[1] and Peng He[2(✉)]

[1] School of Computer and Information Engineering, Hubei University, Wuhan 430062, China
[2] School of Cyber Science and Technology, Hubei University, Wuhan 430062, China
penghe@hubu.edu.cn

Abstract. Most performance enhancements in software engineering practice are made by optimizing a few key codes. A key class identification method for software systems based on a graph neural network is proposed to identify these key codes in software systems effectively. First, a directed weighted class dependency network with classes as nodes and dependencies as edges is constructed based on the dependency degree in the software system and the dependency relationship between classes. Second, the initial feature of class nodes is the embedding vector generated by network embedding learning for each class node. Then, the GraphSAGE model, which is used to learn the hidden features of class nodes, is further explained. Moreover, a multi-layer perceptron turns the learned feature vector into a scalar score. By descending sorting the node scores, the key classes are then identified. Two open datasets are used to test the proposed approach. The experimental results show that our proposed method can improve the recall and accuracy of the top 10% of key node classes by more than 10%, compared with the current work and three graph neural network methods.

Keywords: Class dependency networks · Network embedding · Graph neural networks · Key class recognition · Multi-layer perceptron

1 Summary

It is well known that the "80/20 Rule" are prevalent in software development 1. For instance, optimizing 20% of the code improves performance for 80% of users, 80% of users typically only use 20% of the system's functions, and 20% of detected bugs typically cause 80% of problems. As a result, over 20% of code in a critical software system necessitates focusing on the classes involved in these codes, which can be considered key classes in the software system.

Researchers have successfully applied the node importance measurement method in complex networks to identify node centrality [3], K-shell [4, 5], and other key classes of software systems because the system has been demonstrated to have complex network characteristics. As a result, researchers are accustomed to abstracting the software system as a complex network [2], a class-dependent network with classes acting as nodes and various dependencies between classes acting as edges.

Z. Tari et al. (Eds.): ICA3PP 2023, LNCS 14489, pp. 174–190, 2024.
https://doi.org/10.1007/978-981-97-0798-0_11

With the advancement of deep learning, graph neural networks have gained recognition for their advantages in processing graph data in recent years. Graph analysis is also widely used in many fields [11]. The complex network method needs to identify key nodes based on the calculation of artificial measurement indicators. For example, although Betweenness Centrality has been proved that the intermediary centrality has an excellent ability to describe the importance of nodes, the index will increase due to the scale of the network, and the calculation cost is high, even leading to the problem of uncalculation [7]. With solid characterization ability, graph neural networks can automatically learn the deep features of nodes and obtain global structure information by aggregating features of neighbor nodes, which can effectively overcome a series of problems encountered in complex network methods.

The author uses GAE, a graph auto-encoder, to learn the class-dependent network and successfully identifies key classes in the literature [37]. In this work, the author discusses an undirected class-dependent network. That is, the directivity of dependencies between classes is not considered. Moreover, GAE is a graph neural network model of direct inference, which can only learn fixed graphs rather than new nodes not in the graph. However, in the actual software system, the dependencies and dependencies between classes are different, and the software system will evolve dynamically with the requirements change. For example, a Product class is used as a parameter for an "add" operation in a Cart class. Dependencies point from the Cart class to the Product class in a class diagram. The Cart class may change if the Product class is changed, but not the other way around.

Therefore, based on previous work, this paper attempts to introduce a graph neural network model with an inductive framework (GraphSAGE [45]). While considering the dependency directions between classes, it also uses the attribute information of nodes to learn the characteristics of new nodes and further models the software system more accurately to improve the identification accuracy of key classes.

The following is a summary of the main contributions made in this paper:

(1) Based on the GraphSAGE graph neural network model of the inductive framework, an improved key class identification method suitable for processing directed, class-dependent networks is proposed.
(2) The recall rate and accuracy rate of the top 10% candidate key class identification can be improved by more than 10% compared to the existing benchmark method, which was verified on two public data sets.

The following is the structure of the remainder of this paper: The current state of research in related fields is outlined in Sect. 1. Section 2 introduces several standard models of direct inference graph neural networks. Section 3 explains the method in detail. Section 4 shows the experiment and result analysis. The final section summarizes the entire paper and anticipates future work.

2 Related Works

2.1 Research on the Importance of Nodes in Complex Networks

There are four main categories of importance ranking methods for complex network nodes. The first method uses degree centrality [3] or semi-local centrality algorithms [9] to determine the importance of nodes based on information from their neighbors. The second method evaluates the importance of individual nodes based on the path that network information travels. Random Walk Median Centrality [11] and Median Centrality [10] are the most widely used algorithms. The third type is to judge the importance of nodes based on their location in the network, and the K-shell decomposition algorithm [12] is the leading representative. The fourth category is based on the eigenvectors of nodes, primarily HITS [13], PageRank [14], and their variants [15].

The following drawbacks exist with the above methods: (1) They only pay attention to the structure of nodes in the network topology, like the K-shell decomposition algorithm [12], without considering the attribute information of nodes. (2) It is impossible to evaluate the influence of nodes on the entire network, such as degree centrality [16], by only considering the local or specific attributes of nodes. Methods based on global attributes, such as median centrality, need to find the graph's shortest path between two pairs of nodes. However, the time cost of this indicator will increase exponentially with network size, making it difficult to use it in large networks [46]. (3) Setting the random walk probability of neighbor nodes is required for a method like the LeaderRank method [23] to calculate the importance of nodes based on their characteristics.

2.2 Research on Key Node Identification in Software Network

A software network is an artificial complex network (such as a class-dependent network) that abstracts software systems based on complex network theory. Pan et al. [32] utilized package granularity to create a software network. They introduced the PageRank algorithm to identify key packages in the software system. Considering nodes' excess and call time, the weighted class dependency network is used by Singh et al. [33] to determine the importance of the nodes in the network. Wang et al. [21] proposed a method to automatically identify key classes in software systems, combining the complexity measurement of classes, the interaction with neighbors, and the class's control over software information flow. Node importance identification can also be considered a sorting problem. Srinivasan et al. [34] proposed a core software-based sorting model for locating key software system modules. Pan et al. [35] proposed the ElementRank algorithm to sort multilayer software networks to identify critical nodes. Furthermore, Pan et al. [36] identified key classes in software systems in a directed weighted class-dependent network using generalized k-kernel decomposition. The paper [8] proposes a software system-based key class recognition method based on the graph autoencoder and the complex network research theory to solve current key class recognition issues based on software networks. By fitting reference objects with the intermediate center index, the graph auto-coding is continuously optimized to achieve a more efficient recognition of important nodes in large networks, thus achieving key class recognition.

It is easy to find that the existing methods have achieved good results in practical applications. However, some things could still be improved: (1) The software system's network modeling is incomplete, and the direction and weight are too few. (2) Most of the work still uses the complex network or social network analysis method, and the graph neural network model has not been applied to this topic. Its effectiveness yet to be fully verified.

3 Basic Theory

3.1 Node Feature Extraction Based on Direct Extrapolation Graph Neural Network

Node Feature Extraction Based on Graph Convolutional Neural Network
According to GCN (Graph Convolutional Networks) [42], the dimension of the embeddedness vector represents the initial feature of the node. Represents the features of nodes after updating the layer, expressed as:

$$h_v^{(l)} = \sigma\left(\widetilde{D}^{-1} \widetilde{A} h_v^{(l-1)} W^{l-1}\right) \tag{1}$$

where A is the class-dependent network's adjacency matrix, $\widetilde{A} = A + I_N$, D is the degree matrix, $\widetilde{D}_{ii} = \sum_j \widetilde{A}_{ij}$, and W is the weight matrix. After two levels of iterative learning, Z, the final feature representation vector of the nodes, can be obtained if $\widehat{A} = \widetilde{D}^{-1} \widetilde{A}$.

$$Z = f(X, A) = \text{softmax}\left(\widehat{A} \text{ReLU}(\widehat{A} X W^{(0)}) W^{(1)}\right) \tag{2}$$

Node Feature Extraction Based on Graph Attention Network
A self-attention mechanism based on GCN is introduced in GAT (Graph attention networks) [43] to implement the GAT network by stacking attention layers. A set of node features constitute the attention layer's input: $\mathcal{H} = \{h_1, h_2, \cdots, h_N\}, h_i \in \mathbb{R}^F$. N represents the number of nodes, and F is the dimension of each node's attributes. A set of features derived from attention level learning: $\mathcal{H}' = \{h_1', h_2', \cdots, h_N'\}, h_i' \in \mathbb{R}^{F'}$. F' is the new feature dimension of the node. Using the weight matrix $W \in \mathbb{R}^{F' \times F}$ for the linear transformation of each node in the figure and then calculating e, the attention coefficient, for each node is necessary to transform the input features into higher-order features.

$$e_{ij} = a(Wh_i, Wh_j) \tag{3}$$

where e_{ij} represents the importance of node j to node i, a is a single-layer feed-forward neural network, and the output scalar value represents the correlation between two nodes. Normalize the correlation between target and neighbor nodes to distinguish the weights between nodes:

$$\alpha_{ij} = \text{softmax}(e_{ij}) = \frac{\exp(e_{ij})}{\sum_{k \in N(i)} \exp(e_{ik})} \tag{4}$$

The linear combination of the corresponding features is calculated following the acquisition of the normalized attention factor α. The final output eigenvector of each node is, using the nonlinear activation function, as follows:

$$h_i' = \sigma(\sum\nolimits_{j \in N(i)} \alpha_{ij} W h_j) \tag{5}$$

To stabilize the self-attention learning process, GAT employs multiple attention layers, or K groups of mutually independent attention mechanisms, and then splices the results of K-layers:

$$h_i' = \|_{k=1}^{K} \sigma\left(\sum\nolimits_{j \in N(i)} \alpha_{ij}^{(k)} W^k h_j\right) \tag{6}$$

where $\|$ is splicing, $\alpha_{ij}^{(k)}$ and W^k are the normalized attention and weight coefficients of the kth attention layer, respectively. The average value of the results at layer k is used to reduce the dimension of the feature vector to get h_i':

$$h_i' = \sigma\left(\frac{1}{K}\sum\nolimits_{k=1}^{K} \sum\nolimits_{j \in N(i)} \alpha_{ij}^{(k)} W^k h_j\right) \tag{7}$$

Node Feature Extraction Based on Graph Autoencoder
The integration of GCN into Auto-Encoders (AE) is known as GAE (Graph Autoencoders) [44]. Deep representation vectors of nodes can be obtained by employing a two-layer GCN as the encoder (see formula (2)). The inner product is used as a decoder to reconstruct the class dependency network, which can be expressed as:

$$\overline{A} = \sigma\left(ZZ^T\right) \tag{8}$$

where $\overline{A}$ is the reconstructed adjacency matrix. The better it is, the more similar $\overline{A}$ and A are. As a result, cross-entropy is utilized by GAE as the training loss function:

$$\mathcal{L} = -\frac{1}{N}\sum y \log \overline{y} + (1-y)\log(1-\overline{y}) \tag{9}$$

where y represents the element's value in A, and $\overline{y}$ represents the value in $\overline{A}$.

3.2 Node Feature Extraction Based on Inductive Graph Neural Network

A well-known inductive framework, GraphSAGE (Graph SAmple and aggreGatE) [45] consists of sampling and aggregation in three steps:

1. A fixed number of random samples are taken from the neighbors of each node in the graph.
2. A multi-layer aggregation function is used to aggregate the information of the neighboring nodes.
3. Learning is performed on the nodes based on the aggregated information, as shown in Fig. 1, with a detailed description in Algorithm 1.

Algorithm 1: GraphSAGE algorithm for generating node features input:

Input: $dWCDN = G(V, E, W)$,Initial feature vector X of the node, Target depth K , weight matrix $W^k, \forall k \in \{1, \cdots, K\}$.

Output: Node feature matrix Z.

1　Initialize node representation vector $h_v^{(0)} = X_v, \forall v \in V$;

2　**If** $k = 1 \cdots K$, **then**

3　Perform the following operation on each node $v \in V$:

4　$h_{N(v)}^k = AGGREGATE_k(\{h_u^{k-1}, \forall u \in N(v)\})$;

5　$h_v^k = \sigma(W^k \cdot CONCAT(h_{N(v)}^k, h_v^{k-1}))$;

4　$h_v^k = h_v^k / \left\| h_v^k \right\|_2$; //Normalize the h_v^k obtained for each layer

5　**End**

6　$z_v = h_v^K, \forall v \in V$;

7 The final output node feature matrix Z.

In the algorithm, h_v^k is the representation vector of node v in the k-th layer, $N(v)$ represents the set of neighbor nodes of node v, σ is a non-linear activation function, and AGGREGATE($\cdot$) and CONCAT($\cdot$) are the neighborhood aggregation function and the concatenation function, respectively.

The three types of aggregation functions that GraphSAGE suggests are Long Short-Term Memory (LSTM) aggregator, Mean aggregator, and Pooling aggregator. The Mean aggregator concatenates the node's and its neighbor's feature vectors and then takes the average of each dimension of the feature without requiring any new parameters. The Pooling aggregator connects each neighbor's representation vector with a fully connected neural network, then applies a maximum pooling operation to the new representation vector. The LSTM aggregator randomly sorts the neighbors and inputs them into an LSTM.

GCN and GAT are batch trained and only use information from one-hop neighbors without using higher-order neighbor information because push-forward models have limitations. GraphSAGE, on the other hand, makes mini-batch training based on sub-graph training possible and samples neighbor nodes to increase computational efficiency, making it better suited for applications in the real world. As a result, this paper proposes an improved key class recognition method using the inductive graph neural network model GraphSAGE and the Mean aggregator for neighborhood aggregation.

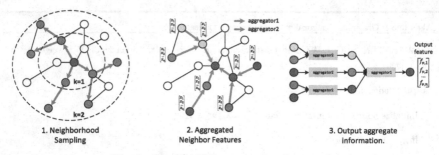

1. Neighborhood   2. Aggregated   3. Output aggregate
   Sampling   Neighbor Features   information.

Fig. 1. GraphSAGE's aggregation framework

4 Research methods

This paper expands on previous research [37] in two ways: First, it builds a directed weighted class dependency network and improves the software system's network modeling. Second, it demonstrates node feature learning from a vertex's nearby neighbors, trains aggregation functions, and introduces the inductive GraphSAGE graph neural network model.

The overall research framework of this paper is depicted in Fig. 2: (1) Using software source code analysis, construct a directed weighted class dependency network with classes as nodes and dependency relationships as edges by extracting the relationships between classes. (2) Class node embedding vectors are created using network embedding learning in the class dependency network. (3) To learn the feature representation of class nodes, the obtained embedding vectors and the class dependency network are fed into the GraphSAGE model as initial features. (4) After the class node features are turned into a scalar and sorted in descending order by a multi-layer perceptron, the model is trained using the labels in the dataset. Finally, key classes are effectively identified.

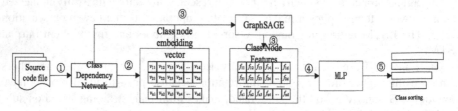

Fig. 2. Overall framework of the method presented

4.1 Modeling Class dependency network

This paper aims to identify the key classes in a software system. The software network model is constructed at the class granularity, creating a class dependency network (CDN) represented by $CDN = V, E, W$, where V is the set of nodes in the network and node $v_i(v_i \in V)$ means a class or interface in the software system. E is the set of network edges, connecting edge $e_{ij}(e_{ij} = (v_i, v_j) \in E)$ represents node v_i depending on node

v_j. W represents the set of weights, and weight $w_{ij}(w_{ij} = (v_i, v_j) \in W)$ represents the weight on edge e_{ij}.

The three main categories of dependency relationships between class nodes in this paper's CDN modeling are as follows:

(1) Aggregation dependency: An edge e_{ij} exists between nodes v_i and v_j when class i contains the attribute of class j;

(2) Inheritance dependency: An edge e_{ij} exists between nodes v_i and v_j when class i inherits from class j or implements the interface of class j;

(3) Calling dependency: An edge e_{ij} exists between nodes v_i and v_j when class i calls class j.

The number of dependencies between classes determines the weight on edge between them. Whenever two classes have any of the above three types of dependencies (aggregation, inheritance, calling), the dependency weight between the two classes is increased by one. The greater the weight, the closer the dependency relationship between the two classes [38].

4.2 Network Embedding Learning

Mapping a network into a vector space, representing nodes as low-dimensional vectors, and obtaining representation vectors simultaneously preserving node information and the network's topology is called "network embedding learning." It makes it possible to represent and reason in vector space. For a given network $G(V, E)$, a function f maps each node v ($v \in V$) in the network to a d-dimensional representation vector X_v ($f : v_i \rightarrow X_v \in \mathbb{R}^d, d \ll |V|$). The learned node representation vectors are used as input for graph neural network models, which extract deeper features of nodes. Common network embedding methods include node2vec [40], the most widely used one.

In node2vec, the random walk strategy is guided by the parameters p and q, defined as follows:

$$\partial_{pq}(t, x) = \begin{cases} \frac{1}{p} & d_{tx} = 0 \\ 1 & d_{tx} = 1 \\ \frac{1}{q} & d_{tx} = 2 \end{cases} \tag{10}$$

where $\alpha_{pq}(t, x)$ is the unnormalized transition probability. The probability of returning to the walking node of the previous step is represented by the number p. Depending on whether the walk is close to or far from the node of the previous step, q is referred to as the away or close probability. The distance between nodes t and x is represented by d_{tx}. $d_{tx} = 0$ Indicates that nodes t and x are the same nodes. $d_{tx} = 1$ Indicates that nodes t and x are neighboring nodes, and $d_{tx} = 2$ indicates that node t does not have an edge directly connected to node x.

If p is relatively small, the random walk will have breadth-first characteristics, and the sampling will wander around node t. The random walk will exhibit depth-first characteristics if the value of q is relatively small, and the sampling will gradually deviate from node t. Through this method, different walking strategies can be adopted for different graphs so that the learning quality can be improved. For more information about node2vec, please refer to the literature [23].

4.3 Key Class Identification

A multi-layer perceptron is used to turn the GraphSAGE node feature representation into a scalar, which is then prioritized based on the scalar. Therefore, it allows for the identification of key classes in software systems.

This paper uses a three-layer perceptron with only one hidden layer to transform the node feature vector. The fully connected layer links the hidden layer and the input layer, and the input is the GraphSAGE model's node feature vector z_v. The output produced by the output layer is f(WZ + b), where f is an activation function that is not linear. In this paper, the LeakyReLU function is utilized because it effectively addresses the issue of gradient disappearance. W is the connection coefficient, b is the bias, and the scalar y_v is shown as:

$$y_v = \text{LeakyReLU}(Wz_v + b) \tag{11}$$

The obtained node scalar value y_v is sorted in descending order, and the top-k nodes are marked as candidate key classes and then compared with the actual true key classes. The model is trained by maximizing the recognition hit rate.

5 Experimental Analysis

5.1 Experiment Data

This paper chooses two datasets with expert annotation information, Ant and JMeter, each with 10 and 14 key classes, as shown in Table 1.

Table 1. Dataset information in this article

Name	Key class count	Version	Node count	Edge count
Ant	10	1.6.1	906	4046
JMeter	14	2.0.1	277	1058

5.2   Experiment Setup

Evaluation Indicators

The conventional metrics of precision and recall, which are defined as follows, are used to evaluate this paper:

$$\text{Precision} = \frac{TP}{TopK\%} \tag{12}$$

$$\text{Recall} = \frac{TP}{Sum} \tag{13}$$

where True Positive (TP) represents the number of key classes in candidate TopK% annotated by experts, and Sum represents the total number of key classes annotated by experts in the dataset.

Standard Method

In addition to comparing with the three direct-push formula graph neural network models in Sect. 3.3, two methods based on complex network theory are selected for comparison and analysis to confirm the method's efficacy:

(1) GCN [42]: This paper uses undirected graphs as inputs for the GCN because the GCN in the literature [42] only works with undirected graphs.
(2) GAT [43]: Since GAT can deal with directed graphs, the directed weighted network serves as the input.
(3) GAE [37]: The node contraction method and GAE are utilized in the paper for undirected class dependency networks to achieve key class identification.
(4) ICOOK [36]: The weighted K-Core decomposition method is used in the literature to identify key classes in software systems. When a node has the same core number as other nodes, the current node is more important if its weighted degree is higher.
(5) ElementRank [35]: The PageRank method is used in the literature to identify key software system classes. After being calculated, the PageRank value of a node increases in proportion to its network importance.

Some changes were made to the GCN and GAT models used in this paper to make a fair comparison because they were not directly used to identify key classes. The MLP model was made into a scalar by incorporating the features of the final hidden layer. Then priority sorting was performed according to the obtained scalar.

Experimental Environment

This paper's software network modeling section was completed in a Java environment on a Windows 10 system (JDK 1.9). At the same time, the rest of the experiments were conducted on a rented cloud server using a Linux system (Red Hat 4.8.5–28) with 32 GB of memory and 4 11GB GTX1080ti GPU. The TensorFlow framework was used to carry out every experiment, and the model parameters were set, as shown in Table 2.

Table 2. Partial model parameter settings

Parameter name	Parameter value
Optimizer	Adam
Learning rate	0.0002
The dimension of the node2vec output embedding vectors d	128
The parameter q (departure probability) in Node2vec	3
GraphSAGE batch-size	8
Number of neighbor nodes sampled	5
Aggregate function level	2

5.3 Analysis of Experimental Results

The recall and accuracy outcomes of various approaches are presented in Table 3. In the Ant project, among the top 10% candidate key classes, the GraphSAGE method proposed in this paper has a better key class recognition effect than the GCN, GAT, and GAE methods, in which the recall rate reaches 90%, and the accuracy rate reaches 10%. The three methods all have the same effect on recognizing the top 15% candidate key categories, and their recall rate and accuracy rate have reached 100% and 7.35%, respectively. For the JMeter project, the recall rate and accuracy rate of the GraphSAGE method in this paper are the best among the top 10% candidate key classes, 71.43% and 35.71%, respectively, while the recall rate and accuracy rate of GCN, GAT, and GAE are the same, reaching 64.29% and 32.14% respectively. GraphSAGE and GAT methods outperform GCN and GAE methods for the top 15% candidate key classes, with recall rates of 85.71% and accuracy rates of 29.27%, up 9.1%, respectively.

According to Table 3, it is easy to see that those key class identification methods based on the graph neural network model perform better than those based on conventional complex network theory. Particularly in project Ant, the recall and accuracy rates significantly increase.

In addition, from the results of several direct GNN models, the inputs of GCN and GAE are undirected graphs. GAT and GraphSAGE adopted in this paper can process directed graphs. Key class identification necessitates taking into account the direction information of the class-dependent network because these two data sets' outcomes are significantly superior to those of GCN and GAE.

Table 3. Comparison of recall and accuracy of various methods

Project	Method	10%		15%	
		Recall	Precision	Recall	Precision
Ant	GCN	80%	8.89%	100%	7.35%
	GAT	80%	8.89%	100%	7.35%
	GAE	80%	8.89%	100%	7.35%
	GraphSAGE	90%	10%	100%	7.35%
	K-Core	70%	7.78%	70%	5.15%
	PageRank	60%	6.67%	70%	5.15%
JMeter	GCN	64.29%	32.14%	78.57%	26.83%
	GAT	64.29%	32.14%	85.71%	29.27%
	GAE	64.29%	32.14%	78.57%	26.83%
	GraphSAGE	71.43%	35.71%	85.71%	29.27%
	K-Core	64.29%	32.14%	71.43%	23.80%
	PageRank	35.71%	18.05%	50%	16.6%

The specific situation regarding the key classes in Ant's recognition is depicted in Table 4. Table 3 from the Ant project demonstrates that the GraphSAGE method recognizes 9 out of 10 expert-labeled key classes with slightly better accuracy than the other 3 GNNs, especially when recognizing ElementHandler and TaskContainer. All methods failed to recognize the class Main, which seems to be mismatched with the intuitive result. One possible explanation is that the current methods only extract information from the structure between classes without considering the semantic information of the classes themselves. The Main class should be more precisely identified if the semantic information of the classes is considered. The recognition performance of the four GNN-based methods is comparable in the top 15% of key classes, and all expert-labeled key classes can be recognized and processed. To put it another way, when the number of candidate key classes that are returned reaches a certain scale, the recognition accuracy of several methods is similar. However, from the perspective of software evolution, the GraphSAGE method proposed in this paper is more adaptable. It only needs to update the representation of new nodes; no need to retrain the entire graph. The computation cost is also significantly lower.

Table 4. Key classes identified by various methods in Ant

Top-n%	Key classes	GCN	GAT	GAE	GraphSAGE	K-Core	PageRank
Top-10%	Project	√	√	√	√	√	√
	UnknownElement	√	√	√	√	√	√
	Task	√	√	√	√	√	√
	Main	×	×	×	×	×	×
	IntrospectionHelper	√	√	√	√	√	√
	ProjectHelper	√	√	√	√	√	×
	RuntimeConfiguration	√	√	√	√	√	√
	Target	√	√	√	√	√	√
	ElementHandler	×	√	×	√	×	×
	TaskContainer	√	×	√	√	×	×
Top-15%	Project	√	√	√	√	√	√
	UnknownElement	√	√	√	√	√	√
	Task	√	√	√	√	√	√
	Main	√	√	√	√	×	×
	IntrospectionHelper	√	√	√	√	√	√
	ProjectHelper	√	√	√	√	√	×
	RuntimeConfiguration	√	√	√	√	√	√
	Target	√	√	√	√	√	√
	ElementHandler	√	√	√	√	×	×
	TaskContainer	√	√	√	√	×	√

As shown in Table 5, for JMeter, in the top 10% of the key class recognition, the three direct graph neural network recognition methods have the same effect, and 9 of the 14 classes marked by experts have been identified. However, the GraphSAGE method proposed in this paper can still identify one additional class, TestCompiler, and the other four unrecognized key classes are consistent with the three previous GNN methods. The recognition effect of the GraphSAGE and GAT methods is the same when the proportion is increased to 15%, and more TestListeners are recognized than GCN and GAE methods.

Consideration of the direction of the dependency relationship between classes in the software system aids in identifying the key classes, as demonstrated by the preceding experimental results in Ant and JMeter projects. In addition, the graph neural network based on induction can better represent the node structure characteristics than the one based on direct inference.

Table 5. Key classes identified by various methods in JMeter

Top-n%	Key classes	GCN	GAT	GAE	GraphSAGE	K-Core	PageRank
Top-10%	AbstractAction	✓	✓	✓	✓	×	✓
	JMeterEngine	×	×	×	×	×	×
	JMeterTreeModel	✓	✓	✓	✓	✓	✓
	JMeterThread	✓	✓	✓	✓	✓	×
	JMeterGUIComponent	✓	✓	✓	✓	✓	✓
	PreCompiler	×	×	✓	✓	×	×
	Sampler	✓	✓	✓	✓	✓	✓
	SampleResult	✓	✓	✓	✓	✓	×
	TestCompiler	✓	✓	×	✓	✓	×
	TestElement	✓	✓	✓	✓	✓	✓
	TestListener	×	×	×	×	×	×
	TestPlan	✓	✓	✓	✓	✓	×
	TestPlanGui	×	×	×	×	×	×
	ThreadGroup	×	×	×	×	✓	×
Top-15%	AbstractAction	✓	✓	✓	✓	×	✓
	JMeterEngine	×	×	×	×	×	×
	JMeterTreeModel	✓	✓	✓	✓	✓	✓
	JMeterThread	✓	✓	✓	✓	✓	×
	JMeterGUIComponent	✓	✓	✓	✓	✓	✓
	PreCompiler	✓	✓	✓	✓	×	×
	Sampler	✓	✓	✓	✓	✓	✓
	SampleResult	✓	✓	✓	✓	✓	✓
	TestCompiler	✓	✓	✓	✓	✓	×
	TestElement	✓	✓	✓	✓	✓	✓
	TestListener	×	✓	×	✓	✓	✓
	TestPlan	✓	✓	✓	✓	✓	×
	TestPlanGui	×	×	×	×	×	×
	ThreadGroup	✓	✓	✓	✓	✓	×

6   Conclusion

This paper attempts to abstract the software system into a directed weighted software network from the perspective of the network to address the issue of key class recognition in software systems. It then introduces the GraphSAGE graph neural network model for network node feature learning and applies the learning results to key class recognition.

In order to verify the effectiveness of the method, an empirical analysis was carried out on two public data sets, and the results showed that:

(1) In the process of building a software network, the direction and weight information of the dependency relationship between classes are considered, which is helpful to describe the software system structure better;
(2) The graph neural network model based on induction can better represent the node structure characteristics than that based on direct push. The recall and accuracy rate of the top 10% candidate key class recognition can be improved by more than 10%.

This paper's proposed technique has generally achieved great recognition results compared to several benchmark strategies. However, considering the rising necessity for accuracy in the recognition of key categories, there is still much opportunity to get better in the subsequent work, primarily involving the following aspects:

(1) To test this method's generalizability, this paper uses two open-source software systems based on Java. These systems should complement the software systems used to verify the identification of key classes, such as software systems written in other languages.
(2) The classes' dependencies are not differentiated by the directed weighted software network constructed in this paper. Various dependencies' impact on identifying key classes is contrasted and analyzed using multiple weights.
(3) There is still much room for improvement, even though the approach presented in this paper has produced satisfactory results to a certain extent. The updated graph neural network model should be used to extract node features to improve further the recognition effect of key classes in the software system.

References

1. Xiao, P.: Analysis and exploration of software testing technology. Comput. CD Softw. Appl. **18**(02), 44–45 (2015)
2. Wang, S., Liu, T., Nam, J., et al.: Deep semantic feature learning for software defect prediction. IEEE Trans. Softw. Eng. **46**, 1267–1293 (2018)
3. Li, J., He, P., Zhu, J., et al.: Software defect prediction via convolutional neural network. In: 2017 IEEE International Conference on Software Quality, Reliability and Security (QRS), pp. 318–328. IEEE (2017)
4. Qu, Y., Liu, T., Chi, J., et al.: node2defect: using network embedding to improve software defect prediction. In: 2018 33rd IEEE/ACM International Conference on Automated Software Engineering (ASE), pp. 844–849. IEEE (2018)
5. Huang, C., Liu, X., Deng, M., et al.: A survey on algorithms for epidemic source identification on complex networks. Chin. J. Comput. **41**(06), 1156–1179 (2018)
6. Kabir, K.A., Kuga, K., Tanimoto, J.: Analysis of SIR epidemic model with information spreading of awareness. Chaos Solitons Fractals **119**, 118–125 (2019)
7. Firth, J.A., Hellewell, J., Klepac, P., et al.: Using a real-world network to model localized Covid-19 control strategies. Nat. Med. **26**, 1616–1622 (2020)
8. Rahmani, H., Blockeel, H., Bender, A.: Using a human drug network for generating novel hypotheses about drugs. Intell. Data Anal. **20**(1), 183–197 (2016)

9. Gu, Q., Ju, C., Wu, G.: Knowledge communication model of social network with user cooperation and leadership encouragement. Telecommun. Sci. **36**(10), 172–182 (2020)

10. Yada, K., Motoda, H., Washio, T., Miyawaki, A.: Consumer behavior analysis by graph mining technique. New Math. Natural Comput. **2**(01), 59–68 (2006)

11. Ma, S., Liu, J., Zuo, X.: Survey on graph neural network. J. Comput. Res. Develop. **59**(01), 47–80 (2022)

12. Zhan, W., Guan, J., Zhang, Z.: Advance in the research of complex network: model and application. J. Chin. Comput. Syst. **32**(2), 193–202 (2011)

13. Watts, D.J., Strogatz, S.H.: Collective dynamics of 'small-world' networks. Nature **393**, 440–442 (1998)

14. Barabási, A.L., Albert, R.: Emergence of scaling in random networks. Science **286**, 509–512 (1999)

15. Li, X., Chen, G.R.: A local-world evolving network model. Physica A: Stat. Mech. Appl. **328**(1–2), 274–286 (2003). https://doi.org/10.1016/s0378-4371(03)00604-6]

16. Yook, S.H., Jeong, H., Barabasi, A.L., et al.: Weighted evolving networks. Phys. Rev. Lett. **86**(25), 5835–5838 (2001). https://doi.org/10.1103/PhysRevLett.86.5835]

17. Wu, Z., Chen, Y.: Link prediction using matrix factorization with bagging. In: 2016 IEEE/ACIS 15th International Conference on Computer and Information Science (ICIS). IEEE (2016)

18. Koene, J.: Applied network analysis : a methodological introduction. North-Holland **17**(3), 422–423 (1984)

19. Chen, D.B, Lul, Y., Shang, M.S., et al.: Identifying influential nodes in complex networks. Physica A: Stat. Mech. Appl. **391**(4), 1777–1787 (2012)

20. Freeman, L.C.: Centrality in social networks conceptual clarification. Soc. Netw. **1**(3), 215–239 (1978–1979)

21. Wang, J., Ai, J., Yang. Y., et al.: Identifying key classes of object-oriented software based on software complex network. In: International Conference on System Reliability & Safety, pp. 444–449. IEEE (2017)

22. Newman Me, J.: A measure of betweenness centrality based on random walks. Soc. Netw. **27**(1), 39–54 (2005)

23. Kitsak, M., Gallos, L.K., Havlin, S., et al.: Identification of influential spreaders in complex networks. Nat. Phys. **6**(11), 888–893 (2010)

24. Kleinberg, J.M.: Authoritative sources in a hyperlinked environment. J. ACM **46**(5), 604–632 (1999)

25. Brin, S., Page, L.: The anatomy of a large-scale hypertextual web search engine. Comput. Netw. **56**(18), 3825–3833 (2012)

26. Opsahl, T., Agneessens, F., Skvoretz, J.: Node centrality in weighted networks: generalizing degree and shortest paths. Soc. Netw. **32**(3), 245–251 (2010)

27. Osman, M.H., Chaudron, M.R.V., Putten, P.V.D.: An analysis of machine learning algorithms for condensing reverse engineered class diagrams. In: Proceedings of the 2013 IEEE International Conference on Software Maintenance (ICSM 2013), Eindhoven, The Netherlands, pp. 140–149 (2013)

28. Li, Q., Zhou, T., LüL, C.D.: Identifying influential spreaders by weighted LeaderRank. Phys A **404**, 47–55 (2014)

29. Yin, L., Deng, Y.: Toward uncertainty of weighted networks: an entropy-based model. Physica A **508**, 176–186 (2018)

30. Valverde, S., Cancho, R.F., Solé, R.V.: Scale free networks from optimal design. Europhys. Lett. **60**(4), 512–517 (2002)

31. Ding, Y.: Research on measurement method in open software ecosystem based on complex network. Wuhan University (2017)

32. Pan, W., Li, B., Ma, Y., et al.: Identifying the key packages using weighted PageRank algorithem. Acta Electronica Sinica **42**(11), 2174–2183 (2014)
33. Singh, S., Jha, R.K.: A survey on software defined networking: architecture for next generation network. J. Netw. Syst. Manage. **25**(2), 321–374 (2017)
34. Srinivasan, S.M., Sangwan, R.S., Neill, C.J.: On the measures for ranking software components. Innovations Syst. Softw. Eng. **13**, 161–175 (2017)
35. Pan, W., Ming, H., Chang, C.K., Yang, Z., Kim, D.-K.: ElementRank: ranking Java Software classes and packages using multilayer complex network-based approach. IEEE Trans. Software Eng. (2019). https://doi.org/10.1109/TSE.2019.2946357
36. Pan, W., Song, B., Li, K., Zhang, K.: Identifying key classes in object-oriented software using generalized k-core decomposition. Futur. Gener. Comput. Syst. **81**, 188–202 (2018)
37. Zhang, J., Song, K., He, P., Li, B.: Identification of key classes in software systems based on graph neural networks. Comput. Sci. **48**(12), 149–158 (2021)
38. Ma, Y., Cheng, G., Liang, X., Li, Y., Yang, Y., Liu, Z.: Improved SDNE in weighted directed network. Comput. Sci. **47**(04), 233–237 (2020)
39. Perozzi, B., Al-Rfou, R., Skiena, S.: Deepwalk: online learning of social representations. In: Proceedings of the 20th ACM SIGKDD International Conference on Knowledge Discovery and Data Mining, pp. 701–710 (2014)
40. Grover, A., Leskovec, J.: node2vec: scalable feature learning for networks. In: Proceedings of the 22nd ACM SIGKDD International Conference on Knowledge Discovery and Data Mining, pp. 855–864 (2016)
41. Figueiredo, D.R., Ribeiro, L.F.R., Saverese, P.H.P.: struc2vec: learning node representations from structural identity. In: Proceedings of the 23rd ACM SIGKDD International Conference on Knowledge Discovery and Data Mining, Halifax, NS, Canada, pp. 13–17 (2017)
42. Kipf, T.N., Welling, M.: Semi-supervised classification with graph convolutional networks. In: Proceedings of ICLR (2017)
43. Velikovi, P., Cucurull, G., Casanovam A., et al.: Graph Attention Networks (2017)
44. Kipf, T.N., Welling, M.: Variational graph auto-encoders. In: NIPS Workshop on Bayesian Deep Learning (2016)
45. Hamilton, W., Ying, Z., Leskovec, J.: Inductive representation learning on large graphs. In: Proceedings of NIPS, pp. 1024–1034 (2017)
46. Fan, C., Zeng, L., Ding, Y., et al.: Learning to identify high betweenness centrality nodes from scratch: a novel graph neural network approach. arXiv:1905.10418v1 (2019)

Spatio-Temporal Fusion Based Low-Loss Video Compression Algorithm for UAVs with Limited Processing Capability

Qianyuan Zhang, Desheng Wan, Hao Chen, Lianghua Cheng, Jiayi Chen, and Chaocan Xiang[✉]

College of Computer Science, Chongqing University, Chongqing 400044, China
xiangchaocan@cqu.edu.cn

Abstract. Real-time urban crowd surveillance is essential for riot supervision, epidemic prevention, and urban emergency management. Unmanned aerial vehicles (UAVs) provide a promising way for real-time crowd surveillance due to their convenient deployment and flexible mobility. However, the limited wireless transmission bandwidth and the large capacity of high-definition video pose great challenges to the real-time transmission of UAV-captured videos. Although existing edge computing-based video compression algorithms can partially solve this dilemma, the complexity of these algorithms makes them inapplicable for edge devices with limited processing capacity. To this end, we propose a lightweight spatiotemporal fusion based low-loss video compression algorithm, which consists of two parts: feature clustering-based temporal sampling and dynamic encoding-based spatial sampling. The first module clips the video content from a temporal perspective by identifying inter-frame redundancy. The second module compresses the video content from a spatial perspective by examining regions of interest (RoIs) within each frame and utilizing background filtering to analyze intra-frame encoding. This lightweight algorithm effectively reduces the size of the video file while maintaining high-quality output, which is compatible with edge devices' constrained process power. The experimental results demonstrate that the proposed algorithm maintains minimal loss in crowd detection accuracy while reducing transmission latency by 31.3%.

Keywords: Urban crowd surveillance · UAVs · Constrained processing capabilities · Low-loss video compression · Spatio-temporal fusion

1 Introduction

Urbanization has led to a rapid increase in population density, increasing the risk of stampedes and riots in urban areas, which poses a serious threat to public

This work is supported by the National Natural Science Foundation of China (62172063).

safety [10]. Therefore, crowd surveillance gains paramount importance in urban safety emergency management. Currently, crowd surveillance mainly relies on fixed cameras deployed in cities [19], but they suffer from blind spots, poor mobility, as well as high deployment and maintenance costs [6]. Other crowd surveillance methods based on WiFi or millimeter-wave radar have also been studied widely. However, they have low accuracy in detecting mobile crowds and limited detection range, making them insufficient to use in practical scenarios.

Fortunately, Unmanned aerial vehicles (UAVs) have become increasingly popular in civil and commercial applications due to their low cost, flexible mobility, and wide range of video capturing capabilities, which provides a promising way for large-scale urban crowd surveillance [24,28]. Although some works have studied UAV-based crowd surveillance systems, they either improve high-precision detection networks that require high-resolution videos, leading to high latency [7–9,33], or only consider real-time performance and use lossy compression on the video, resulting in reduced accuracy [27,32]. Relevant experiments are also conducted to verify the problems:

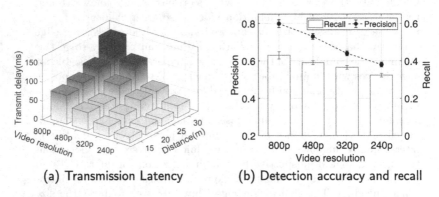

(a) Transmission Latency   (b) Detection accuracy and recall

Fig. 1. Analysis of UAV transmission latency and detection accuracy at different resolutions and transmission distances.

i) High-precision drone surveillance systems are difficult to achieve real-time requirements. As shown in Fig. 1a, When transmitting an 800p high-definition resolution 10-minute video from a drone to a ground station, only 30m away, the latency exceeds 30 min, causing severe lag in crowd surveillance information and making it unable to ensure the real-time safety of urban crowds.

ii) Compressed video leads to a drop in crowd detection accuracy. As illustrated in Fig. 1b, when the transmission distance is fixed, compressing the video frames from 800p to 320p reduces the transmission latency by 65.7%. However, the compression also leads to a sharp drop in detection accuracy from 80% to 64%, making it incapable of meeting the high precision requirements of crowd target detection.

Table 1. Comparisons of edge-computing devices [1].

		Lightweight devices			High-performance devices	
Devices		RP3	RP3+ICS	JetsonNano	OrinNX	AGXOrin
AI Performance		-	-	0.472TFLOPs	70TOPS	200TOPS
Models (fps)	InveptionV4	-	-	11	593	1337
	TinyYoloV3	0.5	-	25	1156	2611
	Unet	-	5	-	183	387
	DashcamNet	-	-	11	689	1482
Weight (kg)		0.1	0.2	0.241	0.76	1.5
Power (watt)		2	2	5	20	40

To address the above issues, UAV-mounted lightweight edge devices is utilized to extract effective crowd information from drone-captured video and remove redundant content to achieve low transmission latency and high detection accuracy in drone-based crowd surveillance. However, achieving this goal confronts the following two challenges:

- **Designing lightweight algorithms adapted to edge devices with limited processing capability is challenging.** As depicted in Table. 1, high-performance edge devices have large weight and high power consumption, while the drone (such as S500 [4]) has a mass of only 1.3 kg and a power consumption of 50 W. If equipped with AGXOrin (1.5 kg, 40 W), the UAV's endurance will be reduced by 75% at least. However, the processing power of lightweight devices (e.g., RP3 and JetsonNano) is severely constrained. Therefore, designing lightweight video processing algorithms for limited capabilities is challenging.
- **It is hard to extract effective crowd information and filter out invalid content in drone-captured videos.** Drone footages contain redundant information, including inter-frame repeated content and intra-frame background regions. However, conventional methods for lightweight video compression either fail to utilize image and crowd features related to crowd surveillance tasks, leading to compressed crowd areas inside frame and a decline in detection accuracy [19,29], or only clip between frames or compress inside frames, resulting in a still relatively large processed video volume [22,23]. Hence, efficiently extracting crowd information while filtering out invalid content to minimize latency and achieve high-precision crowd surveillance is a huge challenge.

To address the above challenges, this paper proposes the **Spatiotemporal Fusion based Low-loss Video Compression Algorithm** tailored to edge devices with limited computational capabilities. Specifically, the algorithm consists of two modules: 1) **Feature Clustering based Temporal Video Sampling module.** The crowd and scene features are initially extracted from the UAV-captured videos to comprehensively characterize the frames' information relevant to the crowd detection task. Then, using a feature similarity-based clustering algorithm with a sliding window, we clip the redundant video frames and retain keyframes, minimizing the inter-frame similarities. **2) Dynamic Encoding based Spatial Video Sampling module.** For the temporally sampled keyframes, the regions of interest (RoIs) are extracted for video encoding based on lightweight background filtering and target detection in the spatial domain. Subsequently, dynamic video encoding is executed to maximize the reduction of redundant non-RoIs data while maintaining the visual quality of the crowd region. As a result, our algorithm retains only useful data related to crowd surveillance in intra-frames and inter-frames, while irrelevant information is excluded, which satisfies the capability limitations of UAV-end lightweight devices and ensures the minimum video transmission delay and loss rate of target detection.

In summary, this paper makes the following contributions:

1. We present an intelligent UAV-based edge computing system to realize low-latency and high-accuracy crowd surveillance through intelligent collaborative processing between UAVs and ground stations at the edge side.
2. We propose a lightweight spatio-temporal fusion based low-loss video compression algorithm for computing capacity limitation of the UAV end, using crowd and scene features and edge detection information to compress surveillance video from the temporal and spatial aspects to ensure the minimization of video transmission latency and loss rate of target detection.
3. We build the system prototype and give an extensive performance evaluation. The experimental results show that the proposed algorithm can reduce the transmission delay by 31.3% while losing only within 2% of the crowd-counting accuracy.

The rest of the paper is organized as follows. First, the system architecture is presented in Sect. 2. Then we design the spatio-temporal fusion based low-loss video compression algorithm in Sect. 3. In Sect. 4, we conduct extensive experiment evaluations, while implementing the system prototype for real-world application in Sect. 5. Finally, we review related work in Sect. 6 and conclude this paper in Sect. 7.

2   System Overview

In this section, we present an overview of the intelligent UAV-based edge computing system for real-time surveillance of urban large-scale crowd gathering. As shown in Fig. 2, it consists of the following two components:

- **UAV module.** This module includes drones and the mounted lightweight edge processing devices, such as Jetson Nano [2]. It performs mobile surveillance and video capturing of crowd areas. The captured videos are then compressed by the lightweight edge device with spatiotemporal low-loss video compression techniques and transmitted to the ground control station in real-time.

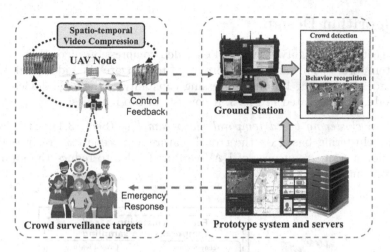

Fig. 2. System architecture.

- **Ground Control Station module.** This module comprises a ground control station (GCS) and a user interface for crowd supervision, responsible for controlling the UAV's flight trajectory and processing the compressed videos from the UAV node for the relevant crowd analysis applications. The analyzed results are finally transmitted to the cloud server for data storage and aggregation.

The workflow of the system is as follows: i) Supervisors deploy the ground control station in crowded areas (e.g., commercial plazas and tourist spots, etc.) and set drones' patrol trajectories. ii) The drones capture videos of the gathering crowds while the mounted edge device processes the videos with the spatio-temporal fusion based video compression method proposed in this paper and transmits them to the ground control station with low latency and high precision. iii) The ground control station receives the UAV-captured videos for crowd surveillance applications such as crowd counting [20] and social distancing analysis [24]. Meanwhile, the crowd analysis results are uploaded to the cloud server for data visualization. Besides, the ground control station feedbacks the crowd analysis results (e.g., crowd detection counts, proposed bounding boxes, confidence level, tracking trajectory, etc.) to the UAV node through wireless networks so that the UAV node can refine the detection results for more accurate spatio-temporal video compression.

The key of the intelligent UAV-based edge computing system is the spatio-temporal fusion based low-loss video compression algorithm on the UAV node, satisfying the computational constraints of lightweight edge devices while minimizing the video transmission latency under a specific loss rate of target detection. Hence, the algorithm design will be specified in Sect. 3.

3   Algorithm Design

In this section, the lightweight low-loss video compression algorithm based on spatio-temporal fusion sampling is proposed to mitigate the video transmission delay and adapt to the limited processing capability of the edge devices at the UAV node. As illustrated in Fig. 3, it consists of two modules:

1) *Feature clustering based temporal video sampling* (Sect. 3.1): Utilizing the frame clustering based on the crowd feature and scene feature, we clip the redundant frame sequences of UAV-captured video to reduce the number of video frames.

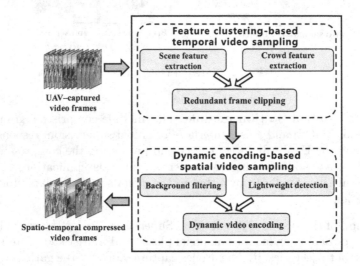

Fig. 3. Low-loss video compression framework based on spatio-temporal sampling fusion.

2) *Dynamic encoding based spatial video sampling* (Sect. 3.2): To further reduce the inefficient information inside a frame, we distinguish regions of interest (RoIs) of the frame and dynamically encode the frame according to the division of the encoding RoIs.

3.1   Feature Clustering Based Temporal Video Sampling

This module clips the drone-captured video based on the feature similarity between frames to significantly remove the redundant frames without affecting the accuracy of crowd detection. As depicted in Fig. 4, the crowd and scene features of video frames are first extracted, and then feature clustering and redundant frames are cut according to the feature similarity of the video frames. Finally, the number of redundant frames in the output video frames is significantly reduced compared with the original input video.

Table 2. Frequently used notations.

Notations	Descriptions
I_k	The k_{th} frame
N_k	The number of bounding box of k_{th} frame
d_i^k	The center coordinate of i_{th} bounding box in k_{th} frame
c_i^k	The confidence levels of i_{th} bounding box in k_{th} frame
$\boldsymbol{m}_k$	The crowd density map of k_{th} frame
$\boldsymbol{P}_k$	The crowd feature of k_{th} frame
$\boldsymbol{F}_k, \boldsymbol{L}_k, \boldsymbol{Y}_k$	The frequency domain, luminance and contrast feature of k_{th} frame
$\boldsymbol{H}_k^R, \boldsymbol{H}_k^G, \boldsymbol{H}_k^B$	The distribution ratio for R, G, B channels of k_{th} frame
$\boldsymbol{S}_k$	The scene feature of k_{th} frame
$\mathcal{V}_{orig}, \mathcal{V}_{out}$	The input and output frames of temporal video sampling
I_{orig}, I_{out}	The input and output frame of spatial video sampling
X_{k_1, k_2}	The spectral value of two-dimensional discrete cosine transform
$\mathcal{R}_{edge}, \mathcal{R}_{det}$	The output RoIs of the background filtering and lightweight detection

Crowd Feature Extraction. Considering the detection speed and accuracy, we utilize a lightweight network model to extract the crowd features of the image (e.g., the number of people and their location distribution, etc.) as the basis for the similarity determination of the temporal sampling.

Specifically, to adapt to the extremely limited computational power of UAV edge devices (e.g., Jetson Nano), the lightweight network yolov4-tiny [5] is used for crowd detection of the UAV-captured videos. The number of bounding boxes of k_{th} image I_k is denoted as N_k, and the center coordinates and the confidence levels of the i_{th} bounding box in the frame are denoted as d_i^k and c_i^k, and $c_i^k \in (0, 1]$. Then a Gaussian kernel is applied to convolve with the above crowd detection results to obtain the crowd density map $\boldsymbol{m}_k$ of the k_{th} frame as follows:

$$\boldsymbol{m}_k = \sum_{i=1}^{N} c_i^k \cdot \delta(x - d_i^k) * G_{\delta_i}(x) \tag{1}$$

where $G_{\delta_i}(x)$ is the two-dimensional Gaussian kernel and $\delta(x - d_i^k)$ is the impulse function. Finally, to alleviate the computation burden of the lightweight edge device, the crowd density map $\boldsymbol{m}_k$ is uniformly pooled to obtain the compressed crowd density map $\boldsymbol{P}_k$ with reduced image size. $\boldsymbol{P}_k$ is then used as the crowd feature of that frame. The frequently used notations are shown in Table 2.

Scene Feature Extraction. We utilize the image's frequency-domain and structural information as its scene features to determine the variations in the video background during crowd surveillance. When extracting scene features, the frame's low-frequency information is retained to depict the scene contours of UAV-captured videos. In particular, we apply two-dimensional Discrete Cosine Transform (2D-DCT) [13] for each frame according to Eq. (2) and retain the low-frequency region where the spectral energy is concentrated as the low-frequency spectrum.

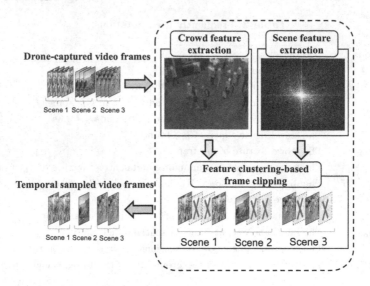

Fig. 4. Temporal video sampling.

$$X_{k_1,k_2} = a_{k_1} a_{k_2} \sum_{n_1=0}^{N_1-1} \sum_{n_2=0}^{N_2-1} x_{n_1,n_2} \cos\left[k_1 \frac{\pi}{N_1}\left(n_1 + \frac{1}{2}\right)\right]$$
$$\cos\left[k_2 \frac{\pi}{N_2}\left(n_2 + \frac{1}{2}\right)\right] \tag{2}$$

where N_1 and N_2 denote the height and width of the original frame, respectively, and k_1 and k_2 are the frequency domain coordinates of the 2D-DCT and satisfy $0 \le k_1 \le N_1 - 1$ and $0 \le k_2 \le N_2 - 1$. When $k_1 = 0$, the coefficient $a_{k_1} = 1/\sqrt{N_1}$; when $1 \le k_1 \le N_1 - 1$, $a_{k_1} = \sqrt{2/N_1}$. When $k_2 = 0$, the coefficient $a_{k_2} = 1/\sqrt{N_2}$; when $1 \le k_2 \le N_2 - 1$, $a_{k_2} = \sqrt{2/N_2}$. According to whether the spectral DCT value is greater than the mean value, each DCT value is quantized to 0 or 1 (i.e., above the DCT mean value is assigned as 1; otherwise, it is assigned as 0). Finally, the quantized result is taken as the frequency domain feature F_k of the k_{th} frame.

Furthermore, the structural information is extracted to capture the structure changes in the scene of the continuous UAV-captured frames. Specifically, the k_{th} frame is uniformly divided into rectangular blocks, and the variance and mean of grayscale values of each block are calculated as the local luminance and contrast. Then the local luminance and contrast values of all blocks are concatenated to form the luminance feature L_k and contrast feature Y_k. Next, the pixel distribution ratios of different grayscale levels for the R, G, B channels of the k_{th} frame is denoted as H_k^R, H_k^G, H_k^B, which serve as the color distribution features. Finally, the frequency-domain, luminance, contrast, and color distribution features of the k_{th} frame are combined as the scene feature S_k.

Feature Clustering Based Frame Clipping. The image's crowd and scene features are jointly employed as the combined features, and feature similarity-based clustering is used to extract keyframes and remove redundant ones. The temporal video sampling process is demonstrated in Algorithm 1, mainly involving the following two steps:

Step 1: Calculation of inter-frame similarity based on sliding window. Since video frames captured by UAV have similarity within a continuous time range , we use a sliding window to cover consecutive video frames and calculate the cosine similarity as the inter-frame similarity based on the combined features for all frames within the window coverage, according to Eq. (3).

$$\text{Similarity}(X, Y) = \frac{X \cdot Y}{\|X\| \times \|Y\|} \tag{3}$$

Step 2: Video frame clustering and redundant frame removal. Based on the inter-frame similarity within the sliding window, the DBSCAN algorithm [16] is applied to feature-based clustering of video frames. Then, based on the clipping threshold, we determine whether to remove each cluster of video frames in the clustering result. Finally, the sliding window is moved in steps, and the feature clustering and redundant frame removal process continues until the sliding window traverses all input frames.

Algorithm 1: Feature Clustering based Temporal Video Sampling.

Input: UAV-captured continuous video sequence with N frames
$\qquad \mathcal{V}_{orig} = \{I_1, \cdots, I_N\}$
Output: Sampled video sequence with K frames
$\qquad \mathcal{V}_{out} = \{I_i, \cdots, I_j\}, 1 \leq i, j \leq N.$
1   Initialize: Clipping threshold $\varepsilon = \varepsilon_0$; Sliding window length $L_S = L_{S_0}$; Sliding
    step $d = \lfloor L_S/2 \rfloor$; $\mathcal{V}_{out} = \varnothing$.
2   **for** $i \leftarrow 0$ **to** $\lfloor (N - d)/d \rfloor - 1$ **do**
3   **for** $j \leftarrow 1$ **to** L_S **do**
4   **for** $k \leftarrow 1$ **to** L_S **do**
5   Calculate the inter-frame similarity $sim_{i+j,i+k}$ according to Eq. (3)
6   **end**
7   **end**
8      $Cls = \text{DBSCANcluster}(\{sim_{i+1,i+1}, \cdots, sim_{i+L_S,i+L_S}\}, \mathcal{V}_{orig})$
9      $\mathcal{V}_{keep} = \text{FrameCut}(Cls, \varepsilon)$
10     $\mathcal{V}_{out} = \mathcal{V}_{out} \cup \mathcal{V}_{keep}$
11 **end**
12 **return** $\mathcal{V}_{out}$.

3.2 Dynamic Encoding Based Spatial Video Sampling

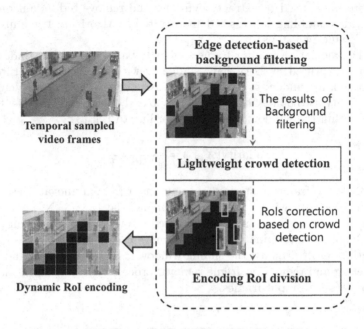

Fig. 5. Spatial video sampling.

Although the duplicate content between frames is filtered out via temporal video sampling, each frame still contains many inefficient background regions (such as streets, sky, and other non-crowd areas in crowd surveillance), occupying a large file volume. Therefore, we design the spatial video sampling module based on dynamic encoding to filter out the non-regions of interest (non-RoIs) while maintaining the visual quality of the crowd areas. As shown in Fig. 5, the background regions are filtered out with edge detection and then lightweight crowd detection results are utilized to correct the misclassification of crowd regions as non-RoIs in the edge detection, thus obtaining an accurate division of encoding RoIs. Finally, the frames are dynamically encoded according to the RoI division, i.e., the task-related RoIs are losslessly encoded, while the non-RoIs are lossy encoded.

Background Filtering Based on Edge Detection. Compared to the texture and contours in the crowd regions, the edge texture information in the background regions of images is relatively simple. Hence, we utilize the edge texture information of different objects to distinguish and filter the background regions in the image. Lightweight edge detection utilizes the Prewitt operator [13] (Eqs. (4) and (5)). Specifically, for the input image I_k, the horizontal and vertical gradients at pixel (x, y) is calculated according to Eqs. (6) and (7):

$$d_x = \begin{bmatrix} 1 & 1 & 1 \\ 0 & 0 & 0 \\ -1 & -1 & -1 \end{bmatrix} \tag{4}$$

$$d_y = \begin{bmatrix} -1 & 0 & 1 \\ -1 & 0 & 1 \\ -1 & 0 & 1 \end{bmatrix} \tag{5}$$

$$G_x = d_x * I_k \tag{6}$$

$$G_y = d_y * I_k \tag{7}$$

where $*$ denotes the 2D convolution operator. Then each element's value $g_{xy}(i,j)$ of the gradient magnitude matrix G_{xy} is calculated as follows:

$$g_{xy}(i,j) = \sqrt{g_x(i,j)^2 + g_y(i,j)^2} \tag{8}$$

where $g_x(i,j)$ and $g_y(i,j)$ represent the values of the horizontal and vertical gradient matrices at position (i,j). If $g_{xy}(i,j) < \lambda_{edge} \cdot \overline{g_{xy}}(i,j)$, the pixel at position (x,y) is labeled as RoI, otherwise it is labeled as non-RoI. The final output RoIs is defined as $\mathcal{R}_{edge} = \{(x,y)|g_{xy}(i,j) \geq \lambda_{edge} \cdot \overline{g_{xy}}(i,j)\}$.

Moreover, to adapt to the limited computation capabilities of the drone nodes, this lightweight edge detector has a time complexity of only $O(HW \cdot \log(HW))$ for edge detection with the image size of $H \times W$, equivalent to a single convolutional layer in a neural network. The algorithm's execution time on actual lightweight edge devices (i.e., Jetson Nano) is less than $3\,\mathrm{ms}$ for 800p frame. Hence, this algorithm incurs minimal computational overhead and negligible processing latency at the edge node.

Lightweight RoI Correction. Since background filtering based on edge detection relies solely on the intensity of edge textures to distinguish background regions without considering the crowd distribution in the image, it is a common issue of misclassifying crowd areas as background regions [21]. To reduce the computational workload on lightweight edge devices, the crowd detection results in the temporal sampling process is reused to reclassify crowd areas erroneously labeled as background into non-background regions. For the crowd density map m_k obtained in temporal sampling, if an element value in m_k is greater than the maximum value of the detection threshold e_0 and the mean value of m_k, it is designated as RoI; otherwise, it is labeled as non-RoI. The final output RoIs of the lightweight detection is $\mathcal{R}_{det} = \{(x,y)|m_k(x,y) \geq \max\{e_0, \overline{m_k}\}\}$.

The RoIs from lightweight crowd detection are used to complement and revise the background filtering results, thus the final RoI division is defined as $\mathcal{R}_f = \mathcal{R}_{edge} \cup \mathcal{R}_{det}$.

Dynamic Quality Encoding. Upon obtaining the final partition of the video frames into RoIs, we utilize a video encoder for dynamic quality encoding of the frames. The encoding quality map for each frame is dynamically computed based on the RoI partition of the frame, defining the encoding quality for each block within the frame (i.e., RoIs are set to lossless encoding, while non-RoIs are set to lossy encoding). Then, based on the encoding quality map of each frame, the video encoder performs corresponding lossless or lossy encoding for each image block.

Algorithm 2: Dynamic Encoding based Spatial Video Sampling.

Input: Coloured video frame I_{orig} with pixel size $m \times m$
Output: Spatially sampled video frame I_{out}; Encoding RoI set $\mathcal{R}$.

1 Initialize: Edge detection threshold $\varepsilon_d = \varepsilon_{d_0}$; $\mathcal{R} = \varnothing$; $I_{\text{out}} = I_{\text{orig}}$.
2 $\mathcal{B} = \text{Division}(I_{\text{orig}})$
3 **for** *box* in $\mathcal{B}$ **do**
4 | $\mathcal{R}_{edge} = \text{EdgeDetect}(box, \varepsilon_d)$
5 | $\mathcal{R}_{det} = \text{CrowdDetect}(box)$
6 | **if** $|\mathcal{R}_{edge}| > 0$ *or* $|\mathcal{R}_{det}| > 0$ **then**
7 | | $\mathcal{R} = \mathcal{R} \cup \{\mathcal{R}_{edge} \cup \mathcal{R}_{det}\}$
8 | | $I_{\text{out}} = \text{HighQualityEncode}(I_{\text{out}}, box)$ /* lossless encoding*/
9 | **else**
10 | | $I_{\text{out}} = \text{LossyQualityEncode}(I_{\text{out}}, box)$ /* lossy encoding*/
11 | **end**
12 **end**
13 **return** I_{out} and $\mathcal{R}$.

The spatial video sampling algorithm based on dynamic encoding is illustrated in Algorithm 2. Its input is $m \times m$ colored video frame I_{orig}. Line 1 initializes the parameters. Then we obtain the results of RoI division (line 2). In line 3–12, the edge detection, lightweight crowd detection and dynamic video encoding are executed. At last, the algorithm outputs the dynamically encoded video frame I_{out} with the RoI set $\mathcal{R}$.

To sum up, by jointly incorporating the results of edge filtering and lightweight crowd detection, dynamic encoding-based spatial video sampling achieves lossless encoding for crowd areas while applying lossy compression to the background regions of non-surveillance targets. This module can further reduce video transmission volume without compromising crowd surveillance accuracy.

4   Evaluation

4.1   Experimental Setup

Datasets Description. The drone crowd surveillance datasets used in this paper include The Oxford Town Centre (OTC) [14], Group-detection (GD) [26], the self-labeled drone-captured Multi-scenario Crowd Dataset (MCD), Drone

Vision Challenge dataset VisDrone2019 [12]. All the datasets are compatible with the crowd detection tasks and contain over 20,000 UAV-captured crowd-gathering video frames with resolutions up to 1280×720 and above, including multiple scenarios such as crowd riots, large-scale gatherings and city street parades.

Experimental Methodology and Settings. Jetson Nano is deployed as lightweight edge processing device on the UAV side and a GPU-equipped laptop is used as a ground base station for data communication via wireless wifi. Deep network inference and video processing is based on TenorRT [3] and FFmpeg. Yolov5 is trained as crowd detection network for GCS with VisDrone2019 dataset on the training server equipped with NVIDIA 3090 GPU.

To show the effectiveness of our algorithm, we evaluate the overall latency, detection accuracy and recall at different crowd scene density and image resolution. The communication distance between the UAV end and the GCS end is set to 30m. Besides, to simulate the scene switching of the real drone-captured video, the video frames of different scenes are mixed and sampled to form the testing set.

Baseline Methods. To comprehensively evaluate the performance of edge computing systems, four typical comparison methods with different performances are used in this paper:

- **Un-CVP**: Uncompressed Video Processing method directly transmits the whole UAV-captured video frames to GCS without any compressing. Thus, this method achieves the highest detection accuracy but incurs the largest overall latency.
- **CVP** [19]: Compressed Video Processing method processes all the UAV-captured video frames with lossy compression.
- **CFRC** [30]: Clipped Fixed-RoI Compressing method first clips the UAV-captured video with structural similarity and then compresses frames with fixed RoIs.
- **REMIX** [18]: This algorithm compresses video frames based on adaptive image region segmentation in different compression qualities according to the crowd density of each region.

Furthermore, to evaluate the performance of each module of our algorithm, the following two comparison algorithms are used:

- **TSO**: Temporal Sampling Only algorithm compresses the UAV-captured videos via the temporal sampling module in this paper.
- **SSO**: Spatial Sampling Only algorithm compresses the UAV-captured videos by the spatial sampling module in this paper.

Evaluation Metrics. For the experimental studies, we adopt three metrics to evaluate the algorithm performance: 1) **Average Processing Time(APT)**, which is the sum of the average processing delay of the UAV-captured video at the edge of the UAV and the average transmission delay of the compressed video to the ground station per frame; 2) **Precision(P)**, which refers to the proportion of the detection results in the relevant categories to the total returned results, as in Eq. (9); 3) **Recall(R)**, the proportion of relevant categories to the total relevant categories in the detection results, as in Eq. (10).

$$P = \frac{|\mathcal{T}_P|}{|\mathcal{T}_E|} \tag{9}$$

$$R = \frac{|\mathcal{T}_P|}{|\mathcal{T}_G|} \tag{10}$$

where $\mathcal{T}_P$, $\mathcal{T}_E$ and $\mathcal{T}_G$ denote the sets of the correct detection results, the whole detection results, and the ground-truth, respectively.

4.2   Experimental Results

Algorithm Performance Evaluation. First, we assess the impact of different video resolutions on crowd detection performance and system processing latency. As depicted in Fig. 6a, the video frame resolution gradually decreases from 800p to 240p, and the experiment results show that our algorithm achieves 98.6% detection accuracy and 95.7% recall of the Un-CVP method. Compared to our algorithm, the detection accuracy of the REMIX, CFRC, and CVP methods is, on average, reduced by 2.2%, 17.2%, and 23.2%, respectively. This significant difference in accuracy demonstrates that only our algorithm and REMIX consider crowd information during video processing, ensuring crowd detection accuracy and recall. Moreover, as shown in Fig. 6b, our algorithm reduces the average APT of the Un-CVP and the REMIX methods by 31.3% and 33.25%, respectively.

Next, we evaluate the impact of crowd-gathering scenes in different densities on crowd-detection performance and system processing latency. As shown in

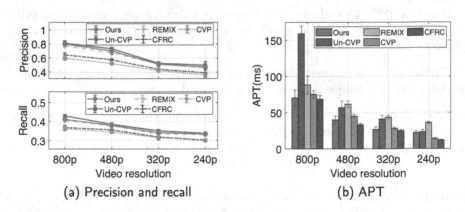

(a) Precision and recall   (b) APT

Fig. 6. Evaluation of crowd detection performance with different video resolutions, in terms of precision, recall, and latency.

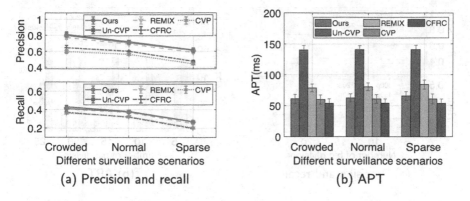

(a) Precision and recall   (b) APT

Fig. 7. Evaluation of crowd detection performance in different crowd density scenarios, in terms of precision, recall, and latency.

Fig. 7a, when the resolution of different crowd density scenes is set to 800p, our algorithm improves the accuracy by 23.6% and 32.6%, and the recall by 21.6% and 20.5% compared to the CVP and CFRC, while achieving crowd detection accuracy and recall close to the Un-CVP. Furthermore, as shown in Fig. 7b, compared with low-density scenes, our algorithm and the REMIX increase the APT by an average of 5.1% and 4.6%, respectively, in medium and high-density scenes due to the increased delay in processing crowd features. In all crowd density scenes, our algorithm reduces average processing latency by 55.2% and 22.3% compared to the Un-CVP and REMIX methods, respectively, with an average reduction of 77.6 ms and 18.0 ms.

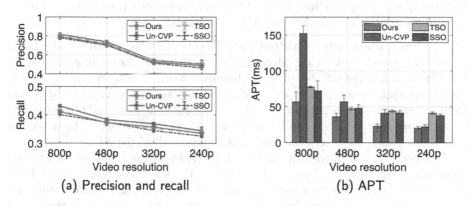

(a) Precision and recall   (b) APT

Fig. 8. Evaluation of crowd detection performance with different algorithm modules, in terms of precision, recall, and latency.

Algorithm's Module Evaluations via Ablation Studies. We conduct ablation experiments to evaluate the effectiveness of each module in our algorithm.

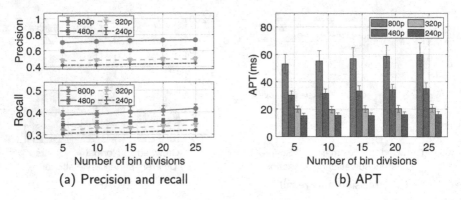

Fig. 9. Evaluation of crowd detection performance with different numbers of image divisions, in terms of precision, recall, and latency.

As shown in Fig. 8a, the accuracy of our algorithm, the TSO, and SSO reached 96.7%, 98.6%, and 94.3% of the Un-CVP method, and the recalls reached 96.7%, 96.6%, and 94.6% of the Un-CVP method, respectively. Then, the experimental results of system processing delay, as shown in Fig. 8b, illustrate that our algorithm reduces the average APT of the Un-CVP method by 38.2%. The delay of the TSO and SSO algorithms at higher resolutions (800p and 480p) is reduced by an average of 34.3% and 33.5%, compared to the Un-CVP, while at lower resolutions (320p and 240p), the delay is increased by 36.3% and 47.2%, respectively. This difference in delay can be attributed to the fact that both the TSO and SSO use neural networks, resulting in inherent processing delay that cannot be effectively reduced at lower resolutions. On the other hand, our algorithm's temporal and spatial sampling modules enable the neural network's time consumption to be evenly distributed, reducing the APT remarkably.

Algorithm's Parameter Evaluations. We evaluate the impact of the block division parameter *bin* on the algorithm's performance. As depicted in Fig. 9a and Fig. 9b, when the number of image block division *bin* increases from 5 to 25, the experimental results show that with the increase of the bin, the detection accuracy and recall are improved by 2.6% and 2.1%, and the APT is increased by 3.4 ms on average. As a result, increasing the *bin* can improve the detection performance to some extent but also lead to a growth of the APT.

Additionally, our algorithm is tested on video frames of four large-scale gatherings scenarios. As illustrated in Fig. 10, the scenes in column 1 to 4 are crowded city streets, election rallies, square clusters, and mass crowd gatherings of marathons, respectively. The detection results demonstrate that the crowd detection performance of compressed images with the CVP method significantly decreases, while the detection performance of video frames processed by our algorithm is similar to the uncompressed video frames (Un-CVP) in all scenarios.

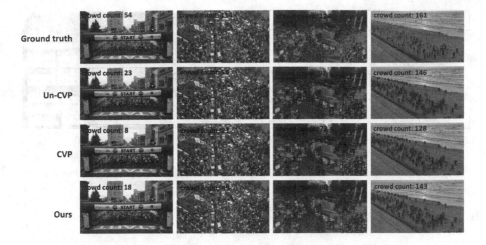

Fig. 10. Illustration of crowd detection results in large-scale gatherings scenarios.

In summary, the above experimental results further validate the effectiveness of our algorithm for real-time high-precision video compression in crowd surveillance. In comparison with baseline methods, our algorithm achieves remarkable improvement in reducing video transmission latency by 31.3% while maintaining the detection accuracy loss within 2%. Also, the performance of our algorithm is stable and robust, making it adaptable to crowd supervision in various surveillance scenarios.

5   Implementation

To validate the effectiveness and feasibility of our design, we build a prototype UAV-based urban crowd surveillance system. As shown in Fig. 11(a), we utilize an S500 drone [4] equipped with a Jetson Nano lightweight processing device as an edge node. The drone also carries a 2k motion camera, flight control, and a wireless transceiver to capture and stream HD video. And a GPU-equipped laptop is used as the ground station, communicating with the drone in real time via wireless wifi.

The prototype system practically executes crowd surveillance tasks in the following steps: 1) we deploy the ground station near crowded places in urban areas and set up patrol areas for UAVs; 2) the UAVs fly around to capture videos of gathering crowds while the edge devices on board UAVs perform the spatio-temporal low-loss video compression and transmit the compressed videos to the ground side with low latency; 3) the ground station conducts crowd surveillance applications, such as crowd counting and density analysis on the UAV-captured videos, and uploads the results to a cloud server for visualization, as depicted in Fig. 11(b), the user interface of the system displays real-time streaming video, detection results of crowds, the UAVs' flight trajectories and status, specific person recognition results, and crowd density information.

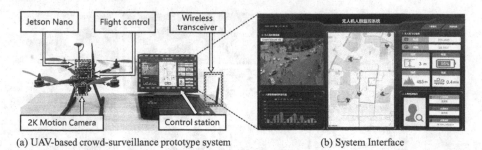

(a) UAV-based crowd-surveillance prototype system   (b) System Interface

Fig. 11. Intelligent UAV-based real-time crowd surveillance prototype system.

6   Related Work

Some work has been done on UAV-based crowd surveillance systems and lightweight video compression algorithms.

UAV-Based Crowd Surveillance Systems. UAVs are often combined with crowd-surveillance methods for urban security applications [17,24,25] due to their small size, low cost, and high safety. Singh *et al.* [25] used ScatterNet for human pose estimation to identify violent behavior from the UAV-captured footage. Castellano *et al.* [9] embedded cameras and GPUs in UAVs for real-time crowd density estimation. Woźniak *et al.* [31] used specially designed UAVs with large edge devices to run deep neural networks for crowd surveillance, reducing endurance and difficulty applying to conventional UAVs. As UAV technology continues to advance, its applications in various fields have further expanded. However, due to the limitations of endurance and size, UAVs cannot carry large-scale processing devices, making it difficult to handle complex computing tasks. Therefore, solving the problem of limited UAV edge computing capabilities has become the focus of current research.

Lightweight Video Compression Algorithms. Due to edge devices' extremely limited computing resources, lightweight video compression algorithms have been widely studied to solve such problems. Lu *et al.* [22] utilized optical flow motion information and autoencoder networks for efficient video encoding. Cohen *et al.* [11] proposed a modified entropy-constrained quantizer design algorithm for lightweight computing between clouds and edges. He *et al.* [15] designed the over-fitted repair neural network (ORNN) to obtain overfitted images. Park *et al.* [23] proposed a fast decision scheme for lightweight neural networks using multi-type trees (MTT) to reduce encoding complexity. Although lightweight video compression algorithms partly solves the video transmission latency, some still have high-performance requirements making it difficult to run on lightweight devices. In addition, the lightweight video compression methods currently are not adapted to crowd detection tasks. Accordingly, achieving high-precision and low-latency drone-captured video compression is still an open problem.

Unlike previous studies, this paper proposes a spatio-temporal low-loss video compression algorithm running on a capability-limited UAV node. It performs inter-frame comprehensive feature extraction and similarity clustering based on video clipping and RoI-based dynamic encoding of videos within frames, ensuring the accuracy of crowd detection and reducing video transmission latency.

7   Conclusion

This paper proposes an intelligent UAV-based edge computing system for real-time urban crowd surveillance. The system aims to address the high latency in video transmission and the low accuracy of compressed videos in UAV-based crowd surveillance with limited processing capabilities. Specifically, we propose a spatio-temporal fusion algorithm for low-loss video compression tailored for UAVs with limited edge capabilities. By integrating feature-based clustering for temporal sampling and dynamic encoding for spatial sampling, the algorithm achieves low-loss compression of UAV-captured videos while minimizing transmission latency under a certain detection information loss rate. The compressed videos are transmitted with low latency from the UAV to the ground station, enabling real-time and accurate UAV-based crowd surveillance. Furthermore, based on the actual UAV, ground station, and cloud server, we build a prototype UAV-based edge intelligent system and conduct comprehensive and in-depth experimental evaluations. The experimental results demonstrate that the proposed approach effectively reduces transmission latency while meeting high accuracy requirements.

References

1. Jetson benchmarks (2023). https://developer.nvidia.com/embedded/jetson-benchmarks
2. Nvidia jetson nano (2022). https://www.nvidia.cn/autonomous-machines/embedded-systems/jetson-nano/
3. Nvidia tensorrt (2022). https://developer.nvidia.com/tensorrt
4. S500 v2 kit (2022). https://www.holybro.com/product/pixhawk4-s500-v2-kit/
5. Yolov4-tiny (2022). https://github.com/AlexeyAB/darknet
6. Cctv: Too many cameras useless, warns surveillance watchdog tony porter (2019). https://www.bbc.com/news/uk-30978995
7. Bai, H., Wen, S., Gary Chan, S.H.: Crowd counting on images with scale variation and isolated clusters. In: 2019 IEEE/CVF International Conference on Computer Vision Workshop (ICCVW), pp. 18–27 (2019)
8. Castellano, G., Castiello, C., Cianciotta, M., Mencar, C., Vessio, G.: Multi-view convolutional network for crowd counting in drone-captured images. In: Bartoli, A., Fusiello, A. (eds.) ECCV 2020. LNCS, vol. 12538, pp. 588–603. Springer, Cham (2020). https://doi.org/10.1007/978-3-030-66823-5_35
9. Castellano, G., Castiello, C., Mencar, C., Vessio, G.: Crowd counting from unmanned aerial vehicles with fully-convolutional neural networks. In: 2020 International Joint Conference on Neural Networks (IJCNN), pp. 1–8. IEEE (2020)

10. CNN: South Korean authorities say they had no guidelines for halloween crowds, as families grieve 156 victims (2022). https://edition.cnn.com/2022/10/31/asia/ seoul-itaewon-halloween-mourning-memorial-intl-hnk/index.html

11. Cohen, R.A., Choi, H., Bajić, I.V.: Lightweight compression of neural network feature tensors for collaborative intelligence. In: 2020 IEEE International Conference on Multimedia and Expo (ICME), pp. 1–6. IEEE (2020)

12. Du, D., et al.: VisDrone-DET2019: the vision meets drone object detection in image challenge results. In: 2019 IEEE/CVF International Conference on Computer Vision Workshop (ICCVW), pp. 213–226 (2019)

13. Gonzalez, R.C.: Digital Image Processing. Pearson Education (2009)

14. Harvey, A., LaPlace, J.: Megapixels: origins, ethics, and privacy implications of publicly available face recognition image datasets. Megapixels 1(2), 6 (2019)

15. He, G., et al.: A video compression framework using an overfitted restoration neural network. In: Proceedings of the IEEE/CVF Conference on Computer Vision and Pattern Recognition Workshops, pp. 148–149 (2020)

16. Ionescu, R.T., Smeureanu, S., Popescu, M., Alexe, B.: Detecting abnormal events in video using narrowed normality clusters. In: 2019 IEEE Winter Conference on Applications of Computer Vision (WACV), pp. 1951–1960. IEEE (2019)

17. Jha, S., et al.: Visage: enabling timely analytics for drone imagery. In: Proceedings of the 27th Annual International Conference on Mobile Computing and Networking, pp. 789–803 (2021)

18. Jiang, S., Lin, Z., Li, Y., Shu, Y., Liu, Y.: Flexible high-resolution object detection on edge devices with tunable latency. In: Proceedings of the 27th Annual International Conference on Mobile Computing and Networking, pp. 559–572 (2021)

19. Jiang, Y., Miao, Y., Alzahrani, B., Barnawi, A., Alotaibi, R., Hu, L.: Ultra large-scale crowd monitoring system architecture and design issues. IEEE Internet Things J. 8(13), 10356–10366 (2021)

20. Liu, J., Gao, C., Meng, D., Hauptmann, A.G.: Decidenet: counting varying density crowds through attention guided detection and density estimation. In: 2018 IEEE/CVF Conference on Computer Vision and Pattern Recognition, pp. 5197–5206. IEEE (2018)

21. Liu, L., Li, H., Gruteser, M.: Edge assisted real-time object detection for mobile augmented reality. In: The 25th Annual International Conference on Mobile Computing and Networking, pp. 1–16 (2019)

22. Lu, G., Zhang, X., Ouyang, W., Chen, L., Gao, Z., Xu, D.: An end-to-end learning framework for video compression. IEEE Trans. Pattern Anal. Mach. Intell. 43(10), 3292–3308 (2020)

23. Park, S.H., Kang, J.W.: Fast multi-type tree partitioning for versatile video coding using a lightweight neural network. IEEE Trans. Multimedia 23, 4388–4399 (2020)

24. Rezaee, K., Mousavirad, S.J., Khosravi, M.R., Moghimi, M.K., Heidari, M.: An autonomous UAV-assisted distance-aware crowd sensing platform using deep shuffleNet transfer learning. IEEE Trans. Intell. Transp. Syst. 23(7), 9404–9413 (2022)

25. Singh, A., Patil, D., Omkar, S.: Eye in the sky: Real-time drone surveillance system (dss) for violent individuals identification using scatternet hybrid deep learning network. In: 2018 IEEE/CVF Conference on Computer Vision and Pattern Recognition Workshops (CVPRW), pp. 1629–1637 (2018)

26. Solera, F., Calderara, S., Cucchiara, R.: Socially constrained structural learning for groups detection in crowd. IEEE Trans. Pattern Anal. Mach. Intell. 38(5), 995–1008 (2015)

27. Sun, J., Li, B., Jiang, Y., Wen, C.Y.: A camera-based target detection and positioning UAV system for search and rescue (SAR) purposes. Sensors 16(11), 1778 (2016)
28. Wang, Y., Su, Z., Xu, Q., Li, R., Luan, T.H.: Lifesaving with RescueChain: energy-efficient and partition-tolerant blockchain based secure information sharing for UAV-aided disaster rescue. In: IEEE INFOCOM 2021-IEEE Conference on Computer Communications, pp. 1–10. IEEE (2021)
29. Wang, Z., Bovik, A.C., Sheikh, H.R., Simoncelli, E.P.: Image quality assessment: from error visibility to structural similarity. IEEE Trans. Image Process. 13(4), 600–612 (2004)
30. Wang, Z., Simoncelli, E.P., Bovik, A.C.: Multiscale structural similarity for image quality assessment. In: The Thrity-Seventh Asilomar Conference on Signals, Systems and Computers, 2003, vol. 2, pp. 1398–1402. IEEE (2003)
31. Woźniak, M., Siłka, J., Wieczorek, M.: Deep learning based crowd counting model for drone assisted systems. In: Proceedings of the 4th ACM MobiCom Workshop on Drone Assisted Wireless Communications for 5G and Beyond, pp. 31–36 (2021)
32. Xiao, L., Ding, Y., Huang, J., Liu, S., Tang, Y., Dai, H.: UAV anti-jamming video transmissions with QoE guarantee: a reinforcement learning-based approach. IEEE Trans. Commun. 69(9), 5933–5947 (2021)
33. Yang, Y., Li, G., Wu, Z., Su, L., Huang, Q., Sebe, N.: Reverse perspective network for perspective-aware object counting. In: Proceedings of the IEEE/CVF Conference on Computer Vision and Pattern Recognition, pp. 4374–4383 (2020)

CRAFT: Common Router Architecture for Throughput Optimization

Jiahua Yan, Mingyu Wang$^{(\boxtimes)}$, Yao Qin, and Zhiyi Yu

School of Microelectronics Science and Technology, Sun Yat-sen University,
Guangzhou 510275, China
wangmingyu@mail.sysu.edu.cn

Abstract. Agile development is considered as an emerging trend in
hardware design, and NoC-based multi-IP integrated system has become
the mainstream solution. The most critical challenge of this solution is to
design a common efficient NoC with ultra-high throughput and low-cost
routers to meet the communication characteristics of varied applications.
Usually, NoC can not achieve both the gains from high throughput and
low cost because high-throughput router architectures need complex virtual channel control logic to ensure that the power consumption and area
of buffers will not be too large. Meanwhile, the low-cost router architecture is difficult to achieve higher throughput by increasing the number
of buffers due to its simple control circuits. Moreover, most routers in
NoC are optimized only for specific applications. To address the problem, this paper presents a "viaduct theory" for throughput optimization
and realizes a common NoC router architecture for different scenario
requirements (CRAFT router). Firstly, we improve the top boundary of
the router throughput with the fully-connected crossbar switch, which is
an expensive solution for high throughput. Secondly, we use a method
similar to the ring-decouple theory in low-cost routers to reduce the
number of router buffers while maintaining high throughput. The experimental results show that the throughput for the low-cost program is
15.4% higher than the conventional router with the same overhead. For
the high-performance program, our router saves 50% hardware overhead
compared with fully-connected routers with the same throughput; The
maximum throughput that can be achieved is 26.3% higher than conventional routers, and 14.2% higher than the other state-of-the-art designs.

Keywords: Common router · Network on chip · High throughput ·
Low cost

1 Introduction

With the requirements of diversified application scenarios, processors need to
be more customizable. The DARPA Circuit Realization At Faster Timescales
(CRAFT) program has shown that rapid integration and agile development have

© The Author(s), under exclusive license to Springer Nature Singapore Pte Ltd. 2024
Z. Tari et al. (Eds.): ICA3PP 2023, LNCS 14489, pp. 212–229, 2024.
https://doi.org/10.1007/978-981-97-0798-0_13

become new methods of hardware development and design [1], which is consistent with the idea of scalable design for networks on chip (NoC). NoC is an interconnection network composed of distributed micro routers on a chip. High throughput and low hardware cost are the two most important design goals in NoC.

A router typically comprises two significant components: buffers and switches. The size and structure of router buffers are crucial in both NoC performance and its size. According to previous studies, buffers account for approximately half of the area and 80% of the power consumption in a router, establishing them as its most expensive component [2,3]. The switch's primary role is to transfer data packets from an input port to an output port based on routing calculations. Buffers temporarily store packets that cannot be forwarded when the output switch competes. In an ideal scenario, a router should function like a connector for two interconnected pipes, allowing for smooth traffic flow. Due to the competition from the switching function, a buffer needs to be of a specific size to maintain traffic stability. Additionally, the buffer structure will have an impact on switch efficiency.

Regarding buffer organization, Tamir and Frazier have defined a typical family of router architectures in their work [4]. This includes First-In-First-Out (FIFO) routers, Multi-Queue (MQ) routers, and Fully-Connected (FC) routers. When designing a network on chip, designers initially select one of these architectures based on their estimation of network traffic and chip area budget. Then, the buffer size is determined based on the actual traffic's saturated throughput and chip area limitations. Generally, the throughput will gradually increase to the top boundary as the buffer size increases, while the power consumption and area overhead also increase. However, if the buffer size cannot be adjusted to meet these conditions, designers must choose another router architecture, which significantly increases the period and workload of chip design. Additionally, some high-performance router designs are so complex that they only exist in theory or simulators. For example, the output buffer router architecture has a high throughput, but its crossbar switch needs to function at a frequency far faster than the system clock. This results in additional challenges concerning actual project implementation [5].

To achieve high-performance router architecture, many input virtual channels and complex virtual channel control logic are typically required. Multiple virtual channels are used to boost crossbar switch utilization and prevent it from bottlenecking network throughput. Complex virtual channel control logic is also essential to ensure efficient use of the buffers. In situations where throughput meets system requirements, a significant reduction in chip area and power consumption is obtained when the least number of buffers are used. High-performance routers perform well when using a large number of buffers. However, when the number of buffers is small, the complexity of the control logic far outstrips the buffer overhead. This makes high-performance routers unsuitable for low-overhead scenarios.

Low-overhead router architectures simplify control logic to ensure buffer efficiency. Such simple control logic can only effectively handle a small number of buffers, comparable to the efficient management of a family by parents, where additional labor can notably enhance the family's living conditions. But effective management of a country necessitates complex administrative systems. Hence, when an adequate budget is available, low-overhead router architectures cannot achieve higher throughput by merely adding more buffers. In such instances, designers must select a new router architecture and design a new one, which could be disastrous for rapid integration and agile development.

In this paper, we propose a "viaduct theory" that can optimize throughput and buffer efficiency by comparing the flit transmission in NoC and the travel mode of vehicles in the urban traffic network. Based on this theory, we realized a simple low-cost but ultra-high performance router with the Common Router Architecture For Throughput-optimization (CRAFT). This architecture is simple in both circuit implementation and code design, allowing developers to implement a router quickly. At the same time, the CRAFT architecture shows good performance in both buffer efficiency and throughput, which will enable us to achieve higher throughput with the same hardware overhead. Finally, the top boundary of router throughput in this architecture is ultra-high. It means that as long as we are willing to spend more area and power, we can achieve higher performance than others. This is also essential in high-performance computation (HPC) design.

2   Related Work

The FIFO router (Fig. 1(a)) represents the traditional architecture used to achieve the First Come First Serve (FCFS) mechanism in communication networks. Incoming packets are placed in each input queue in the order of their arrival. If several packets from different input queues are scheduled to be routed to the same output port, switch allocation (SA) organizes competition among the packets at the head of each queue for the output order. Ultimately, the winning packets are linked to their corresponding output port through a crossbar switch, configured based on the allocation results. In a conventional k-port FIFO router, a maximum of k packets from the input queue will compete in each cycle, while the same number of output ports will be waiting for matching. This often results in a low matching rate and a bottleneck in the switch allocation, consequently impacting the overall throughput of FIFO routers.

Multi-Queue router(Fig. 1(b)), which is generally considered as a router architecture with high performance and buffer efficiency, including Statically-Allocated-Multi-Queue (SAMQ) router, and Dynamically-Allocated-Multi-Queue (DAMQ) router. It allows the incoming packets to use any input queue by the virtual channel allocation mechanism, which improves the utilization of the input queue's storage space and provides more chances for switch allocation. Without pre-routing, incoming packets can be stored in free queues like FIFO routers. The switch allocation approach is divided into two stages: the first stage

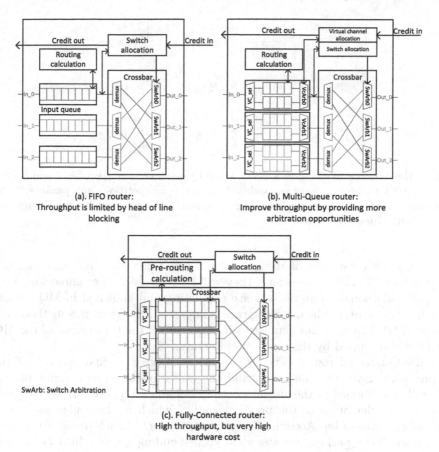

(a). FIFO router:
Throughput is limited by head of line blocking

(b). Multi-Queue router:
Improve throughput by providing more arbitration opportunities

SwArb: Switch Arbitration

(c). Fully-Connected router:
High throughput, but very high hardware cost

Fig. 1. Micro-architecture of three common mainstream routers.

reduces the number of requests for an output port to one request per input port by a v:1 arbiter, since the number of VCs is v. The second stage produces a winner for each output port among all the competing input ports using a k:1 arbiter in the k-port router [10]. The second stage is the same as the switch allocation in FIFO routers, while the first stage can reduce the competition of the second stage by adjusting the priority, which is called virtual channel allocation (VCA). Previous studies [6,7] have shown that the more virtual channels, the less competition. Some researchers also try to dynamically adjust the number of virtual channels when the network is running and result in further complex circuits, like the regulator in ViChaR [10] or the linked-list-table in EDVC [11].

The biggest bottleneck of throughput in the MQ router is the head-of-line (HOL) blocking problem. As shown in Fig. 2, in conventional MQ router architecture, packets are received and stored in the input queue. Only the flit at the head of the queue can get a chance of arbitration, and then transmit by a multiplexor to the output port if granted. When the head flit fails in arbitration,

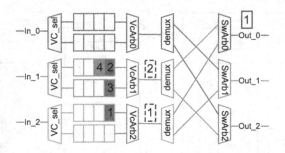

Fig. 2. Head-of-line blocking in a 2-VC Multi-Queue router: Packets 1, 2 and 3 need to be output to port 0. Assume packet 1 wins the competition, and packets 2 and 3 occupy heads of each queue, resulting in packet 4 being blocked, even though the output port of packet 4 is idle

it still occupies the head of the queue, and other flits after the head can not get a chance of arbitration even if they will be routed to other unblocking output ports. Although more packets are competing and allocated in MQ routers than in FIFO routers, the matching rate is improved. However, this method only reduces HOL blocking, not eliminates it. As a result, the throughput of the MQ router is still limited by the router architecture.

Fully-Connected router (Fig. 1(c)) consists of separate lines to each of the output ports from an input port, and the buffer queue in each input port is statically partitioned to the output port [4]. Before each packet arrives, it needs pre-routing calculation in the previous router, which is also called lookahead routing mechanism [9]. According to the output port of each packet calculated previously, these packets are stored in corresponding queues. In a k-port FC router, each input port requires $k - 1$ queues. The minimum length of queues that can maintain the continuous transmission of data flow is determined by the recovery time of credits. Specifically, when a packet is sent to the next router, it needs to consume a credit of the input port corresponding to the next router. Assuming that there is only one credit, the next packet in this queue can only be sent after the credit is recovered. Therefore, we need enough credits to maintain the continuity of data flow. For a lossless NoC, the buffer length of the input queue must be greater than or equal to the number of credits [2]. For fully-connected switch routers, all queue lengths of each port need to meet the above rule, which causes massive waste because only one of the input queues is available as potential storage for any given packet [4].

Low-Cost (LC) router (Fig. 3) evolved from ringbus network router [13]. Ringbus network is generally considered to be a simple, low-overhead and bufferless interconnect network. Its router is composed of just pipeline registers and multiplexers, which can be seen as a special FIFO router (single buffer FIFO router). The main idea of the low-cost network is to decouple the 2D-mesh network into X-direction and Y-direction ringbus network. And the LC router is partitioned or sliced into two separate routers: one for X-ring and the other for Y-ring. Data

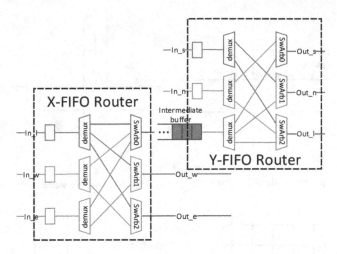

Fig. 3. Block diagram of Low-Cost router: composed of X-ring router and Y-ring router in 2D-mesh.

can be exchanged through an intermediate buffer between two ringbus routers of the same node.

From the perspective of network, the way to reduce the cost is to decouple the complex network into a simple network. In this paper, we will further optimize this method to make our NoC and router simple, versatile and realizable.

3   The Proposed Viaduct Theory in Throughput Optimization

From a macro point of view, we can think of routers as crossroads in the urban traffic network. Conventional FIFO routers are regarded as crossroads with traffic lights and packets passing through the router need to wait for the traffic lights to turn green. At this type of crossroads, there is a waiting relationship between packets going in different directions. For example, vehicles going straight from the north can only pass this crossroads after the vehicles turning left from the south (Fig. 4(a)), which makes the road from the north to the South idle, even though there are vehicles that need to use this road. This is a waste of road resources and reduces the throughput. Urban designers believe that the most essential way to improve traffic flow is to improve road utilization. Therefore, the most effective way to solve this problem is the "viaduct" (Fig. 4(b)).

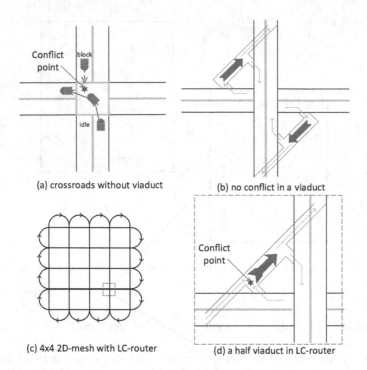

Fig. 4. Demonstration of the viaduct in crossroad conflict resolution and viaduct theory in the low-cost router.

3.1   Viaduct Theory in Throughput Optimization

The viaduct theory comprises the following aspects: Viaduct has two basic functions, connection and temporary storage; The viaduct should be as complete as possible to ensure that there are independent connections between different turns and there is almost no conflict; The viaduct in routers needs to be capable of temporarily storing a certain number of packets. Otherwise these diverted packets will block the straight packets and cause coupling relationship between different turns; The final decision of throughput is the road utilization, so packets going to the same road can share the viaduct buffer while waiting in connection; Viaduct is the most efficient way to change direction, high-throughput routers should be like crossroads with viaducts. In summary, high-throughput routers should ensure that if the packets in the router need to go to a port, the port should not be idle.

3.2   Comparison of Ring Decouple Theory and Viaduct Theory in Low-Cost Router

The theory of Low-Cost router is based on Ring-Decouple Theory. This theory consists of three parts: Firstly, the traffic of the ring network is unidirectional, just like the pipeline of the processor. Secondly, in a unidirectional pipeline, only one register between pipeline stages is required to ensure data continuity, resulting in a low-overhead data transmission path. Finally, a 2D-mesh network can be decoupled into several separate ring networks, and data is exchanged between networks through an intermediate buffer. The greater the buffer depth, the greater the independence of the ring network and the higher the throughput of the entire network. By leveraging this mechanism, a low-throughput 2D-mesh network can be transformed into several high-throughput ring networks as depicted in Fig. 4(c).

From the theory of "viaduct", this design separates the horizontal and vertical roads, and primarily ensures the smooth flow of straight vehicles on each road. The horizontal and vertical roads are connected by a "viaduct" (Fig. 4(d)), which serves as the intermediate buffer in the Low-Cost router design. The "viaduct" in the LC router reduces the congestion during turning and improves the network throughput. However, this can only be called a "half viaduct" and some conflicts remain, because both the forward and reverse directions of X turn through this viaduct in the LC router. Obviously, the competition for the viaduct will be the bottleneck of the network at high throughput.

4   A Common Router Architecture for Throughput Optimization Based on Viaduct Theory

Through viaduct theory, the micro-architecture of the CRAFT router is illustrated in Fig. 5. Our proposed CRAFT router consists of k input port (representing different roads), each with $k-1$ queues for eliminating conflict between different turns, an arbiter, and a shared queue for decoupling near the output port. In fact, it is similar to the FC router architecture we presented in Fig. 1. However, the shared queue of the CRAFT router significantly improves the hardware efficiency in buffering.

4.1   Flow Control Optimization with Viaduct Theory

In the viaduct theory, a router acts as a crossroads for vehicles to turn in different directions. Obviously, a small car can turn more easily than a long truck. Thus, in the CRAFT router, the single flit packet is used as the Basic Transmission Unit (BTU). This is a special kind of wormhole flow control mechanism that is widely used in modern many-core processors. A BTU consists of these three parts: destination of this packet, output port information and the payload. The arriving packets select different queues according to the current output direction as well as calculating the output direction of the next router, then the pre-routed packets are remixed and stored in the selected queue. Hence the output direction of all packets in the same queue is the same. Next, packets from different input queues to the same output port begin to compete, the winner enters the shared queue and releases a credit to the corresponding front router. Finally, the head flit of share queue checks the credit of its next routing direction, and if the number of credits is not zero, the flit transmits.

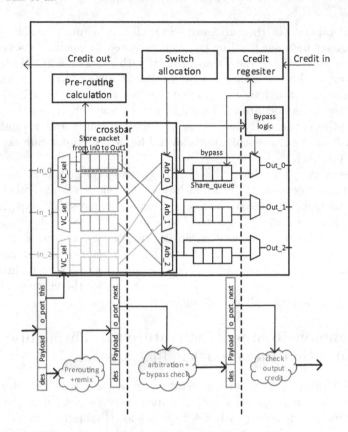

Fig. 5. An example of three port CRAFT router and pipeline stage division.

4.2 Pipeline Stage and Latency Optimization

The pipeline of router is usually divided into these stages: buffer selection and write, routing calculation, virtual channel allocation, switch allocation, and transmission. A set of registers is required between each stage of the pipeline to store data. In CRAFT router, the routing calculation can be set at the same stage of buffer selection because they are parallel. The stage of virtual channel allocation is removed because we statically allocate each virtual channel to an output port. And the function of registers between switch allocation and transmission stage can also be replaced by the share queue, which simplifies the design of routers and reduces latency. As shown in Fig. 5, the pipeline stages can be reduced to 2 or 3 stages, and registers between stages of the pipeline are replaced by the existing buffers, thus the latency and overhead are reduced.

4.3   High Throughput Optimization with Viaduct Theory

For general NoC, each output port can only transmit one flit per cycle, so the bottleneck of network throughput is the utilization of switches. Head flits that fail in competition must wait for the next round of switch arbitration, leading to Head of Line (HOL) blocking. This blocking hinders packets behind the queue from delivery to the head of the queue. When an output port is idle and the packets routing to this port are blocked behind the head of the queue, the utilization of the switch associated with that port is greatly reduced.

According to the viaduct theory, our proposed CRAFT router solves this blocking problem by fully-connected switches, which is the same as the FC router. Fully-connected switches provide independent connections for each turn similar to that of a viaduct. Therefore, the assumption made in the above paragraph does not occur in fully-connected crossbar switches because of the separating queues of the input port. Each separate queue of the input port is connected to a corresponding output port independently. If an output port were idle, all queues connected to it should be empty. In other words, a packet will only be blocked by packets routing to the same direction, which will not reduce the utilization of switches and output ports.

4.4   Low Hardware Cost with Viaduct Theory

The main problem of FC router is the excessive hardware overhead. Within the viaduct theory, the way to reduce the hardware overhead in our CRAFT router is to establish a shared buffer in the same output direction, just like the decoupling of LC routers (Fig. 6) in different directions. From the perspective of packet flits, in FC routers, the packet at the head of the queue needs to meet two conditions simultaneously before it can be transmitted: get the highest priority in arbitration, and get a credit from the next queue. The output shared queue in the CRAFT router decouples these two conditions. Packet with the highest priority will first enter the shared queue and release a credit to the corresponding previous router. Next, the packet waits in the shared queue for transmission until it gets credit from the next router.

Figure 7 depicts the credit recovery and consumption process in different routers. Four flits consumed credits in the last router and arrived. In the FC router, another four buffer slots are required in each queue if we want to release credit to the last router right now, resulting in $k - 1$ times extra buffer slots in a k-port router. However, we only need four extra buffer slots in the share queue for the CRAFT router to release credit. The decoupling of credit recovery and consumption greatly reduces the number of buffers required for each queue.

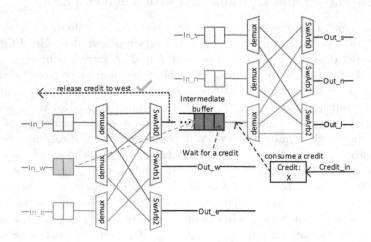

Fig. 6. Detail diagram and flow control of Low-Cost router: Packets from X to Y need to pass through the intermediate buffer, so the release of credit in X direction is earlier than the consumption of credit in Y direction. The larger the buffer, the weaker the correlation

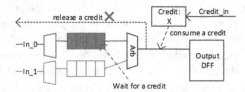

(a) FC router: more buffer is required for each queue

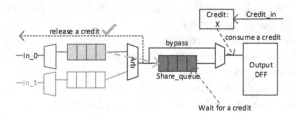

(b) CRAFT router: packets are stored in share queue
and credits have been release

Fig. 7. Credit recovery and consumption process in (a) FC and (b) CRAFT router. The credit release method of the CRAFT router is the same as that of the LC router (Fig. 6), regardless of whether the next router returns a credit.

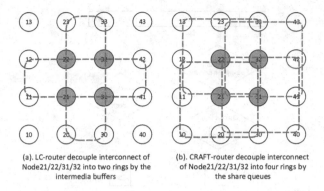

(a). LC-router decouple interconnect of Node21/22/31/32 into two rings by the intermedia buffers

(b). CRAFT-router decouple interconnect of Node21/22/31/32 into four rings by the share queues

Fig. 8. Decoupling diagram of LC network and CRAFT network.

In terms of networks, the Low-Cost router decouples the 2D-mesh network into two ring networks, vertical and horizontal, as shown in (Fig. 8(a)); our proposed CRAFT router decouples the 2D-mesh network into four or more ring networks as the same. The network decoupling operation enables us to achieve higher throughput with fewer buffers (Fig. 8(b)) like the Low-Cost network.

4.5 Bypass Strategy for Low Workload

In the previous section, we mentioned that packets need to enter the shared buffer queue before reaching the output port. This method improves the hardware efficiency and maintains high throughput under high workload. However, this method will increase the latency under low workload. In order to adapt to the situation of low workload, we propose a bypass strategy for the shared queue illustrated in (Fig. 9): If the shared queue is empty, check whether the packet to be stored in the queue meets the output conditions. If so, the packet will be transmitted to the output port through a short bypass way without being stored in the shared queue.

5 Experiment Results

In this section, we will be evaluating the proposed CRAFT router architecture and comparing it with the conventional FIFO router architecture (Fig. 1(a)), the Low-Cost router architecture (Fig. 6), the Fully-Connected architecture (Fig. 1(c)) and the state-of-the-art Multi-Queue router architecture. We use the 5-ports routers in a 64 node, 8x8 2D-mesh network with the uniform random traffic workload, of which the maximum throughput is 0.5 flit/node/cycle [2]. We randomly inject data packets into the network according to the setting injection rate, and record the time spent from the generation of packets to being

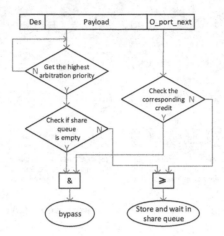

Fig. 9. Bypass strategy of share queue.

received, which is the latency-throughput curve. Generally, the network saturation throughput is defined as the injection rate when the average network latency is three times that of zero-load latency [8].

In Fig. 10, we evaluated the performance of different router architectures under different buffer sizes. For FIFO routers and FC routers, we can change the length of each input queue (b); For LC routers, according to the author's design, the length of each input queue is fixed to 2, and we can change the size of the intermediate buffer (bi). For our CRAFT router, we can change the length of the input queue (b) and the buffer size of the shared queue (bs) respectively. In general, without changing the router architecture, we can adjust the number of buffers to meet the throughput, area and power consumption requirements of the NoC. When the number of buffers is small, the throughput increases rapidly with the increase of the buffer; When the number of buffers is large, the throughput increases slowly to the top boundary. As shown in Fig. 10, the top boundary of throughput is determined by the architecture of the router.

5.1 For Low-Cost Program

When NoC is used in embedded systems or very large-scale small-core interconnected systems, the area budget of routers is usually small. Previous studies [14,15] have shown that routers spend about 50% of their area on buffers. Suppose that we can only put down no more than totally 26 buffers in a router at this configuration. For the baseline conventional FIFO router, 5 buffer slots are set for each port, 25 buffer slots in total. Besides, all the pipeline registers in this router are removed. For the Low-Cost router, 2 buffer slots are set for each input port and the size of the intermediate buffer is 16, thus 26 buffer slots in total. Our CRAFT router is configured with 1 buffer per input queue, and the depth of each shared queue is 2. In fact, due to paths do not exist in the X-Y

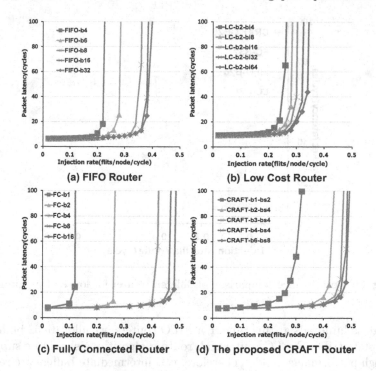

Fig. 10. Latency vs. offer load comparison of (a) the conventional FIFO router architecture, (b) the low cost router, (c) the fully connected router, and (d) the proposed CRAFT router for uniform random traffic in 8x8 2D mesh network as the amount of buffer is varied.

routing algorithm, some queues are removed from this router such as the queue from south to east or west. The remaining 16 input queues and 5 shared queues cost a total of 26 buffers. For the FC router, our configuration is 2 buffers per queue. The 16 effective queues under the X-Y routing algorithm spend 32 buffers in total.

Figure 11 shows the latency-throughput curves of the above routers. Under random traffic, the saturated throughput of the LC router and CRAFT router is nearly 15.4% higher than that of the FIFO router and FC router. Comparing the FC router with the CRAFT router, although the FC router uses six more buffers than the CRAFT router, its throughput is still only 86% of that of the CRAFT router. This is because the shared queue of the CRAFT router improves the buffer efficiency, just like the role of the intermediate buffer in the LC router.

5.2   For High-Performance Program

In high-performance computation and applications, data throughput is the primary indicator. Moreover, we must consider reducing the area and power consumption as much as possible. Suppose that the buffer budget of each router can

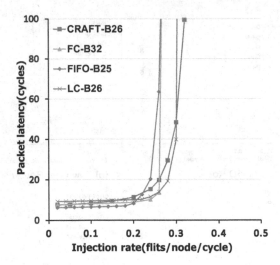

Fig. 11. Performance comparison of different router for low-cost program.

be as high as hundreds at this time. For FIFO routers, we allocate 32 buffers per port, a total of 160. For LC routers, all routing algorithms should be supported in the high-performance mode. Therefore, two intermediate buffers are required to ensure that all turning from the X direction to the Y direction and from the Y direction to the X direction are supported. We allocate two buffers to each port. The size of the intermediate buffer is 64, a total of 138. For FC routers, the number of effective queues supporting all routing algorithms is 20. We allocate 8 buffers and 16 buffers per port to evaluate their performance, respectively. For our proposed CRAFT router, the length of each input queue is configured as 6, the depth of the shared queue is 8, and the total number of buffers is 160.

For other state-of-the-art throughput optimized routers, we evaluate the following routes: The Swap buffer router (SwB) [16] proposed last year, uses a swap buffer to manage traffic, just like the traffic police at crossroads. The remaining two are 5-stage multi queue routers with 4 and 8 virtual channels respectively. The total buffer size of these routers is all about 160.

As shown in Fig. 12, The CRAFT router can achieve about 95% of the maximum throughput when it has enough buffers. To achieve this throughput, the FC router needs twice as many buffers as the CRAFT router. The throughput of other routers can not be further improved under more buffers. The performance of the LC router is similar to that of the CRAFT router in low configuration, but it cannot effectively use these buffers in high configuration mode. Hence, its max throughput is much lower than that of the CRAFT router, about 66.7%. It can also be seen from the figure, the maximum throughput of the CRAFT router is 26.3% and 14.2% higher than that of the FIFO router and MQ router, respectively.

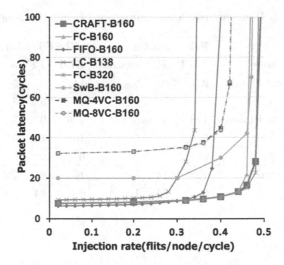

Fig. 12. Performance comparison of different router for the high-performance program.

The zero-load latency of a conventional MQ router is much higher than that of a CRAFT router with at least two-stage pipelines because it usually requires five-stage pipelines in MQ routers. The SwB router improves some of the throughput by the management of the "traffic police" and reduces the latency, but it is still a certain distance from the CRAFT router. The single cycle FIFO router and LC router both have two-stage pipelines. Under low workload, the latency of the CRAFT router is as low as that of the two routers because of the bypass strategy.

5.3 Hardware Overhead

To evaluate the hardware overhead, we synthesize the FIFO router and the CRAFT router under VIVADO with the 64-bit buffer and 160 buffer slots in total. Table 1 compares the Look-Up Table (LUT) and Flip-Flop (FF) resources consumed by the main components of these two routers. It can be seen that the crossbar and buffer are the two components that consume the most resources. Compared with the FIFO router, the CRAFT router has not much additional overhead except for requiring more queue control logic.

Table 1. Hardware overhead comparison of main components

	FIFO router			CRAFT router		
Components	*LUT*	*FF*	*amount*	*LUT*	*FF*	*amount*
Routing calculation	2	0	5	11	0	5
Switch allocation	25	5	5	24	5	5
Crossbar multiplexer	128	0	5	128	0	5
Credit	9	7	5	10	15	5
Bypass logic	0	0	0	5	0	5
Queue control	12	14	5	10	8	25
Buffer	0	64	160	0	64	160

6   Conclusion

In this paper, by comparing the blocking problem in network transmission on chip with the crossroad conflict problem in traffic network, we propose a "viaduct theory" for throughput optimization on the basis of optimizing the "ring-decouple theory" of Low-Cost router. This theory shows that the viaduct is the most efficient crossing structure of crossroads. Routers in 2D-mesh network with a more similar architecture to the viaduct have higher throughput and buffer efficiency. Through this theory, we designed a CRAFT architecture that can adapt to various situations. In embedded and other low-cost programs, our CRAFT router performs similarly to Low-Cost routers, and the throughput is 15.4% higher than that of conventional routers. In the high-performance program, the CRAFT router can provide 95% of the maximum network throughput, which is 26.3% and 14.2% higher than the conventional FIFO router Multi-Queue virtual channel router respectively, and saves 50% of the buffer than the Fully-Connected router with the same throughput. Based on the CRAFT router architecture, we can simply adjust the number of router buffers to meet the needs of different NoC scenarios without re-designing a router, which provides effective support for rapid integration and celerity development.

Acknowledgements. This work is supported by the National Natural Science Foundation of China (NSFC) under Grant 62204271, the Key-Area Research and Development Program of Guangdong Province under Grant 2021B0101410004, and the Guangdong Basic and Applied Basic Research Foundation under Grant 2022A1515011708.

References

1. Ajayi, T., Al-Hawaj, K., Amarnath, A., et al.: Celerity: an open source RISC-V tiered accelerator fabric. In: Symposium on High Performance Chips (Hot Chips) (2017)
2. Jerger, N.E., Krishna, T., Peh, L.-S.: On-chip networks. Synthesis Lectures on Computer Architecture **12**(3), 1–210 (2017)

3. Dally, W.J., Towles, B.: Bufferd Flow Control, in Principles and Practices of Inter-connection Networks. Morgan Kaufmann, San Francisco (2003)
4. Tamir, Y., Frazier, G.L.: Dynamically-allocated multi-queue buffers for VLSI communication switches. IEEE Trans. Comput. **41**(06), 725–737 (1992)
5. Ramanujam, R.S., Soteriou, V., Lin, B., et al.: Design of a high-throughput distributed shared-buffer NoC router. In: 2010 Fourth ACM/IEEE International Symposium on Networks-on-Chip, pp. 69–78. IEEE (2010)
6. Boura, Y.M., Das, C.R.: Performance analysis of buffering schemes in wormhole routers. IEEE Trans. Comput. **46**(6), 687–694 (1997)
7. Jiang, N., Becker, D.U., Michelogiannakis, G., et al.: A detailed and flexible cycle-accurate network-on-chip simulator. In: 2013 IEEE International Symposium on Performance Analysis of Systems and Software (ISPASS), pp. 86–96. IEEE (2013)
8. Ma, S., Enright Jerger, N., Wang, Z.: DBAR: an efficient routing algorithm to support multiple concurrent applications in networks-on-chip. In: Proceedings of the 38th Annual International Symposium on Computer Architecture, pp. 413–424 2011
9. Xin, L., Choy, C.: A low-latency NoC router with lookahead bypass. In: Proceedings of 2010 IEEE International Symposium on Circuits and Systems, pp. 3981–3984. IEEE (2010)
10. Nicopoulos, C.A., Park, D., Kim, J., et al.: ViChaR: a dynamic virtual channel regulator for network-on-chip routers. In: 2006 39th Annual IEEE/ACM International Symposium on Microarchitecture (MICRO 2006), pp. 333–346. IEEE (2006)
11. Oveis-Gharan, M., Khan, G.N.: Efficient dynamic virtual channel organization and architecture for NoC systems. IEEE Trans. Very Large Scale Integr. (VLSI) Syst. **24**(2), 465–478 (2015)
12. Dally, S.: Deadlock-free message routing in multiprocessor interconnection networks. IEEE Trans. Comput. **100**(5), 547–553 (1987)
13. Kim, J.: Low-cost router microarchitecture for on-chip networks. In: Proceedings of the 42nd Annual IEEE/ACM International Symposium on Microarchitecture, pp. 255–266 (2009)
14. Zimmer, H., Zink, S., Hollstein, T., et al.: Buffer-architecture exploration for routers in a hierarchical network-on-chip. In: 19th IEEE International Parallel and Distributed Processing Symposium. IEEE (2005). 4 pp
15. Petrisko, D., Zhao, C., Davidson, S., et al.: NoC Symbiosis (Special Session Paper). In: 2020 14th IEEE/ACM International Symposium on Networks-on-Chip (NOCS) 1–8. IEEE (2020)
16. Katta, M., Ramesh, T.K., Plosila, J.: SB-Router: a swapped buffer activated low latency network-on-chip router. IEEE Access **9**, 126564–126578 (2021)

A Cross-Chain System Supports Verifiable Complete Data Provenance Queries

Jingyi Tian[1], Yang Xiao[2(✉)], Enyuan Zhou[3], and Qingqi Pei[1(✉)]

[1] State Key Laboratory of Integrated Services Networks and Shaanxi Key Laboratory of Blockchain and Secure Computing, Xidian University, Xi'an, China
qqpei@mail.xidian.edu.cn
[2] State Key Laboratory of Integrated Service Networks, School of Cyber Engineering, and also with the Engineering Research Center of Trusted Digital Economy, Universities of Shaanxi Province, Xidian University, Xi'an, China
yxiao@xidian.edu.cn
[3] The Hong Kong Polytechnic University, Hung Hom, Kowloon, Hong Kong

Abstract. In recent years, blockchain has been widely used as a decentralized database. However, its limited scalability makes it unsuitable for large-scale applications. To overcome this challenge, high scalability cross-chain technologies have become crucial for implementing blockchain applications. While existing cross-chain technologies focus on implementing cross-chain logic, ignoring the issue of incomplete results that may arise when querying data provenance in cross-chain systems. In this work, we propose a cross-chain system that supports efficient and verifiable complete data provenance queries. We achieve this by designing an index based on a linear list and adjacent linked list to store related transactions, enabling us to perform a complete query. We also utilize vector commitments to generate verification objects indicating whether transactions are included in our index entries, ensuring verifiable integrity of query results. Furthermore, we construct an index structure based on B+ tree and Merkle tree to enhance the system's availability and query efficiency. Our experimental results demonstrate that our scheme not only has excellent performance in terms of the cost of verification object, but it also improves query efficiency by approximately two times.

Keywords: Blockchain · Cross-chain · Verifiable Query · Efficient Query

1 Introduction

Blockchain is a decentralized ledger composed of multiple data blocks linked by hash pointers, originally utilized for recording Bitcoin [1] transactions. The hash algorithm and consensus protocol make the blockchain possesses characteristics

---

J. Tian and E. Zhou—These authors contributed equally to this work.

such as security, trustworthiness, and tamper resistance. These features drew more attention to the blockchain's potential to support applications beyond digital currencies. Consequently, blockchain platforms that support general applications emerged, such as Ethereum [2] and Hyperledger [3]. Today, the blockchain is increasingly used as a distributed database in various fields, including supply chain [4], finance [5], and medical care [6].

However, due to the decentralized nature of blockchain, all members of the blockchain network must maintain all data on the entire chain. This presents a challenge as some users may choose not to participate due to the significant storage burden. Furthermore, achieving consensus among all users in the network for storing each piece of data results in limited throughput when the blockchain is used as a database, which significantly restricts a single blockchain's scalability. To address these issue, large-scale applications require multiple blockchains to work together to support a substantial amount of data. This requirement has led to the emergence of cross-chain technology that enables interconnectedness between multiple blockchains, improving interoperability and scalability. However, current cross-chain schemes focus on implementing cross-chain logic, such as breaking data islands and ensuring atomicity [7], with few addressing issues related to data provenance query in the cross-chain system.

Specifically, when a user on a blockchain wants to query the complete provenance of a transaction in the cross-chain system, for example, a user wants to find the complete circulation information of a product on all blockchains in a cross-chain system, the user's blockchain may intentionally or unintentionally ignore the fact that the transaction is part of cross-chain transactions with other chains. The user might only receive related transactions queried on their own chain, without finding the related transactions on other chains. Therefore, in this scenario, users are highly likely to obtain incomplete query results and are unable to verify the integrity of the outcomes. Consequently, the current cross-chain system falls short of achieving a verifiable complete data provenance query, which significantly undermines the blockchain's original design intention, that is, trusted and verifiable.

In addition, unlike other traditional databases have built-in index structure, blockchain's query process requires comparison items one by one, resulting in significant inefficiency [8]. Additionally, blockchain employs levelDB [9], a key-value database that stores random hash strings as keys, making the blockchain only support queries based on hash values and unable to carry out complex queries based on attribute values [10]. These concerns also arise during cross-chain queries, which will greatly limit the wide application of blockchain in various fields. Therefore, we outline the challenges faced by current cross-chain systems when conducting queries:

– When querying the complete provenance of a certain data in all blockchains in the cross-chain system, incomplete query results may be returned because the system cannot perceive the cross-chain relationship between transactions.
– Users can only verify the authenticity of the results obtained by the complete provenance query, but cannot verify its integrity.

- Queries are inefficient and do not support other types of complex queries except hash-based.

To address the above challenges, this work designs a cross-chain system supports efficient and verifiable complete data provenance query based on the most popular cross-chain technology, relay chain. Our work fills the gap in research on cross-chain system queries, and our contributions are outlined as follows:

- We design an index stores related transactions to achieve complete cross-chain queries, improving the reliability of cross-chain systems during queries. This is not considered by previous research on cross-chain systems, which promotes the landing application of cross-chain technology.
- Based on the complete query, we use the vector commitment [11] for the index to generate verification objects (VOs), implements the integrity verification function of query results. This ensures the trustworthiness of the query results.
- In order to improve the practicability of cross-chain system, we also design an index to speed up the query efficiency, and extend the block structure to realize the query based on attribute value. Make the query performance of cross-chain system towards traditional database.

The rest of this work is organized as follows: Sect. 2 presents the related work of our research, Sect. 3 presents the system components and some preliminary knowledge, Sect. 4 presents our scheme, Sect. 5 evaluates our system through experiments and we concludes this work in Sect. 6.

2   Related Work

In this section, we introduce the related works of our research from three aspects: query function, verifiable query and cross-chain technology.

2.1   Query Function

Researches on the query function of blockchain mainly fall into two categories: outbound database and built-in index structure. Outbound database methods, such as ChainSQL [14], configure a relational database management system(RDBMS) on top of the Ripple [15] blockchain and synchronise data from the blockchain to the database, giving users the option of querying the blockchain network directly or accessing the database for diverse queries. BigchainDB [16] combines MongoDB [17] for fast querying of blockchain data and adds consensus and cryptographic signatures to ChainSQL so that data is only uploaded to MongoDB if a majority of complete nodes in the blockchain system agree. EtherQL [18] synchronises data from the most dominant blockchain platform, Ethernet, into an RDBMS and adds a query layer with a defined efficient query primitive. Users can initiate analytical queries such as range queries and TOP-K queries. The drawback of this outreach database approach is that there would be a centralised RDBMS, which goes against the security and trust that comes with the decentralised nature of the blockchain.

Built-in indexing methods, such as SEBDB [19], add relational semantics to blockchains, where each transaction is stored in a block as a row of records with multiple attribute values, and provide SQL-like language for users to perform complex types of queries; Lineagechain [20] adds a runtime traceability function for smart contracts on Hyperledger based on the Forkbase [21] storage engine, which makes up for the defect that blockchain can only analyze data offline, and sets up a novel index structure based on skip list in the blockchain to achieve fast and efficient traceability queries, and a simple smart contract interface is provided for users to use; Blockchain cannot store huge amount of data due to its decentralized nature, so current blockchain-based systems adopt a hybrid on-chain and off-chain storage architecture, storing core transactions on-chain and the rest of data off-chain. MST [22] then establishes the connection between on-chain and off-chain data by extracting their mapping relationships through a modified TF-IDF model, and constructs an index structure called MST based on inverted indexes [23], hash pointers and B+ trees [12] to support range queries and fuzzy queries. The problem with the built-in indexing method is that the index structure set up inside the block can further increase the storage burden imposed by the blockchain on the full node.

2.2 Verifiable Query

For the study of verifiable queries, Xu C et al. proposed vChain [24], which constructs a built-in authentication data structure (ADS) for each block by aggregating the data within the block with the data between the blocks and using a multiset accumulator to generate a VO for the light nodes to achieve verifiable integrity of the blockchain Boolean range query. Secondly, vChain sets up intra-block index based on tree structure with skip list to improve the efficiency of light node query, it also adds bulk subscription query and authentication function based on IP prefix tree index.

Wang H et al. found the defect in vChain that the inter-block index set up by vChain does not work in 80% of the cases when performing query, and vChain converts all numeric attributes to set-valued attributes for processing and can only support fixed-point numbers and integers, which greatly limits the application of vChain. Therefore, Wang H et al. proposed vChain+ [25], which uses a sliding window accumulator to solve the problem of inter-block index failure and applies a B+ tree to support range queries for floating-point numbers, making vChain more flexible query types and improves the query performance of vChain by a factor of 913. Verifiable queries in blockchains make the most closely related to the research in this work, but existing schemes are considered based on individual blockchains, and no work has focused on some of the problems that exist when performing verifiable data provenance queries in cross-chain systems.

2.3 Cross-Chain Technology

The research on cross-chain technology is currently divided into three technical solutions: side/relay chains [26], hash locking [27], and notary mechanisms [28] in academia. The mainstream solution currently used by academia and industry is the relay chain. The most popular cross-chain systems Cosmos [29] and Polkadot [30] are based on relay chains. Cosmos consists of a Cosmos Hub and several parallel blockchains, and the different parallel blockchains communicate and interoperate through the Cosmos Hub based on cross-chain communication protocols. Another popular relay chain project Polkadot aims to solve the interoperability problem between multiple heterogeneous blockchains and puts most of the complexity on the relay chain, so that heterogeneous chains can be more easily accessed into the system with better scalability.

It can be seen that the research on query function and verifiable query processing of blockchain has made some progress in recent years, but these works are considered based on a single blockchain without considering the query processing in cross-chain systems; and in the research on cross-chain technology has become more and more mature, the existing solutions can already break the data silos between blockchains and realize the atomicity transactions. However, there is still a gap in the research of verifiable cross-chain data provenance query.

3   Preliminaries

In this section, we introduce the components of the system and the preparatory knowledge necessary to understand the solution presented in this work.

3.1   Preliminary Knowledge

Blockchain Structure. All data on the blockchain are stored in blocks as the smallest unit, and each block is divided into two parts: the block body and the block header [1]. The block header stores some short information to maintain the structure of the blockchain, and the block body stores specific transaction information.

The meanings of each field in the block header are as follows. ParentHash records the unique hash value obtained by hashing all the contents of the previous block, which can make the block into a chain and provide tamper-proof function; VersionID records the version number of the block header; Timestamp records the time information when the block is chained; Difficult to record the difficulty of the generation of the block; Nonce records a random number that changes the hash value of the block when the block is generated; MerkleRoot records the root of the Merkle tree [13] consisting of transactions in the block body, which is used for verification of the authenticity of transactions by light nodes [24].

It is worth mentioning that there are usually two types of nodes in blockchain, *i.e.*, full node and light node. Full node needs to maintain all the data on the blockchain, including the block header and the block body. Light node only needs to synchronize the block header part, does not participate in the consensus uploading of data, and only initiates transaction and query requests to the full node.

Accumulator. The accumulator is a cryptographic primitive. One of its applications is to prove whether a set contains an element. We use vector commitment(VC) [11], a position-bound accumulator, which allows us to quickly locate the required items in the set. A VC usually contains the following algorithms:

- $Paramgen(1^k, q) \rightarrow pp$: Given the security parameter k and the vector size q, output the public parameter pp.
- $Com(m) \rightarrow C, aux$: Input vector $m = m_1, m_2, \ldots, m_q$ and pp, output commitment C and auxiliary information aux.
- $Open(m, i, aux) \rightarrow \theta_i$: Input message vector m and position i, and generate proof θ_i of i position information m_i.
- $Verify(C, m, i, \theta_i) \rightarrow 0, 1$: Verify θ_i is the proof created on m_i, if it is, it outputs 1, otherwise it outputs 0.
- $ComUpdate(C, m, m', i) \rightarrow C', U$: Update the $i - th$ message of m, the new message becomes m', and output a new commitment C' and update-related information U.
- $ProofUpdate(C, \theta_j, m', i, U) \rightarrow \theta'_j$: After the message changes from m to m', update the proof of the j position message from θ_j to θ'_j.

RSA Assumption. Assuming two very large prime numbers $p, q, N = p \times q$, and $m = (p - 1)(q - 1)$, given an exponent $e(1 < e < m)$ coprime to m such that $gcd(m, e) = 1$. Given a random number y, compute x such that $y = x^e \bmod N$, and the RSA assumption holds if any adversary computes x with negligible probability that $x^e = y \bmod N$ [31]. In this work, the algorithm in vector commitment is implemented based on the RSA assumption.

3.2   System Components

Figure 1 illustrates the architecture of the proposed cross-chain system supporting efficient and verifiable complete data provenance queries in this work. The workflow of the system is roughly as follows: full node or light node users initiates cross-chain transactions to the parallel chain. The relay chain will listen for these cross-chain transactions and update the index structure and block headers based on their relationship. When light node users has a query need, they can initiate a request to the relay chain to achieve a complete query in an efficient manner and verify the integrity of the result after it is obtained. In the following text, we will introduce the functions of key components in the system.

Relay Chain. A relay chain is a channel for complete cross-chain transactions between blockchains, also responsible for reaching consensus on cross-chain transactions and storing them on the chain. In our system, the relay chain is also responsible for updating the index and generating the verification object(VO) based on the relationship between cross-chain transactions.

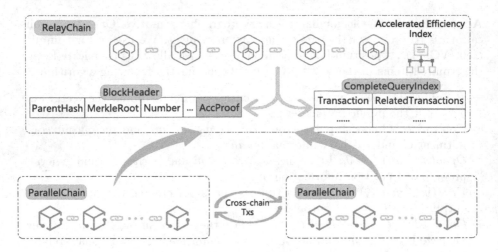

Fig. 1. Illustration of our scheme.

Parallel Chain. Parallel chains are blockchains in the traditional sense and are responsible for conducting ordinary transactions and cross-chain transactions. Ordinary transactions are only stored on the parallel chain; cross-chain transactions will be stored on both the parallel and relay chains after the relay chain completes consensus on them.

AccProof. We extend the *AccProof* field for the block header of the relay chain. This field stores the cryptographic proof generated by vector commitment [11] as VO. Light nodes can use this field to verify the integrity of the query result.

Complete Query Index (CQI). Complete Query Index is designed based on linear list and adjacent linked list for complete data provenance queries and is maintained in a relay chain. When a new cross-chain transaction pair occurs, the relay chain updates the index with it.

Accelerated Efficiency Index (AEI). The Accelerated Efficiency Index is designed based on B+ tree [12] and Merkle tree [13], which exist in relay chain and parallel chains. The characteristics of B+ trees ensure efficient querying and Merkle trees ensure verifiable authenticity of the queried data. The detailed construction process of the above components will be described in Sect. 4.

It's worth mentioning that *AccProof*, CQI, and AEI are all maintained within the block by miners, like transactions, these contents will be verified and stored on the blockchain after consensus by miners.

4 Solution

In this section, we will introduce how we address the challenges of cross-chain systems when performing queries in two parts: verifiable complete data provenance query and efficient query.

4.1 Verifiable Complete Data Provenance Query

The relay chain stores the cross-chain transactions after listening to them and extracts the link between them, then constructs a Complete Query Index (CQI) based on this link for integrity queries. After obtaining the entries in the CQI, we can move to the phase of generating VO. The relay chain will use the algorithm in vector commitment to generate the VO. Since the VO is provided to light nodes for verification, it is also necessary to expand an *AccProof* field for the block header and store the VO in this field. Light nodes can then perform integrity verification through *AccProof*.

To help readers understand our solution, we first define cross-chain transactions and related transactions as follows:

Cross-Chain Transactions. If there are blockchains $chain_1, ...chain_m$, and transactions $tx_1, ...tx_n$, tx_i belongs to $chain_x$ and tx_j belongs to $chain_y$ ($1 \leq i, j \leq n, 1 \leq x, y \leq m$, and $i \neq j, x \neq y$), tx_j is initiated by tx_i on $chain_i$ or tx_i is initiated by tx_j on $chain_j$, then tx_i and tx_j are a pair of cross-chain transactions.

Related Transactions. If tx_i and tx_j are a pair of cross-chain transactions, and tx_j and tx_k are a pair of cross-chain transactions, then tx_i, tx_j, and tx_k are related transactions.

CQI Construction. When the relay chain first listens to the occurrence of a cross-chain transaction, it uses the pair of cross-chain transactions to initialize the CQI. It adds all the cross-chain transactions to the linear list, and analyzes the transactions that are cross-chain or related to the entries in the linear list, then adds them to the adjacent linked list of that entry. When another new cross-chain transaction arrives, the latest state of CQI needs to be read out in the previous block first, and then the CQI is updated according to the relationship of these transactions. The update process is the same as the initialization process. In addition, The blockchain information that the transaction is on is also added to the CQI when the transaction is added.

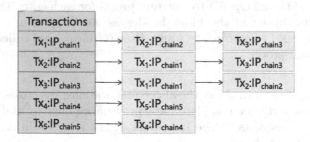

Fig. 2. Complete query index illustration.

Figure 2 shows an illustration of CQI. Assume the following cross-chain transaction pairs are sent to the relay chain: $tx_1(chain_1)$ $\longleftrightarrow$ $tx_2(chain_2), tx_2(chain_2)$ $\longleftrightarrow$ $tx_3(chain_3), tx_4(chain_4)$ $\longleftrightarrow$ $tx_5(chain_5)$. The index construction process is: tx_1 and tx_2 are added to the linear list and appended to each other's adjacent linked list; when listening to the cross-chain transaction relationship between tx_2 and tx_3, tx_3 is added to the linear list and they are appended to each other's adjacent linked list. Since tx_1 and tx_3 are related transactions, they will also be appended to each other's adjacent linked list. The other cross-chain transactions will update to CQI in the same process.

VO Generation. The algorithm for vector commitment needs to be constructed before generating the VO. The algorithm included in vector commitment has been introduced in Sect. 3, and here we present the specific implementation of RSA-based vector commitment [11] with the following algorithm:

- $Paramgen(1^k, l, q) \rightarrow pp$: Select two random prime numbers p_1 and p_2 of $\frac{k}{2}$ bits, and $N = p_1 \times p_2$, $m = (p_1 - 1)(p_2 - 1)$, and select the $q(l+1)$-bit prime numbers $e_1, e_2, \ldots, e_q$ that are relatively prime to m, and q is The length of the vector, calculate $S_i = a^{\prod_{j=1, j \neq i}^{q} e^j} (1 \leq i \leq q)$. Then $(N, a, S_1, \ldots, e_1, \ldots, e_q)$ will be used as a public parameter pp.
- $Com(m) \rightarrow C, aux$: Compute $C = S_1^{m_1} \times \ldots \times S_q^{m_q}$, output commitment C and auxiliary information $aux = (m_1, \ldots, m_q)$.
- $Open(m, i, aux) \rightarrow \theta_i$: Compute $\theta_i = \sqrt[e_i]{\prod_{j=1, j \neq i}^{q} S_j^{m_j}} \mod N$, output θ_i.
- $Verify(C, m, i, \theta_i) \rightarrow 0, 1$: If $C = S_i^m \times \theta_i^{e_i} \mod N$ then output 1, otherwise output 0.
- $ComUpdate(C, m, m', i) \rightarrow C', U$: Output $C' = C \cdot S_i^{m'-m}$ and $U = (m, m', i)$.
- $ProofUpdate(C, \theta_j, m', i, U) \rightarrow \theta_j'$: If $i = j$, then the proof remains the same θ_j; otherwise compute $\theta_j' = \theta_j \times \sqrt[e_i]{S_i^{m'-m}}$ and output θ_j'.

We can use these algorithm to generate *AccProof*. The process of generating *AccProof* is as follows: (1) Call $Paramgen()$ to generate public parameters; (2) Input all entries in the index linear table part as messages to $Com()$ to generate commitment C; (3) Call $Open()$ to generate proofs for each entry, the generated proofs It can be stored in the block header as *AccProof*. When an entry in the index is changed, $ComUpdate()$ and $ProofUpdate()$ are called to update commitments and proofs.

Query and Verification. After the above preparations, the cross-chain system already supports verifiable complete data provenance queries, and light nodes can launch query requests to the full nodes of the relay chain and verify the integrity of the results locally.

Query Process. When the light node wants to trace the complete provenance of a transaction on a certain parallel chain, the relay chain will search the linear list of CQI. If this transaction does not exist, it means that this transaction has no related transaction in other parallel chains, so it only needs to search in the parallel chain of this transaction; If it exists, it means that the transaction has related transactions on other parallel chains. It is necessary to read all related transactions and their parallel chain information in the adjacent linked list corresponding to this transaction, then search in parallel on each parallel chain to find the complete transaction provenance.

Verification Process. After the query results are obtained by the light node, it can verify the integrity of the results by using the verification object *AccProof* and the *Verify*() method of vector commitment. The function of *AccProof* is to verify whether a transaction exists in CQI. The light node needs to call *Verify*() for the transactions it has obtained. If the output is 0, it means that the transaction has no related transaction in other parallel chains and the complete result has been obtained. If the output is 1, it means that the CQI contains the transaction, that is, the transaction has related transactions in other parallel chains. At this time, CQI can be used to check all related transactions. If it does not get these transactions, the result is incomplete.

4.2   Efficient Query

The improvement of query efficiency in this section mainly lies in two parts: cross-chain query and parallel chain query.

The first part is cross-chain query. When querying the complete data provenance of a transaction, the parallel chain to which the transaction belongs must be searched first. After finding the earliest related transaction in this chain, it is necessary to judge whether this transaction is a cross-chain transaction. If not, the query is complete; If so, go to the other chain that corresponds to the cross-chain transaction, repeat the above process until all related transactions are found. This is a serial process. If the information of all parallel chains to be searched can be quickly obtained, the parallel search can be conducted on these parallel chains at the same time, which will greatly improve the query efficiency. Therefore, this work designed an accelerated efficiency index(AEI) for the relay chain based on B+ tree [12] and Merkle tree [13]. The leaf nodes store all parallel chain information related to cross-chain transactions, so that the parallel chain information can be quickly obtained for parallel query. The Merkle tree has the same function as it in the existing blockchain platform, providing light nodes with simple payment verification (SPV) [1].

The second part is the parallel chain query. Since the transaction is stored in the parallel chain, it is necessary to query the transaction in the parallel chain. There are dozens to tens of thousands of transactions in each block, parallel chain needs to start from the latest block, transaction by transaction comparison, which is the slowest query process. So we used the same method to set AEI for parallel chains. Different from the relay chain, we extended the transaction structure of the parallel chain block, so that the transaction can be stored in the block according to the attribute value. The leaf node of AEI stores the information of the transaction, while the branch node stores the ID attribute that uniquely identifies a transaction. This way can locate transactions quickly.

Algorithm 1: Node insert algorithm

Input : B+tree root: nodePointer, insert data: entry, child node: newChildEntry

Output: NewNode

1 **Function** INSERT(*nodePointer, entry, newChildEntry*)
2   create M = maxKeyCountOfOneNode
3   **if** *nodePointer isNotLeaf* **then**
4   create Node N = nodePointer
5   find nodei which node.firstKey <entry.key <node.lastKey
6   insert(nodei, entry, newChildEntry)
7   **if** *newChildEntry == null* **then**
8   return null
9   **else**
10   **if** *N.keyCount <M* **then**
11   add *newChildEntry to N
12   return newChildEntry = null
13   **else**
14   create a new Node N2
15   move last M/2 keys and pointers of N to N2
16   newChildEntry = &Node(N2.firstKey, &N2)
17   **if** *N isRootNode* **then**
18   create newNode = Node(&N, *N)
19   N.pointer = &newNode
20   return newChildEntry
21   **else**
22   create Node L = nodePointer
23   **if** *L.keyCount <M* **then**
24   add entry to L
25   return newChildEntry = null
26   **else**
27   create a new Node L2
28   move last M/2 keys and pointers of L to L2
29   newChildEntry = &Node(L2.firstKey, &L2)
30   L.pointer = &L2
31   return newChildEntry

AEI Construction. In the relay chain, AEI is mainly to quickly locate the related transactions and its chain information with a certain transaction. We extract every entry in the linear list of CQI as the key value of AEI, and take all the information in the adjacent linked list corresponding to each entry as the leaf node of the key value, and insert the node according to the rules of B+ tree. Algorithm 1 describes the algorithm of node insertion. After the construction of B+ tree, it is also necessary to hash from the leaf node in two upwards until the root of Merkle tree is generated to provide SPV function.

In the CQI fragment shown in Fig. 3, there are eight cross-chain transactions, $tx_1 - tx_8$, in the relay chain. We take the hash value of $tx_1 - tx_8$ in the linear List part as the key value of the AEI, insert nodes into the index shown in Fig. 4 according to Algorithm 1, and use the information of the adjacent linked list as the leaf node. When the leaf node where tx_1 is located is located, cross-chain transactions related to tx_1 can be read, including tx_1 on $parallelchain_1$ and tx_5 on $parallelchain_2$. Then related transactions can be searched on these two chains in parallel. The improvement of query efficiency by AEI in the relay chain is reflected in two aspects. On the one hand, using B+trees can quickly locate the parallel chain information of all related transactions. On the other hand, after obtaining these information, parallel queries can be performed on these parallel chains simultaneously, further improving retrieval efficiency.

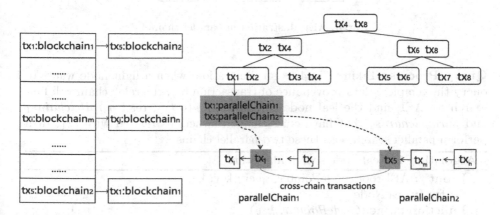

Fig. 3. CQI fragment.   **Fig. 4.** AEI illustration in relay chain.

In parallel chains, the function of AEI is to quickly find complete information about intra-block transactions based on their ID values. The construction process is the same as AEI in the relay chain, which is not detailed here. The difference is that the branch node of AEI in parallel chain stores the transaction ID value that can uniquely identify a transaction, while the leaf node points to the complete information of the transaction corresponding to a transaction ID. Figure 5 shows an illustration of AEI in a parallel chain. Several transactions are stored in the block with expanded transaction structure in the form of attribute values. Then, the value $key_1 - key_8$, which uniquely identifies a transaction, is used to construct the B+ tree part of AEI according to Algorithm 1. Each key value in the leaf node points to its corresponding complete transaction information. Similarly,

AEI also needs to perform a bottom-up hash operation in pairs until the Merkel tree root is generated, so that light nodes can verify the authenticity of the transaction.

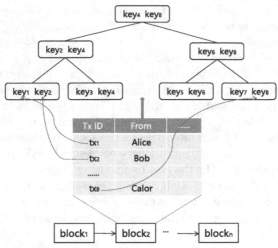

Fig. 5. AEI illustration in parallel chain.

Query Process. Taking Fig. 4 as an illustration, when a light node wants to query the complete data provenance of transaction tx_1, the relay chain will first search for AEI, find the leaf node corresponding to tx_1, read $parallelchains_1$ and $parallelchains_2$ that have cross-chain transactions related with tx_1, then perform parallel searches on these two parallel chains.

Algorithm 2: Leaf node search algorithm

    Input : AEI root: nodePointer, query key: key
    Output: LeafNode
1 **Function** SEARCH(*nodePointer, key*)
2   **if** *nodePointer isLeaf* **then**
3   return nodePointer
4   **else**
5   **if** $key <$ **nodePointer.nextKey* **then**
6   return search(*nodePointer.firstPointer,key)
7   **else if** $key >$ **nodePointer.lastKey* **then**
8   return search(*nodePointer.lastPointer, key)
9   **else**
10   **for** *each item in *nodePointer* **do**
11   **if** $item \leq key < nextItem$ **then**
12   return search(itemPointer, key)

The process of searching for leaf nodes is shown in Algorithm 2. Firstly, it is necessary to start searching from the root node. If the first keyword in the root node is larger than the keyword to be searched for, the subtree pointed to by the pointer represented by the first keyword should be removed to continue searching downwards; If the last keyword in the root node is smaller than the keyword to be searched for, then go to the subtree pointed to by the pointer represented by the last keyword to continue searching down; Otherwise, it is necessary to find a pointer where the left keyword is less than or equal to the pending keyword and the right keyword is greater than the pending keyword, and then repeat the above search process to the subtree it points to until the leaf node containing the pending keyword is found.

5   Experiment

In this section, we analyze the performance of our proposed schemes via evaluating its temporal and spatial expression on real datasets.

5.1   Experimental Setup

Settings. The experiment was run on the Intel Core i7-11700 2.50 GHz CPU. We built a cross-chain system with two parallel chains and one relay chain on Ubuntu 20.04. The function of parallel chain is to conduct common and cross-chain transactions, while the function of relay chain is to reach consensus on cross-chain transactions. We chose this structure because it's the most main-stream cross-chain structure, and parallel chains are independent of each other, even if the number of parallel chains is increased, cross-chain transactions still occur between two parallel chains in the system, so setting more parallel chains will not have a significant difference from our test results. We built the most popular blockchain platform - Ethereum client geth1.8 [2] as parallel blockchains, and a relay chain implemented using nodejs. Based on this work's solution, we have improved these blockchains to test the performance of our solution.

Dataset. In order to better match the actual requirements, we selected a supply chain dataset published by DataCo Global to test the system performance. This dataset contains more than 180,000 pieces of information about the circulation of real commodities in the supply chain. Each piece of information has 53 attributes. In this work, some key attributes are selected as transaction data and stored in the Ethereum blockchain we built in the format of <productName, productPrice, shippingDate, daysForShipping, orderId, orderCity, customerId, customerCity>. By inquiring the complete provenance of a certain commodity in the cross-chain system, instead of tracing the currency transaction information, it is closer to the common application logic, which can better demonstrate our solution's support for non-digital currency applications.

Performance Index. The core of this work is efficient verifiable query, so the performance of the verification object(VO) and query efficiency are important measurement criteria, including the following indicators:

(1) VO cost: Since VO is stored in the relay chain block, it's necessary to measure the impact of the time spent generating VO on block packaging and whether the space taken up by the VO imposes a storage burden on the blockchain node.
(2) CQI and AEI cost: Similar to VO, it is also necessary to measure the time and storage space occupied by CQI and AEI construction.
(3) Verification efficiency: The time taken by light nodes to verify the integrity of the results is also a factor affecting system availability.
(4) On-chain time: It is necessary to know whether the write performance of the system is affected. We take the time that the transaction is stored on the blockchain as the measurement standard, namely the on-chain time.
(5) Query efficiency: Finally, it is also necessary to measure how much query efficiency has improved after the introduction of indexes.

5.2   Results and Discussion

VO Cost. We evaluate the time taken to generate VO and the space occupied by VO in the block header when sending 2000, 4000, 6000, 8000, and 10000 cross-chain transactions.

From Fig. 6, when sending 2000–10000 cross-chain transactions, the time for generating VO only takes up about 1/5 of the time for generating blocks, is not the main part of the time for packaging blocks.

From Fig. 7, the upward trend of VO occupying space in the block header is relatively stable, and there will be no significant growth trend when increasing the number of transactions. Even when sending 10000 transactions, the size of VO is only 45 MB, which is relatively small compared to other common applications, and the vast majority of users will not feel storage pressure. And in practical applications, the number of cross-chain transactions is much smaller than the number of common transactions, which generally does not reach the number we tested. Therefore, the storage burden caused by adding VO in the block header on light nodes is also acceptable.

CQI and AEI Cost. We have set up CQI and AEI in the system, which require all nodes to build and update, inevitably increasing the time and storage overhead of the blockchain consensus on the chain, which may reduce the throughput of the blockchain and cause an explosion in block data volume. Therefore, we tested the time and space costs of generating CQI and AEI when sending different numbers of transactions, And compare the cost of generating blocks to verify whether our system is practical.

In terms of time cost, Fig. 8 and Fig. 9 show that the time for generating CQI is negligible compared to the time for generating blocks. When the number of transactions increases exponentially, this time only increases slightly; The time

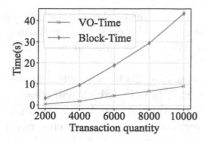

Fig. 6. VO time cost.

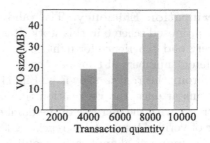

Fig. 7. VO space cost.

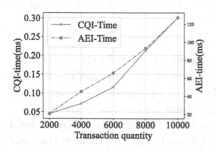

Fig. 8. Index time cost.

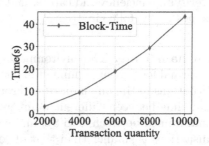

Fig. 9. Block time cost.

to generate AEI is slightly higher than CQI, because the process of building a B+tree itself is much more complex than building a Linear List, but it only accounts for about 0.4% of the time to generate blocks, which will not have too much impact on the throughput of the cross-chain system.

In terms of space cost, Fig. 10 shows CQI occupies approximately 0.3% of the space occupied by the block, while AEI occupies larger space than CQI, accounting for about 15% of the block space, both not the main part of the block. Moreover, as the number of transactions increases, this value will correspondingly decrease, with little impact on the storage burden of the full node.

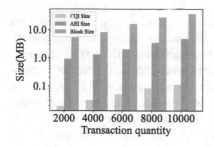

Fig. 10. Index space cost.

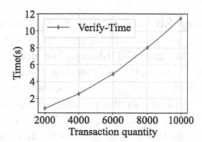

Fig. 11. Verification efficiency.

Verification Efficiency. The validation efficiency reflects the practicality of the proposed scheme in this work. If the validation time is too long, it will cause a very bad experience for light node. Therefore, We test the time taken to verify different numbers of transactions.

From the results shown in Fig. 11, when verifying 2000 cross-chain transactions at once, the verification time is approximately 0.8 s, which will steadily increase with the increase of the number of verified transactions. When the number of verified transactions reaches 10000, the verification time will reach around 11.5 s. In practical applications, only single digit transactions will be verified, and the verification time will be very small. So our scheme performs relatively well in verifying efficiency indicators, and we will compare this indicator with other schemes in the following text.

On-Chain Time. The environment cannot always be stable in the same state, it's normal for the same number of common and cross-chain transactions to have fluctuations in the on-chain time. From Table 1, we can see the difference in on-chain time between different numbers of cross-chain and common transactions is maintained between 1 to 8 s, and sometimes the on-chain time of cross-chain transactions is smaller than that of common transactions, It indicates that our system will not have a significant impact on the on-chain time.

Table 1. Comparison of on-chain time between common transaction and cross-chain transaction.

Unit: second(s)		Transaction quantity				
		2000	4000	6000	8000	10000
$Parallelchain_1$ on-chain time	common transaction	12	36	55	85	116
	cross-chain transaction	11	38	61	85	121
$Parallelchain_2$ on-chain time	common transaction	9	35	51	83	120
	cross-chain transaction	14	42	59	79	120

Query Efficiency. Due to the uncontrollable number of transactions in each block, we tested the query time of two parallel chains while searching for 10, 50, 100, and 150 blocks. We first initiate transactions with certain filtering conditions (such as querying transactions with productName = "apple"). Subsequently, cross-chain queries were conducted with and without AEI. As shown in Fig. 13, when the number of search blocks is the same, the time spent using AEI for multi chain parallel traceability is much smaller than when no AEI is used. As the number of blocks increases, using AEI can shorten the query time by nearly half, and it is already close to the time spent querying a parallel chain in Fig. 12. This proves that AEI significantly improves query performance.

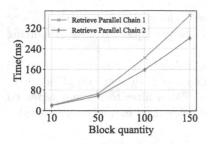

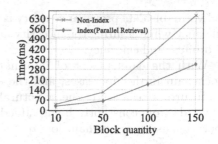

Fig. 12. Parallel chain query time.

Fig. 13. Cross-chain query time.

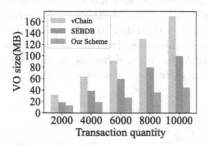

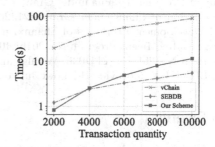

Fig. 14. VO size comparison.

Fig. 15. Verification time comparison.

Comparative Experiment. At last, we compare the space cost and validation time of VO in our scheme with representative works in this field, vChain [24] and SEBDB [19]. Similarly, in the case of sending different numbers of cross-chain transactions, the size of VO in vChain, SEBDB, in this work, as well as the time it took for the light nodes to verify these cross-chain transactions were tested.

From Fig. 14, it can be seen that the size of VO used in this work is approximately 30% of vChain and 45% of SEBDB, indicating a significant improvement. Figure 15 shows the comparison results of the time spent on verification. Compared with vChain, the verification time of our scheme has been reduced by about 90%. Although when the transaction number reaches 4000, there is a slight increase compared to SEBDB, this is determined by the nature of the accumulator. The accumulator used in this work will experience a greater increase compared to SEBDB when the transaction number increases. But the VO in SEBDB are used for authenticity queries, the VO used in this work can provide integrity queries that SEBDB does not have, our functionality is superior to SEBDB.

6 Conclusion

This work shows that the current cross-chain system may have incomplete query results and users cannot verify the integrity of query results. Combined with the low retrieval efficiency in blockchain, a cross-chain system supporting efficient and verifiable data provenance query is proposed. The scheme can improve query efficiency while ensuring the integrity and verifiability of query results. Balancing credibility and efficiency.

With the development of information technology, the amount of data on blockchain further increases, and the query demand is further complicated. In the future, we can study the structural such as Bloon filter to prevent repeated keywords from constantly being added to the index, and extend more complex operations such as on-chain join queries.

Acknowledgement. This work is supported by the National Key Research and Development Program of China under Grant 2022YFB3102700, the National Natural Science Foundation of China under Grant 62132013, 62102295, 62202358, the Key Research and Development Programs of Shaanxi under Grants 2021ZDLGY06-03, the Guangdong Leading Talent Program No. 2016LJ06D658 and the Guangdong Innovation Team Program No. 2018KCXTD030. We also appreciate Alibaba Cloud Intelligent Computing LINGJUN to provide the powerful computation ability in the experiments.

References

1. Bitcoin, N.S.: A peer-to-peer electronic cash system. Decentral. Bus. Rev. 21260 (2008)
2. Buterin, V.: A next-generation smart contract and decentralized application platform. White Paper **3**(37), 2–1
3. Androulaki, E., Barger, A., et al.: Hyperledger fabric: a distributed operating system for permissioned blockchains. In: Proceedings of the Thirteenth EuroSys Conference, pp. 1–15 (2018)
4. Biswas, D., Jalali, H., Ansaripoor, A.H., et al.: Traceability vs. sustainability in supply chains: the implications of blockchain. Eur. J. Oper. Res. **305**(1), 128–147 (2023)
5. Patel, R., Migliavacca, M., Oriani, M.: Blockchain in Banking and Finance: is the best yet to come? A bibliometric review. Res. Int. Bus. Fin. 101718 (2022)
6. Taherdoost, H.: Blockchain-based internet of medical things. Appl. Sci. **13**(3), 1287 (2023)
7. Belchior, R., et al.: A survey on blockchain interoperability: past, present, and future trends. ACM Comput. Surv. (CSUR) **54**(8), 1–41 (2021)
8. Peng, Z., Wu, H., Xiao, B., et al.: VQL: providing query efficiency and data authenticity in blockchain systems. In: 2019 IEEE 35th International Conference on Data Engineering Workshops (ICDEW), pp. 1–6. IEEE (2019)
9. Wood, G.: Ethereum: a secure decentralised generalised transaction ledger. Ethereum Project Yellow Paper **2014**(151), 1–32 (2014)
10. Zhu, Y., Zhang, Z., Jin, C., et al.: Towards rich Query blockchain database. In: Proceedings of the 29th ACM International Conference on Information & Knowledge Management, pp. 3497–3500 (2020)
11. Catalano, D., Fiore, D.: Vector commitments and their applications. In: Kurosawa, K., Hanaoka, G. (eds.) PKC 2013. LNCS, vol. 7778, pp. 55–72. Springer, Heidelberg (2013). https://doi.org/10.1007/978-3-642-36362-7_5
12. Graefe, G.: Modern B-tree techniques. Found. Trends® Databases **3**(4), 203–402 (2011)

13. De Ocáriz Borde, H.S.: An Overview of Trees in Blockchain Technology: Merkle Trees and Merkle Patricia Tries (2022)
14. Muzammal, M., Qu, Q., Nasrulin, B., et al.: A blockchain database application platform. arXiv preprint arXiv:1808.05199 (2018)
15. Armknecht, F., Karame, G.O., Mandal, A., Youssef, F., Zenner, E.: Ripple: overview and outlook. In: Conti, M., Schunter, M., Askoxylakis, I. (eds.) Trust 2015. LNCS, vol. 9229, pp. 163–180. Springer, Cham (2015). https://doi.org/10.1007/978-3-319-22846-4_10
16. McConaghy, T., Marques, R., Müller, A., et al.: Bigchaindb: a scalable blockchain database. white paper, BigChainDB (2016)
17. Bradshaw, S., Brazil, E., Chodorow, K.: MongoDB: The Definitive Guide: Powerful and Scalable Data storage. O'Reilly Media, Sebastopol (2019)
18. Li, Y., Zheng, K., Yan, Y., Liu, Q., Zhou, X.: EtherQL: a query layer for blockchain system. In: Candan, S., Chen, L., Pedersen, T.B., Chang, L., Hua, W. (eds.) DASFAA 2017. LNCS, vol. 10178, pp. 556–567. Springer, Cham (2017). https://doi.org/10.1007/978-3-319-55699-4_34
19. Zhu, Y., Zhang, Z., Jin, C., et al.: SEBDB: semantics empowered blockchain database. In: 2019 IEEE 35th International Conference on Data Engineering (ICDE), pp. 1820–1831. IEEE (2019)
20. Ruan, P., Dinh, T.T.A., Lin, Q., et al.: LineageChain: a fine-grained, secure and efficient data provenance system for blockchains. VLDB J. **30**, 3–24 (2021)
21. Wang, S., Dinh, T.T.A., Lin, Q., et al.: Forkbase: an efficient storage engine for blockchain and forkable applications. arXiv preprint arXiv:1802.04949 (2018)
22. Zhou, E., Hong, Z., Xiao, Y., et al.: MSTDB: a hybrid storage-empowered scalable semantic blockchain database. IEEE Trans. Knowl. Data Eng. (2022)
23. Mahapatra, A.K., Biswas, S.: Inverted indexes: types and techniques. Int. J. Comput. Sci. Issues (IJCSI) **8**(4), 384 (2011)
24. Xu C, Zhang C, Xu J.: vchain: enabling verifiable Boolean range queries over blockchain databases. In: Proceedings of the 2019 International Conference on Management of Data, pp. 141–158 (2019)
25. Wang, H., Xu, C., Zhang, C., et al.: vChain+: optimizing verifiable blockchain Boolean range queries. In: 2022 IEEE 38th International Conference on Data Engineering (ICDE), pp. 1927–1940. IEEE (2022)
26. Frauenthaler, P., Sigwart, M., Spanring, C., et al.: ETII relay: a cost-efficient relay for ethereum-based blockchains. In: 2020 IEEE International Conference on Blockchain (Blockchain), pp. 204–213. IEEE (2020)
27. Poon, J., Dryja, T.: The bitcoin lightning network: scalable off-chain instant payments (2016)
28. Hope-Bailie, A., Thomas, S.: Interledger: creating a standard for payments. In: Proceedings of the 25th International Conference Companion on World Wide Web, pp. 281–282 (2016)
29. Kwon, J., et al.: A network of distributed ledgers. Cosmos. Accessed 1–41 2018
30. Wood, G.: Polkadot: vision for a heterogeneous multi-chain framework. White Paper **21**(2327), 4662 (2016)
31. Milanov, E.: The RSA algorithm. RSA Laboratories, 1–11 (2009)

Enabling Traffic-Differentiated Load Balancing for Datacenter Networks

Jinbin Hu, Ying Liu, Shuying Rao, Jing Wang, and Dengyong Zhang(✉)

School of Computer and Communication Engineering,
Changsha University of Science and Technology,
Changsha 410004, China
{jinbinhu,znwj_cs,zhdy}@csust.edu.cn,{yingliu,shuyingrao}@stu.csust.edu.cn

Abstract. In modern datacenter networks (DCNs), load balancing mechanisms are widely deployed to enhance link utilization and alleviate congestion. Recently, a large number of load balancing algorithms have been proposed to spread traffic among the multiple parallel paths. The existing solutions make rerouting decisions for all flows once they experience congestion on a path. They are unable to distinguish between the flows that really need to be rerouted and the flows that potentially have negative effects due to rerouting, resulting in frequently ineffective rerouting and performance degradation. To address the above issues, we present a traffic-differentiated load balancing (TDLB) mechanism, which focuses on distinguishing flows that necessarily to be rerouted and employing corresponding measures to make optimize routing decisions. Specifically, TDLB detects path congestion based on queue length at the switches, and distinguishes the traffic that must be rerouted through the host pair information in the packet header, and selects an optimal path for rerouting. The remaining traffic remains on the original path and relies on congestion control protocols to slow down to alleviate congestion. The NS-2 simulation results show that TDLB effectively reduces tailing latency and average flow completion time (FCT) for short flows by up to 45% and 46%, respectively, compared to the state-of-the-art load balancing schemes.

Keywords: Datacenter networks · Load balancing · Traffic differentition · Multiple paths

1 Introduction

With the stringent requirements of low latency and high throughput for various datacenter applications like high-performance computing, big data analytics, and

This work is supported by the National Natural Science Foundation of China (62102046, 62072056), the Natural Science Foundation of Hunan Province (2023JJ50331, 2022JJ30618, 2020JJ2029), the Hunan Provincial Key Research and Development Program (2022GK2019), the Scientific Research Fund of Hunan Provincial Education Department (22B0300), the Changsha University of Science and Technology Graduate Innovation Project (CLSJCX23101).

Z. Tari et al. (Eds.): ICA3PP 2023, LNCS 14489, pp. 250–269, 2024.
https://doi.org/10.1007/978-981-97-0798-0_15

Internet of Things (IOT), providing high-performance network services in datacenters is of paramount importance [1–5]. Modern datacenter networks (DCNs) widely adopt a multi-rooted tree network topology to establish multiple parallel paths among host pairs, providing high-bandwidth network transmission capabilities [6,7]. In recent years, numerous outstanding network processing technologies have emerged to fully leverage the multipath characteristics of datacenter networks for achieving high-performance network transmission. Among these technologies, load balancing mechanisms play a critical role in datacenter networks by effectively distributing the workload across multiple paths, mitigating link failures, and maximizing the utilization of parallel paths, thereby improving link utilization and system performance [8–10].

In order to enhance performance in multiple paths transmission scenarios, various load balancing mechanisms have been proposed to improve the transmission performance of the network. ECMP (Equal Cost Multi-Path) [11], which is presently most widely used load balancing mechanism, balances traffic poorly under multiple paths. ECMP uses a hash algorithm to randomly select forwarding paths for traffic, balancing traffic to multiple parallel paths to reduce path load and network congestion, and achieve high chain utilization and low latency transmission. However, ECMP fails to take advantage of the congestion information and traffic characteristics of the network, and hash routing can easily cause hash conflicts between different data flows on the transmission path, resulting in load imbalance and exacerbating network congestion hotspots, wasting network bandwidth.

Therefore, many enhanced load balancing mechanisms have been proposed to address issues such as hash conflicts and congestion hotspots in ECMP. In contrast, other solutions adopt useful information on the network. CONGA [12] selects the optimal transmission path for data flows by collecting global congestion information in real-time. DRILL [13] uses the queue information of the local switch and selects the forwarding port based on the queue length. LetFlow [14] detects path congestion and randomly forwards traffic along different paths at fixed time intervals. Hermes [15] detects global congestion and performs rerouting based on path conditions and data flows status. They effectively resolve hash collision issues, to a certain extent reduce path congestion, and improve network performance through making full use of information on multiple paths.

However, despite the excellent performance of load balancing mechanisms in many aspects, previous solutions have had a few significant shortcomings. The existing load balancing mechanism only triggers rerouting when congestion is perceived, without distinguishing the traffic that truly needs to be rerouted [16–18]. As a result, it does not perform differentiation processing and directly reroutes all traffic on the congested path. These solutions have to some extent alleviated the problem of link workload, but coarse-grained rerouting solutions cannot solve path congestion. Specifically, a small amount of congested traffic has not effectively solved the original problem through load balancing and rerouting [19–21]. This is because when these congested flows are rerouted, they may cause congestion to spread and affect other transmission paths, unable to solve congestion hotspots. In addition, it will increase the queuing latency of some flows, leading to an increase in flow completion time. Therefore, these congested flows can only be alleviated by reducing the transmission rate of the link through congestion control mechanisms [22,23].

To address these issues, we aim to provide a more intelligent and efficient load balancing solution to further optimize network performance and improve user experience [24–26]. We propose a traffic-differentiated load balancing mechanism (TDLB) primarily based on the principle of distinguishing congested flows and prioritizing traffic. Firstly, we introduce a traffic recognition mechanism that accurately identifies the traffic that truly needs to be rerouted and performs special processing on it. Secondly, we have designed a priority scheduling algorithm that allocates network resources reasonably based on the importance and priority of traffic, ensuring that high priority traffic is processed with priority. Among these aspects, we focus on how to distinguish traffic that truly needs to be rerouted and how to minimize the disruption caused by rerouting operations [15,16]. In addition, we conduct performance comparisons between symmetric and asymmetric network topologies to evaluate the performance of TDLB.

1) We conduct in-depth research and analysis on two main issues during path congestion, including the increased tailing latency caused by undifferentiated traffic that truly needs to be rerouted, and the longer flow completion time caused by reordering resulting from rerouting.
2) We determine the path status by evaluating the queue length of the switch and design a traffic recognition mechanism to differentiate between flows that contribute to congestion and flows that truly need to be rerouted.
3) We propose a fine-grained load balancing mechanism TDLB that effectively separates traffic on congested paths and implements different scheduling schemes for flows with different characteristics.
4) We implement TDLB using NS-2 simulation, which effectively improves transmission performance under congested conditions and significantly reduces tailing latency and average flow completion time (AFCT).

In Sect. 2, we introduce the main motivation of this paper. In Sect. 3, we provide a detailed overview of the design and its details. In Sect. 4, we conduct experimental simulations and analysis, including the analysis of relevant comparative results. In Sect. 5, we introduce some representative related work. In the final section, we summarize the work of this paper.

2   Motivation

However, current load balancing mechanisms (such as ECMP, CONGA, and Hermes) have a common problem in that they all reroute all flows on congested paths without distinguishing between those that truly need to be rerouted. In this scenario, there are two serious issues. Firstly, the traffic that truly needs to be rerouted may still choose the same path as the traffic that causes congestion, leading to an increase in tail latency. Secondly, rerouting increases the chance of

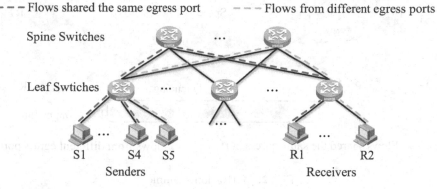

- - - Flows shared the same egress port   - - - Flows from different egress ports

Spine Switches

Leaf Swtiches

S1   S4   S5   R1   R2

Senders   Receivers

Fig. 1. Leaf-spine topology

disorder, which not only fails to reduce the flow completion time of congested flows but also leads to longer FCT [27,28].

Study Case: In order to better illustrate the above issues in existing load balancing mechanisms, we will use a simple example to illustrate. As shown in Fig. 1, this case adopts a common leaf-spine topology, where two leaf switches L1 and L2 are connected to two sending terminals (S1, S2) and two receiving terminals (R1, R2), respectively. The two leaf switches are fully connected to three spine switches, meaning there are three identical parallel paths between L1 and L2, which have the same link bandwidth and latency. S1 sends 4 flows ($f1-f4$) to R1 at T1 moment, and S2 subsequently sends 1 flow $f5$ at T2 moment. We use five typical load balancing mechanisms to handle this network transmission scenario.

As depicted in Fig. 2, whether it is ECMP at the flow level, LetFlow and CONGA also at the flow level, or DRILL and Hermes at the packet level, they may all be directed to the same path when handling flows ($f1-f5$) on leaf switch L1. Moreover, a significant number of flows become congested in the queues of L2 switches, with 80% of the traffic being sent to the receiving terminal R1 and the remaining 20% to the receiving terminal R2. Consequently, this link experiences congestion, prompting the aforementioned load balancing mechanisms to redirect these flows and packets to the upper-level switch L1, employing different processing schemes to reroute this portion of traffic.

We analyz the impact of congestion on the completion time and tailing latency of short flows through NS-2 simulation testing [25]. We use a leaf-spine topology with three equal-cost paths between host pairs. The bottleneck bandwidth is 40 Gbps, and the round-trip propagation latency is 10 μs. Each sender sends a DCTCP flow to the receiver through leaf and spine switches [23]. In this sample experiment, a mixture of a large number of flows from the same receiver and a small number of flows from other receivers are randomly generated under different network loads.

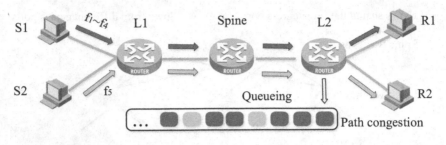

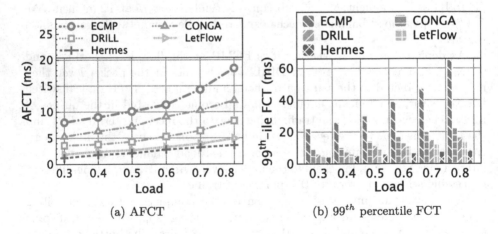

Fig. 2. Motivation example

A large number of flows (*f1–f4*) with the same outbound port and a small number of flows *f5* with different outbound ports are crowded on the same transmission path, where flows (*f1–f4*) are sent from leaf switch L2 to receiver R1, while flow *f5* is sent from leaf switch L2 to receiver R2. Obviously, this may lead to network performance bottlenecks and increases transmission latency. This is because a large amount of traffic shares the same transmission path, which may not be able to handle all traffic simultaneously. If not processed, a small amount of flows *f5* sent to different receiving ports will be blocked in the output queue of leaf switch L2, seriously increasing the FCT of *f5*. Therefore, this situation requires a load balancing mechanism to schedule its flows.

(a) AFCT

(b) 99th percentile FCT

Fig. 3. Performance under different workloads

Large Tailing Latency. The existing load balancing mechanisms first perceive high load or congested paths and propagate all information on that path back to the upstream switch. Then, according to the corresponding load balancing mechanism, this portion of traffic is rerouted to other parallel paths to reduce flow completion time and improve link utilization. As shown in Fig. 3 (a), as the network load increases, AFCT continues to increase. Figure 3 (b) show the 99th percentile FCT for different load balancing mechanisms as network load increases. We observed that the tailing latency is approximately twice the average flow completion time. It can be inferred that some flows have encountered severe congestion, despite load balancing and rerouting of congested paths, congestion still exists after switching paths. It is worth emphasizing that the tailing latency of ECMP is much higher than other load balancing mechanisms. This is because ECMP uses a hash method and does not have a congestion aware mechanism, only spreading the flow to different paths for forwarding. For paths that have already generated congestion, it is likely to exacerbate the congestion of the path, and generating hash collisions will also increase the possibility of link congestion. Overall, these congested flows are also rerouted and forwarded to other paths, which is likely to continue to cause congestion on other paths and ultimately increase the tailing latency of the flow.

Large Average Latency. The completion time of the flow actually increases. In Fig. 2, the flows ($f1$–$f4$) sent to the same receiving terminal experience congestion on both the receiving terminal and the last hop switch. In this case, these flows are processed through load balancing scheduling, and the traffic on the entire congested path is rerouted and forwarded to other parallel paths. Figure 3 (a) shows the average flow completion time under different load balancing mechanisms as the load increases. Despite rerouting the flow on congested paths, the overall average flow completion time has significantly increased. When the network load is 0.3, the AFCT of Hermes is 1.2 ms, which has reached a higher latency level compared to the link latency of 10 μs. ECMP mechanism based on hash routing has achieved an average flow completion time of 8 ms at a load of 0.3. This is because the ECMP does not utilize any information from the link, and the re-selected path is random, which may result in choosing a worse path. In addition, if the flow on a congested path is rerouted and redirected to another path, it may still face congestion, increasing the possibility of disorderly transmission [6,14]. The flow that causes congestion cannot reduce the FCT by rerouting, on the contrary, disorderly arrival actually leads to an increase for the FCT.

In brief, we analyze the impact of existing load balancing mechanisms on network performance under congested conditions and come to the following conclusions: (i) Flows that truly need rerouting can still potentially choose the same path as the congesting traffic, resulting in increasing tail latency. (ii) Congestion-induced flows, after rerouting, increase the chances of packet reordering and are unable to reduce flow completion time, leading to longer completion times. These conclusions drive us to propose a congestion-aware load balancing mechanism that distinguishes between flows that genuinely require rerouting and congested flows, aiming to achieve low-latency, high-performance transmission.

3   TDLB Design

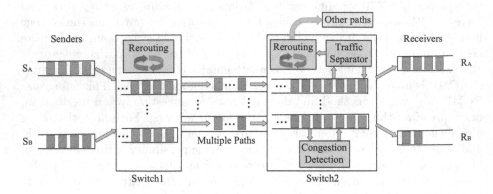

Fig. 4. The architecture of TDLB

3.1   Design Overview

In this section, we provide an overview of the design of TDLB. The key to TDLB is to distinguish the traffic that needs to be rerouted, flexibly select the ideal route for this portion of the traffic, and effectively adjust the traffic on congested links. Specifically, on the one hand, identifying flows that truly need to be rerouted and rerouting them to other better parallel paths may reduce serious queuing and tailing latency. On the other hand, to prevent the traffic that causes congestion from being rerouted to other paths, causing congestion spread and affecting the normal transmission of other paths, and to prevent rerouting from increasing the chance of disorder and increasing the FCT. Figure 4 demonstrates the framework of TDLB, which comprises of three components.

1) **Congestion Detection:** TDLB performs congestion detection on the switch that is the last hop from the receiving terminal. Firstly, determine whether congestion has occurred on the path according to the queue length of the switch. Subsequently, the switch sends link congestion notifications to prevent the path congestion from becoming more serious [17, 20].
   Specifically, switches usually establish a queue for each port to store incoming packets. Through monitoring the queue length of the switch (i.e., the total amount of packets in the queue), it is possible to inadvertently figure out whether congestion has occurred on this path. When the queue length exceeds the threshold set in advance, congestion can be regarded as having occurred.
2) **Traffic Separation:** The traffic separator is employed to separate congested traffic and other harmless traffic on congested paths, and to forward innocent traffic to other parallel paths through the rerouting mechanism on the local switch.

When the switch detects congestion, it indicates that some traffic has affected the normal transmission of the path, which is the flow causing congestion. Only when congestion occurs in the link, TDLB will employ the traffic separator to distinguish between the congested traffic and other innocent traffic, and deal with them in a fine-grained method to alleviate path congestion.

3) **Truly Necessary Rerouting:** For innocent traffic on congested paths, selects a better transmission path based on the current network state and the queue length of the switch forwarding port.

   To further alleviate network congestion, TDLB transforms its innocent flows from congested paths to other available non-congested paths. This can reduce the workload on network congestion, improve overall network performance and service quality.

3.2 Design Rationale

In the current load balancing mechanism, there is no distinguishing flows on the congested path that truly needs to be rerouted. They are forwarded to other parallel paths in the same method, resulting in increasing congestion and tailing latency. Specifically, when a path becomes congested, not all traffic is negative. Only a portion of the traffic is actually causing congestion, while the remaining traffic is innocent normal traffic. If no processing is carried out, the traffic causing congestion will continue to block the queue exit, and innocent traffic waiting in the queue will also become a criminal flow causing congestion. This will undoubtedly increase the FCT of traffic on the path, as well as the tailing latency. The load balancing mechanism for this case will forward every flow on the congested path to the higher-level switch, which will subsequently reroute depending on a specific routing algorithm. The existing load balancing mechanisms reroutes all flows without distinguishing between congested and innocent flows, which can cause congestion spread in other paths and increase the possibility of packet disorder, increasing the FCT of congested traffic.

We address these issues from different perspectives focusing on the topic of whether it is possible to explore a novel load balancing mechanism with the following design goals: (i) Detecting path congestion in a more effective and quicker manner; (ii) Designing an efficient flow separator to distinguish traffic that truly need to be rerouted; (iii) Making sure rerouting to better parallel paths to achieve efficient network load balancing.

3.3 Design Detail

In this section, we provide a detailed explanation of the key designs in each part of TDLB.

Congestion Detection: The switch maintains a queue for storing packets, and whenever a packet arrives at the switch and cannot be immediately forwarded, it will be added to the queue [30,31]. When the path becomes congested, the switch ports become blocked, and the queue length gradually increases. At the same time, the path delay also increases. If relying solely on queue length to perceive congestion, it may lead to misjudgment or omission. Therefore, the exchange opportunity regularly checks the length of the queue and sets a threshold of Φ for the queue length in advance. When the length of the queue exceeds this threshold, we believe it may cause congestion. To ensure the accuracy of congestion detection, we also need to measure whether the path delay changes when the queue length exceeds the threshold. When the threshold is exceeded and the path delay increases, we determine that congestion has occurred. We set Φ to the half of queue length. It is worth noting that this article adopts DCTCP [23] as the congestion control algorithm.

Traffic Separation: The existing load balancing mechanism is unable to differentiate traffic in congested cases, and coarse-grained routing leads to an increase in FCT and tail latency [28,32,33]. Inspired by the distinction between long and short flows to reduce FCT of short flows, we differentiate between congested and innocent flows and perform fine-grained processing based on the characteristics of the traffic. When congestion is detected, distinguish the flow that truly needs to be rerouted to achieve the separation of the congested flow from the innocent traffic.

Specifically, we identify the traffic causing congestion by analyzing traffic characteristics and utilizing traffic statistics data. Firstly, it is necessary to establish classification rules for the traffic in the switch and classify it based on the characteristics of the traffic. Generally speaking, traffic can be distinguished by five tuples (source IP address, destination IP address, source port number, destination port number, protocol type). In TDLB, we have developed a hash array for distinguishing congested traffic to record the hash value and quantity of traffic quintuples. At the same time, a congestion traffic flag has been added to the header of the data packet.

Algorithm 1: Distinguishing congested flows algorithm

Input: A packet belongings to flow f
Output: Congestion flow tag T

1   **if** *cur_queue_length $>$ queue_threshold and cur_path_delay $>$ default_delay* **then**
2 | Record current congestion flows into the **array**;
3 | Mark congestion flow tag $T \to$ True ;
4   **end**
5   **for** *every packet* **do**
6 | Assume the corresponding flow is f;
7 | **if** *hash(f) exists in array* **then**
8 | Mark congestion flow tag $T \to$ True ;
9 | **else**
10 | Mark congestion flow tag $T \to$ False
11 | **end**
12 | return T
13   **end**

The mechanism for distinguishing congested flows is shown in Algorithm 1. When the switch detects congestion, the hash array will treat all recorded traffic as causing the congestion. After other subsequent traffic through hash operations, check if the hash value exists in the array. If it does, mark the data packet with congestion, and increase the count corresponding to the hash value by one; Otherwise, it is considered innocent traffic and truly needs to be rerouted.

Truly Necessary Rerouting: The rerouting rules play a crucial role in the performance of load balancing mechanisms [34,35]. It ensures high reliability transmission of the network through reselecting other parallel paths in the event of link congestion or network topology changes. The switch maintains a routing table for storing routing information in the network. When a link failure or network topology change occurs, the switching opportunity updates the routing table accordingly to reflect the new network state.

The rerouting operation is used by local switches experiencing congestion to handle a small amount of innocent traffic on congested paths. In order to better handle the differences between congested and non congested flows, a congestion separation based rerouting algorithm is proposed. TDLB uses a traffic separation mechanism to create congestion markers for congested and innocent flows, which can clearly distinguish congested flows on the path. Congestion flows will be assigned lower priority and kept on the path waiting for congestion to end, in order to limit their transmission rate and reduce the impact on network congestion; Innocent flows will be assigned higher priority to obtain better network resources. TDLB reroutes these innocent flows without congestion markers and forwards them to good paths to ensure better quality of service.

The rerouting mechanism of TDLB is deployed on the incoming port of the switch, determining the forwarding path based on the current queue length and link delay of the switch, in order to route traffic that truly necessary rerouting to other better path. Specifically, when data packets enter the switch, the detected innocent traffic will trigger the rerouting mechanism to be forwarded to other better parallel paths. For the selection of parallel paths, it is determined based on the queue length of the switch's output port, scanning to find the output port with the smallest queue length, which is the current optimal path. It is worth noting that using only queue length for rerouting cannot effectively determine the optimal path. Therefore, when the queue length of different outgoing ports in the switch is the same, it is also necessary to select a path with smaller latency through link latency. By comparing the queue length and link delay of multiple paths, a better forwarding path can be obtained and congestion on that path can be effectively alleviated and eliminated.

4   Evaluation

We conduct experiments with two classic workloads, Web Search and Data Mining, to evaluate the performance of TDLB in large-scale scenarios [6,36]. In the above workloads, the distribution of flow size is heavy-tailed, and the specific flow size distribution is shown in Table 1. In general, flows smaller than 100 KB are regarded as short flows, while others are considered long flows. Due to the small size of short flows, special attention is paid to the AFCT, while long flows are more concerned with higher throughput. In terms of traffic size distribution, under Web Search workload, approximately 20% of traffic is greater than 1 MB, while in Data Mining workload, less than 5% of traffic is greater than 35 MB. Web Search and Data Mining have 62% and 83% of short flows smaller than 100 KB, respectively, resulting in a heavy-tailed distribution.

Table 1. Flow size distribution of two workloads

	Web Search	Data Mining
0-100KB	62%	83%
100KB-1MB	18%	8%
>1 MB	20%	9%
Average flow-size	1.6 MB	7.4 MB

The flow size distribution of Web Search and Data Mining usually presents a long tail distribution. For Web Search, this means that a few popular search terms will account for the majority of traffic, while most search terms only contribute a small proportion of traffic. Although the proportion of long wake flow is relatively low, they still have certain importance. Long wakes may play a crucial role in specific scenarios or problems. Therefore, it is necessary to pay special attention to the transmission quality of long wakes in the network to avoid the negative impact of severe tailing on network performance as much as possible.

We compare TDLB with five state-of-the-art load balancing mechanisms, including ECMP, CONGA, DRILL, LetFlow, and Hermes. ECMP uses static hashing to allocate each flow to an equivalent multipath. CONGA selects the optimal transmission path for data flows through real-time feedback of global congestion information. DRILL selects the port for packet forwarding based on the length of the local queue. Letflow classifies flowlets using fixed time intervals and randomly selects forwarding ports for each flowlet. Hermes reroutes based on path conditions and data flows status.

We still use the same leaf-spine topology as the motivation as the experimental network topology, which consists of 8 leaf switches and 9 spine switches. Each leaf switch connects 32 terminals, so the entire network has 256 independent terminals connected through a 40 Gbps bandwidth link, with a round-trip propagation latency of 10 us. The traffic in the network is generated randomly through the Poisson distribution between random host pairs. We adjust the workload from 0.3 to 0.8 to thoroughly evaluate the performance of TDLB.

In the following two experiments, we use symmetric and asymmetric network topologies for performance evaluation. A symmetric topology structure means that the connection modes between nodes in the network are the same, that is, the bandwidth and latency on each link are the same. Therefore, a symmetric topology structure is easier to achieve load balancing in the network. On the contrary, the connections and characteristics between network nodes in asymmetric topology structures are different, and the bandwidth and latency on different paths may vary. Therefore, if you want to achieve high-quality transmission performance in an asymmetric topology network, it requires a more flexible and scalable load balancing mechanism.

4.1 Performce Under Symmetric Topology

In this section, we test the performance of TDLB on the AFCT, 95th percentile FCT and 99th percentile FCT of short flows in network transmission. Compared with five load balancing mechanisms under Web Search and Data Mining workloads.

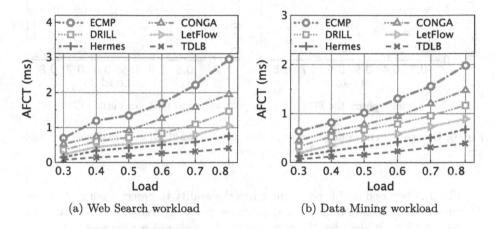

(a) Web Search workload   (b) Data Mining workload

Fig. 5. Comparison of AFCT in different workloads under symmetric topology

Figure 5 (a) and Fig. 5 (b) show the average FCT of short flows under Web Search and Data Mining workloads, respectively. Compared with the other five load balancing mechanisms, TDLB significantly reduces AFCT, especially in high workload situations. Specifically, under a 0.8 workload, in the Web Search scenario, TDLB decreases by ~86%, ~79%, ~72%, ~61%, and ~46%, compared to ECMP, CONGA, DRILL, LetFlow, and Hermes, respectively. In the Data Mining scenario, compared to ECMP, CONGA, DRILL, LetFlow, and Hermes, TDLB decreases by ~80%, ~73%, ~67%, ~56%, and ~44%, respectively.

These test results indicate that TDLB performs well in low latency transmission. Within the same switch, as workload increases, more mixed flows queue up in the output queue, leading to path congestion. Consequently, this results in the blocking of additional traffic that has just arrived at the path, causing it to experience significant queuing latency and packet reordering. Specifically, ECMP operates at the flow level and experiences long queuing latency in cases of path congestion. Similarly, LetFlow operates solely at the flow level but utilizes random routing. The feedback latency of CONGA in obtaining congestion information increases, leading to the creation of long queues. While DRILL can handle sudden congestion in a timely manner, it struggles to accurately perceive global congestion information and avoid the retransmission cost caused by disorder. The rerouting mechanism triggered by Hermes is overly conservative, often leading to low link utilization.

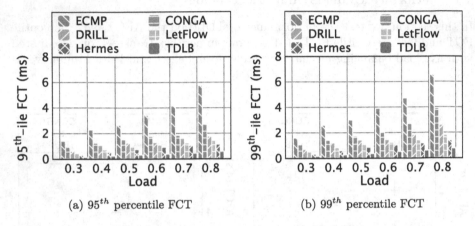

(a) 95th percentile FCT   (b) 99th percentile FCT

Fig. 6. Comparison of tailing latency in Web Search workload under symmetric topology

TDLB, proposed in this paper, first has the ability to perceive congestion and detect sudden congestion in a timely manner. Secondly, based on the characteristics of the flow, it identifies the flow that truly needs redirection and forwards it to a better path. This effectively solves the original path congestion problem and prevents congestion diffusion. The congested flows continue to queue and wait on the original path to avoid rerouting causing disorder and increasing the completion time of the flow. Non-congested innocent flows truly need to be rerouted to find better transmission paths, accelerate the arrival time of the flow, and reduce the completion time of the flow. Therefore, TDLB can achieve a lower average flow completion time.

In addition, we find that the short flows in the Web Search workload is larger than the FCT in Data Mining. The reason is that in Web Search, there are more flows ranging in size from 100 KB to 1 MB, accounting for 18% of the total traffic, resulting in longer queue lengths for switch queues and increasing opportunities for packet disorder. In Data Mining, flows smaller than 100KB account for about 83%, and a small amount of long flows will reduce the latency of short flows queuing. Under the Web Search workload, the proportion of long flows is heavier, and during transmission, long flows occupy the transmission path, while short flows experience longer queuing latency, resulting in a larger overall AFCT.

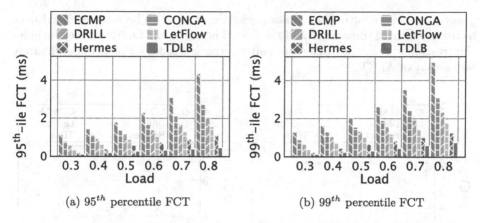

(a) 95th percentile FCT   (b) 99th percentile FCT

Fig. 7. Comparison of tailing latency in Data Mining workload under symmetric topology

Figure 6 (a), (b) and Fig. 7 (a), (b) show the 95th percentile FCT and 99th percentile FCT of short flows under Web Search and Data Mining workloads, respectively. Obviously, compared to the other five solutions, TDLB significantly improves the tail FCT, especially in high workload situations. Specifically, under a 0.8 workload, in the Web Search scenario, TDLB decreases by ∼88%, ∼80%, ∼70%, ∼58%, and ∼45% compared to ECMP, CONGA, DRILL, LetFlow, and Hermes, respectively. In the Data Mining scenario, compared to ECMP, CONGA, DRILL, LetFlow, and Hermes, TDLB decreases by ∼84%, ∼76%, ∼68%, ∼58%, and ∼40% respectively.

Severe tailing refers to a situation where the latency of a small portion of the flow is significantly higher than the average latency. This phenomenon may have a negative impact on network performance. ECMP randomly select forwarding paths, without considering the congestion situation of other parallel paths, and transmit at the flow level. Therefore, ECMP has the greatest tailing effect. When the load balancing mechanism is unable to evenly distribute the load onto available servers, it may lead to some servers being overloaded, leading to serious latency. TDLB can truly redirect the flow that needs to be redirected and reroute it to a more optimal parallel path. This enables those innocent flows to complete transmission as early as possible, reducing overall flows latency.

Furthermore, we are surprised to find that under the same workload, the 99th percentile FCT is approximately twice that of AFCT. We conduct a detailed theoretical analysis of the experimental results. Firstly, the 99th percentile FCT represents the portion of the flow with the highest tailing latency in transmission. Under normal case, the traffic is transmitted at the maximum transmission rate of the transmission path, and its FCT is less than or equal to AFCT. However, when the network load increases, it is inevitable that congestion will occur, resulting in some queuing latency and increasing the FCT. Meanwhile, prolonged waiting or excessive congestion can lead to packet loss and disorder. In general congestion situations, data packets only need to be retransmitted once to restore normal transmission of the flow. Therefore, the tailing latency includes the transmission latency before and after retransmission, which is approximately twice that of AFCT.

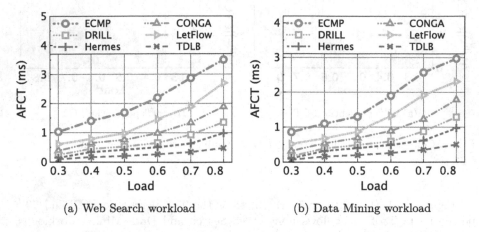

(a) Web Search workload   (b) Data Mining workload

Fig. 8. Comparison of AFCT in different workloads under asymmetric topology

4.2 Performance Under Asymmetric Topology

In this section, we test the performance of TDLB under asymmetric network topology. Firstly, we introduce the characteristics of asymmetric network topology, and then conducted a detailed analysis of the experimental results.

Figure 8 (a) and Fig. 8 (b) show AFCT under Web Search and Data Mining workloads, respectively. From the experimental results, ECMP is still the worst, followed by LetFlow, while TDLB still performs best and can achieve excellent performance in asymmetric topology scenarios.

In asymmetric topology, due to the differences between nodes, some nodes need to carry more traffic. In the case of unequal link bandwidth, when the network data volume increases, the node with unequal bandwidth between the uplink and downlink links will cause serious congestion at that node due to the significant difference in acceptance rate and transmission rate [37,38]. Due to ECMP and LetFlow randomly selecting paths through hash scattering and fixed time intervals, the uncertainty caused by random path selection may be forwarded to paths with lower link bandwidth, resulting in excessive congestion and FCT elongation. CONGA, DRILL, and Hermes utilize the congestion information of the path to alleviate the sudden congestion on the current path, but the feedback latency of global congestion perception was too large. The switch is unable to sense the congestion status of other paths in a timely manner, and cannot reroute traffic on congested paths to appropriate paths, which may increase

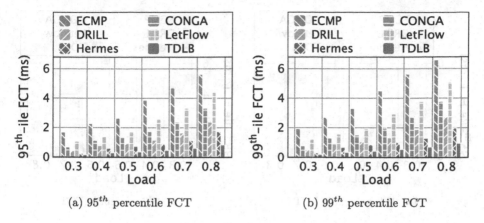

(a) 95th percentile FCT   (b) 99th percentile FCT

Fig. 9. Comparison of tailing latency in Web Search workload under asymmetric topology

the load on instantaneous congested paths and ultimately lead to longer FCT. At the same time, the unequal bandwidth and latency of asymmetric topology links can lead to more severe packet disorder due to partial rerouting, which affects the original transmission latency.

In addition, they do not distinguish the traffic that needs to be rerouted before rerouting, which can cause congestion when the traffic that causes congestion flows through rerouting and is forwarded to other paths to continue to generate congestion. In congested case, TDLB distinguishes innocent flows that truly need to be rerouted and forwards them to better paths based on link congestion information, thereby reducing the FCT of innocent flows. The congested flow continues to queue on the original path, replacing rerouting with speed reduction to avoid disorder and significantly reduce the FCT of the congested flow. Figure 9 (a), Fig. 9 (b), Fig. 10 (a), and Fig. 10 (b) respectively show the 95th percentile FCT and 99th percentile FCT under the Web Search and Data Mining workloads with asymmetric topology. When the workload in the Web Search scenario is 0.8, the 99th percentile FCT of TDLB is 0.96 ms, which increases TDLB decreases by ~85%, ~75%, ~65%, ~81%, and ~52% compared to ECMP, CONGA, DRILL, LetFlow, and Hermes, respectively. Overall, TDLB can also exhibit excellent performance in asymmetric topological structures.

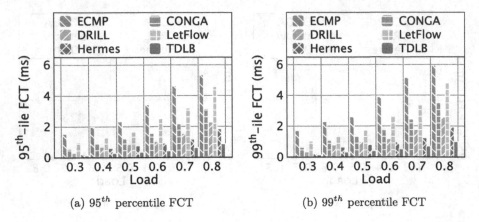

(a) 95th percentile FCT   (b) 99th percentile FCT

Fig. 10. Comparison of tailing latency in Data Mining workload under asymmetric topology

5   Related Work

There are many literature on datacenter load balancing [6,11–16,35,36,39,40]. Among them, we only discuss and summarize some representative work related to this paper.

ECMP [11] achieves load balancing by distributing network traffic across multiple paths with equal costs. When a packet reaches the load balancer or router, the ECMP mechanism utilizes predefined load balancing algorithms, such as hash or round-robin algorithms, to select an optimal path from the available options for transmitting the packet. While ECMP effectively achieves load balancing, it is not without limitations. It may suffer from line-rate congestion and tail latency issues, particularly when dealing with short and long flows. Additionally, the presence of hash collisions can lead to suboptimal utilization of available bandwidth. In comparison, RPS [16] offers a straightforward packet-level load balancing mechanism by randomly distributing packets among all available transmission paths on the switch. This technique enhances link utilization and ensures better distribution of traffic. However, it also has inherent drawbacks. Packet reordering issues may arise, potentially impacting the order of packet delivery. Moreover, RPS lacks the ability to proactively detect congestion, which can result in packet loss and subsequent retransmissions.

CONGA [12] achieves global congestion-aware load balancing by leveraging end-to-end path feedback information. It effectively detects congestion and failure in paths, making it suitable for asymmetric networks. However, the implementation of CONGA presents challenges due to the substantial storage requirements for path information and the reliance on custom switches, limiting its scalability. Additionally, the accuracy of feedback obtained from remote switches may be compromised, leading to potential discrepancies in real-time path conditions. HULA [40] tackles the scalability issues of CONGA by employing programmable switches for congestion-aware load balancing. Nevertheless, HULA's strategy of selecting solely the optimal next-hop path may give rise to herd effects, resulting in congestion along the chosen path. Moreover, the update rate of the best path depends on the probing frequency, which, if excessively used, can degrade network efficiency.

To mitigate the overhead associated with switch-based congestion detection, LetFlow [14] autonomously senses path congestion based on intrinsic packet characteristics. Compared to global congestion-aware schemes like CONGA, LetFlow eliminates the need for comprehensive global congestion information, enhancing its scalability. However, the stochastic nature of LetFlow scheduling prevents it from achieving optimal load balancing performance. To efficiently alleviate congestion caused by bursty traffic, DRILL [13] introduces a load balancing strategy specifically designed for micro-bursts. By leveraging local queue lengths in switches, DRILL swiftly performs packet-level forwarding decisions, enabling microsecond-level load balancing and effectively resolving congestion issues arising from micro-bursts. Nevertheless, DRILL exhibits limitations when confronted with high-speed bursts. Hermes [15] utilizes ECN, latency, and probing packets to detect path congestion and failure states, estimating the benefits of switching paths to determine if rerouting should be executed. This approach mitigates the negative consequences associated with blind path switching, such as packet reordering, while also circumventing the mismatch between link states and congestion windows. Although Hermes is applicable to asymmetric networks, its conservative rerouting decision-making process can lead to suboptimal link utilization.

6 Conclusion

We present a traffic-differentiated load balancing mechanism called TDLB to solve the problem of large tailing latency caused by coarse-grained rerouting in existing load balancing mechanisms. TDLB aims at distinguishing traffic that truly needs to be rerouted and taking different measures to optimize the routing and transmission of traffic on congested paths. Firstly, the switch senses whether the current path is congested based on queue length and path delay. Secondly, when the path becomes congested, TDLB distinguishes between the flows causing congestion and the ones that truly need to be rerouted. Finally, the flow that truly needs to be rerouted will select a better transmission path based on the congestion information and queue length of other parallel paths. The experimental results of NS-2 simulation show that compared with existing mechanisms, TDLB can decrease the tailing latency and AFCT by to 45% and 46%, respectively, supplying more effectively network performance.

References

1. Li, W., Chen, S., Li, K., Qi, H., Xu, R., Zhang, S.: Efficient online scheduling for coflow-aware machine learning clusters. IEEE Trans. Cloud Comput. **10**(4), 2564–2579 (2020)
2. Wang, J., Liu, Y., Rao, S., Zhou, X., Hu, J.: A novel self-adaptive multi-strategy artificial Bee Colony algorithm for coverage optimization in wireless sensor networks. Ad Hoc Netw. **150**, 103284 (2023)
3. Li, H., Zhang, Y., Li, D., et al.: URSA: hybrid block storage for cloud-scale virtual disks. In: Proceedings of the Fourteenth EuroSys Conference, pp. 1–17 (2019)
4. Wang, J., Liu, Y., Rao, S., et al.: Enhancing security by using GIFT and ECC encryption method in multi-tenant datacenters. Comput. Mater. Continua **75**(2), 3849–3865 (2023)
5. Wang, Y., Wang, W., Liu, D., et al.: Enabling edge-cloud video analytics for robotics applications. IEEE Trans. Cloud Comput. **11**(2), 1500–1513 (2023)
6. Wang J., Rao S., Liu Y., et al.: Load balancing for heterogeneous traffic in datacenter networks. J. Netw. Comput. Appl. **217** (2023)
7. Hu, J., Zeng, C., Wang, Z., et al.: Enabling load balancing for lossless datacenters. In: Proceedings of IEEE ICNP (2023)
8. Xu, R., Li, W., Li, K., Zhou, X., Qi, H.: DarkTE: towards dark traffic engineering in data center networks with ensemble learning. In: Proceedings of IEEE/ACM IWQOS, pp. 1–10 (2021)
9. Li, W., Yuan, X., Li, K., Qi, H., Zhou, X.: Leveraging endpoint flexibility when scheduling coflows across geo-distributed datacenters. In: Proceedings of IEEE INFOCOM, pp. 873–881 (2018)
10. Bai, W., Chen, K., Hu, S., Tan, K., Xiong, Y.: Congestion control for high-speed extremely shallow-buffered datacenter networks. In: Proceedings of ACM APNet, pp. 29–35 (2017)
11. Hopps, C.E.: Analysis of an equal-cost multi-path algorithm (2000)
12. Alizadeh, M., et al.: CONGA: distributed congestion-aware load balancing for datacenters. In Proceedings of ACM Conference on SIGCOMM, pp. 503–514 (2014)
13. Ghorbani, S., Yang, Z., Godfrey, P.B., Ganjali, Y., Firoozshahian, A.: DRILL: micro load balancing for low-latency data center networks. In: Proceedings of ACM SIGCOMM, pp. 225–238 (2017)
14. Vanini, E., Pan, R., Alizadeh, M., Taheri, P., Edsall, T.: Let it flow: resilient asymmetric load balancing with flowlet switching. In: Proceedings of USENIX NSDI, pp. 407–420 (2017)
15. Zhang, H., Zhang, J., Bai, W., Chen, K., Chowdhury, M.: Resilient datacenter load balancing in the wild. In: Proceedings of ACM SIGCOMM, pp. 253–266 (2017)
16. Dixit, A., Prakash, P., Hu, Y.C., Kompella, R.R.: On the impact of packet spraying in data center networks. In: Proceedings of IEEE INFOCOM, pp. 2130–2138 (2013)
17. Hu, J., Huang, J., Li, Z., Wang, J., He, T.: A receiver-driven transport protocol with high link utilization using anti-ECN marking in data center networks. IEEE Trans. Netw. Serv. Manag. **20**(2), 1898–1912 (2023)
18. He, X., Li, W., Zhang, S., Li, K.: Efficient control of unscheduled packets for credit-based proactive transport. In: Proceedings of ICPADS, pp. 593–600 (2023)
19. Kabbani, A., Vamanan, B., Hasan, J., Duchene, F.: FlowBender: flow-level adaptive routing for improved latency and throughput in datacenter networks. In: Proceedings of CoNEXT, pp. 149–160 (2014)

20. Wang, J., Yuan, D., Luo, W., et al.: Congestion control using in-network telemetry for lossless datacenters. Comput. Mater. Continua **75**(1), 1195–1212 (2023)
21. Wen, K., Qian, Z., Zhang, S., Lu, S.: OmniFlow: coupling load balancing with flow control in datacenter networks. In: Proceedings of ICDCS, pp. 725–726 (2016)
22. Shafiee, M., Ghaderi, J.: A simple congestion-aware algorithm for load balancing in datacenter networks. In: Proceedings of INFOCOM, pp. 1–9 (2016)
23. Alizadeh, M., Greenberg, A. et al.: Data center TCP (DCTCP). In: Proceedings of ACM SIGCOMM, pp. 63–74 (2010)
24. Munir, A., et al.: Minimizing flow completion times in data centers. In: Proceedings of INFOCOM, pp. 2157–2165 (2013)
25. Li, Z., Bai, W., Chen, K., et al.: Rate-aware flow scheduling for commodity data center networks. In: Proceedings of IEEE INFOCOM, pp. 1–9 (2017)
26. David, Z., Tathagata, D., Prashanth, M., Dhruba, B., Randy, K.: DeTail: reducing the flow completion time tail in datacenter networks. In: Proceedings of the ACM SIGCOMM, pp. 139–150 (2012)
27. Benson, T., Akella, A., Maltz, D.: Network traffic characteristics of data centers in the wild. In: Proceedings of ACM IMC, pp. 267–280 (2010)
28. Hu, C., Liu, B., Zhao, H., et al.: Discount counting for fast flow statistics on flow size and flow volume. IEEE/ACM Trans. Network. **22**(3), 970–981 (2013)
29. The NS-2 network simulator. http://www.isi.edu/nsnam/ns
30. Bai, W., Hu, S., Chen, K., Tan, K., Xiong, Y.: One more config is enough: saving (DC) TCP for high-speed extremely shallow-buffered datacenters. IEEE/ACM Trans. Network. **29**(2), 489–502 (2020)
31. Liu, Z., et al.: Enabling work-conserving bandwidth guarantees for multi-tenant datacenters via dynamic tenant-queue binding. In: IEEE INFOCOM 2018-IEEE Conference on Computer Communications, pp. 1–9 (2018)
32. Hu, C., Liu, B., Zhao, H., Chen, K., et al.: Disco: memory efficient and accurate flow statistics for network measurement. In: Proceedings of IEEE ICDCS, pp. 665–674 (2010)
33. Wei, W., Gu, H., Deng, W., Xiao, Z., Ren, X.: ABL-TC: a lightweight design for network traffic classification empowered by deep learning. Neurocomputing **489**, 333–344 (2022)
34. Wei, W., et al.: GRL-PS: graph embedding-based DRL approach for adaptive path selection. IEEE Trans. Netw. Serv. Manag. (2023)
35. Hu, J., He, Y., Wang, J., et al.: RLB: reordering-robust load balancing in lossless datacenter network. In: Proceedings of ACM ICPP (2023)
36. Hu, J., Zeng, C., Wang, Z., Xu, H., Huang, J., Chen, K.: Load balancing in PFC-enabled datacenter networks. In: Proceedings of ACM APNet (2022). Wang, J., Rao, S., Liu, Y., et al.: Load balancing for heterogeneous traffic in datacenter networks. J. Netw. Comput. Appl. **217** (2023)
37. Zhao, Y., Huang, Y., Chen, K., Yu, M., et al.: Joint VM placement and topology optimization for traffic scalability in dynamic datacenter networks. Comput. Netw. **80**, 109–123 (2015)
38. Zheng, J., Du, Z., Zha, Z., et al.: Learning to configure converters in hybrid switching data center networks. IEEE/ACM Trans. Network. 1–15 (2023)
39. Liu, Y., Li, W., Qu, W., Qi, H.: BULB: lightweight and automated load balancing for fast datacenter networks. In: Proceedings of ACM ICPP, pp. 1–11 (2022)
40. Katta, N., Hira, M., Kim, C., Sivaraman, A., Rexford, J.: HULA: scalable load balancing using programmable data planes. In: Proceedings of the Symposium on SDN Research, pp. 1–12 (2016)

Deep Reinforcement Learning Based Load Balancing for Heterogeneous Traffic in Datacenter Networks

Jinbin Hu, Wangqing Luo, Yi He, Jing Wang, and Dengyong Zhang[✉]

School of Computer and Communication Engineering,
Changsha University of Science and Technology, Changsha 410004, China
{jinbinhu,znwj_cs,zhdy}@csust.edu.cn, {luowangqing,heyi}@stu.csust.edu.cn

Abstract. Modern high-speed datacenter networks (DCNs) employ multi-tree topologies to provide large bisection bandwidth. Load balancing is crucial for making full use of parallel equal-cost paths and ensuring high link utilization. In the past decades, a large number of heuristic load balancing mechanisms have been proposed to alleviate congestion. However, they cannot be resilient to the concurrent explosive growth and unpredictable dynamic traffic scenarios. Recently, deep reinforcement learning methods have become very powerful techniques for reacting to dynamically changing networks, but the existing learning-based load balancing schemes are agnostic to the heterogeneous traffic generated by diverse applications. Thus, the latency-sensitive short flows suffer from large tail delays due to the long pre-learning period. To address the above issues, we propose a deep reinforcement learning-based load balancing mechanism called DRLB. Specifically, DRLB differentiates heterogeneous traffic by using Deep Reinforcement Learning (DRL) in conjunction with the Distributed Distributional Deterministic Policy Gradients (D4PG) algorithm to make the optimal (re)routing for long flows while adopting the Weighted Cost Multipathing (WCMP) mechanism for short flows. The NS-3 experimental results show that, DRLB increases the throughput of long flows by up to 47% and reduces the flow completion time (FCT) of short flows by up to 58% compared to the state-of-the-art load balancing mechanisms.

Keywords: Datacenter Networks · Deep Reinforcement Learning · Load Balancing · Heterogeneous Traffic

1 Introduction

In recent years, as the stringent demands of high throughput and low latency for applications like cloud computing, big data, and artificial intelligence have

This work is supported by the National Natural Science Foundation of China (62102046, 62072056), the Natural Science Foundation of Hunan Province (2023JJ50331, 2022JJ30618, 2020JJ2029), the Hunan Provincial Key Research and Development Program (2022GK2019), the Scientific Research Fund of Hunan Provincial Education Department (22B0300).

Z. Tari et al. (Eds.): ICA3PP 2023, LNCS 14489, pp. 270–289, 2024.
https://doi.org/10.1007/978-981-97-0798-0_16

continued to expand, modern datacenter networks have been gradually scaling up with rising bandwidth and traffic [1–3]. As a result, the dynamic network environment and real-time traffic changes have become more pronounced [4–6]. For instance, during each round of model training iterations in the present large-scale distributed training, the parameter synchronization updating procedure generates a significant amount of traffic interactions at the network endpoints [7].

To support large bisection bandwidth, production datacenter networks usually employ multi-rooted tree architectures. Existing load balancing mechanisms, however, confront emerging challenges in utilizing these equivalent multipaths, enhancing link utilization, and decreasing Flow Completion Time (FCT) to satisfy the requirements of varied applications [8]. As a result, developing advanced load balancing mechanisms has become an important research direction for datacenter networks [9].

Recently, a number of load balancing mechanisms have been proposed, but they aren't performing effectively in this constantly shifting environment. In particular, the inability to switch paths flexibly in ECMP and Freeway [10, 11], two flow-based scheduling granularity mechanisms, results in ineffective link utilization. Two packet-based scheduling granularity mechanisms, DRILL and RPS [12], might worsen overall performance by increasing the frequency of disorder in dynamic network scenarios. Hermes and TLB mechanisms also attempt to tackle the challenge by distinguishing heterogeneous traffic and utilizing routes with various granularities [13, 14], but they also struggle to adapt to the dynamically changing network environment.

The problem of load balancing offers novel solutions due to the development of artificial intelligence techniques [15–17]. Deep reinforcement learning is used in datacenter networks to continuously interact with the network environment and learn the best load balancing strategy in order to adapt to the dynamically changing network environment and increase network performance and reliability [18, 19].

We propose a heterogeneous traffic load balancing mechanism (DRLB) based on Deep Reinforcement Learning (DRL) to tackle the aforementioned problem. This mechanism distinguishes heterogeneous traffic by employing DRL for long flows and a straightforward Weighted Cost Multipathing (WCMP) mechanism for short flows [20]. It is important to note that in this paper, the long flows to be processed by us are accumulated according to the sending rate over the cycle time of the host-side parameter updates. We build a network model applicable to DRLB mechanism that can observe the network state in real time as input to the Distributed Distributional Deterministic Policy Gradient (D4PG) algorithm [21] and dynamically adjust the Link State Metric (LSM) to guide the forwarding of long flows through DRL agent, where a lower LSM value indicates a better link. By distinguishing heterogeneous traffic, we ensure optimal routing for all varieties of traffic, ensuring low latency for short flows and high throughput for long flows. The main contributions of this paper are as follows:

- We highlight the inadequacies and issues of the load balancing mechanisms presently employed in datacenters to deal with network changes and congestion. We devise a novel load balancing mechanism by combining machine learning techniques and investigate the problems of DRL pre-learning and short flow mismatch in order to address these issues. Experiments verify the rationality and scientific validity of these problems.
- We propose a deep reinforcement learning-based heterogeneous traffic load balancing mechanism called DRLB that applies DRL on long flows and a WCMP mechanism on short flows. DRLB reduces the 99th percentile FCT of short flows and increases the throughput of long flows. We develop a network model applicable to DRLB mechanism that monitors the network state in real time as the input to D4PG algorithm and dynamically modifies LSM via DRL agent to direct the forwarding of long flows. In addition, a plausible state action space and reward function are proposed.
- We conduct a series of experiments on the NS-3 simulator involving diverse traffic workloads and symmetric and asymmetric topology scenarios. We compared the experimental results to the existing load-balancing mechanisms, including Freeway, Hermes, and TLB. The experimental results demonstrate that DRLB substantially increases the throughput of long flows and decreases the FCT of short flows at wide-range traffic load.

The rest of this paper is organized as following. In Sect. 2, we discuss the context and rationale for this paper. Section 3 describes DRLB mechanism's design and implementation in detail. Section 4 provides an experimental evaluation of DRLB mechanism and a comparison to other mechanisms. Section 5 examines load balancing mechanisms in datacenter networks and DRL-related research. In Sect. 6, we summarize the paper's contributions, key findings, and prospective research directions.

2   Background and Motivation

2.1   Background

Load Balancing. In recent years, numerous load balancing mechanisms have been proposed to enhance the performance of datacenter networks. When designing these mechanisms, numerous considerations must be made, such as selecting the appropriate granularity for load balancing. A smaller granularity offers more routing options, enhancing network utilization and load balancing performance. Granularity can be packet (DRILL mechanism) [12], flow (WCMP mechanism) [20], flowlet (CONGA mechanism), or adaptive (TLB mechanism) [14] among the existing mechanisms [22–24]. In this paper, DRLB mechanism handles heterogeneous traffic differently, with short flows at the granularity of flows and long flows at the granularity of adaptive flows formed by the host-side sending rate accumulated over DRL agent iteration time. In addition to granularity, network congestion-aware policies, mechanism control structure, deployment difficulty, and scalability must be considered when devising a load balancing mechanism [25, 26].

However, traditional load balancing mechanisms continue to encounter numerous challenges in the face of growing datacenter network size and complexity. Traditional mechanisms are incapable of adapting to the heterogeneous devices and complex topologies encountered in datacenter networks, nor can they precisely assess network state and traffic distribution [27–29]. In addition, highly unpredictable and bursty traffic in datacenter networks frequently causes network failures, while traditional load balancing mechanisms are unable to modify their policies in a timely manner, exacerbating the issues of congestion and performance degradation. This paper proposes DRLB mechanism, which combines deep reinforcement learning techniques to adaptively adjust the load balancing policy by learning the characteristics of network state and traffic distribution, which can better adapt to the dynamic network environment and improve the performance and reliability of the data center network.

Deep Reinforcement Learning. DRL has emerged as a significant branch of artificial intelligence technology in recent years, achieving remarkable results in natural language processing, computer vision, speech recognition, and other fields. DRL combines the information extraction ability of deep learning models with the decision-making ability of reinforcement learning to achieve abstraction and feature extraction of complex data through multi-layer nonlinear mapping and to provide excellent solutions for complex tasks driven by the reinforcement learning objective [30, 31].

DRL has robust generalization and adaptability to meet the challenges of complex environments and large-scale tasks, which is one of the reasons why we are combining it with datacenter network problems [32–34]. DRL excels at coping with high-dimensional, complex environments, extracting high-level features from raw input data, and modeling the environment effectively. Moreover, DRL is able to learn complex policies and performs exceptionally well when confronted with complex tasks. Due to its ability to train end-to-end, it can effectively address difficult-to-model problems in datacenter networks, such as complex and variable traffic and enormous scope.

However, DRL is not entirely devoid of disadvantages. Its training requires vast quantities of data and computational resources, a lengthy training period, and an unstable training process that requires careful calibration of the network structure and hyperparameters to assure convergence and stability. In order to solve the problem of network load balancing in a datacenter, the appropriate methods and algorithms must be chosen based on the specific mission requirements and constraints [35, 36]. This paper's DRLB mechanism utilizes D4PG algorithm, which combines the Deep Deterministic Policy Gradient (DDPG) algorithm and distributed experience replay technology to efficiently utilize large-scale data for accurate and stable policy learning. Dynamically, D4PG algorithm is able to intelligently and dynamically modify the load balancing policy to adapt to network environment changes as well as diverse traffic distribution and device heterogeneity [21]. By applying D4PG algorithm to the design of load balancing mechanisms, network performance and dependability can be enhanced while maintenance costs and the need for manual intervention are reduced.

2.2 Motivation

In this subsection, we experimentally validate the load balancing mechanisms described in the introduction and the learning period problem for DRL.

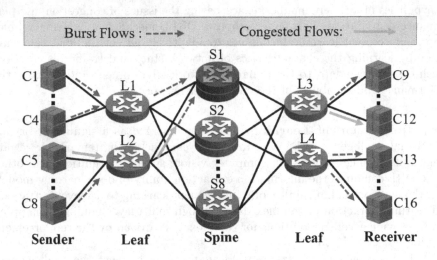

Fig. 1. Leaf-Spine topology

Load Balancing Problems. Using the NS-3 simulation platform, we conduct experiments to address the problems of existing load balancing mechanisms (Freeway, Hermers, and TLB) in adapting to changing network environments. In the leaf spine topology shown in Fig. 1, the sender C_5 transmits numerous data flows to the receiver C_{12}. According to the experimental findings, the spine switch S_1 is overloaded by a large number of bursty long flows when the three load balancing mechanisms tend to transmit smoothly and the throughput is largely stable. Specifically, Fig. 2 demonstrates that the long flow throughput of the three load-balancing mechanisms increases by approximately 35% at approximately 30 ms. We offer a comprehensive analysis of these three load balancing mechanisms.

First, for TLB mechanism, which performs best in Fig. 2, it can only sense local congestion information and cannot acquire entirely accurate, real-time, and comprehensive congestion information. In the case that S_1 becomes congested due to a large number of burst long flows, L_2 is unable to detect this and may continue to allocate traffic to S_1, resulting in a performance degradation. Although

Hermes can detect global congestion and understand that a large number of congested long flows occur in S_1, its overly conservative rerouting mechanism prevents it from resolving the S_1 congestion problem in a timely manner, resulting in throughput increases but actual throughput decreases. Although Freeway mechanism senses the change in path congestion and detects the congestion problem in S_1, its algorithm to reclassify paths to select high-throughput paths is not timely, and the network congestion may intensify during this period and new congested links may appear, resulting in poor overall performance.

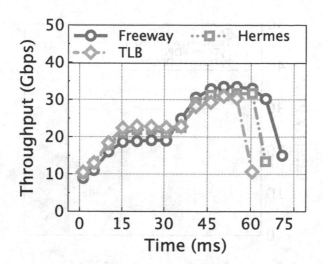

Fig. 2. Throughput of long flows under three load balancing mechanisms

Deep Reinforcement Learning Problems. The fact that the learning period of DRL is too lengthy for delay-sensitive short flows is one of the challenges associated with the application of DRL algorithms in the field of load balancing. Since DRL algorithms need to interact with the environment in order to learn the optimal policy, it takes time for these interactions to accumulate sufficient experience and optimise the policy. However, delay-sensitive short flows have stringent requirements for timeliness and need to be communicated quickly, whereas the learning process of DRL algorithms is lengthy and may not be able to accommodate the short delay in time. When the short flows arrive, the algorithm may still be in the process of learning and cannot instantaneously assign the optimal action.

To demonstrate our point, we conduct the DPG and DDPG algorithms using TensorFlow and PyTorch, respectively, as machine learning frameworks. In order to devise an appropriate neural network model, we created a multilayer perceptron network with two hidden layers, each containing 256 neurons. Such a network structure can provide sufficient expressive capacity for illustrating and learning complex load balancing policies in datacenter networks. To simulate real-world traffic conditions, we set up short flows that are sensitive to delay and large-scale long flows. The average sending rate for short flows is set to 100 Mbps, and the traffic size is generated randomly between 10 KB and 100 KB. The average sending rate for long flows is 1 Gbps, and the generated traffic size is randomly distributed between 1 MB and 10vMB. This type of traffic is typically very sensitive to latency, necessitating quick and precise load balancing mechanisms to ensure timely processing and delivery.

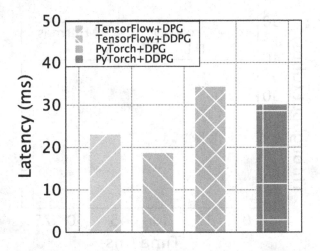

Fig. 3. Latency of two algorithms in different frameworks

Based on the experimental results depicted in Fig. 3, we observe a minimum processing latency of 23 ms when conducting the most advanced machine learning algorithms on both frameworks. This indicates that the short flows have already completed transmission before DRL algorithm implements the load-balancing guidance action. Under the current configuration, this result indicates that DRL algorithm has some latency issues when processing delay-sensitive short flows. Since the transmission time of short flows is relatively short, DRL algorithm requires some time to sense the network state, make decisions, and perform load balancing actions, resulting in some short flows having completed transmission before the algorithm becomes active.

3   DRLB Design

3.1   Design Overview

In this section, we briefly describe DRLB mechanism's guiding principles. Each server in the leaf-spine topology, as depicted in Fig. 4, contains a DRL agent responsible for network data collection, data processing, feature selection, model training, and output command. On top of each server, DRL algorithm module deploys a distributed actor-critic. Using WCMP mechanism, the actor obtains real-time network data (e.g., throughput, link utilization, and latency) from the switch and dispatches short flows to the link with the lowest link weight.

For long flows, the distributed actor stores its generated experience in the replay buffer, while the critic samples the replay buffer and modifies the priority. Then, based on the sample information, the critic assesses the performance and load of the links and outputs LSM values via the learner to direct the distributed actor to choose the optimal link for forwarding long flows. Utilizing the reinforcement learning capability of D4PG algorithm, the forwarding policy of extended flows is dynamically amended based on the real-time network state and traffic characteristics, resulting in improved load balancing.

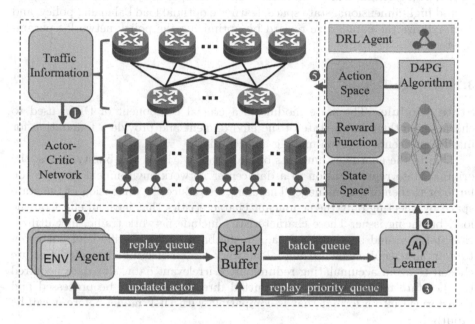

Fig. 4. DRLB overview

DRL agent is composed of four components:

- **State module.** A suitable state representation is devised in order for DRL algorithm to comprehend the current state of the datacenter network. This module aggregates real-time network information from the switches, including throughput, link utilisation, and latency, for use as input to DRL in order to inform the agent's decisions.
- **Action module.** The decision made by DRL agent based on the current condition is an action. In the problem of load balancing, the action consists of selecting the optimal link for forwarding long flows. Since the action space is

continuous, we employ D4PG algorithm to address the difficulty of continuous action space.

- **Reward module.** Reward is the feedback an agent receives based on their actions. In the load balancing problem, we construct an appropriate reward function based on network performance metrics such as latency to help agents learn the correct load balancing policy. The objective is to use rewards to encourage agents to learn load-balancing policies that can enhance network performance and reliability. An appropriate reward structure can motivate agents to take measures to enhance network performance.

- **DRL algorithm module.** By learning and optimizing their policies, the algorithm is used to train agents to progressively improve their performance on load balancing problems. To solve the load balancing problem, we employ D4PG algorithm, which can handle the challenges of continuous action space and high-dimensional state space, learn the optimal load balancing policy, and maximize the cumulative reward by optimizing the policy network's parameters.

3.2   Design Detail

State Module. The state module is a crucial component of DRL, used to characterize the present state of the environment and provide a foundation for intelligent agents' decision-making and learning.

Due to the enormous amounts of traffic and variable and bursty states, the original state representation in a datacenter network environment may be too large or challenging to manage. Therefore, we reduce the dimension of the state space by selecting only the significant characteristics that have an effect on the load balancing issue. These characteristics include network traffic, link utilization, latency, and bandwidth data. On the basis of the specific load balancing targets and requirements, essential network state-reflecting characteristics are chosen to avoid accumulating redundant or irrelevant data. The data acquired by the state module cannot be inputted directly and must be processed and transformed into a format that is compatible with DRL model before it can be inputted.

In this paper, these data are transformed into a vector matrix. This paper describes the deployment of DRLB mechanism on an $N * M$ leaf spine topology with K servers connected to each leaf switch. We collect queue length (QL), throughput (T), link utilization (LU), and latency (L) information at the switches and design them as state vectors for DRLB:

$$S = \begin{bmatrix} S_N \\ S_M \end{bmatrix} \tag{1}$$

$$S_N = \begin{bmatrix} S_{N_1} \\ S_{N_2} \\ \cdots \\ S_{N_N} \end{bmatrix} = \begin{bmatrix} QL_{N1}, T_{N_1}, LU_{N_1}, L_{N_1} \\ QL_{N2}, T_{N_2}, LU_{N_2}, L_{N_2} \\ \cdots \\ QL_{N_N}, T_{N_N}, LU_{N_N}, L_{N_N} \end{bmatrix} \tag{2}$$

where $S_{Ni} = [QL_{Ni}, T_{Ni}, LU_{Ni}, L_{Ni}]$ represents the real-time queue length and throughput information collected on the $i - th$ leaf switch. The vectors S_M and S_N on the M spine switches are obtained identically, so we will not list them individually here.

Action Module. Another essential component of DRL is the action module, which defines the set of actions that an agent can choose from. When designing the action module in DRLB mechanism, the agent-selectable actions are defined by inputting the calculated state into DRL agent in accordance with the load balancing objectives and requirements of DRLB mechanism. The action module transmits LSM values to the host side, instructing it to forward long flows to the finest integrated link. The action module defines each action's meaning and operation explicitly, allowing the agent to select the action with the highest reward value. Since the actions for load balancing are continuous, DRL agent needs to select a continuous set of parameter values to execute the action operations. In this instance, the action space is defined as a contiguous parameter space, and the agent chooses the optimal action to accomplish load balancing through LSM.

Similarly to the state module, the action space module should be able to be dynamically updated to accommodate environmental changes. In a datacenter network environment, the network state and traffic distribution may change, so the action space module has to be able to reflect these alterations and promptly update the collection of optional actions. By appropriately designing and optimizing the action space module, DRL agent can direct the host side to achieve the load balancing goal based on the current state information and the output LSM value and continuously learn and improve its decision strategy to enhance the performance and effectiveness of the system, as described below for DRLB action space:

$$A = [A_1, A_2, \ldots, A_K] \tag{3}$$

where A_i indicates that the i_{th} host end at this moment is guided by LSM value and the current state to select the action with the greatest reward value, thereby optimizing the transmission link for the long flow.

Reward Module. DRLB's reward function incorporates the round-trip latency (RTT) of all links and calculates the total reward value by aggregating them. This enables the case of all equivalent links on the leaf-spine topology to be considered, as opposed to only individual links. By comparing the RTT of each link to the average RTT, the reward function can determine the congestion level on the link. When a link's RTT exceeds the average RTT, the reward value becomes negative, indicating a higher level of congestion. The reward function that includes a logarithmic transformation can restrict the reward value to an appropriate range. The logarithmic function can make the change in reward value more gradual and prevent extreme values. Also, the average RTT is used as a reference value in the reward function so that the reward function can adapt to changes in the network and thus guide DRL agent to make adjustments and decisions accordingly. DRLB reward function is depicted by Eq. (4):

$$R = \frac{1}{N \times M} \log \left(\sum_{i=1}^{N \times M} \left(-\frac{RTT_i - \overline{RTT}}{N \times M} \right) \right) \tag{4}$$

N*M represents the total number of equivalent links, where N and M represent the number of leaf and spine switches, respectively. RTT_i is the RTT of the i_{th} link, whereas $\overline{RTT}$ is the average of all connections' RTTs. The reward function is computed by first normalizing and then aggregating the difference between each link's RTT and the average RTT. The value of the normalized difference is divided by $(N \times M)$ to determine the average contribution to all links. The reward value is determined by taking the logarithm of the preceding sum. The condition of the system's load balancing is determined by calculating the difference between the RTT of each link and the average RTT. When all RTTs are close to the average, the value of the reward will be close to zero. And the reward value will become increasingly negative when the RTT of some links deviates significantly from the average. The logarithmic operation of the reward function can limit the incentive value to a narrow range, making gradient optimization and learning simpler to perform.

DRL Algorithm Module. In this paper, DRLB mechanism selects the policy gradient-based D4PG algorithm. The convergence and stability of D4PG algorithm can accommodate the dynamic and complex nature of the network environment in a datacenter, resulting in significant changes to the reward signal. Using a parameterized policy network, D4PG algorithm generates continuous action values that are appropriate for continuous action selection in load balancing.

In addition, D4PG algorithm has a robust policy search capability; when the network state and load conditions in load balancing problems change frequently, D4PG algorithm is able to modify the policy based on real-time environment information to accommodate varying load demands. D4PG algorithm's replay caching mechanism permits sampling and utilization of historical experience during training. This is crucial for the load balancing problem because it enables DRL agent to learn the policy under a variety of network states and load conditions, thereby enhancing overall performance and adaptability.

The objective of DRLB mechanism is to optimize the performance of the datacenter network, including throughput, latency, and resource utilization metrics. D4PG algorithm discovers policies by maximizing reward value, which can be optimized for these performance indices. Through proper design of the reward function and optimization of the policy network, D4PG algorithm can help DRLB mechanism attain improved network performance.

D4PG algorithm is selected as DRL algorithm for DRLB mechanism. When using D4PG algorithm for load balancing design, it is also necessary to consider network topology, state modeling, reward function design, policy network architecture, and training parameter optimization for improved performance and results. Table 1 displays the primary parameters involved in D4PG algorithm.

Table 1. Main parameters involved in D4PG algorithm.

Parameters	Meaning of parameters
$s \in S, a \in A, r \in R$	States, Actions, Rewards
$(s_{i:i+V}, a_{i:i+V-1}, r_{i:i+V-1})$	Status , Action, Reward Sequence
P_i	Sampling priority
D	Replay buffer size
W	The number of distributed actors
V	Sample time length
U	Sample size
ς	Degree of agent exploration
ρ_0 and ϱ_0	Initial learning rate value
γ	Reward discount factor
Z_μ	Random variables
d	Difference between value and target value distribution

Here, we parse D4PG algorithm of DRLB, initializing the network parameters (θ, μ) and designating them to the target network (θ', μ'), as well as some hyperparameters, and deploy W actors by duplicating the network weights for each actor. (lines 1–4). During the training process, the policy network is kept up-to-date by making training blocks using this sampling method. Then, by constructing the target distribution, we can obtain the target Q values for each step, compare them with the predicted values of the value function network, and calculate the loss function of the value function network.

Finally, by calculating the updates of actors and critics, we can update the parameters of the strategy network and the value function network. In lines 10 and 11, we update the target network and propagate the network weights to the actors to ensure that the target network is periodically updated and that the actor and learner weights are in sync for an efficient DRL training process. When the predetermined number of training cycles have been completed, the current policy parameter is returned as the final policy parameter.

In this paper, LSM value is introduced to characterize the advantages and disadvantages of the current real-time link, and DRL agent outputs LSM value to assist the host side in performing the most reasonable load balancing operation on the long flow. Since D4PG needs to maximize LSM value through extensive learning, we have to improve D4PG algorithm's exploration and speed up its training by presetting an optimal action based on LSM value, which we refer to as alsm. As shown in line 15 of the code, $p_{(t)}$ is a decreasing function; as t increases, the effect of the a_{lsm} value on the agent's choice of action diminishes, and the policy network's action is more likely to be chosen. DRLB executes the action and interacts with the environment based on the selected action in the subsequent phase. After the action is performed, the reward r and the subsequent state s' are observed. The experience replay buffer maintains a tuple with the current state s, action a, reward r, and next state s. Past experiences are stored and sampled in the experience replay buffer for optimal learning. D4PG algorithm for DRLB is described below:

Algorithm 1: DRLB's D4PG algorithm

Initialization stage:
    Hyper-parameters: D, W, V, U, ς, ρ_0 and ϱ_0
    Random network weights(θ, μ),target weights $(\theta', \mu') \leftarrow (\theta, \mu)$
    Initiate W actors while replicating (θ, μ) to each actors.
Learning Stage:
for $t = 1,...,T$ **do**
    Sample U transitions $(s_{i:i+V}, a_{i:i+V-1}, r_{i:i+V-1})$ of length V with priority
    P_i from replay buffer D
    Target distributions: $Y_i = \sum_{v=0}^{V-1} \gamma^v r_{i+V} + \gamma^V Z_{\mu'}(s_{i+V}, \pi_{\theta'}(s_{i+V}))$
    Update the actors and critics:
        $\delta_\mu = \frac{1}{U} \sum_i \nabla_\mu (Dp_i)^{-1} d(Y_i, Z_\mu(s_i, a_i))$
        $\delta_\theta = \frac{1}{U} \sum_i \nabla_\theta \pi_\theta(\mathbf{s}_i) \mathbb{E} [\nabla_{\mathbf{a}} Z_\mu(\mathbf{s}_i, \mathbf{a})]|_{\mathbf{a}=\pi_\theta(\mathbf{s}_i)}$
    Update network parameters:$\theta \leftarrow \theta + \rho_t \delta_\theta, \mu \leftarrow \mu + \varrho_t \delta_\mu$
    If $t = 0 \bmod t_{target}$, updating target network $(\theta', \mu') \leftarrow (\theta, \mu)$
    If $t = 0 \bmod t_{actors}$, replication of network weights to actors
end
Return θ
Actor

Repeat
    Sample random action $a = (1 - p(t)) * \pi_\theta(s) + \varsigma V(0, 1) + p(t) * a_{lsm}$
    Execute Sample action a, then observe reward r and the next state s'
    Store (s, a, r, s') in replay buffer
Until learner completes

Through distributed learning, distribution-based value function estimation, distributed empirical playback, prioritized empirical playback, and model-free policy optimization, DRL algorithm of DRLB provides an efficient, stable, and accurate method for solving the load balancing problem in continuous action space.

4 Experimental Results and Evaluation

4.1 Experimental Setup

In this subsection, using the NS-3 simulator, we conduct a series of large-scale experiments to evaluate the performance of DRLB mechanism under various traffic and topology scenarios. We compare DRLB mechanism to several existing load-balancing algorithms (Freeway, Hermers, and TLB). The 99th percentile FCT of short flows and the throughput of long flows are choosed as the evaluation metric for a comprehensive performance evaluation of each load balancing algorithm.Table 2 displays the remaining experimental settings:

Table 2. Key parameters of DRLB in experiments.

Parameters	Value/Description
Number of leaf switches	8
Number of spine switches	12
Number of end-hosts	40
Number of links	1
Spine-Leaf Capacity	40 Gbps
Leaf-Server Capacity	40 Gbps
Link Latency	2us
Realistic datacenter workloads	Web Sever and Data Mining Communication
Pattern	All-to-all
Transport	DCTCP
Switch buffer	512 KB

4.2   Evaluation of Experimental Results

In this paper, we compare the performance evaluation of the Freeway, Hermers, TLB, and DRLB load balancing mechanisms under two actual datacenter workloads, Web server (64 KB) and datamining (7.4 MB), in a symmetric leaf spine topology. Figures 5 and 6 demonstrate that as the traffic load increases, the 99th percentile FCT of the short flows increases considerably, whereas the throughput of the long flows decreases as the traffic load increases. Figures 5(a) and 6(a) demonstrate that DRLB mechanism reduces the FCT of short flows by approximately 54%, 38%, and 12% on average compared to the Freeway, Hermers, and TLB mechanisms, respectively, when the traffic load reaches 80%, and Figs. 5(b) and 6(b) demonstrate that for the throughput of long flows, DRLB mechanism reduces the FCT by approximately 38% compared to the Freeway, Hermers, and TLB mechanisms.

As shown in Fig. 5 and Fig. 6, we can observe that the performance of Hermes mechanism decreases by approximately 30% when the traffic workload increases, and we conclude that this is because Hermes mechanism is not timely enough based on end-system-aware congestion, and the path-aware and re-routing mechanisms are too conservative. In contrast, the performance of DRLB mechanism is much more stable, with a loss of only about 4% in both FCT and throughput on average.

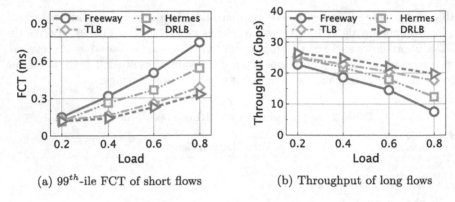

(a) 99th-ile FCT of short flows   (b) Throughput of long flows

Fig. 5. Web server workload under symmetrical topology

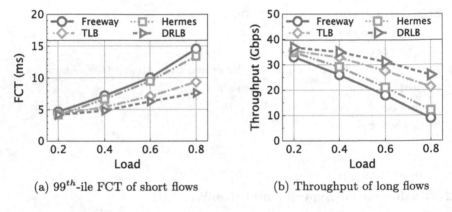

(a) 99th-ile FCT of short flows   (b) Throughput of long flows

Fig. 6. Data mining workload under symmetrical topology

Topologies frequently exhibit asymmetric link congestion in real-world application scenarios due to datacenter network traffic increases, multiple dynamic changes, and other factors. We set the capacity of multiple links in the leaf spine topology on NS-3 to be only 1/5 of other links in order to simulate the asymmetric topology for experiments, and Web server and datamining traffic remain the primary burden. According to Fig. 7(a), DRLB mechanism reduces the 99th percentile FCT of short flows by 63%, 42%, and 29% compared to the Freeway, Hermers, and TLB mechanisms, respectively, under the webserver load. Figure 8(a) reveals that DRLB mechanism optimizes the performance metric of FCT by more than 28% compared to the other three mechanisms in the datamining scenario. By comparing Fig. 7(b) and Fig. 8(b), we can see that DRLB does a superior job of ensuring high throughput requirements for long flows than Freeway, Hermers, and TLB, where the average throughput of long flows decreases by 49%, 30%, and 25%, respectively.

From the above experimental results, we observe that the asymmetric topology affects the performance of DRLB mechanism by only 3% in both traffic load

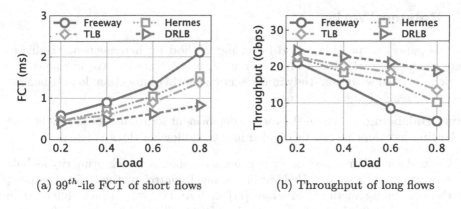

(a) 99th-ile FCT of short flows   (b) Throughput of long flows

Fig. 7. Web server workload under asymmetrical topology

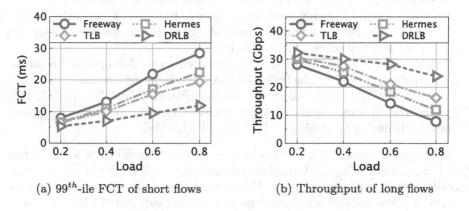

(a) 99th-ile FCT of short flows   (b) Throughput of long flows

Fig. 8. Data mining workload under asymmetrical topology

scenarios, whereas TLB mechanism suffers a performance loss of up to 37%. In the symmetric topology scenario, TLB mechanism performs 27% better than Hermes mechanism under both workloads, but in the asymmetric topology, TLB handles short flows at packet granularity because its mechanism must adaptively adjust the reroute granularity of long flows based on the strength of short flows, while it only senses local congestion and cannot detect the problem in time when congestion occurs in other links. The performance degradation is severe due to the high number of packet retransmissions, which is only about 12% better than Hermes.

By conducting simulation experiments under two traffic loads and two leaf spine topology scenarios to evaluate the two performance metrics, we can conclude that DRLB improves the long flow throughput by 47% and reduces the short FCT by 58% when compared to the existing Freeway, Hermers, and TLB mechanisms. It demonstrates the efficacy and progression of its precise differentiation of heterogeneous traffic into short flows using WCMP mechanism and long flows combined with DRL for load balancing.

5   Related Works

In this paper, we provide a load balancing method for heterogeneous traffic in data center networks based on deep reinforcement learning. We introduce the work associated with this study in two sections in this subsection: load balancing and deep reinforcement learning.

Load Balancing. Based on the various deployment locations, the prevalent load balancing mechanisms can be divided into the following three categories:

- Central controller-based load balancing mechanisms, which primarily include re-routing methods like Hedera; fine-grained control methods like Fastpass; the scheduling methods Freeway [11] and AuTO, etc., which differentiate between short and long flows; and the LBDC [37] mechanism, which is used to solve the problem of load imbalance among multiple controllers.
- Host-based load balancing mechanisms include Hermes [13], and other rerouting techniques; Presto, and other traffic slicing protocols; and CAPS [38], and other fine-grained scheduling mechanisms.
- Switch-based load balancing mechanisms need to be distinguished from the scheduling granularity, including flow-level scheduling methods like Dart and WCMP [20]; packet-level scheduling methods like RPS, DRILL [12]; packet-cluster-level scheduling methods like CONGA; and adaptive granularity scheduling methods like AG [39] and TLB [14].

The central controller-based load balancing mechanism can precisely detect link failures and designate the optimal transmission path for data transfers using global network state information. However, obtaining and maintaining global information necessitates a certain deployment overhead, and the large feedback and control delays hinder load balancing performance under dynamic surge traffic. The host-based load balancing mechanism is more scalable than the centralized scheduling scheme because its hosts are independent and each host can modify the load balancing mechanism to meet its own needs. However, the end-to-end feedback latency of host-based load balancing prevents it from adapting to extremely dynamic surge traffic. The switch-based load balancing mechanism can detect the link load in the network in real time and quickly balance the network traffic on the path connected to the switch out ports, but it is difficult to obtain the end-to-end path accurately and quickly, thereby affecting the accuracy of traffic transfer.

Deep Reinforcement Learning. DRL has become a research hotspot in the field of machine learning in recent years, and DRL algorithms are primarily divided into the following categories:

- Value function-based algorithms employ iterative updates to optimize the value function. These algorithms evaluate the value of each state using the value function, and they update the value function's parameters based on the most recent forecasts and reward signals. Among these, Deep Q-Network

(DQN) [19] is a traditional approach that stabilizes training by using experience replay and target networks and a deep neural network to approximate the value function. By addressing the issue of high valuation functions in DQN to enhance performance, Double DQN (DDQN) improves on DQN.
- Function approximation is used by policy gradient-based algorithms to build policy networks. These algorithms choose actions, retrieve the related reward values via the policy network, and then optimize the policy by maximizing the reward value by adjusting the policy network's parameters in the gradient direction. An actor-based algorithm known as DDPG combines the concepts of value functions with policy gradients. By restricting the size of policy updates, the Proximal Policy Optimization (PPO) algorithm ensures the stability and convergence of policy optimization. These algorithms are built on the concept of policy gradients and enhance the algorithm's performance through various enhancements and optimizations.

In addition, DRL can improve the policy search process by incorporating artificial supervision and integrating the policy evaluation and value estimation capabilities of deep neural networks with the policy search and exploration capabilities of Monte Carlo Tree Search (MCTS) [40]. Hierarchical reinforcement learning methods can also be used to decompose the overall goal into multiple subtasks and solve the problem of low efficiency of directly optimising the strategy of the overall goal in complex (DRL) tasks by learning hierarchical strategies and combining the strategies of multiple subtasks to form an effective overall strategy. This technique is commonly referred to as hierarchical DRL based on spatio-temporal abstraction and intrinsic motivation.

6   Conclusion

In this paper, we propose DRLB, a DRL-based load balancing mechanism for heterogeneous traffic in datacenter networks. DRLB distinguishes heterogeneous traffic and employs DRL technology for load balancing. Specifically, DRLB uses D4PG algorithm to learn the optimal load balancing policy, and dynamically adjusts LSM value via DRL agent to guide long flows to select the least congested forwarding paths. The most suitable link for forwarding addresses the issue that the traditional load balancing mechanism cannot adapt well to change traffic scenarios and employs WCMP mechanism to safeguard delay-sensitive short flows against the issue of a lengthy pre-learning period for DRL. Using the 99th percentile FCT of short flows and the throughput of long flows as performance metrics, our experimental results demonstrate that DRLB significantly improves the overall performance and stability of the datacenter network when compared with the existing Freeway, Hermes, and TLB mechanisms. The algorithm can be further enhanced and validated in real-world scenarios, and DRLB can be refined in future research.

References

1. Li, Z., Bai, W., Chen, K.: Rate-aware flow scheduling for commodity data center networks. In: Proceedings of IEEE INFOCOM, pp. 1–9 (2017)
2. Hu, J., He, Y., Wang, J., Luo, W., Huang. J.: RLB: reordering-robust load balancing in lossless datacenter network. In: Proceedings ofACM ICPP (2023)
3. Ma, X., et al.: Error tolerant address configuration for data center networks with malfunctioning devices. In: Proceedings of IEEE INFOCOM, pp. 708–717 (2012)
4. Jing, Q., Wang, W., Zhang, J., Tian, H., Chen, K.: Quantifying the performance of federated transfer learning. arXiv preprint arXiv:1912.12795 (2019)
5. Hu, C., Chen, K., Chen, Y., Liu, B.: Evaluating potential routing diversity for internet failure recovery. In: Proceedings of IEEE INFOCOM, pp. 1–5 (2010)
6. Wang, Y., Wang, W., Liu, D., Jin, X., Jiang, J., Chen, K.: Enabling edge-cloud video analytics for robotics applications. Proc. IEEE Trans. Cloud Comput. (2022)
7. Li, W., Yuan, X., Li, K., Qi, H., Zhou, X.: Leveraging endpoint flexibility when scheduling coflows across geo-distributed datacenters. In: Proceedings of IEEE INFOCOM, pp. 873–881 (2018)
8. Li, W., Chen, S., Li, K., Qi, H., Xu, R., Zhang, S.: Efficient online scheduling for coflow-aware machine learning clusters. IEEE Trans. Cloud Comput. 10(4), 2564–2579 (2020)
9. Zeng, G., Bai, W., Chen, G., Chen, K., Han, D., Zhu, Y.: Combining ECN and RTT for datacenter transport. In: Proceedings of the First Asia-Pacific Workshop on Networking, pp. 36–42 (2017)
10. Hopps, C.: Analysis of an equal-cost multi-path algorithm. RFC 2992 (2000)
11. Wang, W., Sun, Y., Zheng, K., Kaafar, M.A., Li, D., Li, Z.: Freeway: adaptively isolating the elephant and mice flows on different transmission paths. In: Proceedings of the IEEE International conference on network protocols, pp. 362–367 (2014)
12. Ghorbani, S., Yang, Z., Godfrey, P.B., Ganjali, Y., Firoozshahian, A.: DRILL: micro load balancing for low-latency data center networks. In: Proceedings of ACM SIGCOMM, pp. 225–238 (2017)
13. Zhang, H., Zhang, J., Bai, W., Chen, K., Chowdhury, M.: Resilient datacenter load balancing in the wild. In: Proceedings of ACM SIGCOMM, pp. 253–266 (2017)
14. Hu, J., Huang, J., Lv, W., Li, W., Wang J., He, T.: TLB: trafficaware load balancing with adaptive granularity in data center networks. In: Proceedings of ACM ICPP, pp. 1–10 (2019)
15. He, X., Li, W., Zhang, S., Li, K.: Efficient control of unscheduled packets for credit-based proactive transport. In: Proceedings of ICPADS, pp. 593–600 (2023)
16. Wang, J., Rao, S., Ying, L., Sharma, P.K., Hu, J.: Load balancing for heterogeneous traffic in datacenter networks. J. Netw. Comput. Appl. 217 (2023)
17. Hu, J., et al.: Enabling load balancing for lossless datacenters. In: Proceedings of IEEE ICNP (2023)
18. Hu, J., Huang, J., Li, Z., Wang, J., He, T.: A receiver-driven transport protocol with high link utilization using anti-ECN marking in data center networks. IEEE Trans. Netw. Serv. Manag. 20(2), 1898–1912 (2023)
19. Mnih, V., Kavukcuoglu, K., Silver, D.: Playing Atari with deep reinforcement learning. In: Proceedings of Workshops at the 26th Neural Information Processing Systems, pp. 201–220 (2013)
20. Zhou, J.L., et al.: WCMP: weighted cost multipathing for improved fairness in data centers. In: Proceedings of ACM SIGCOMM, pp. 1–14 (2014)

21. Barth-Maron, G., Hoffman, M.W., Budden, D.: Distributed distributional deterministic policy gradients. arXiv preprint arXiv:1804.08617, (2018)
22. Wang, J., Yuan, D., Luo, W., Rao, S., Sherratt, R.S., Hu, J.: Congestion control using in-network telemetry for lossless datacenters. CMC-Comput. Mater. Continua **75**(1), 1195–1212 (2023)
23. Zhao, Y., Huang, Y., Chen, K.: Joint VM placement and topology optimization for traffic scalability in dynamic datacenter networks. Comput. Netw. 109–123 (2015)
24. Wei, W., Gu, H., Wang, K., Li, J., Zhang, X., Wang, N.: Multi-dimensional resource allocation in distributed data centers using deep reinforcement learning. In: IEEE TNSM, pp. 1817–1829 (2023)
25. Wang, J., Liu, Y., Rao, S., Sherratt, R.S., Hu, J.: Enhancing security by using GIFT and ECC encryption method in multi-tenant datacenters. CMC-Comput. Mater. Continua **75**(2), 3849–3865 (2023)
26. Wei, W., Gu, H., Deng, W., Xiao, Z., Ren, X.: ABL-TC: a lightweight design for network traffic classification empowered by deep learning. Neurocomputing, 333–344 (2022)
27. Hu, C., Liu, B., Zhao, H.: DISCO: memory efficient and accurate flow statistics for network measurement. In: Proceedings of IEEE ICDCS, pp. 665–674 (2010)
28. Li, H., Zhang, Y., Zhang, Z.: URSA: hybrid block storage for cloud-scale virtual disks. In: Proceedings of ACM EuroSys, pp. 1–17 (2019)
29. Hu, J., Zeng, C., Wang, Z., Xu, H., Huang, J., Chen, K.: Load balancing in PFC-enabled datacenter networks. In: Proceedings of ACM APNet (2022)
30. Bai, W., Chen, K., Hu, S., Tan, K., Xiong, Y.: Congestion control for high-speed extremely shallow buffered datacenter networks. In: Proceedings of ACM APNet, pp. 29–35 (2017)
31. Liu, Y., Li, W., Qu, W., Qi, H.: BULB: lightweight and automated load balancing for fast datacenter networks. In: Proceedings of ACM ICPP, pp. 1–11 (2022)
32. Xu, R., Li, W., Li, K., Zhou X., Qi, H.: DarkTE: towards dark traffic engineering in data center networks with ensemble learning. In: Proceedings of IEEE/ACM IWQOS, pp. 1–10 (2021)
33. Wang, J., Liu, Y., Rao, S., Zhou, X., Hu, J.: A novel self-adaptive multi-strategy artificial Bee colony algorithm for coverage optimization in wireless sensor networks. Ad Hoc Netw. **150** (2023)
34. Hu, C., Liu, B., Zhao, H.: Discount counting for fast flow statistics on flow size and flow volume. IEEE/ACM Trans. Network. **22**(3), 970–981 (2014)
35. Wei, W., et al.: GRL-PS: graph embedding-based DRL approach for adaptive path selection. Proc. IEEE Trans. Netw. Serv. Manag. 1 (2023)
36. Zheng, J., Du, Z., Zha, Z., Yang, Z., Gao, X., Chen, G.: Learning to configure converters in hybrid switching data center networks. IEEE/ACM Trans. Network. 1–15 (2023)
37. Gao, X., Kong, L., Li, W., Liang, W., Chen, Y., Chen, G.: Traffic load balancing schemes for devolved controllers in mega data centers. Proc. IEEE Trans. Parallel Distrib. Syst. 572–585 (2017)
38. Hu, J., Huang, J., Lv, W., Zhou, Y., Wang, J., He, T.: CAPS: coding-based adaptive packet spraying to reduce flow completion time in data center. In: Proceedings of IEEE INFOCOM, pp. 2294–2302 (2018)
39. Liu, J., Huang, J., Li, W., Wang, J.: AG: adaptive switching granularity for load balancing with asymmetric topology in data center network. In: Proceedings of IEEE ICNP, pp. 1–11 (2019)
40. Silver, D., Huang, A., Maddison, C. J.: Mastering the game of Go with deep neural networks and tree search. Nature 484–489 (2016)

Adaptive Routing for Datacenter Networks Using Ant Colony Optimization

Jinbin Hu, Man He, Shuying Rao, Yue Wang, Jing Wang, and Shiming He[✉]

School of Computer and Communication Engineering, Changsha University of
Science and Technology, Changsha 410004, China
{jinbinhu,znwj_cs,smhe_cs}@csust.edu.cn,
{heman,shuyingrao,wyyy}@stu.csust.edu.cn

Abstract. Modern datacenter networks (DCNs) employ Clos topologies
that providing sufficient cross-sectional bandwidth, various load balancing mechanisms are proposed to make full use of multiple parallel paths
between end-hosts. Faced with a large number of heterogeneous flows,
existing load balancing schemes cannot work well and cause performance
degradation, such as latency-sensitive short flows experiencing large tail
delay and severe link bandwidth waste due to random rerouting. To
solve these issues, we propose an adaptive routing mechanism based
on ant colony optimization algorithm (RACO), which adopts different
(re)routing strategies for heterogeneous flows. Specifically, RACO uses
the improved ant colony optimization algorithm to make optimal rerouting decisions to obtain high throughput and low latency for both elephant
flows and mice flows, respectively. The experimental results based on
Mininet simulation show that RACO effectively increases the throughput of long flows and reduces the average flow completion time (FCT)
of short flows by up to 42% and 61%, respectively, compared with the
state-of-the-art load balancing mechanisms.

Keywords: Datacenter networks · Routing · Ant colony
optimization · Heterogeneous flows

1 Inroduction

With the rise of cloud computing and big data, datacenters have become an
indispensable part of network computing infrastructure [1–4]. In datacenters, a
large number of commercial switches and servers form numerous high-speed clusters through large-scale network interconnection, providing rich computing and
storage resources [5,6]. In order to cope with the increasing bandwidth demand,

This work is supported by the National Natural Science Foundation of China
(62102046, 62072056), the Natural Science Foundation of Hunan Province
(2023JJ50331, 2022JJ30618, 2020JJ2029), the Hunan Provincial Key Research and
Development Program (2022GK2019), the Scientific Research Fund of Hunan Provincial Education Department (22B0300).

datacenters typically adopt a tree topology [7] to provide multiple parallel transmission links between nodes, such as Dcell [8], Fat-tree [9], and Bcube [10]. At the same time, various excellent load balancing schemes [11–18] have been proposed to fully utilize multiple parallel paths between terminal hosts.

Load balancing mechanism is one of the important mechanisms to achieve low latency, high bandwidth requirements, and improve network transmission quality in datacenter networks. Through load balancing mechanism, the traffic in datacenter networks can be dispersed across multiple paths, thereby improving the bandwidth utilization and network transmission quality of the network. There are a large number of heterogeneous traffic [19–27] with different transmission requirements [28–30] in datacenter networks, such as latency-sensitive mice flows and throughput-sensitive elephant flows. Therefore, it is crucial for the load balancing mechanism to fully utilize the rich link resources and meet the transmission needs of long and short flows in datacenter networks.

However, existing load balancing mechanisms often encounter problems such as large tail latency and severe link bandwidth waste when facing a large amount of heterogeneous traffic. ECMP [31] is a commonly used standard load balancing mechanism in datacenter networks, which mainly utilizes static hashing mechanism, that is, by utilizing network information to forward traffic to various paths in the network. However, ECMP is vulnerable to hash collisions, in which numerous elephant flows are routed to the same path, causing chain congestion and squandering a significant percentage of the network's available bandwidth. In response to the shortcomings of ECMP, Dixit A et al. proposed the random packet scattering mechanism RPS [32]. It reduces the number of hash conflicts while improving link utilization by randomly sending packets to various available paths in the link. Meanwhile, due to the serious disorder caused by packet granularity transmission, LetFlow [33] based on flowlet transmission has emerged. LetFlow divides the elephant flow by setting time intervals and randomly sends the divided flow to each available path, effectively alleviating the disorder problem and improving link utilization. However, the above methods ignore the transmission characteristics of network traffic and do not adopt reasonable forwarding strategies for heterogeneous flows, resulting in the inability to meet the transmission requirements of throughput-sensitive elephant flows and latency-sensitive mice flows. And due to the randomly selected rerouting method, traffic cannot be accurately routed to the optimal path.

The Ant Colony Optimization (ACO) algorithm, as a heuristic optimization algorithm, has been widely used in network routing [34] and other combinatorial optimization problems [35]. The ant colony optimization algorithm can avoid the congestion problem in datacenter networks through the guidance of pheromone and the path selection strategy, disperse the traffic to different paths, and improve the overall performance and throughput of the network. By using the ant colony optimization algorithm, we can greatly increase the accuracy of traffic routing.

In response to the above analysis, this paper proposes an adaptive routing mechanism called RACO based on ACO algorithm, which selects paths for flows according to their different needs. We improve the heuristic function that guides ant routing, taking bandwidth utilization and transmission delay as influencing conditions to guide elephant flows in selecting paths with low bandwidth utilization and low latency for mice flows. Through this approach, RACO can effectively improve link utilization and reduce transmission delay, thus meeting the transmission requirements of heterogeneous traffic.

The main contributions of this paper are as follows:

- We conduct in-depth research to analyze the problems caused by ignoring the needs of different types of traffic and randomly selecting rerouted paths: short flows experiencing long tail delays lead to an increase in flow completion time, and random routing of traffic leads to a significant waste of link resources.
- We propose RACO, an adaptive routing mechanism based on ACO algorithm. RACO uses ACO algorithm to select the path rules, redefines the relevant heuristic information, takes the bandwidth utilization and transmission delay as the influencing factors for ants to select the next node, and then selects the optimal transmission path for long and short flows in the network transmission.
- We conduct Mininet simulation under symmetric and asymmetric scenes to test the effectiveness of RACO. The results show that RACO can effectively avoid the long tail congestion and the waste of link resources. Compared with the state-of-the-art schemes, RACO effectively increases the throughput of long flows and reduces the average FCT of short flows by up to 42% and 61%, respectively.

The rest of the paper is organized as following. We describe the background and motivation in Sect. 2. In Sect. 3, we describe the overview. In Sect. 4, we show the Mininet simulation results. In Sect. 5, we present the related works and then conclude the paper in Sect. 6.

2   Background and Motivation

2.1   Background

Intelligent optimization algorithms are widely applied in the field of networking [36]. The Ant Colony Optimization (ACO) algorithm is an intelligent optimization algorithm that simulates the foraging behavior of ants. At present, it is widely used in path planning, network routing, resource allocation, and other fields. It solves various combinatorial optimization problems by simulating information exchange and cooperation among ants.

The ant colony optimization algorithm is adaptive and robust, and can automatically adapt to the changes and dynamics of the datacenter networks. When the network topology changes, the link fails, or there is a new traffic demand, the ant colony optimization algorithm can adapt to the new situation through the update of pheromones and the adjustment of path selection, without explicit adjustment or re-optimization of the algorithm. The adaptability and robustness of the ant colony optimization algorithm provide significant advantages in dynamic datacenter network environments. In addition, the ant colony optimization algorithm has relatively low computational and communication overhead. Since the ant colony optimization algorithm only needs each ant to select the path and update the pheromone according to the local information, it does not need the global state information, so the calculation and communication overhead of the algorithm are relatively small. This enables the ant colony optimization algorithm to efficiently perform path selection in datacenter networks with high real-time requirements.

2.2 Motivation

In this section, we first investigate why existing load balancing schemes cannot meet the transmission requirements of heterogeneous traffic in datacenter networks with multiple parallel paths. Then, we conduct extensive simulation experiments to objectively present the performance comparison of existing load balancing schemes when faced with a large amount of heterogeneous traffic transmission.

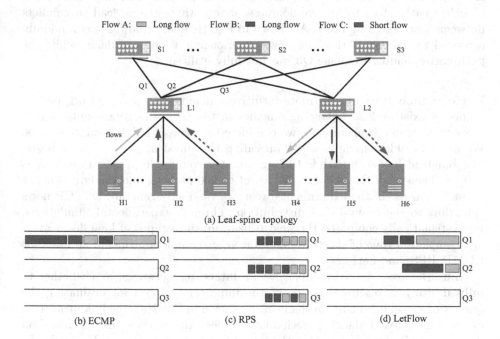

Fig. 1. Queueing under different load balancing mechanisms

Problem Description. In datacenter networks, blindly selecting rerouting paths without considering the needs of different traffic may cause damage to network transmission performance. We choose three typical load balancing schemes, ECMP, RPS, and LetFlow, to illustrate this issue and demonstrate its impact.

We show how ECMP, RPS, and LetFlow work in Fig. 1. In Fig. 1(a), we have sent three flows from the senders (H1, H2, H3) to the receivers (H4, H5, H6), consisting of two long flows (A, B) and one short flow C. These flows from switch L1 arrive at switch L2 using three parallel paths, each aligned with the three output port queues (Q1, Q2, Q3) of switch L1.

ECMP selects a fixed forwarding path for the flow based on the five tuples, so it is inevitable that different flows will be clustered on the same path, resulting in long tail latency and hash conflicts. As illustrated in Fig. 1(b), it can be seen that short flow C ranks behind long flows (A, B), while long flows (A, B) are all routed to a single path, and the paths corresponding to the remaining queues (Q2, Q3) are not fully utilized. The RPS scheme effectively improves link utilization by randomly forwarding packets to various available paths. However, due to unknown path conditions, it may lead to issues such as long tail delay and disorder. As shown in Fig. 1(c), the short flow C is still randomly placed after the long flows (B, C), and there are out-of-order packets. LetFlow distinguishes grouping clusters based on time interval thresholds and randomly sends them to other paths. Due to the randomness of rerouting, it may lead to conflicts between short and long flows. As shown in Fig. 1(d), short flow C was randomly rerouted to the end of the long flow, experiencing a long tail delay, while the path corresponding to queue Q3 was not fully utilized.

Performance Impact. To more intuitively demonstrate the network performance of existing load balancing schemes in the face of large amounts of heterogeneous traffic transmission, we conducted a series of NS-3 simulation tests. We use a 2×10 leaf-spine topology providing 10 equivalent paths between hosts. The bandwidth of each path is 10Gbps, and the round-trip propagation delay is 100 μs. The size of the switch buffer is set to 200 packets, and the flow timeout is set to 150 μs. In the experimental test, we randomly generate DCTCP flows according to the heavy-tailed distribution. Through experimental simulations, we systematically compared the link utilization, throughput of long flows, average completion time of short flows, and average completion time of all flows for ECMP, RPS, and LetFlow.

Link resources are very expensive in datacenter networks, and whether to fully utilize link resources is one of the important criteria for evaluating the quality of a load balancing mechanism. As shown in Fig. 2(a), the link utilization under various load balancing mechanisms varies. Among them, ECMP based on flow granularity transmission has the lowest link utilization rate (only one-third), which is due to the adverse consequences of a large number of hash collisions leading to multiple flows be sent to the same path for transmission. Although LetFlow has improved performance compared to ECMP, the use of flowlet granularity to randomly select rerouted paths still results in low link utilization. In addition, although the link utilization of RPS transmitted in packet granularity is significantly higher than the other two schemes, it only reaches 80%, and there is still a utilization inefficiency in link resources.

As shown in Fig. 2(b), consistent with the comparison of link utilization, the throughput of ECMP's long flows is still at its lowest level. ECMP is prone to forwarding different long flows to the same path, leading to network congestion and reducing the throughput of long flows. We can also observe that the throughput of long flows with high link utilization RPS is higher than that of ECMP and LetFlow. This is because RPS can divide long flows into multiple data packets for transmission on different paths, improving the throughput of long flows. But it is precisely because of the random transmission of multiple packets that the problem of out-of-order retransmission caused by RPS is very serious and has a negative impact.

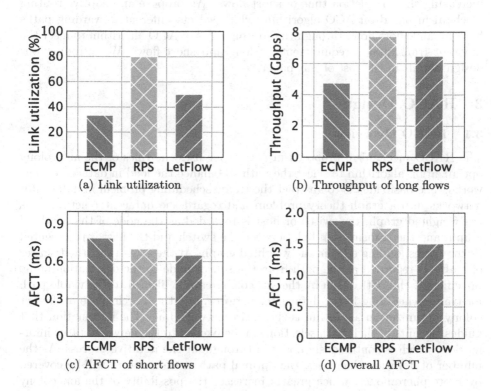

Fig. 2. Performance under different load balancing mechanisms

As shown in Fig. 2(c) and Fig. 2(d), LetFlow has significantly lower average completion time of short flows and average FCT for overall flows compared to ECMP and RPS. This is because ECMP may forward both long and short flows to the same path, causing short flows to experience significant long tail delays, thereby increasing the average flow completion time of short flows. The waste of link resources and the low throughput of long flows fully validate the fact that the average FCT of overall flows of ECMP is high. At the same time, RPS randomly sends packets to various paths through the switch, which can lead to adverse effects such as disorderly retransmission or even packet loss retransmission of each data packet, resulting in an increase in FCT and affecting network transmission performance.

2.3 Summary

Therefore, based on the observations of the above experiments, we conclude that (1) random rerouting leads to the failure to fully utilize the resources of each link, resulting in serious bandwidth waste and low link utilization, thereby affecting the throughput of long flows. (2) Random rerouting results in long and short flows sharing the same path, and short flows experience large tail delays, increasing the completion time of short flows. We propose an adaptive routing mechanism based on ACO algorithm, which selects different forwarding paths for long and short flows through the routing rules of ACO algorithm to meet the different transmission requirements of long and short flows. We introduce our design details in the rest of this paper.

3   RACO Design

3.1   RACO Overview

The RACO mechanism proposed in this paper uses the rules of the ant colony optimization algorithm to select the path to balance the load in datacenter networks. The basic idea is to abstract the traffic scheduling problem in datacenter networks into a graph theory problem and regard the datacenter networks as the weighted graph, the switch or host is regarded as the node of the weighted graph, and the transmission link between the switch and the host is represented as the corresponding edge in the weighted graph. At the same time, the transfer of traffic is regarded as the path from the source node to the destination node. Specifically, the path taken by the ant colony can be likened to the viable path for traffic scheduling in the datacenter network, and the search space of the ant colony optimization algorithm is the entire network link. The information that guides the ants in their path selection can be likened to the network status information, which is continuously updated throughout the iterative process. As the number of iterations increases, the optimal path in the network will be covered by more pheromones, which greatly increases the possibility of the ant colony choosing the optimal path. Promoted by positive feedback, the ant colony optimization algorithm can effectively point to the best path in the network, so as to realize the effective scheduling of convection. RACO consists of two modules, as shown in Fig. 3.

(1) **Statistics Module.** In RACO, we first distinguish between traffic types, dividing the flow into elephant and mice flows based on the number of bytes sent by the flow. Then according to the different requirements of two kinds of traffic, the relevant network state information is collected.

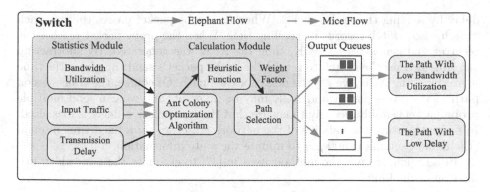

Fig. 3. RACO architecture

(2) **Calculation Module.** The calculation module is mainly responsible for deciding different forwarding strategies based on the flow size. Because of the need to ensure as fast as possible, the mice flow chooses the path with lower delay to avoid long tail delay. The long flow selects the path with low bandwidth utilization to make full use of the abundant link resources.

3.2 Statistics Module

The adaptive routing mechanism based on ACO algorithm proposed in this paper requires distinguishing between elephant flows and mice flows, and implementing different path selection strategies to meet the requirements of high bandwidth for elephant flows and low latency for mice flows. So this paper distinguishes traffic based on the number of bytes sent by the flow. When the number of bytes received exceeds the set threshold (set to 100KB in this paper), it is determined that the sent flow is an elephant flow; Otherwise, it's a mice flow. In addition, we use graph to represent the topology of datacenter network. A set $S=\{s1, s2, ..., s_n\}$ is used to represent a switch node, and the set $L=\{l1, l2, ..., l_n\}$ represents a link in the network, where l_{ij} is represented as a link between l_i node and l_j node.

In order to meet the needs of high throughput of elephant flows and low latency of mice flows in datacenter networks, we need to divide the statistical network state information into transmission delay and bandwidth utilization, because it has a great impact on the performance of long and short flows. We choose $band(l_{ij})$ to represent inter-link bandwidth utilization and $delay(l_{ij})$ represent inter-link transmission delay. in datacenter networks, we take ants as packets of data. We record the network state information through the switch, which is used as the influence factor for the ants to choose the available path for different types of flows. We maintain a link state estimation table on the switch port that records the addresses, bandwidth utilization, and transmission delays of all the next switch nodes. We use the method used in hula to estimate the bandwidth utilization of the link. In addition, we record the transmission delay by sending the probe packet. When the probe packet passes through each switch, each switch records its sending time. When the probe packet reaches the receiver and generates an ACK, time the ACK was generated. By subtracting the time it takes the switch to record the probe packet's send time from the time it produces the ACK, we get the delay for each path. Of course, the transmission path of the probe packet and ACK may be different, and we can send multiple probe packets simultaneously to ensure that the transmission delay is measured. In order to ensure the validity of the network state, we set a fixed time interval (the default setting is 500 μs) to update the state information.

In addition, we use the method used in [37] to estimate the bandwidth utilization of the link:

$$U = D + U \times (1 - \frac{\Delta t}{\tau})$$

where U is the estimated bandwidth utilization of the link, D represents the size of the data packet, Δt represents the time elapsed since the last update of bandwidth utilization, τ is a time constant and should be set to at least twice the sending frequency of the probe packet.

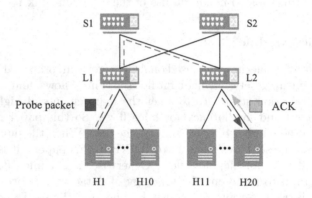

Fig. 4. Calculation of transmission delay

As shown in the Fig. 4, we use a 2×2 leaf-spine topology to illustrate how to measure the transmission delay. We send a probe packet from H1 to H2. When passing through switches L_1, S_1, L_2, we record the sending time of this probe packet on each switch. When the probe packet arrives at H2 to generate ACK, we record the time of generating ack. In this way, when the switch receives the ACK, we can subtract the sending time of each switch's recorded probe packet from the time of generating the ack to obtain the path delay from each switch to the destination node. Then we can subtract the path delay from two adjacent switches to the destination node to obtain the path delay of the adjacent switch. For example, $delay(S_1 - L_2) = delay(S1 - H2) - delay(L_2 - H_2)$.

3.3   Calculation Module

The calculation module mainly makes different path calculation decisions for elephant flows and mice flows to meet their different transmission needs. The rule for calculating the path selection is based on the transition probability P_{ij} to select the next node. The calculation formula is shown in equation (2):

$$P_{ij} = \begin{cases} \frac{(\tau_{ij})^\alpha \times (\eta_{ij})^\beta}{\sum_{u \in ad_k} (\tau_{iu})^\alpha \times (\eta_{iu})^\beta}, & u \in ad_k \\ 0, & others \end{cases} \tag{2}$$

Among them, P_{ij} represents the probability of ants moving from node i to node j, ad_k represents k paths that ants can move, and α represents the importance of pheromone. β represents the importance of the heuristic factor, τ_{ij} represents the concentration of pheromone on path l_{ij}, η_{ij} refers to heuristic function on path l_{ij}.

We can see from the above formula that the transition probability P_{ij} that affects the ant to select the next node is mainly affected by pheromone concentration and heuristic function. In this paper, we mainly improve the heuristic function to meet the needs of long and short flows. Specifically, considering the impact of transmission delay and bandwidth utilization on traffic transmission, we propose a corresponding improved heuristic function to guide ants to select the next node. Path l_{ij} of the heuristic function on is shown in equation (3):

$$\eta = \frac{\varepsilon}{band(l_{ij})} + \frac{\theta}{delay(l_{ij})} \tag{3}$$

Among them, ε and θ are weight factors, $0 < \varepsilon < 1$, $0 < \theta < 1$, and the values of varepsilon and theta are determined based on the current type of traffic. When the forwarded traffic in the network is elephant flow, it is more sensitive to link bandwidth utilization, and the value of ε should be greater than θ. On the contrary, when the traffic forwarded in the network is mice flow, it requires a higher transmission delay of the link, and the value of θ is greater than ε. The larger the $band(l_{ij})$ and $delay(l_{ij})$ values, the higher the load and delay of this link. Therefore, the shorter the value of the heuristic function, the less likely the ant is to choose the path, in order to guide the ant in heuristic routing.

In addition, in order to ensure the accuracy of the algorithm, we will adjust the pheromone of the path the ants go through to obtain the optimal result after the ants complete a routing. The update of pheromone can be realized by the following formula:

$$\tau_{ij}(t+1) = (1 - P) \times \tau_{ij}(t) + \nabla\tau_{ij}(t) \tag{4}$$

Among them, $\tau_{ij}(t+1)$ is the pheromone concentration on the path l_{ij}, $\nabla\tau_{ij}(t)$ is the increment of pheromone on the path l_{ij}, and P is the volatilization factor of pheromone.

When the ant reaches its destination, we record its path in the Path Table PathList. The PathList= $[P1, P2, ..., P_n]$. We use the SelectBestPath function to select the most suitable path for mice and elephant flows. Specifically, when the next transmission of the data flow for the mice flow, we can choose the optimal path Ph in the path of the least delay. When the next data flow is an elephant flow, the path with the lowest bandwidth utilization in the optimal path Ph is chosen. In addition, RACO updates the optimal path at regular intervals (the default is 500 μs) to ensure the efficiency of the flow.

3.4 Algorithm Pseudocode

This paper proposes an adaptive routing mechanism based on ACO algorithm, which improves the heuristic function according to the different needs of long and short flows. By allocating size and network state information through weight factors, the number of link congestion is reduced, throughput and link utilization are improved. The RACO algorithm processing pseudocode is as follows:

Algorithm 1: RACO Algorithm

Input: The first node on the path src, the last node on the path dst,$G(S, L)$
Output: The optimal path Ph
1 **while** $iter < M$ **do**
2   **for** $i = 1$ to k **do**
3   Set curNode=src;
4   **while** $curNode! = src$ **do**
5   Choose next node with P_{ij};
6   path.append(curNode);
7   **end**
8   PathList(path) update τ_{ij}
9   **end**
10 **end**
11 Ph=SelectBestPath(PathList);
12 return Ph

---

4   Performance Evaluation

4.1   Experimental Setup

Choices of Basedline. We choose four load balancing mechanisms, ECMP, RPS, DRILL, and LetFlow, to compare with RACO. ECMP is a common load balancing mechanism in datacenter networks, which distributes flows to each path by hashing. RPS is based on packets, which are sent randomly to each available path. Drill selects the port with the smallest queue length as the packet forwarding port by comparing the two currently randomly selected ports with the one with the smallest load in the previous round. LetFlow divides the flow according to a set time interval and randomly selects a forwarding port for the divided flow.

Simulation Setup. We build a K=4 Fat-tree topology using Mininet network simulation platform. The topology consists of 20 switches, 16 hosts and 48 links. In all experiments, we set the link bandwidth to 1 Gbps, the link propagation delay to $100\mu s$, and the buffer size of the switch to 256 packets. All traffic generation follows a Poisson distribution. In addition, we adjusted the workload from 0.4 to 0.7 to evaluate RACO performance. In order to reduce the adverse effect of random error and guarantee the fairness of the experiment, We test each load balancing mechanism twenty times, and take the average of twenty times as the final result.

At the same time, the parameter settings of the RACO algorithm also affect its performance. Based on multiple simulation tests, this paper sets the following parameters, as shown in Table 1.

Table 1. RACO algorithm parameters.

Parameter	Value
Iterations M	10
Pheromone volatile factor p	0.2
Pheromone weight α	2
Heuristic factor β	3
weight factors ε	mice flows: 0.3, elephant flows: 0.7
weight factors θ	mice flows: 0.7, elephant flows: 0.3

4.2 Performance Under Symmetrical Topology

In order to evaluate the effectiveness of the adaptive routing mechanism RACO based on ACO algorithm, link utilization, throughput, round-trip delay and average FCT are tested and analyzed under different network loads in a symmetric network topology, in conjunction with ECMP, RPS, DRILL, and LetFlow. The experimental results are as follows:

As shown in Fig. 5, RACO consistently maintains the highest link utilization compared to ECMP, RPS, DRILL, and LetFlow as network load increases. As a whole, with the increasing network load, the link utilization of all load balancing mechanisms is increasing, but the growth rate is different. For example, with a network load of 0.6, RACO has increased link utilization by about 12%, 14%, 26%, compared to DRILL, LetFlow, and ECMP, respectively. This is because RACO can choose the path with low bandwidth utilization for elephant flows, avoid the waste of link resources, and improve link utilization. RPS and DRILL based on packet granularity also have relatively high link utilization,

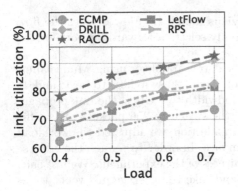

Fig. 5. Link utilization

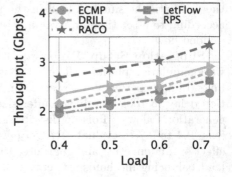

Fig. 6. Throughput of long flows

while ECMP has the lowest link utilization under different loads. This is because ECMP uses hash mechanism to schedule flows, different flows may choose the same fixed path forwarding, easy to cause a link in idle state for a long time.

Figure 6 shows that RACO consistently maintains the highest throughput compared to other advanced load balancing mechanisms as the network load increases. Specifically, RACO increases throughput by about 34%, 15%, 22%, and 33%, compared to ECMP, RPS, DRILL, and LetFlow at network load of 0.6. This is because RACO can route elephant flows to low bandwidth utilization links based on heuristic information, which greatly improves link utilization and overall throughput. The flow-based load balancing routing strategy, ECMP, will lead to the waste of link resources. Due to the inevitable hash collision and the design flaw of not being able to sense the congestion, which will also aggravate the congestion of the link, this will reduce network throughput.

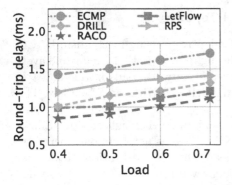

Fig. 7. Round-trip delay

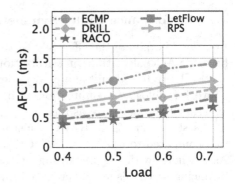

Fig. 8. AFCT of all flows

As shown in Fig. 7, the round trip delay increases as the load increases, but the round trip delay of RACO is always at the lowest level. Specifically, at a network load of 0.6, RACO reduces round-trip latency by 35%, 21%, 16%, and 8% compared to ECMP, RPS, DRILL, and LetFlow, respectively. This is because RACO distinguishes heterogeneous traffic and routes elephant traffic to transport paths with low link utilization. This effectively equalizes the transmission pressure of each path in the network, avoids the link congestion, and completes the transmission faster. In addition, RACO makes the mice flow always route to the switch port with the lowest delay, effectively reducing the probability of long and short flows routing together, avoiding the long tail delay of short flows, and effectively reducing the transmission delay.

Figure 8 shows that RACO significantly reduces the average FCT as network load increases compared to ECMP, RPS, DRILL, and LetFlow. The performance of fine-grained solutions such as RPS and DRILL decreases as the network load increases. This is because the congestion path increases and the packet disorder becomes more serious, causes more packet retransmissions to occur. However, ECMP and Letflow, as coarse-grained schemes, do not fully utilize the link resources in the network and do not take into account the transmission demand of heterogeneous traffic. So the average FCT is higher. RACO takes into account the needs of different traffic, and combined with the ACO algorithm iterative fast characteristics, can be more accurate and faster for the flow to find the best link path condition. More specifically, compared to ECMP, RPS, DRILL, and LetFlow at a load of 0.6, RACO can reduce the average FCT by 56%, 45%, 31%, and 12%, respectively.

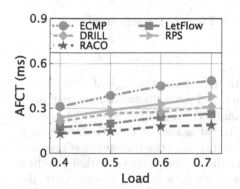
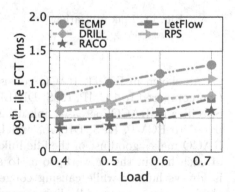

Fig. 9. AFCT of short flows   **Fig. 10.** 99th-ile FCT of short flows

Figures 9 and 10 show the mean FCT and 99th percentile FCT of the short flow from load 0.4 to 0.7, respectively. We notice that RACO is able to effectively reduce AFCT and tail FCT compared with the remaining four schemes. Specifically, at network load of 0.7, RACO reduces ACT and 99th percentile FCT by 61%, 50%, 40%, 29% and 53%, 44%, 27%, 23% compared to ECMP, RPS,

DRILL, and LetFlow, respectively. This is because when the workload becomes larger, more short flows will experience queuing delay and out-of-order retransmission. RACO can select the path with low transmission delay for mice flows, effectively improving the transmission performance of short flows and reducing queuing delays and the impact of retransmissions.

4.3 Performance Under Asymmetric Topology

In order to evaluate the effectiveness of the adaptive routing mechanism RACO based on ACO algorithm, link utilization, throughput, round-trip delay and average FCT are tested and analyzed under different network loads in an asymmetric network topology, in conjunction with ECMP, RPS, DRILL, and LetFlow.

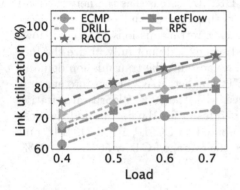

Fig. 11. Link utilization   **Fig. 12.** Throughput of long flows

As shown in Fig. 11, RACO consistently maintains the highest link utilization compared to ECMP, RPS, DRILL, and LetFlow as network load increases. Specifically, RACO increases link utilization by 34%, 15%, 22%, and 33%, compared to ECMP, RPS, DRILL, and LetFlow at a load of 0.6. This is because RACO makes good use of the idle links, and disperses the transmission pressure of each link in the network, so as to avoid the extreme situation that one link is always hungry while causing congestion. In this way, RACO can route the elephant flows to the idle link according to the heuristic information even in the asymmetric network transmission scenario, which makes full use of the link resources and improves the link utilization.

Figure 12 shows that RACO consistently maintains the highest throughput compared to other advanced load balancing mechanisms as the network load increases. Specifically, with a network load of 0.7, RACO increases throughput by approximately 46%, 19%, 24%, and 30%, compared to ECMP, RPS, DRILL, and LetFlow, respectively. RACO searches and optimizes continuously through heuristic information, and makes rational use of link resources to a large extent.

Even under the fast changing network load, elephant flows can still be sent quickly to the optimal forwarding path. Therefore, in an asymmetric topology, RACO still performs well with increasing load. Although RPS can improve link utilization and throughput by randomly scattering packets. The serious out-of-order problem can not be ignored.

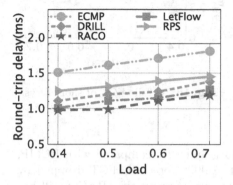

Fig. 13. Round-trip delay

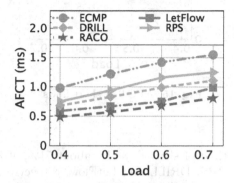

Fig. 14. AFCT of all flows

As shown in Fig. 13, RACO always maintains the lowest round trip latency compared to other advanced load balancing mechanisms under different network loads. This is because RACO fully considers the transport characteristics of heterogeneous traffic, routes the throughput-sensitive elephant flows to the port with low link utilization while routing the delay-sensitive mice flows to the port with low delay. This design can improve the link utilization and reduce the probability of long tail delay, thus reducing the round trip delay. In the asymmetric network scenario, the round trip delay of ECMP and RPS, which are prone to long tail delay and serious packet reorder. Specifically, compared with ECMP, RPS, DRILL, and LetFlow, the round trip delay is decreased by 35%, 20%, 10%, and 9%, respectively, at the load of 0.6.

Figure 14 shows that RACO significantly reduces the AFCT compared to ECMP, RPS, DRILL, and LetFlow as the network load increases. Specifically, at the load of 0.6, RACO reduced the average FCT by 52%, 41%, 31%, and 9% compared with ECMP, RPS, DRILL, and LetFlow, respectively. The results show that RACO still performs well in asymmetric topology. This is because RACO can continuously search and optimize by heuristic information, and can quickly adapt to network load changes, and choose the appropriate forwarding path for the flows in time. Through reasonable path forwarding, RACO can make full use of link resources to avoid long tail delay, thus reducing the average FCT.

As shown in Fig. 15 and Fig. 16, under different network loads, compared with other advanced load balancing mechanisms, the average completion time and 99th percentile FCT of short flows in RACO always maintain a lower level. Specifically, when the network load is 0.7, RACO reduces the average flow completion

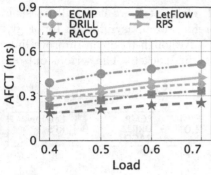
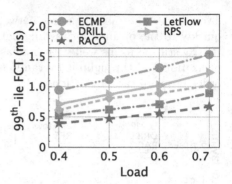

Fig. 15. AFCT of short flows   **Fig. 16.** 99th-ile FCT of short flows

time of short flows by about 51%, 40%, 33% and 32% compared with ECMP, RPS, DRILL and LetFlow, respectively. The 99th percentile FCT dropped by about 56%, 45%, 34%, and 25%. These test results prove that RACO still has advantages in asymmetric topologies. This is because RACO can dynamically select a more appropriate forwarding path for short flows based on the heuristic information, thereby avoiding long-tail delays for short flows and improving the transmission quality of short flows.

5   Related Work

Load Balancing Schemes Based on Flow. ECMP is a load balancing strategy based on flow granularity. In order to increase network capacity usage, a router will distribute various types of traffic to various links when it finds many equivalent paths that lead to the same destination address. ECMP, on the other hand, is unable to detect congestion, and for lines that are already congested, it is likely to make the situation worse. Hedera [38] achieves effective transmission of elephant flows by navigating all open channels on the network to identify the first one that can match user bandwidth requirements. Although this approach reduces network congestion, the chosen path might not be the best one. Mahout [39] monitors host side flows for elephant fluxes, which are then identified and tagged. The controller assigns the elephant flows to lighter loaded channels once the switch node gets the identified elephant flows. Although it requires changes to the host, this technique can significantly lower the switch's overhead.

Load Balancing Schemes Based on Flowlet. DRE technology is primarily used by CONGA [40] to gauge and assess the level of obstruction in the path. The switch keeps a brief flow information table and a congestion information table, and the switch chooses the flow's forwarding path based on the flow and congestion information. Adopting global congestion aware technology can reduce link latency and increase resource utilization, but it still has difficult deployment problems because it needs to store a lot of path information on switches. LetFlow makes use of the inherent properties shared by packets to detect path congestion automatically. Using time interval criteria to detect packet clusters, it sends them to different pathways at random. LetFlow can prevent chaos and successfully handle asymmetric issues. However, due to the randomness of LetFlow scheduling, optimal load balancing performance cannot be achieved.

Load Balancing Schemes Based on Packet. An approach for load balancing at the packet level is RPS. In this approach, multiple equivalent pathways to the same destination address are used for each connection on a per-packet basis, as determined by the router. Although it is straightforward, simple to deploy, and fully utilizes network links, it may cause major disorder issues. Between two random ports and the port with the least queue length from the previous round, DRILL [41] chooses a forwarding path. The basic idea behind path selection is to choose the port closest to them with the shortest wait length, then use that way to transmit the packet. However, because it may assess the status of the path based on local information, it is unable to completely eliminate the disorder issue.

6 Conclusion

In this paper, we propose RACO, an adaptive routing mechanism based on ACO algorithm in datacenter networks. RACO uses ACO algorithm to find the optimal path, and uses bandwidth utilization and transmission delay as heuristic information for ant to select the path. Then, the long and short flows are rerouted to the transmission path with low link utilization and low delay respectively, so as to avoid encountering the path with congestion and improve the network transmission performance. According to the Mininet simulation results, RACO can effectively avoid the long tail congestion and the waste of link resources. Compared with the most advanced solutions, RACO effectively increases the throughput of long flows and reduces the average FCT of short flows by up to 42% and 61%, respectively. In the future work, we plan to test the performance of RACO in a variety of different real scenarios and further improve the mechanism.

References

1. Wei, W., Gu, H., Wang, K., et al.: Multi-dimensional resource allocation in distributed data centers using deep reinforcement learning. IEEE Trans. Netw. Serv. Manage. **20**(2), 1817–1829 (2022)
2. Li, H., Zhang, Y., Li, D., et al.: URSA: hybrid block storage for cloud-scale virtual disks. In: Proceedings of the Fourteenth EuroSys Conference, pp. 1–17 (2019)
3. Zhao, Y., Huang, Y., Chen, K., Yu, M., et al.: Joint VM placement and topology optimization for traffic scalability in dynamic datacenter networks. Comput. Netw. **80**, 109–123 (2015)
4. Wang, J., Yuan, D., Luo, W., et al.: Congestion control using in-network telemetry for lossless datacenters. Comput. Mater. Continua. **75**(1), 1195–1212 (2023)
5. Wang, Y., Wang, W., Liu, D., et al.: Enabling edge-cloud video analytics for robotics applications. IEEE Trans. Cloud Comput. **11**(2), 1500–1513 (2023)

6. Hu, J., Huang, J., Li, Z., Wang, J., He, T.: A receiver-driven transport protocol with high link utilization using anti-ECN marking in data center networks. IEEE Trans. Netw. Serv. Manage. **20**(2), 1898–1912 (2023)
7. Zheng, J., Du, Z., Zha, Z., et al.: Learning to configure converters in hybrid switching data center networks. IEEE/ACM Trans. Netw., 1–15 (2023)
8. Guo, C., Wu, H., Tan, K., Shi, L., Zhang, Y., Lu, S.: DCell: a scalable and fault-tolerant network structure for data centers. In: Proceedings of ACM SIGCOMM, pp. 75–86 (2008)
9. Al-Fares, M., Loukissas, A., Vahdat, A.: A scalable, commodity data center network architecture. ACM SIGCOMM Comput. Commun. Rev. **38**(4), 63–74 (2008)
10. Guo, C., Lu, G., Li, D., et al.: BCube: a high performance, server-centric network architecture for modular data centers. In: Proceedings of ACM SIGCOMM, pp. 63–74 (2009)
11. Hu, J., Huang, J., Lv, W., Zhou, Y., Wang, J., He, T.: CAPS: coding-based adaptive packet spraying to reduce flow completion time in data center. In: Proceedings of IEEE INFOCOM, pp. 2294–2302 (2018)
12. Hu, J., Huang, J., Lv, W., Li, W., Wang J., He, T.: TLB: Trafficaware load balancing with adaptive granularity in data center networks. In: Proceedings of ACM ICPP, pp. 1–10 (2019)
13. Hu, J., He, Y., Wang, J., et al.: RLB: Reordering-robust load balancing in lossless datacenter network. In: Proceedings of ACM ICPP (2023)
14. Hu, J., Zeng, C., Wang, Z., et al.: Enabling load balancing for lossless datacenters. In: Proceedings of IEEE ICNP (2023)
15. Zhang, H., Zhang, J., Bai, W., Chen, K., Chowdhury, M.: Resilient datacenter load balancing in the wild. In: Proceedings of ACM SIGCOMM, pp. 253–266 (2017)
16. Liu, Y., Li, W., Qu, W., Qi, H.: BULB: lightweight and automated load balancing for fast datacenter networks. In: Proceedings of ACM ICPP, pp. 1–11 (2022)
17. Hu, J., Zeng, C., Wang, Z., Xu, H., Huang, J., Chen, K.: Load balancing in PFC-enabled datacenter networks. In: Proceedings of ACM APNet (2022)
18. Xu, R., Li, W., Li, K., Zhou, X., Qi, H.: DarkTE: towards dark traffic engineering in data center networks with ensemble learning. In: Proceedings of IEEE/ACM IWQOS, pp. 1–10 (2021)
19. Li, W., Chen, S., Li, K., Qi, H., Xu, R., Zhang, S.: Efficient online scheduling for coflow-aware machine learning clusters. IEEE Trans. Cloud Comput. **10**(4), 2564–2579 (2020)
20. Wang, J., Rao, S., Liu, Y., et al.: Load balancing for heterogeneous traffic in datacenter networks. J. Netw. Comput. Appl., 217 (2023)
21. Wei, W., Gu, H., Deng, W., et al.: ABL-TC: a lightweight design for network traffic classification empowered by deep learning. Neurocomputing **489**, 333–344 (2022)
22. He, X., Li, W., Zhang, S., Li, K.: Efficient control of unscheduled packets for credit-based proactive transport. In: Proceedings of ICPADS, pp. 593–600 (2023)
23. Li, W., Yuan, X., Li, K., Qi, H., Zhou, X.: Leveraging endpoint flexibility when scheduling coflows across geo-distributed datacenters. In: Proceedings of IEEE INFOCOM, pp. 873–881 (2018)
24. Hu, C., Liu, B., Zhao, H., Chen, K., et al.: DISCO: memory efficient and accurate flow statistics for network measurement. In: Proceedings of IEEE ICDCS, pp. 665–674 (2010)
25. Bai, W., Chen, K., Hu, S., Tan, K., Xiong, Y.: Congestion control for high-speed extremely shallow-buffered datacenter networks. In: Proceedings of ACM APNet, pp. 29–35 (2017)

26. Cho, I., Jang, K., Han, D.: Credit-scheduled delay-bounded congestion control for datacenters. In: Proceedings of ACM SIGCOMM, pp. 239–252 (2017)
27. Hu, C., Liu, B., Zhao, H., et al.: Discount counting for fast flow statistics on flow size and flow volume. IEEE/ACM Trans. Netw. **22**(3), 970–981 (2013)
28. Li, Z., Bai, W., Chen, K., et al.: Rate-aware flow scheduling for commodity data center networks. In: Proceedings of IEEE INFOCOM, pp. 1–9 (2017)
29. Zhang, J., Bai, W., Chen, K.: Enabling ECN for datacenter networks with RTT variations. In Proceedings of ACM the 15th International Conference on Emerging Networking Experiments And Technologies, pp. 233–245 (2020)
30. Wang, J., Liu, Y., Rao, S., et al.: Enhancing security by using GIFT and ECC encryption method in multi-tenant datacenters. Comput. Mater. Continua. **75**(2), 3849–3865 (2023)
31. Hopps, C.E.: Analysis of an equal-cost multi-path algorithm (2000)
32. Dixit, A., Prakash, P., Hu, Y. C., Kompella, R.R.: On the impact of packet spraying in data center networks. In: Proceedings of IEEE INFOCOM, pp. 2130–2138 (2013)
33. Vanini, E., Pan, R., Alizadeh, M., Taheri, P., Edsall, T.: Let it flow: resilient asymmetric load balancing with flowlet switching. In: Proceedings of NSDI, pp. 407–420 (2017)
34. Lv, J., Wang, X., Ren, K., Huang, M., Li, K.: ACO-inspired information-centric networking routing mechanism. Comput. Netw. **126**, 200–217 (2017)
35. Gupta, A., Garg, R.: Load balancing based task scheduling with ACO in cloud computing. In: Proceedings of IEEE ICCA, pp. 174–179 (2017)
36. Wang, J., Liu, Y., Rao, S., et al.: A novel self-adaptive multi-strategy artificial bee colony algorithm for coverage optimization in wireless sensor networks. Ad Hoc Netw., 150 (2023)
37. Katta, N., Hira, M., Kim, C., Sivaraman, A., Rexford, J.: HULA: scalable load balancing using programmable data planes. In: Proceedings of ACM SOSR, pp. 1–12 (2016)
38. Al-Fares, M., Radhakrishnan, S., Raghavan, B., Huang, N., Vahdat, A.: Hedera: dynamic flow scheduling for data center networks. In: Proceedings of NSDI, pp. 89–92 (2010)
39. Curtis, A.R., Kim, W., Yalagandula, P.: Mahout: low-overhead datacenter traffic management using end-host-based elephant detection. In: Proceedings of IEEE INFOCOM, pp. 1629–1637 (2011)
40. Alizadeh, M., Edsall, T., Dharmapurikar, S., et al.: CONGA: distributed congestion-aware load balancing for datacenters. In: Proceedings of ACM SIG-COMM, pp. 503–514 (2014)
41. Ghorbani, S., Yang, Z., Godfrey, P.B., et al.: DRILL: micro load balancing for low-latency data center networks. In: Proceedings of ACM SIGCOMM, pp. 225–238 (2017)

MPC: A Novel Internal Clustering Validity Index Based on Midpoint-Involved Distance

Yating Zuo[1], Zhujuan Ma[2], and Erzhou Zhu[1(✉)]

[1] School of Computer Science and Technology, Anhui University, Hefei 230601,
People's Republic of China
ezzhu@ahu.edu.cn
[2] School of Big Data and Artificial Intelligence, Anhui Xinhua University,
Hefei 230088, People's Republic of China

Abstract. As one of the most import machine learning technique, clustering is widely used in many data classification areas. Due to the unsupervised learning feature, the quality of the clustering results needed to be evaluated. In this paper, the $MPdist$ (midpoint-involved distance) based on the midpoint of centers between two clusters is firstly defined to measure the inter-cluster separation. Then, the MPC ($MPdist$ based clustering validity index), a novel internal clustering validity index based on the combination of the new defined inner-cluster compactness and the inter-cluster separation, is proposed to effectively evaluate the validity of the clustering results of many clustering algorithms. Experimental results on testing many types of datasets have demonstrated that the MPC index proposed in this paper is able to quickly handle datasets like spherical datasets, non-spherical datasets and real large-scale datasets.

Keywords: clustering validity index · clustering algorithm · data mining

1 Introduction

Clustering is one of the most important techniques in machine learning which divides datasets into structural partitions based on similarity or dissimilarity measures. Clustering has been widely studied in many fields, such as pattern recognition, bioinformatics, image segmentation, and so on. At present, many clustering algorithms have developed and they can be roughly classified into 4 categories, the partitional clustering algorithms, the hierarchical clustering algorithms, the density-based clustering algorithms and the hybrid clustering algorithms [1].

Due to the unsupervised machine learning feature, different clustering algorithms or different configurations of the same algorithm may produce different partitions for a single dataset. There is no clustering algorithm can optimally process all applications [2]. Therefore, it is necessary to find an effective method

Z. Tari et al. (Eds.): ICA3PP 2023, LNCS 14489, pp. 310–323, 2024.
https://doi.org/10.1007/978-981-97-0798-0_18

to evaluate the performance of clustering before an algorithm being used to solve the practical problems.

The process of evaluating partitions generated by clustering algorithms is called clustering validation. The common approach for the evaluation is to use clustering validity indices (CVIs). The existing CVIs can be roughly classified into two types: the internal CVIs and the external CVIs [3]. The main difference between the internal and the external CVIs is whether the external information (such as labels indicate which cluster the point belongs to) is known.

At present, many internal CVIs have been proposed in literatures and they work well for datasets with specific structures. However, many of the existing CVIs lose their effectiveness when the complicated datasets, such as datasets composed of linear distributed clusters or clusters with outliers, are encountered. Actually, there is no CVI can optimally process all types of datasets [4].

To evaluate the clustering results stably and efficiently for more types of datasets, this paper proposes the MPC, a novel clustering validity index based on the midpoint involved distance. Generally speaking, the contributions of this paper are as follows:

(1) The robust midpoint-involved distance, $MPdist$. The $MPdist$ is defined to measure the inter-cluster separation among different clusters of datasets. The midpoint between centers of any two clusters is firstly defined. Then, the $MPdist$ of the two clusters is defined by the sum of average distances of all data points of the two clusters to the midpoint.
(2) The novel clustering validity index, MPC. The MPC is constructed based on the inner-cluster compactness and the inter-cluster separation. For a given cluster, the inner-cluster compactness is defined by the average distances of all data points to the center of this cluster; the inter-cluster separation is defined as the minimum $MPdist$ of this cluster to the other clusters. By the combination of the inner-cluster compactness and the inter-cluster separation, the MPC is defined which can stably process many kinds of datasets.

2   MPdist: A Midpoint-Involved Distance

In this section, the midpoint of cluster centers is firstly defined. Then, the midpoint-involved distance is defined to evaluate the distance of clusters. The definitions in this section are under the following assumptions:

In the Euclid space R^m, a m-dimensional dataset $D = \{x_1, x_2, ..., x_n\}$ contains n data points. In this dataset, each data points $x_i = \{x_{i1}, x_{i2}, ..., x_{in}\}, i = 1, 2, ..., n$, has m attributes. For a given clustering algorithm, the dataset D is divided into K clusters $C = \{C_1, C_2, ..., C_k\}$. The corresponding clustering centers are $V = \{V_1, V_2, ..., V_k\}$. $|C_i|$ is the number of data points in the cluster C_i. In this space, the Euclidean distance between data points x_i and $x_j (x_i, x_j \in D)$ can be calculated by $d(x_i, x_j) = \sqrt{(x_{i1} - x_{j1})^2 + (x_{i2} - x_{j2})^2 + ... + (x_{im} - x_{jm})^2}$.

In order to address the unbalance distributed clusters, the midpoint between two clustering centers is firstly defined as follows:

Definition 1. It is supposed that v_i and v_j are the centers of clusters C_i and C_j respectively. Then, the midpoint between centers v_i and v_j is defined as:

$$mp_{ij} = (v_i + v_j)/2 = ((v_{i1} + v_{j1})/2, (v_{i2} + v_{j2})/2, ..., (v_{im} + v_{jm})/2) \quad (1)$$

Definition 2. It is supposed that v_i and v_j are the centers of clusters C_i and C_j respectively; mp_{ij} is the midpoint between centers v_i and v_j; v_p and v_q are the two arbitrary non-center data points of clusters C_i and C_j respectively. Then, the distances from x_p and x_q to mp_{ij} are defined as follows:

$$D_{x_p} = d(x_p, mp_{ij}) \quad (2)$$

$$D_{x_q} = d(x_q, mp_{ij}) \quad (3)$$

In definition 2, the midpoint and the conception of non-center are combined to calculate the distances between two clusters. The centers of two clusters are not directly involved in the calculation of the distance. By this calculation, the unbalance distributed problem caused by directly calculating the distance between the centers of the two clusters is avoided.

Another limitation of many existing CVIs is that, between two different clusters, the definition of the separations may be sensitive to the outliers. Due to sensitive to the outliers of single non-center, the multiple non-centers and the midpoint are combined to define a new distance to evaluate the inter-cluster separation.

Definition 3. It is supposed that C_i and C_j are the two clusters in C; $|C_i|$ and $|C_j|$ are the numbers of data points in the two clusters respectively; mp_{ij} is the midpoint of clusters C_i and C_j; D_{x_p} is the distance from an arbitrary data point x_p in C_j to the midpoint mp_{ij}; D_{x_q} is the distance from an arbitrary data point x_q in C_j to the midpoint mp_{ij}. Then, the midpoint-involved distance, marked as $MPdist$, between C_i and C_j is defined as follows:

$$MPdist_{ij} = \frac{1}{|C_i|}\sum_{p=1}^{|C_i|}D_{x_p} + \frac{1}{|C_j|}\sum_{q=1}^{|C_j|}D_{x_q} \quad (4)$$

Normally, any outlier should not act as a representative for the evaluation of the inter-cluster separation. As defined in Equation (4), the midpoint-involved distance can effectively avoid the influence of outliers.

3   MPC: An Internal CVI Based on MPdist

The definitions introduced in this section are also hold the assumptions described in the second paragraph of Sect. 2. On available of the midpoint-involved distance, the definition of the MPC can be given.

3.1 The Definition of MPC

The MPC index in this section is also based on the combination of measures of the inner-cluster compactness and the inter-cluster separation.

Definition 4. The inner-cluster compactness of the cluster C_i is defined as:

$$f_c(i) = \frac{1}{|C_i|}\sum_{x_p \in C_i} d(x_p, v_i) \tag{5}$$

where, $f_c(i)$ calculates the average similarity between each data point x_p ($x_p \in C_i$) and the center (v_i) of the cluster C_i. As calculated in Eq. (5), the smaller value of $f_c(i)$ is obtained, the better performance on evaluating the inner-cluster compactness. In the special case, i.e. $C_i = 0$, $f_c(i)$ is set to 0.

Definition 5. The inter-cluster separation of the cluster C_i is defined as:

$$f_s(i) = \min_{1 \le j \le K, j \ne i} MPdist_{ij} = \min_{1 \le j \le K, j \ne i} \left\{ \frac{1}{|C_i|}\sum_{p=1}^{|C_i|} D_{x_p} + \frac{1}{|C_j|}\sum_{q=1}^{|C_j|} D_{x_q} \right\} \tag{6}$$

where, $f_s(i)$ calculates the minimum of midpoint-involved distance ($MPdist_{ij}$). As calculated in Eq. (6), the larger value of $f_s(i)$ is obtained, the better performance on evaluating the inter-cluster separation.

Definition 6. Based on the definitions of the inner-cluster compactness and the inter-cluster separation, the influence factor of the cluster C_i is defined as:

$$factor(i) = \frac{f_s(i) - f_c(i)}{f_s(i)} \tag{7}$$

where, $factor(i)$ represents the influence of the cluster C_i on the overall MPC index.

Definition 7. Based on the above definitions, the MPC index based on the midpoint-involved distance is defined by Eq. (8).

$$MPC(K) = \frac{1}{K}\sum_{i=1}^{K} factor(i) = \frac{1}{K}\sum_{i=1}^{K} \frac{f_s(i) - f_c(i)}{f_s(i)} \tag{8}$$

By taking the influences of all clusters into consideration, the average value of all the $factor(i)$, $i = 1, 2, \ldots, K$, is used to define the MPC index.

3.2 The Analysis of Time Complexity

It is supposed that n and m represent the number of data points in the target dataset D and the number of attributes of each data points respectively, meanwhile the time complexity of $d(x_i, x_j)$ is 1, the time complexity the MPC(K) can be computed as follows:

(1) According to Eq. (5), the time complexity of the $f_c(i)$ can be calculated as:

$$T_1 = |C_i| + m|C_i| = (m+1)|C_i|$$

where, $|C_i|$ is the number of data points of the cluster C_i.

(2) According to Eq. (6), the time complexity of the $f_s(i)$ can be calculated as:

$$T_2 = (|C_i| + m|C_i| + |C_j| + m|C_j|)(K-1) = (m+1)(K-1)(|C_i| + |C_j|)$$

(3) According to Eq. (7), the time complexity of the $factor(i)$ can be calculated as:

$$T_3 = 2T_2 + T_1 = 2(m+1)(K-1)(|C_i| + |C_j|) + (m+1)|C_i|$$

(4) According to Eq. (8), the time complexity of the $MPC(K)$ can be calculated as:

$$\begin{aligned}
T_4 &= 2(m+1)(K-1)(|C_1| + |C_j|) + (m+1)|C_1| + 2(m+1)(K-1)(|C_2| + \\
&\quad |C_j|) + (m+1)|C_2| + \ldots\ldots + 2(m+1)(K-1)(|C_K| + |C_j|) + (m+1)|C_K| \\
&= 2(m+1)(K-1)((|C_1| + |C_2| + \ldots\ldots + |C_K|) + K|C_j|) + (m+1)(|C_1| + \\
&\quad |C_2| + \ldots\ldots + |C_K|) \\
&= 2(m+1)(K-1)(n + K|C_j|) + (m+1)n \\
&\leq 2(m+1)(K-1)(n + Kn) + (m+1)n \\
&= 2(m+1)(K-1)(K+1)n + (m+1)n
\end{aligned}$$

In the general cases, the values of K and m are far less than the value of n, they can be taken as constants. So, the time complexity of the $MPC(K)$ can be roughly expressed as $O(n)$.

4   Experimental Results

In this section, two algorithms, the AHC and K-means, from different categories are selected to test the performances of the MPC. Due to different principles on the formations of these algorithms, experiments in this part are organized in different styles. Meanwhile, the performances of the 7 existing CVIs, the CH [5], STR [9], BCVI [10], COP [7], CSP [11], Sil [6] and SMV [8], from different categories are compared with the MPC index. Among these CVIs, the CH, STR and BCVI are the single center CVIs; the COP and CSP are the single non-center CVIs; the Sil is the multiple non-center CVI; the SMV and MPC are the hybrid CVIs. The CH, STR, CSP and Sil get the optimal clustering numbers at the biggest index values and thus they are marked as CH^+, STR^+, CSP^+ and Sil^+ respectively. On the contrary, the BCVI, COP, and SMV get the optimal clustering numbers at the smallest index values and thus they are marked as $BCVI^-$, COP^- and SMV^- respectively. The MPC gets the optimal clustering number at the biggest index value according to Eq. $K_{opt} = \{K | \max_{2 \leq K \leq \sqrt{n}} MPC(K)\}$ and thus is marked as MPC^+.

In the experiments, the search range of the clustering number K is limited to the interval of $[K_{min}, K_{max}]$. According to the empirical rule $K \leq \sqrt{n}$, the K_{min} and K_{max} are set to 2 and $\sqrt{n}$ respectively. The optimal clustering numbers calculated by different CVIs are the basic criteria on evaluating their performances. Based on this criterion, some matrices are defined to evaluate the performances of the tested CVIs detailly.

(1) *Error*. For a given dataset D_i, as defined in Eq. (9), this metrics records the absolute value of difference between the number of clusters computed (marked by K_{opt}) by a CVI and the actual number of clusters of this dataset (marked by *correct_number*).

$$Error_i = |correct\_number - K_{opt}| \tag{9}$$

(2) *Error_rate*. When the clustering algorithm is unstable, for example the K-means, the results of different executions of this algorithm on the same dataset are different. It is unscientific to use the Error to evaluate the performances of CVIs. For this reason, the *Error_rate* is defined. The *Error_rate* of a CVI on the dataset D_i is defined as follows:

$$Error\_rate_i = \frac{1}{T}\sum_{t=1}^{T} Error_{it} \tag{10}$$

where, "T" specifies the times of executions of the clustering algorithm; $Error_{it}$ records value of *Error* computed by the CVI on the dataset D_i when this CVI is used to evaluate the result of the tth execution of the clustering algorithm.

(3) *Average_error*. This metrics records the average *Error* of a CVI on all the n test datasets. Specifically, if the clustering algorithm to be evaluated by the CVI is stable (for example the AHC algorithm), the *Average_error* is defined as follows:

$$Averag\_error_s = \frac{1}{n}\sum_{i=1}^{n} Error_i \tag{11}$$

On the contrary, the $Averag\_error_{us}$ is defined as follows:

$$Averag\_error_{us} = \frac{1}{n}\sum_{i=1}^{n} Error\_rate_i \tag{12}$$

where, n is the number of datasets to be tested; $Error_i$ and $Error\_rate_i$ are defined in Eq. (9) and Eq. (10) respectively.

4.1 Description of the Test Datasets

As listed in Table 1, all the 12 test datasets are classified into 3 categories, the first 4 are the spherical distributed datasets, the middle 4 are the non-spherical distributed datasets; the last 4 are the real datasets. Table 1 is divided into 7 columns by the names, data point numbers, cluster numbers, dimensions, compositions, range of K and the source of the corresponding datasets. In the last column of this table, the symbol of "-" specifies the corresponding dataset is randomly generated by MATLAB. For example, the 8th line of this table tell us that the Parallel4 is the randomly generated artificial dataset which is composed of 4 clusters with 95, 100, 102 and 103 data points in each of them. Since there are 400 data points in this dataset, the range of K is limited to the interval of [2,20] by the empirical rule $2 \le K \le \sqrt{n}$.

Table 1. Discription of the 12 test datasets.

Datasets	Points	Clusters	Dimensions	Compositions	Range of K	Sources
Artificial spherical datasets						
D6	600	6	2	6*100	[2,24]	[35]
R8	320	8	2	8*40	[2,17]	[35]
Unbalance	6000	3	2	3*2000	[2,77]	[38]
N10	20000	10	2	10*2000	[2,141]	-
Artificial non-spherical datasets						
Parallel4	400	4	2	95+100+102+103	[2,20]	-
Spiral	312	3	2	101+105+106	[2,17]	[10]
Compound	399	6	2	16+38+45+50+92+158	[2,19]	[41]
Lsun	400	3	2	100+100+300	[2,20]	[42]
Real datasets						
Seed	210	3	7	3*70	[2,14]	UCI
Movement_Libras	360	15	90	15*24	[2,18]	UCI
Wdbc	569	2	30	212+357	[2,23]	KEEL
Statlog_German	1000	2	24	300+700	[2,31]	UCI

As shown in Fig. 1, the clusters of the 4 artificial spherical distributed datasets are characterized by "within-cluster compactness, between-cluster separable". Meanwhile, there are outliers at the edge of each cluster which are far from the cluster centers. As shown in Fig. 2, most of the 4 artificial non-spherical datasets are unbalance distributed. Among them, the parallel distributed dataset, Parallel4, is randomly generated by MATLAB. In Fig. 3, except the Wdbc dataset,

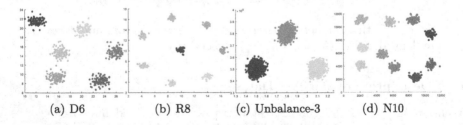

(a) D6　　　　(b) R8　　　　(c) Unbalance-3　　　　(d) N10

Fig. 1. Spatial distributions of the 4 artificial spherical distributed datasets.

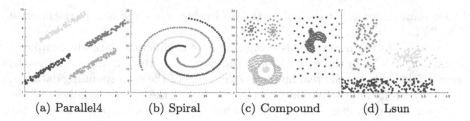

| (a) Parallel4 | (b) Spiral | (c) Compound | (d) Lsun |

Fig. 2. Spatial distributions of the 4 artificial non-spherical distributed datasets.

the other 3 real datasets are the UCI machine learning datasets (http://archive. ics.uci.edu/ml/datasets.php). Since the attributes of data points in the Wdbc dataset from UCI contains characters, the distances among data points cannot be calculated. For this reason, this dataset is downloaded from KEEL (https:// sci2s.ugr.es/keel/datasets.php).

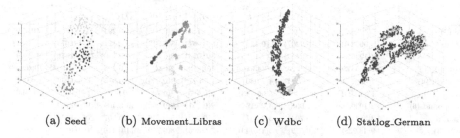

| (a) Seed | (b) Movement_Libras | (c) Wdbc | (d) Statlog_German |

Fig. 3. Spatial distributions of the 4 artificial non-spherical distributed datasets.

As listed in Table 1, some of real datasets are high dimensional. It is needed to reduce the dimensions before displaying them in the low dimensional space.Curr-ently, the dimensionality reduction tools can be divided into two categories, the linear dimensionality reduction tools and the nonlinear dimensionality reduction tools. In this paper, the widely used non-linear dimensionality reduction tool T-SNE [12]is used to preprocess all the high dimensional datasets.

4.2 Performances of CVIs on Evaluating the Results of the AHC

In this section, the 12 datasets listed in Table 1 are clustered by the AHC algorithm at first. Then, the clustering results are evaluated and the optimal clustering numbers are determined by the MPC^+ and the other 7 existing CVIs (the CH^+, STR^+, $BCVI^-$, COP^-, CSP^+, SMV^- and Sil^+). During the processes of utilizing different CVIs to evaluated the clustering results, the ranges of K are limited to the interval of $[2, \sqrt{n}]$ (n is the number of data points in the target dataset) by the empirical rule. For example, since there are 600 data points in the D6 dataset, we only to calculate the CVI values when the K is in the interval of [2, 24].

Table 2 lists the optimal clustering numbers of different datasets computed by different CVIs. In the first column, the number in the parentheses after the names of datasets are the real numbers of clusters of the corresponding datasets. The bold numbers specify the CVIs correctly get the cluster numbers of the corresponding datasets. After each number, the number in parenthese is the corresponding CVI value. For example, the number pair 6(4718.770) at the row #3 and column #2 means the CH^+ index gets the biggest value (4718.770) when the value of K is 6 on the D6 dataset. Meanwhile, the K_{opt} (6) computed by the CH^+ index is the correct number of clusters of the D6 dataset. As can be seen in this table, except Spiral and Compound datasets, the MPC^+ can find all the correct cluster numbers for the other 11 datasets. The performance of the MPC^+ is the best among the 8 CVIs.

Table 2. Optimal clustering numbers (K_{opt}) and index values for all the test datasets.

Datasets	CVIs							
	CH^+	STR^+	$BCVI^-$	COP^-	CSP^+	SMV^-	Sil^+	MPC^+
D6 (6)	6(4718.770)	6(9.198)	5(9.478)	6(0.146)	2(0.895)	23(0.293)	24(−0.897)	6(0.856)
R8 (8)	8(9068.016)	8(92.207)	6(9.226)	8(0.067)	8(0.942)	8(0.162)	17(−0.671)	8(0.934)
Unbalance-3 (3)	3(123048)	2(23.482)	3(2.18E7)	3(0.093)	2(0.967)	29(0.150)	71(−0.976)	3(0.909)
N10 (10)	10(258130.39)	10(26.248)	10(430901.66)	10(0.137)	2(0.935)	24(0.296)	141(−0.994)	10(0.869)
Parallel4 (4)	4(530.250)	16(17.714)	4(2.080)	16(0.278)	2(0.909)	16(0.515)	20(−0.913)	4(0.698)
Spiral (3)	13(10.231)	9(0.158)	3(96.703)	2(0.580)	2(0.729)	5(0.671)	7(−0.985)	2(0.496)
Compound (6)	2(1030.166)	2(1.055)	3(25.192)	2(0.251)	6(0.759)	19(0.499)	19(−0.914)	2(0.192)
Lsun (3)	3(384.439)	3(0.704)	3(1.237)	7(0.313)	5(0.779)	11(0.507)	20(−0.919)	3(0.647)
Seed (3)	9(48.228)	9(0.579)	3(12.443)	3(0.305)	8(0.759)	14(0.410)	2(−0.010)	3(0.699)
Movement_Libras (15)	5(4.509)	17(0.017)	2(2.829)	4(0.441)	14(0.904)	14(0.202)	2(0.133)	15(0.146)
Wdbc (2)	6(46.607)	6(0.387)	2(477728.3)	2(0.105)	3(0.973)	3(0.160)	2(0.801)	2(0.374)
Statlog_German (2)	8(32.250)	5(0.029)	2(1098.735)	3(0.174)	29(0.919)	29(0.142)	3(0.663)	2(0.300)

For different datasets, Table 2 only gives the optimal clustering numbers and the corresponding index values of CVIs (the biggest or smallest index values in the interval of $[2, \sqrt{n}]$) for saving spaces. Actually, for each dataset, all the index values of all CVIs in the interval of $[2, \sqrt{n}]$ are calculated in our experiments.

According to Eq. (9), Table 3 lists the *Error* of different CVIs. For example, the CH^+, STR^+, $BCVI^-$, COP^-, CSP^+, SMV^-, Sil^+ and MPC^+ can get the correct clustering number 6 (as listed in Table 2) for the D6 dataset, the corresponding *Errors* of these CVIs (row #2) are all 0. The clustering number of the D6 dataset computed by the CSP^+ is 2. However, the real number of clusters of this dataset is 6. So, the *Error* of the CSP^+ on the D6 dataset is 4. For the same reason, the *Error* of the Sil^+ on the D6 dataset is 18. The last two rows give the total and average *Error* of the corresponding CVI for all 12

Table 3. Errors of all test datasets by CVIs.

CVIs	Datasets							
	CH^+	STR^+	$BCVI^-$	COP^-	CSP^+	SMV^-	Sil^+	MPC^+
D6	0	0	1	0	4	17	18	0
R8	0	0	2	0	0	0	9	0
Unbalance-3	0	0	0	0	1	26	68	0
N10	0	0	0	0	8	14	131	0
Parallel4	0	12	0	12	2	12	16	0
Spiral	10	6	0	1	1	2	4	1
Compound	4	4	3	4	0	13	13	4
Lsun	0	0	0	4	2	8	17	0
Seed	6	6	0	0	5	11	1	0
Movement_Libras	10	2	13	11	1	1	13	0
Wdbc	4	4	0	0	1	1	0	0
Statlog_German	6	3	0	1	27	27	1	0
Total Errors	**40**	**37**	**19**	**33**	**52**	**132**	**291**	**5**
Average_Errors	**3.333**	**3.083**	**1.583**	**2.75**	**4.333**	**11**	**24.25**	**0.416**

test datasets. As can be seen in Table 3, the MPC^+ generates the smallest total *Errors* and *Average_errors* for the 12 test datasets.

4.3   Performances of CVIs on Evaluating the Results of the K-Means

Due to randomly selection of initial clustering centers, the K-means is unstable. For a single dataset, the clustering results may be different for different executions of the K-means. For this reason, the results in this part are the average value of 20 repeat experiments. Meanwhile, since the K-means cannot precisely partition the non-spherical datasets as the AHC algorithm, only the 4 spherical datasets and the 4 real datasets listed in Table 1 are tested in this part.

Specifically, experiments in this part are organized as follows: (1) For each dataset, the K-means is used to generate the clustering result; (2) all the 8 CVIs are used to evaluate this result; (3) the *Error_rate* (according to Eq. (10)) is computed by repeatedly executing Step (1) and Step (2) 20 times; (4) Eq. (12) is used to compute the *Average_error* for the 8 test datasets.

Table 4. Error_rate of the 8 CVIs on 4 spherical datasets and the 4 real datasets.

CVIs	Datasets							
	CH^+	STR^+	$BCVI^-$	COP^-	CSP^+	SMV^-	Sil^+	MPC^+
D6	0.65	12.00	0.60	2.00	2.85	1.25	13.05	0.20
R8	3.00	7.60	5.60	2.50	4.20	2.60	5.00	2.40
Unbalance	0.00	61.20	1.00	0.00	1.00	0.00	56.6	0.00
N10	15.20	37.00	3.80	58.40	4.00	68.60	15.40	2.80
Seed	0.00	9.00	1.00	0.00	3.90	0.40	9.20	1.00
Movement_Libras	12.60	2.00	9.20	2.00	3.00	1.60	1.60	8.80
Wdbc	5.90	19.80	1.60	0.00	3.00	0.60	19.20	0.00
Statlog_German	0.00	63.80	2.90	0.00	13.50	0.70	27.80	0.00
Total Error_rate	**37.35**	**212.4**	**25.7**	**64.9**	**35.45**	**75.75**	**147.85**	**15.2**
Average_error	**4.668**	**26.55**	**3.212**	**8.112**	**4.431**	**9.468**	**18.481**	**1.9**

Table 4 gives the *Error_rate* of the 8 CVIs on the 8 test datasets. In Table 4, the last two lines list the total *Error_rate* and the *Average_error* of each CVI respectively. As can been seen from this table, the total *Error_rate* and the *Average_error* of the MPC is the smallest among the 8 tested CVIs. The performance of the MPC^+ is the better than the other 7 CVIs.

4.4   Performance Evaluation by Real Large-Scale Datasets

In this part, another 6 real datasets with large number of data points are selected to evaluate the performance of the proposed MPC^+. The detailed descriptions and the spatial distributions of the 6 datasets are given in Table 5 and Fig. 4 respectively.

Table 5. Characteristics of the 6 real large-scale datasets.

Datasets	Points	Clusters	Dimensions	Compositions	Range of K	Sources
Segment	2310	7	19	7*330	[2,48]	KEEL
Thyroid	7200	3	21	166+368+6666	[2,84]	KEEL
Penbased	10992	10	16	4*1055+1056+1142+2*1143+2*1144	[2,104]	KEEL
Magic	19020	2	10	6688+12332	[2,137]	KEEL
Covertype	581012	7	54	2747+9493+17367+20510+35754+211840+283301	[2,762]	UCI
Poker	1025010	10	10	9+17+236+1460+2050 +3978+21634+48828+513701+433097	[2,1012]	KEEL

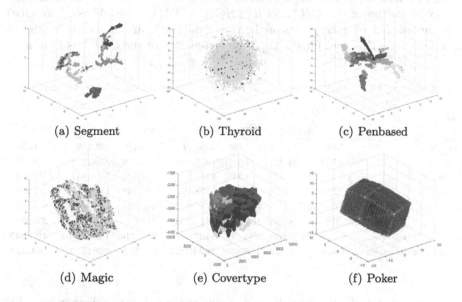

(a) Segment   (b) Thyroid   (c) Penbased

(d) Magic   (e) Covertype   (f) Poker

Fig. 4. Spatial distributions of the 6 real large-scale datasets.

Table 6. Time complexity of different CVIs.

CVI types	Single center			Single non-center		Multiple non-centers	Hybrid	
CVIs	CH+	STR+	BCVI-	COP-	CSP+	Sil+	SMV-	MPC+
Complexity	$O(n)$	$O(n)$	$O(n)$	$O(n^2)$	$O(n^3)$	$O(n^2)$	$O(n)$	$O(n)$

Table 7. *Error_rate* of the 7 CVIs on the 6 real large-scale datasets.

Datasets	CVIs						
	CH^+	STR^+	$BCVI^-$	COP^-	SMV^-	Sil^+	MPC^+
Segment	2.10	21.20	2.40	5.00	27.40	39.70	2.00
Thyroid	0.90	21.70	13.10	22.50	45.60	48.90	0.80
Penbased	6.40	44.90	10.00	12.20	1.60	71.50	4.40
Magic	0.00	78.60	12.90	2.10	9.80	131.50	0.00
Covertype	5.00	4.00	14.00	4.00	1.85	11.55	3.85
Poker	8.00	5.00	117.70	6.00	6.00	508.00	4.50
Total Error_rate	**22.40**	**175.40**	**170.10**	**51.80**	**92.25**	**811.15**	**15.55**
Average_error	**3.733**	**29.23**	**28.35**	**8.63**	**15.38**	**135.19**	**2.59**

Table 6 lists the time complexities of the 8 CVIs. As listed in this table, the time complexities of the CH^+, STR^+, $BCVI^-$, SMV^- and MPC^+ are $O(n)$ (n is the number of data points of the target dataset); the time complexities of the COP^- and Sil^+ are $O(n^2)$; the time complexity of the CSP^+ is $O(n^3)$.

Table 8. Time costs of the 7 CVIs on the 6 real large-scale datasets (ms).

Datasets	CVIs						
	CH^+	STR^+	$BCVI^-$	COP^-	SMV^-	Sil^+	MPC^+
Segment	5.615	10.011	1.198	290.349	1.180	283.178	1.859
Thyroid	9.303	21.451	3.981	1557.006	11.956	2966.181	8.655
Penbased	1.365	16.871	4.943	4449.131	1.471	6466.341	12.091
Magic	22.520	40.124	8.774	6319.565	15.237	14832.508	27.200
Covertype	413.954	1456.216	772.772	37569626	143.793	44094075	854.992
Poker	164.216	626.281	106.337	78457920	217.669	1.058E+8	201.44
Total time costs	616.973	2170.954	898.005	116040162	391.306	1.50E+08	1105.943
Average time costs	102.829	361.826	149.668	19340027	65.218	25015679	184.324

As in Sect. 4.3, the K-means is used to cluster the 8 datasets. Due to unstable of the K-means, the results in this part are the average values of the 20 repeat experiments. The descriptions of experiments in this part are like the second paragraph of Sect. 4.3. According to Eq. (10), Table 7 lists the $Error\_rate$ of the 7 CVIs on the 6 test datasets. Due to high time complexity ($O(n^3)$), the CSP^+ is not suitable to handle large-scale datasets. For this reason, the CSP^+ is not included in this experiment. The last two rows compute the total $Error\_rate$ and the $Average\_error$ respectively. Table 8 lists the time costs of the 7 CVIs. As can be seen from the two tables, the MPC^+ is able to processing real large-scale datasets. Compared with the other 6 existing CVIs, the MPC^+ is more accurate and not time consuming in evaluating the clustering results. Due to more clusters in the Covertype and Poker datasets and the calculation of the $MPdist$ among each clusters pair, the time cost of the MPC index is higher than the other three linear time complexity CVIs (CH^+, $BCVI^-$ and SMV^-).

5   Conclusion and Future Works

In this paper, a new robust distance, $MPdist$, which is based on the midpoint of centers is firstly proposed to measure the inter-cluster separation. Then, based on the combination of the new defined inner-cluster compactness and the inter-cluster separation, a novel interval clustering validity index, MPC, is proposed to effectively evaluate the results of the clustering algorithms. Due to the linear time complexity, the MPC is also able to handle real large-scale dataset. Since the MPC is based on the midpoint-involved distance, the execution time of MPC is mainly caused by the computation on the inter-cluster separation. On facing of datasets with linear separable clusters, the time cost of the MPC may be higher than the other CVIs with linear time complexity. Therefore, further study will be expected to overcome this shortcoming.

Acknowledgments. This study was supported by the Natural Science Foundation of Anhui Province (China)(No. 2008085MF188) and the University Natural Science Research Project of Anhui Province (China) (No. KJ2021A0041).

References

1. Xie, J., Xiong, Z., Dai, Q., Wang, X., Zhang, Y.: A new internal index based on density core for clustering validation. Inf. Sci. **506**, 346–365 (2020)
2. Adolfsson, A., Ackerman, M., Brownstein, N.C.: to cluster, or not to cluster: an analysis of clusterability methods. Pattern Recogn. **88**, 13–26 (2019)
3. Rathore, P., Bezdek, J.C., Erfani, S.M., Rajasegarar, S., Palaniswami, M.: Ensemble fuzzy clustering using cumulative aggregation on random projections. IEEE Trans. Fuzzy Syst. **26**(3), 1510–1524 (2018)
4. Zhu, E., Ma, R.: An effective partitional clustering algorithm based on new clustering validity index. Appl. Soft Comput. **71**, 608–621 (2018)
5. Calinski, T., Harabasz, J.: A dendrite method for cluster analysis. Commun. Stat. **3**(1), 1–27 (1974)
6. Rousseeuw, P.J.: Silhouettes: a graphical aid to the interpretation and validation of cluster analysis. J. Comput. Appl. Math. **22**, 53–65 (1987)
7. Gurrutxaga, I., et al.: SEP/COP: an efficient method to find the best partition in hierarchical clustering based on a new cluster validity index. Pattern Recogn. **43**, 3364–3373 (2010)
8. Yue, S., Wang, J., Wang, J., Bao, X.: A new validity index for evaluating the clustering results by partitional clustering algorithm. Soft. Comput. **20**(3), 1127–1138 (2016)
9. Starczewski, A.: A new validity index for crisp clusters. Pattern Anal. Appl. **20**, 687–700 (2017)
10. Zhu, E., Zhang, Y., Wen, P., Liu, F.: Fast and stable clustering analysis based on grid-mapping K-means algorithm and new clustering validity index. Neurocomputing **363**, 149–170 (2019)
11. Zhou, S., Zhenyuan, X., Liu, F.: Method for determining the optimal number of clusters based on agglomerative hierarchical clustering. IEEE Trans. Neural Netw. Learn. Syst. **28**(12), 3007–3017 (2017)
12. Laurens van der Maaten. t-SNE. Available at: https://lvdmaaten.github.io/tsne
13. Author, F., Author, S.: Title of a proceedings paper. In: Editor, F., Editor, S. (eds.) Conference 2016, LNCS, vol. 9999, pp. 1–13. Springer, Heidelberg (2016). https://doi.org/10.10007/1234567890

HAECN: Hierarchical Automatic ECN Tuning with Ultra-Low Overhead in Datacenter Networks

Jinbin Hu, Youyang Wang, Zikai Zhou, Shuying Rao, Rundong Xin, Jing Wang, and Shiming He[✉]

School of Computer and Communication Engineering,
Changsha University of Science and Technology, Changsha 410004, China
{jinbinhu,znwj_cs,smhe_cs}@csust.edu.cn,
{zhouzikai,shuyingrao,rundongxin}@stu.csust.edu.cn

Abstract. In modern datacenter networks (DCNs), mainstream congestion control (CC) mechanisms essentially rely on Explicit Congestion Notification (ECN) that is widely supported by commercial switches to reflect congestion. The traditional static ECN threshold performs poorly under dynamic scenarios, and setting a proper ECN threshold under various traffic patterns is challenging and time-consuming. The recently proposed Automatic ECN Tuning algorithm (ACC) dynamically adjusts the ECN threshold based on reinforcement learning (RL). However, the RL-based model consumes a large number of computational resources, making it difficult to deploy on switches. In this paper, we present a hierarchical automated ECN tuning algorithm called HAECN, which can fully exploit the performance benefits of deep reinforcement learning with ultra-low overhead. The simulation results show that HAECN improves performance significantly by reducing latency and increasing throughput in stable network conditions. For example, HAECN effectively improves throughput by up to 47%, 34%, 32% and 24% over DCQCN, TIMELY, HPCC and ACC, respectively.

Keywords: Datacenter Network · ECN · Congestion Control · Deep Reinforcement Learning

1 Introduction

In modern DCNs, with the increasingly stringent requirements for diverse services, such as big data processing [1], distributed storage [2], high-performance computing [3], and online services, effective congestion control is crucial to

This work is supported by the National Natural Science Foundation of China (62102046, 62072056), the Natural Science Foundation of Hunan Province (2023JJ50331, 2022JJ30618, 2020JJ2029), the Hunan Provincial Key Research and Development Program (2022GK2019), the Scientific Research Fund of Hunan Provincial Education Department (22B0300).

Z. Tari et al. (Eds.): ICA3PP 2023, LNCS 14489, pp. 324–343, 2024.
https://doi.org/10.1007/978-981-97-0798-0_19

achieveing ultra-low latency and high throughput. Explicit Congestion Notification (ECN) [4] becomes an essential congestion signal for mainstream congestion control mechanisms, which is widely enabled by commercial switches to indicate network congestion due to its simple and effective superior performance. By leveraging ECN, congestion control mechanisms quickly detect queueing building up and perform the corresponding rate adjustment to ensure efficient data transmission in DCNs.

However, a static ECN threshold is not sufficient to cope with dynamic changes in the network environment, leading to suboptimal performance for existing congestion control schemes. Furthermore, the preset static threshold is difficult to adapt to varying traffic patterns, resulting in degradation of application performance. Additionally, network operators need to dedicate significant time and effort to setting suitable ECN thresholds in large distributed networks. Recently, dynamic threshold-setting has gained attraction in both academia and industry for datacenters. Dynamic ECN threshold-setting plays a crucial role in achieving optimal network performance in modern networks, especially in complex topologies with multiple paths. However, finding the right balance is challenging as a low threshold increases packet latency, while a high threshold leads to underutilization. Traditional threshold-setting algorithms do not work well in high-speed networks, further exacerbating underutilization. Moreover, bursty traffic in datacenters complicates dynamic ECN threshold-setting, necessitating adaptive algorithms to maintain optimal network performance. Therefore, advanced algorithms capable of adapting to changing traffic patterns are essential for setting dynamic ECN thresholds in high-speed networks with complex topologies.

To solve the drawbacks of the traditional static ECN threshold-setting, researchers have developed a new solution called ACC [5]. This technology leverages a deep learning-based approach to deploy DRL agents on each switch, allowing for autonomous and dynamic adjustment of ECN tagging thresholds based on real-time information of the buffer and traffic status. ACC significantly outperforms static ECN threshold-settings with zero configuration, achieving lower FCT and higher IOPS for storage services. In essence, ACC represents a breakthrough in addressing the challenges posed by traditional ECN threshold-settings. Although the RL-based model is promising, it requires significant computational resources. This poses a particular challenge for the RL-based automatic ECN threshold adjustment scheme, which operates on switches with limited and valuable CPU resources. When multiple concurrent traffic forwarding cases occur on the same switch, ACC may lead to severe performance degradation due to the large amount of reasoning overhead that interferes with the data path throughput and consumes non-negligible CPU resources. One potential solution to reduce the overhead is to increase the time interval for automatically optimizing ECN configuration decisions, but this can result in serious performance degradation [6,7].

We currently face a dilemma regarding ECN threshold auto-tuning using RL-based techniques. While these techniques are effective in adapting to different network circumstances, they suffer from high overhead issues that hinder their ability to quickly respond to changes in the dynamic network environment. Considering the aforementioned shortcomings of existing solutions, the question arises: Is there a solution that can dynamically provide the appropriate ECN threshold without incurring significant overhead costs?

In this paper, we propose a hierarchical automated ECN tuning mechanism called HAECN that provides a positive answer to this question. HAECN utilizes a flexible hierarchical control architecture comprising decision and policy generator modules. The key design decision involves separating the resource-intensive policy module from the overall agent. HAECN establishes a hierarchical policy structure powered by DRL, which includes a decision module and a policy module. The policy module is activated only when the decision module determines that the current ECN threshold-setting is unsuitable for the environment. It then generates a new ECN threshold that is appropriate for the current situation. With HAECN, the overhead problem no longer poses a significant obstacle to the development of RL-based ECN auto-tuning algorithms.

The main contributions of this paper are as follows:

- This research focuses on congestion control in DCNs, with a specific emphasis on addressing the computational burden associated with automatically adjusting dynamic ECN thresholds on switches. The objective is to minimize computational overhead, thereby enhancing network performance and optimizing congestion control operations in DCNs.
- HAECN is a hierarchical and adaptive approach to network congestion control that effectively addresses the limitations of static thresholds and computational overhead. By leveraging DRL, HAECN optimizes network functionality while minimizing the computational burden associated with dynamically adjusting ECN thresholds.
- The results of the simulation highlight the exceptional performance of HAECN in gradual stabilization network environments, surpassing the ACC solution in both throughput and round-trip time (RTT). For instance, HAECN demonstrates remarkable efficiency by reducing latency and significantly increasing throughput compared to DCQCN, TIMELY, HPCC, and ACC across various leaf nodes, with throughput improvements up to 47%, 34%, 32% and 24%, respectively.

The rest of the paper is organized as follows. We establish our design motivation in Sect. 2. In Sect. 3, we provide an overview of the design and introduce the details of HAECN. Section 4 discusses the implementation. The simulation results are presented in Sect. 5. We present the related works in Sect. 6 and conclude the paper in Sect. 7.

2   Motivation

Static ECN settings are incapable of adapting the network conditions change, which leads to suboptimal utilization of network resources and degraded performance. Tuning static ECN parameters is a complex and time-consuming task, and even if they are set optimally, a static threshold may not be suitable for all traffic types and congestion scenarios. Moreover, the use of static ECN can result in congestion collapse, especially in large datacenters, causing significant performance degradation and even network failure. Hence, an adaptive approach to ECN tuning is necessary to accommodate the dynamic nature of network conditions and traffic patterns, optimize network resource utilization, and ensure optimal performance.

To meet the requirements of low latency and high bandwidth, HPCC [8] employs precise load information acquired from In-network telemetry (INT) to calculate accurate flow rates. Similarly, TIMELY [9] adjusts flow rates based on precise delay measurements using NIC timestamps instead of relying solely on ECN-based signals. These innovative designs have demonstrated significant performance enhancements. However, deploying these solutions in heterogeneous datacenters with legacy devices presents a challenge, as these devices may not support new features like INT.

An adaptive approach to ECN tuning is crucial for effective network congestion management. However, its implementation is constrained by the substantial computational resources it requires. This challenge is further exacerbated in the case of automatic RL-based ECN threshold tuning schemes, as they rely on limited CPU resources and can result in severe performance degradation. Based on these factors, it is evident that several challenging issues need to be addressed to achieve effective ECN tuning in modern datacenters.

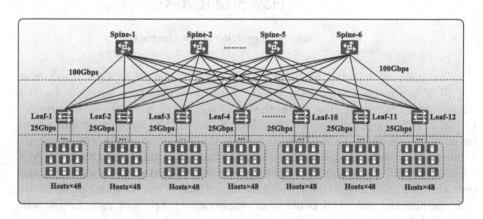

Fig. 1. Leaf-spine topology

Observation 1: Sophisticated model inferences are often responsible for generating substantial computational overhead, which ultimately results in decreased system performance.

To evaluate the performance impact of the computational burden associated with a typical commercial switch with ACC functionality, we perform an extensive evaluation in a simulated network environment. This evaluation included a specific topology configuration featuring a leaf-spine architecture consisting of 12 leaf switches and 6 spine switches, as shown in Fig. 1. In this particular setup, each leaf switch is equipped with a total of 48 links, each with a bandwidth of 25 Gbps, to facilitate connectivity to the server infrastructure. In addition, both leaf and spine switches are interconnected by 6 links, each with a bandwidth of 100 Gbps. Every fourth node linked to each leaf switch was programmed to generate a data flow. These generated flows are subsequently directed towards the server that is specifically connected to leaf switch 1.

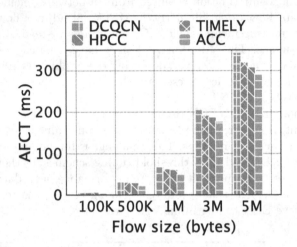

Fig. 2. Performances under bursty scenario

The analysis depicted in Fig. 2 illustrates that the ACC algorithm's average FCT is adversely affected when the network experiences bursty links. It is important to highlight that while the ACC algorithm and the static threshold DCQCN [10] algorithm, as well as TIMELY and HPCC, exhibit certain advantages when dealing with small-scale traffic, their performance becomes comparable as the traffic size increases. In particular, the implementation of ACC with AFCT fails to adequately showcase the anticipated substantial advantages over the aforementioned RDMA control algorithms. These findings demonstrate that despite the potential benefits of employing DCQCN in conjunction with ACC, it introduces notable inference overhead, thereby consuming a significant amount of CPU resources. Consequently, this situation leads to considerable performance degradation, negatively impacting the overall efficiency of data transmission within the network infrastructure.

Observation 2: In stable DCNs environment, frequent adjustments to ECN threshold can disrupt network stability and lead to performance degradation.

In DCNs, it's common for the network environment to gradually stabilize over time. When an appropriate ECN threshold is set, there is often no need to frequently adjust it. However, the current approach for tuning ECN, known as ACC doesn't take full advantage of this fact. Furthermore, ACC is not well-suited to handling the inherent uncertainty and variability in the network environment. The system may overreact to minor changes, resulting in unnecessary adjustments to the ECN threshold. This, in turn, can lead to further instability and congestion. Given these drawbacks, it's clear that a more efficient and effective approach for tuning ECN is needed in DCN environments. By taking advantage of the inherent stability of the network, it may be possible to reduce the overhead of ECN tuning and improve overall network performance.

To assess the effectiveness of the ACC algorithm in a specific scenario, a series of experiments were conducted utilizing the NS-3 simulation framework [11]. The evaluation is carried out within the context of the Observasion-1 scenario, with certain modifications introduced. Initially, the experimental setup involved the random selection of one flow per 12 nodes. However, after achieving network stabilization, the configuration is adjusted to direct one flow per 6 randomly selected nodes to a specific host connected to switch 1.

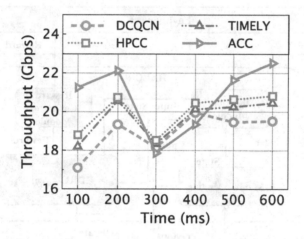

Fig. 3. Performance under gradual stable networks

As illustrated in Fig. 3, the ACC-loaded DCQCN algorithm exhibited a gradual stabilization process between $t_1 = 100ms$ and $t_2 = 200ms$, accompanied by a consistent increase in throughput. However, at time t_2, the ACC algorithm experiences pronounced network fluctuations as the traffic load intensifies. This leads to a continuous degradation in the overall network performance. Notably, within a certain range, the network throughput of the ACC algorithm is lower than that of other congestion control mechanisms such as DCQCN and TIMELY. These findings suggest that while the ACC algorithm effectively manages congestion during periods of network instability, it may generate redundant control decisions during stable periods, thereby resulting in performance degradation.

3   HAECN Design

3.1   HAECN Overview

HAECN is a hierarchical adaptive ECN threshold tuning approach for efficient network congestion control. By leveraging DRL, HAECN addresses the limitations of static threshold and computational overhead. In the architectural framework of HAECN, as is shown in Fig. 4, we need to entail the acquisition and observation of data as input to determine the RL agent's current state. Subsequently, a decision-making process is initiated to evaluate the necessity of activating the policy generator module, followed by appropriate actions being taken accordingly. Furthermore, the RL agent retrieves fresh state and reward values from the environment, facilitating the updating of its knowledge and optimizing its performance. This cyclical process ensures the continuous refinement and adaptation of the RL agent within the HAECN system.

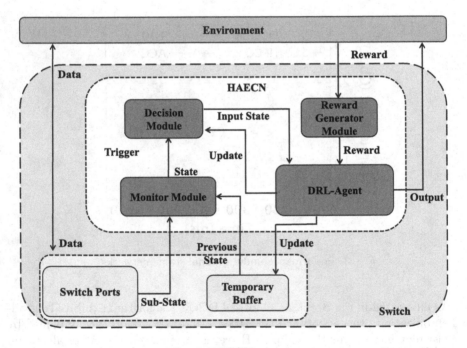

Fig. 4. HAECN overview

3.2   Design Details of HAECN

The architecture of HAECN incorporates a hierarchical control logic, depicted in Fig. 5. Within each time interval, the monitoring module retrieves environmental data as input for the RL agent's current state. Subsequently, this data is transmitted to the decision module, where informed decisions are made, and the policy generator is periodically activated. It is important to highlight that the policy generator dynamically adjusts to triggers by updating the decision module accordingly. By employing this hierarchical control logic, HAECN effectively integrates data collection, decision-making, and policy adaptation, contributing to its overall efficacy in optimizing network performance while minimizing overhead.

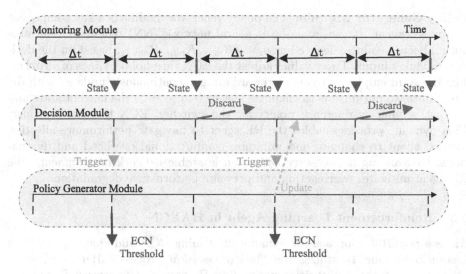

Fig. 5. The hierarchical control logic of HAECN

Consequently, the policy generator remains idle when the ongoing ECN threshold performs satisfactorily, resulting in a considerably lower average activation frequency. Furthermore, owing to its straightforward learning objective, the monitor module is typically smaller in size compared to the policy generator. Consequently, the observer incurs a lesser regular computational cost than previous schemes based on DRL.

Monitoring Module: The monitoring module plays a crucial role in the RL agent by collecting real-time data from the network's switches. It continuously monitors queue length and output data rate for each link. Instead of raw data, we use normalized statistics to provide a more refined approach. The normalization process is detailed in §3.3. It enables the agent to generalize observations from training sessions to unseen environments and improve the model's performance. Normalization also ensures equal treatment of input signals, avoiding the exaggerated impact of large values on the final model. Data collection occurs during consecutive monitoring intervals (Δt), with the monitoring module gathering real-time data, normalizing it to S_t, and transferring it to the decision module.

Decision Module: The decision module assesses the current ECN threshold in a dynamic network using a flag-triggering mechanism. If optimal performance can be achieved without the monitoring module's input, the decision module maximizes efficiency. When an update is needed, it triggers the policy generator for changes. State and action information $\{S^{Save}, a^{Save}\}$ are stored upon activation. After a time interval (Δt), with normalized data S_{t+1} from the monitoring module, the decision module repeats the process with $\{S_{t+1}, S^{Save}\}$.

Policy Generator Module: The policy generator module is a key component of our RL agent. It uses a six-layer neural network (NN) model to determine optimal action values, denoted as $a_t = \{K_{\max}, K_{\min}, P_{\max}\}_t$, based on the decision module's input. These values adjust the ECN threshold-settings of switches. The RL agent employs a dynamic reward system, continuously updated with the latest data, to optimize performance. The system records the current state and actions, serving as reference points for the switches' ECN threshold-settings. This dynamic process enables the RL agent to navigate performance-affecting factors, adapt to changes, and minimize computational overhead and fluctuations. By avoiding unnecessary activation in stable network environments, the RL agent mitigates overreaction and prevents performance degradation.

3.3   Reinforcement Learning Agent in HAECN

RL is a powerful approach for dynamically tuning ECN and enabling adaptive system behaviour. It utilizes the Markov Decision Process (MDP) with state space S, action set A, transition probability P, intermediate reward R, and discount factor γ. In the context of ECN tuning, RL divides time into monitoring intervals. At each slot, the RL agent observes the network state, takes action for ECN configuration, and receives a reward. The goal is to find the optimal policy that maximizes accumulated rewards. Our distributed RL agent design consists of independent agents at each switch, collecting local network information, making ECN configuration decisions, and updating policies. We employ Double-Deep Q-Learning (DDQN) and Deep Q-Learning (DQN) models to enhance ECN configurations, improve performance, and ensure scalability while minimizing computational overhead. By distributing RL agents across switches, we handle large-scale network complexities, adapt to changes, and optimize ECN configurations based on real-time information.

State: In RL, the term S_t refers to the input provided to the agent that describes the current environment. We use three important feature inputs, namely the normalized current port queue length (Q), the normalized link utilization (T), and the current ECN-setting $(Ecn = \{K_{\max}, K_{\min}, P_{\max}\}_t)$ which serve as crucial indicators of the current network state. We further denote the combination of these features as the sub-state $S'_t = \{Q, T, Ecn\}_t$. To generalize the trend of changes across different network environments and evaluate queue length and throughput changes, we combine the sub-states from the past k timestamps and denote it as the current state $S_t = \{S'_{t-k+1}, S'_{t-k+2}, \ldots, S'_t\}$. Our agents utilize the information gathered from the environment to make informed decisions that enable consistently high network performance.

Normalization is crucial for ensuring comparable scale inputs in our approach. We calculate statistics by averaging values from the last report time to the current report time, reducing fluctuations. Link utilization $(T_{(t)})$ is obtained by dividing $txrate_{(t)}$ by the link capacity (B), representing the percentage of link capacity utilized. For queuing length normalization $(ql_{(t)})$, we use a step mapping function to map values to $Q_{(t)}$ between 0 and 1 [5]. Higher values indicate more severe congestion. This enables informed decision-making and improves network performance.

Action: In the paper, we define the action taken at time slot t as the configuration of the ECN setting, which includes the high marking threshold $(K_{\max})$, the low marking threshold $(K_{\min})$, and the tagging probability $(P_{\max})$.

$$a_t = \{K_{\max}, K_{\min}, P_{\max}\}_t \tag{1}$$

To reduce the complexity of the action space, we discretize the ECN adjustment action space and form a template for the ECN configuration at the switch. Specifically, we choose the discretization as the ECN marking threshold. By discretizing the action space, we can reduce the number of possible actions, making it easier for the agent to learn and make decisions. The choice of ECN marking threshold as the discretization allows us to maintain a fine granularity in the ECN adjustment action space, ensuring that we can make accurate and effective adjustments to the network.

Reward: DRL systems have shown great potential for improving the performance of various applications, including computer networks. The reward function is a critical factor that significantly impacts the performance of a DRL system. In the context of network traffic control, the reward function plays a crucial role in defining the optimization objectives. It provides the necessary feedback to the agent on the effectiveness of its actions and helps improve the network's throughput and minimize queue length.

In this regard, we define the reward function for our network traffic control problem as a combination of the normalized link utilization and queue length, weighted by the parameter ω. Specifically, we calculate the reward as follows:

$$r = [\omega \times T_{(t)} + (1 - \omega) \times Q_{(t)}] - \alpha \times \text{trigger} \tag{2}$$

where $T_{(t)}$ symbolizes the normalized representation of link utilization achieved by dividing the port rate by the link bandwidth. Moreover, $Q_{(t)}$ indicates a normalized queue mapping function that gradually decreases from 1 to 0, with lower values representing better queue conditions. The hyperparameter α (set to 0.05) determines the penalty's significance during model training [7]. A higher α reduces policy generator activation frequency, decreasing CPU overhead. The decision module learns to activate the policy generator judiciously based on α penalties. Reducing α allows more frequent trigger activation, achieving finer network control. Proper reward function design is crucial for desired performance in DRL-based network traffic control systems. This particular topic will be covered in more detail in the upcoming sections of the discussion.

3.4   Learning Algorithm in HAECN

ECN optimization can be represented as a DRL problem, which allows us to use advanced techniques to optimize the ECN protocol. Specifically, we use a DDQN to model our policy generator module, and a lightweight DQN network to model our decision module.

Algorithm 1: HAECN's Learning Algorithm

Input: Replay Memory Buffer D, Batch Size N, Temporary Buffer M
Output: $a_t = \{K_{\max}, K_{\min}, P_{\max}\}_t$

1 **for** *every* Δt **do**
2   **Retrieve sub-state** $S'_t = \{Q, T, Ecn\}_t$ **and obtain the current state**
3   The agent retrieves a sub-state $S'_t = \{Q, T, Ecn\}_t$ and obtains the current state;

5     $S_t = \{S'_{t-k+1}, S'_{t-k+2}, \dots, S'_t\}$;
6   The trigger activation is determined based on $\{S, S^{\text{Save}}\}$;
7   **if** *triggering = 1* **then**
8   The action a_t is selected as $\arg\max_a Q(S_t, a, \theta_i)$ and executed;
9   Update the Temporary Memory Buffer with the newest $\{S^{\text{Save}}, a^{\text{Save}}\}$;
10   **else**
11   **continue**;
12   **end**
13   At time step $t+1$, observe S_{t+1}, r_t, and store the transition;
14   Sample N transitions $\{S_j, a_j, r_j, S_{j+1}\}$ from D;
15    $y_j = r_j + \gamma \times Q(S_{j+1}, \arg\max_a Q(S_{j+1}, a, \theta); \theta')$;
16    $L(\theta) = \frac{1}{N} \sum_j (y_j - Q(S_j, a_j; \theta))^2$;
17   Compute the gradient for actors and critics:
    $\nabla_\theta L(\theta) = \frac{1}{N} \sum_j (y_j - Q(S_j, a_j; \theta)) \nabla_\theta Q(S_j, a_j; \theta)$;
18   Update the parameters θ;
19 **end**

At time step j, our monitor module observes the environment and inputs S'_j to the decision module. The decision module integrates several consecutive time segments into S_j and decides whether the current ECN setting still satisfies the current environment, putting the trigger in the corresponding position. If the trigger is 0, the previously stored action a^{Save} is executed. However, if the trigger is 1, S_j is sent to the policy generator module, which generates a new a_j. The tuple (S_j, a_j, r_j, S_{j+1}), which includes the observation reward r_j and the next state S_{j+1}, is called experience. If the trigger is 1, we save the experience in buffer D for experience replay and update the saved state S_j^{Save} and action a_j^{Save} with S_j and a_j, respectively. The network is then trained by uniformly sampling from D. Periodic target updating and experience replay can significantly enhance and stabilize the training process of Q-learning.

This experience is stored in buffer D for experience replay, allowing the network to be trained by uniformly sampling from D. Periodic target updating and experience replay can significantly enhance and stabilize the training process of Q-learning, which is crucial for optimizing ECN. Overall, the use of DRL techniques allows us to formulate the ECN optimization problem in a new and powerful way and offers the potential for significant performance improvements in network congestion control.

4 Implementation

The model architecture of our proposed method adopts a hierarchical policy model constructed using the Pytorch [12] framework. This hierarchical policy model is instrumental in facilitating effective decision-making and control in the network environment. To create a realistic and comprehensive training environment, we leverage the NS-3 network simulator, a widely used tool for network research and development. Additionally, we incorporate ns3-ai [13], a specialized interface, to establish a seamless connection between the NS-3 simulation environment and the agent model, enabling efficient training and evaluation.

Table 1. Training hyperparameters in HAECN

Hyperparameter	Value
Learning Rate	0.005
Gamma (γ)	0.98
Batch Size (N)	64
Model Update Interval	20
Monitoring Time Interval (Δt)	15×RTT
Reward Penalty (α)	0.05

For the training and evaluation of our proposed method, we utilize a Linux computer equipped with an NVIDIA GeForce RTX 3090 GPU. This high performance computing infrastructure enables us to efficiently train and evaluate our HAECN model. Leveraging the computational power of the GPU, we can effectively handle the complex computations and optimizations required during the training process. By employing this state-of-the-art hardware setup, we ensure accurate and reliable results for our HAECN model.

Hyperparameters play a crucial role in training machine learning models as they define the behaviour and performance of the learning algorithm. In the context of HAECN, our proposed method, we carefully select and tune specific hyperparameters to ensure effective training and optimization. The hyperparameters listed in Table 1 provide important insights into the configuration of HAECN during the training process.

5   Evaluation

In this section, our objective is to assess the effectiveness and practical applicability of HAECN by utilizing NS-3 simulations. These simulations serve as a valuable tool for evaluating and comparing various factors pertaining to performance and computational overhead. Through the application of NS-3 simulations, we can accurately quantify the impact of HAECN on service delivery and thoroughly evaluate its relevance in real-world scenarios.

To assess the effectiveness of HAECN in typical DCNs, we conduct a large-scale simulation using the NS-3 framework. The objective is to create a realistic network environment, which is achieved by implementing a two-level leaf-spine topology comprising 12 leaf switches and 6 spine switches. Our evaluation focuses on four distinct network environments, denoted as $SECN_1$, $SECN_2$, $SECN_3$, and $SECN_4$, respectively. These environments share a similar topology but differ in the number of links connected to each leaf switch. Specifically, we vary the number of links per leaf switch, examining scenarios with 12, 24, 36, and 48 links. Each link in the network has a bandwidth capacity of 25 Gbps. Additionally, to ensure efficient communication, we interconnect the leaf and spine switches using 6 links, each supporting a bandwidth of 100 Gbps.

5.1   Performance Under Bursty Networks

To enhance the randomness of our experimental network and ensure generality across the four aforementioned network scenarios, namely $SECN_1$, $SECN_2$, $SECN_3$, and $SECN_4$, we implement specific configurations. These configurations involve programming every fourth node connected to each leaf switch to generate a random data flow. The purpose of this setup is to introduce variability in the network traffic patterns and simulate a realistic data transmission scenarios. Furthermore, the generated flows are directed towards the server connected to leaf switch 1, allowing us to analyze the impact of such data flows on the performance of the network and the effectiveness of the evaluated mechanisms.

As shown in Fig. 6, even with the continuous increase of leaf nodes, the performance of ACC and HAECN far exceeds other advanced load balancing mechanisms in $SECN_1$ and $SECN_2$. Notably, the average FCT achieved by ACC and HAECN is found to be shorter than DCQCN, TIMELY and HPCC. Meanwhile, due to overcoming the high overhead of ACC communication, HAECN has achieved low latency and much higher performance than ACC. Specifically, HAECN reduces the average FCT by about 30%, 24%, 22% and 16% compared

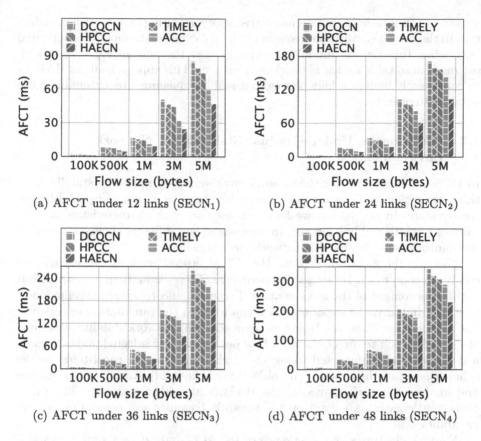

Fig. 6. Performance under different bursty network schemes

to DCQCN, TIMELY, HPCC and ACC at flow size of 5M in the case of 36 leaf nodes.

Meanwhile, it is important to highlight that employing ACC with the DCQCN algorithm introduces a significant computational overhead when the number of links on the leaf switches increases to 48 in SECN$_4$, as illustrated in Fig. 6. This overhead becomes particularly evident when the network operates at high speeds. Consequently, the network experiences an increase in transmission delay and a noticeable decline in overall performance. Specifically, compared to DCQCN, TIMELY, HPCC and ACC at flow size of 5M, HAECN can reduce the average FCT by 33%, 28%, 25% and 21%, respectively. Interestingly, even in SECN$_3$, where ACC is implemented, a relatively high average FCT is observed, surpassing that of algorithms such as DCQCN and TIMELY.

In Fig. 6, it can be seen that HAECN can effectively reduce the average FCT. This is because HAECN adopts a hierarchical design structure, effectively isolating the resource intensive policy generator module from the main agent. This design ensures that the policy generator module remains inactive when the ongoing ECN thresholds are deemed satisfactory, resulting in a significant reduction in the average activation frequency. The policy generato module is activated only when the decision module determines that the current ECN threshold-setting is unsuitable for the network environment. This approach allows HAECN to consistently maintain high throughput and low queuing latency, enabling the network to operate efficiently.

5.2   Performance Under Gradual Stabilization Network Environments

In DCNs, it is common for the network environment to gradually stabilize over time, transitioning from an initial unstable state to a more predictable and steady state. In our paper, we focus on assessing link characteristics, specifically bandwidth and link latency, to understand how different congestion control mechanisms perform in stable network environments.

To assess the effectiveness of the HAECN algorithm in the target scenario, we conduct a comprehensive series of experiments. These evaluations are carried out within the context of the aforementioned scenario, incorporating specific modifications. Initially, the experimental setup involves the randomized selection of one data flow per 12 nodes. However, upon achieving network stability, we adjust the configuration to direct one data flow per 6 randomly selected nodes towards a designated host connected to switch 1, while maintaining consistency in the other aspects of the design. The objective of these adjustments is to observe and analyze the performance of the HAECN algorithm under specific conditions, particularly its impact on the network dynamics and congestion control capabilities.

Figure 7 illustrates the emergence of the ACC, indicating its gradual stabilization and the subsequent increase in throughput. Over time, the ACC algorithm begins to surpass other congestion control mechanisms like DCQCN and TIMELY. Nevertheless, as the traffic load intensifies, the ACC algorithm experiences significant network fluctuations. Consequently, the overall network performance undergoes a continuous oscillatory decline. Notably, the ACC algorithm's throughput remains lower than that of DCQCN and TIMELY within a specific range. These observations imply that while the ACC algorithm may generate unnecessary control decisions during stable periods, leading to performance degradation. In contrast, HAECN exhibits remarkable performance in a stable environment. Its network throughput steadily grows and surpasses state-of-the-art RDMA algorithms such as DCQCN, without significant fluctuations or disruptions. Specifically, compared to DCQCN, TIMELY, HPCC and ACC under different leaf nodes, HAECN increases throughput up to 47%, 34%, 32%, 24%, respectively.

HAECN algorithm demonstrates remarkable efficacy in achieving high throughput while maintaining low queue lengths. Neglecting to promptly adjust the current ECN threshold in response to increasing queue lengths can result in rapid queue accumulation and subsequent latency spikes. In contrast, HAECN

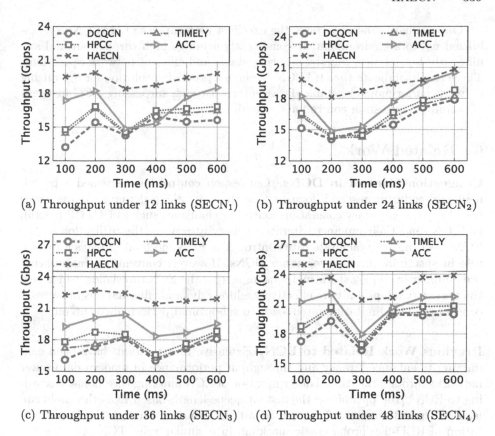

(a) Throughput under 12 links (SECN$_1$)   (b) Throughput under 24 links (SECN$_2$)

(c) Throughput under 36 links (SECN$_3$)   (d) Throughput under 48 links (SECN$_4$)

Fig. 7. Performance under gradual stabilization network environments

proactively responds to queue length and link utilization changes by employing lower ECN thresholds to generate more ECN-tagged packets. Conversely, as the queue length approaches a lower threshold, HAECN applies a higher ECN threshold to prevent potential starvation and ensure optimal throughput performance. This dynamic adjustment of the ECN marker threshold by HAECN ensures the maintenance of short queues and adaptability to prevailing environmental conditions.

The flexible hierarchical control architecture of HAECN proves advantageous as the network environment stabilizes. This architecture effectively separates the policy module from the overall agent, allowing for informed decision-making regarding the network environment before making direct modifications to the ECN thresholds. This approach ensures system stability and prevents misadjustment of the DCN network environment caused by sudden changes in ECN thresholds, which could potentially degrade the performance of the entire network.

Our evaluation demonstrates the excellent performance of HAECN in a stabilized network environment. It consistently achieves high throughput and significantly low latency, thereby ensuring stable and efficient network operations. These results indicate that HAECN presents a promising solution for optimizing network performance in stable data center networks, surpassing ACC in terms of stability, congestion control, and overall performance.

6   Related Work

Congestion Control in DCNs. Congestion control has remained a prominent and enduring research focus within the networking domain for over three decades. Contemporary congestion control mechanisms, such as DCTCP [14,15], DCQCN, and their enhanced iterations, heavily rely on the utilization of the ECN mechanism to facilitate rate control. The ECN mechanism plays a pivotal role in sustaining high-performance DCNs. However, conventional approaches predominantly employ static methods for setting the ECN thresholds, which lack the necessary adaptability to effectively address dynamic fluctuations in network conditions. This limitation often leads to suboptimal performance outcomes.

Previous Work Related to ECN. Extensive investigations have been conducting to enhance latency and throughput performance in modern datacenter networks through the careful determination of ECN marking thresholds. According to ECN* [16], optimizing the instant queue length-based ECN threshold can lead to the attainment of optimal incast performance through the implementation of RED-like probabilistic marking. In a similar vein, TCN proposes the utilization of the sojourn time, which quantifies the duration that packets reside in the queue, as a means to label packets. Moreover, the study presented in ECN# focuses on analyzing the variation of RTT within the datacenter network and marks packets based on both instantaneous and persistent congestion states. It is important to highlight that despite the availability of two threshold parameters (K_{max} and K_{min}) for the ECN switch, a significant number of researchers have commonly opted to assign identical values to both thresholds. Consequently, these studies primarily focus on the examination of a single threshold rather than exploring the potential benefits of leveraging distinct threshold values.

Learning-Based Network Optimization. Learning-based methods have become popular for optimizing network performance and setting parameters for congestion control mechanisms. Remy [17], Indigo [18], Vivace [19], and Aurora [20] are examples of such methods, using techniques like dynamic rate adjustment and DRL. Orca combines traditional TCP Cubic with learning-based approaches to tackle unpredictable traffic patterns. While most learning-based approaches focus on adjusting sending rates based on feedback, ACC proposes using DRL to autonomously adjust ECN parameters. ACC has shown promising results in reducing FCT and maintaining high throughput, but still faces challenges related to overhead and stability in stable network environments.

7 Conclusion

This paper presents HAECN, a novel methodology for addressing congestion control and determining optimal ECN thresholds DCNs. Unlike existing approaches, HAECN leverages DRL techniques to dynamically adapt ECN thresholds based on the evaluation conducted by its decision modules. A key contribution of HAECN lies in its adoption of a flexible hierarchical control architecture, which encompasses decision and policy generation modules to effectively reduce computational overhead. This adaptive framework ensures superior network performance without incurring unnecessary computational burden, while also guaranteeing network stability in stable environments. Experimental evaluations demonstrate the superior performance of HAECN compared to existing methods, showcasing its ability to significantly reduce computational overhead and enhance network functionality. Specifically, HAECN demonstrates remarkable efficiency by reducing latency and significantly increasing throughput compared to DCQCN, TIMELY, HPCC, and ACC across various leaf nodes, with throughput improvements up to 47%, 34%, 32% and 24%, respectively.

References

1. Chen, T., Li, M., Li, Y., Lin, M., Wang, N., Wang, M., et al.: MXNet: a flexible and efficient machine learning library for heterogeneous distributed systems. arXiv preprint arXiv:1512.01274 (2015)
2. Bunnag, C., Jareoncharsri, P., Tantilipikorn, P., Vichyanond, P., Pawankar, R.: Epidemiology and current status of allergic rhinitis and asthma in Thailand-ARIA Asia-Pacific Workshop report. Asian Pac. J. Allergy Immunol. **27**(1), 79–86 (2009)
3. Lu, X., et al.: High-performance design of Hadoop RPC with RDMA over Infini-Band. In: 2013 42nd International Conference on Parallel Processing, pp. 641–650. IEEE (2013)
4. Ramakrishnan, K., Floyd, S., Black, D.: The addition of explicit congestion notification (ECN) to IP. In: No. rfc3168 (2001)
5. Yan, S., Wang, X., Zheng, X., Xia, Y., Liu, D., Deng, W.: ACC: Automatic ECN tuning for high-speed datacenter networks. In: Proceedings of the 2021 ACM SIGCOMM 2021 Conference, pp. 384–397 (2021)
6. Abbasloo, S., Yen, C.Y., Chao, H.J.: Classic meets modern: a pragmatic learning-based congestion control for the internet. In: Proceedings of the Annual Conference of the ACM Special Interest Group on Data Communication on the Applications, Technologies, Architectures, and Protocols for Computer Communication, pp. 632–647 (2020)
7. Tian, H., Liao, X., Zeng, C., Zhang, J., Chen, K.: Spine: an efficient DRL-based congestion control with ultra-low overhead. In: Proceedings of the 18th International Conference on emerging Networking EXperiments and Technologies, pp. 261–275 (2022)

8. Li, Y., Alizadeh, M., Yu, M., Miao, R., Kelly, F.: HPCC: high precision congestion control. In: Proceedings of the ACM Special Interest Group on Data Communication, pp. 44–58 (2019)
9. Mittal, R., Lam, V.T., Dukkipati, N., Blem, E., Wassel, H., Ghobadi, M., et al.: TIMELY: RTT-based congestion control for the datacenter. ACM SIGCOMM Comput. Commun. Rev. **45**(4), 537–550. (2015)
10. Zhu, Y., Eran, H., Firestone, D., Guo, C., Lipshteyn, M., Liron, Y., et al.: ACM SIGCOMM Comput. Commun. Rev. **45**(4), 523–536 (2015)
11. Network Simulator. https://wwwnsnam.org. April 2023
12. Paszke, A, Gross S, Massa F, et al. : Pytorch: an imperative style, high-performance deep learning library. In: Advances in Neural Information Processing Systems, 32 (2019)
13. Yin, H., et al.: ns3-ai: fostering artificial intelligence algorithms for networking research. In: Proceedings of the 2020 Workshop on ns-3, pp. 57–64 (2020)
14. Alizadeh, M., Greenberg, A., Maltz, D.A., Padhye, J., Patel, P., Prabhakar, B., A., et al.: Data center TCP (DCTCP). In: Proceedings of the ACM SIGCOMM 2010 Conference, pp. 63–74 (2010)
15. Alizadeh, M., Javanmard, A., Prabhakar, B.: Analysis of DCTCP: stability, convergence, and fairness. ACM SIGMETRICS Perform. Eval. Rev. **39**(1), 73–84 (2011)
16. Wu, H., Ju, J., Lu, G., Guo, C., Xiong, Y., Zhang, Y.: Tuning ECN for data center networks. In: Proceedings of the 8th International Conference on Emerging Networking Experiments and Technologies, pp. 25–36 (2012)
17. Winstein, K., Balakrishnan, H.: TCP ex machina: computer-generated congestion control. ACM SIGCOMM Comput. Commun. Rev. **43**(4), 123–134 (2013)
18. Yan, F.Y., et al.: Pantheon: the training ground for Internet congestion-control research. In: 2018 USENIX Annual Technical Conference (USENIXATC 18), pp. 731–743 (2018)
19. Dong, M., et al.: PCC Vivace: online-learning congestion control. In: 15th USENIX Symposium on Networked Systems Design and Implementation (NSDI 18), pp. 343–356 (2018)
20. Jay, N., Rotman, N., Godfrey, B., Schapira, M., Tamar, A.: A deep reinforcement learning perspective on internet congestion control. In: International Conference on Machine Learning, pp. 3050–3059. PMLR (2019)
21. Xu, R., Li, W., Li, K., Zhou, X., Qi, H.: DarkTE: towards dark traffic engineering in data center networks with ensemble learning. In: Proceedings of IEEE/ACM IWQOS, pp. 1–10 (2021)
22. Liu, Y., Li, W., Qu, W., Qi, H.: BULB: lightweight and automated load balancing for fast datacenter networks. In: Proceedings of ACM ICPP, pp. 1–11 (2022)
23. Li, W., Yuan, X., Li, K., Qi, H., Zhou, X.: Leveraging endpoint flexibility when scheduling coflows across geo-distributed datacenters. In: Proceedings of IEEE INFOCOM, pp. 873–881 (2018)
24. Li, W., Chen, S., Li, K., Qi, H., Xu, R., Zhang, S.: Efficient online scheduling for coflow-aware machine learning clusters. IEEE Trans. Cloud Comput. **10**(4), 2564–2579 (2020)
25. He, X., Li, W., Zhang, S., Li, K.: Efficient control of unscheduled packets for credit-based proactive transport. In: Proceedings of ICPADS, pp. 593–600 (2023)
26. Wang, J., Rao, S., Ying, L., Sharman, P.K., Hu, J.: Load balancing for heterogeneous traffic in datacenter networks. J. Netw. Comput. Appl. **217**, 103692 (2023)
27. Hu, J., Huang, J., Li, Z., Wang, J., He, T.: A receiver-driven transport protocol with high link utilization using anti-ECN marking in data center networks. IEEE Trans. Netw. Serv. Manage. **20**(2), 1812–1898 (2022)

28. Hu, J., et al.: Enabling load balancing for lossless datacenters. In Proceedings IEEE ICNP (2023)
29. Hu, J., He, Y., Wang, J., Luo, W., Huang. J.: RLB: reordering-robust load balancing in lossless datacenter network. In: Proceedings ACM ICPP (2023)
30. Hu, J., Zeng, C., Wang, Z., Xu, H., Huang, J., Chen, K.: Load balancing in PFC-enabled datacenter networks. In: Proceedings of ACM APNet (2022)
31. Floyd, S., Jacobson, V.: Random early detection gateways for congestion avoidance. IEEE/ACM Trans. Netw. **1**(4), 397–413 (1993)
32. Cardwell, N., Cheng, Y., Gunn, C.S., Yeganeh, S.H., Jacobson, V.: BBR: congestion-based congestion control. Commun. ACM **60**(2), 58–66 (2017)
33. Chung, J., Ahn, S., Bengio, Y.: Hierarchical multiscale recurrent neural networks. arXiv preprint arXiv:1609.01704. (2016)
34. Gawłowicz, P., Zubow, A.: ns3-gym: extending openAI gym for networking research. arXiv preprint arXiv:1810.03943 (2018)
35. Abbasloo, S., Yen, C. Y., & Chao, H. J.: Wanna make your TCP scheme great for cellular networks? Let machines do it for you!. IEEE J. Sel. Areas. Commun. **39**(1), 265–279 (2020)
36. Wang, J., Yuan, D., Luo, W., Rao, S., Sherratt, R.S., Hu, J.: Congestion control using in-network telemetry for lossless datacenters. CMC-Comput. Mater. Continua **75**(1), 1195–1212 (2023)
37. Wang, J., Liu, Y., Rao, S., Sherratt, R.S., Hu, J.: Enhancing security by using GIFT and ECC encryption method in multi-tenant datacenters. CMC-Comput. Mater. Continua **75**(2), 3849–3865 (2023)
38. Hu, C., Liu, B., Zhao, H.: DISCO: memory efficient and accurate flow statistics for network measurement. In Proceedings IEEE ICDCS, pp. 665–674 (2010)
39. Li, H., Zhang, Y., Zhang, Z.: Ursa: hybrid block storage for cloud-scale virtual disks. In: Proceedings ACM EuroSys, pp. 1–17 (2019)
40. Bai, W., Chen, K., Hu, S., Tan, K., Xiong, Y.: Congestion control for high-speed extremely shallow buffered datacenter networks. In: Proceedings ACM APNet, pp. 29–35 (2017)
41. Wang, Y., Wang, W., Liu, D., Jin, X., Jiang, J., Chen, K.: Enabling edge-cloud video analytics for robotics applications. In: Proceedings IEEE INFOCOM, pp. 1–10 (2021)
42. Li, Z., Bai, W., Chen, K.: Rate-aware flow scheduling for commodity data center networks. In: Proceedings IEEE INFOCOM, pp. 1–9 (2017)
43. Zhao, Y., Huang, Y., Chen, K.: Joint VM placement and topology optimization for traffic scalability in dynamic datacenter networks. Comput. Netw. **80**, 109–123 (2015)
44. Hu, C., Liu, B., Zhao, H.: Discount counting for fast flow statistics on flow size and flow volume. IEEE/ACM Trans. Netw. **22**(3), 970–981 (2014)
45. Hu, J., et al.: Load balancing with multi-level signals for lossless data center networks. IEEE/ACM Trans. Netw., 1–13 (2024). https://doi.org/10.1109/TNET.2024.3366336
46. Wang, J., Liu, Y., Rao, S., Zhou, X., Hu, J.: A novel self-adaptive multi-strategy artificial bee colony algorithm for coverage optimization in wireless sensor networks. Ad Hoc Netw. **150**, 103284 (2023)

An Android Malware Detection Method Based on Metapath Aggregated Graph Neural Network

Qingru Li[1,2], Yufei Zhang[2], Fangwei Wang[1,2(✉)],
and Changguang Wang[1,2(✉)]

[1] Key Laboratory of Network and Information Security of Hebei Province,
Hebei Normal University, Shijiazhuang 050024, China
{fw_wang,wangcg}@hebtu.edu.cn
[2] College of Computer and Cyberspace Security, Hebei Normal University,
Shijiazhuang 050024, China

Abstract. Android system is facing an increasing threat of malware. Most of the current malware detection systems need to use large-scale training samples to get high accuracy. However, it is difficult to get a lot of samples. To solve this problem, we proposed a MAAMD (Metapath Aggregated Android Malware Detection) to obtain more useful information under the condition of a limited number of samples. In our method, the two most critical information types, namely API and Permission, are extracted as nodes, and the heterogeneous graph is constructed together with the APP nodes. The embedding of APP nodes is completed through two modules: intra-metapath aggregation and inter-metapath aggregation, then the classification of APP nodes is implemented by the SVM classifier. Each selected pathway is aggregated in the intra-metapath aggregation module, which aggregates the features of the intermediate node on the metapath and the features of the destination node itself. Therefore, the aggregated metapath obtains more semantic information which is helpful for malware detection. In the experiment, we achieved 99.68% accuracy only by 2,505 malicious APPs and 2,431 benign APPs. Compared with other methods that use large-scale training samples, MAAMD achieves higher accuracy by small-scale samples.

Keywords: Android · Heterogeneous graph neural network · Metapath aggregation · Malware detection

1 Introduction

Nowadays, the Android system has a huge number of users, devices, and a wealth of applications. For its scalability and openness, more and more mobile smart devices are choosing Android as their operating system. Meanwhile, Android devices are facing serious security threats. Android users are facing various risks such as privacy theft [1], data leakage [2], and spam [3].

Millions of malware are created every year, and malware is constantly evolving. Malicious code can bypass the detection system through obfuscation and

Z. Tari et al. (Eds.): ICA3PP 2023, LNCS 14489, pp. 344–357, 2024.
https://doi.org/10.1007/978-981-97-0798-0_20

other methods. Currently, in Android malware detection, graph neural networks [4,5] show great potential and have become a research hotspot. Heterogeneous graph neural networks analyze rich nodes and edges to extract their semantic information and obtain deeper implicit relationships by learning the features of nodes themselves and the topological structure of the graph. And these implicit relationships are not easily hidden by some malicious behaviors. This characteristic makes heterogeneous graphs very suitable for Android malware detection. However, the heterogeneous graph neural networks that have been applied to malware detection, such as HAN [6], determines the neighbor nodes (In graph neural networks, a node's neighbor nodes are the same type of nodes that this node is connected to based on metapath) through the metapath, which only aggregates the features of the neighbor nodes. This network model ignores the features of the intermediate node on the metapath and the features of the destination node itself. Based on this hypothesis, we imitate the MAGNN [7] and propose a MAAMD (Metapath Aggregated Android Malware Detection) model. In MAAMD, we select a certain number of pathways under each kind of metapath for each APP node and aggregate each pathway into a specific vector representation, then fuse the aggregated results into the APP node. Thus, more comprehensive semantic information in the heterogeneous graph is obtained. We use MAAMD to detect Android malware and achieve better results. Our main contributions are as the follows.

1) A new Android malware detection method MAAMD based on metapath aggregation is proposed. It not only aggregates the features of metapath-based neighbor APP nodes, but also includes the features of intermediate nodes along the metapath and the features of the destination APP node itself, which can provide more comprehensive semantic information and help to obtain more accurate detection results.
2) Setting a parameter that limits the number of each kind of metapath to ensure the experimental effect while reducing the consumption of system resources and improving efficiency.
3) The proposed method with small-scale training samples overperforms the effect obtained by other methods with large-scale training samples. In our experiment, we achieve an accuracy of 99.68% only by 2,431 benign APPs and 2,505 malicious APPs.

The remainder of this paper is organized as follows: Sect. 2 introduces the related work on Android malware detection. We describe the construction of a heterogeneous graph and the details of our method in Sect. 3. Section 4 presents the detailed experimental setup. Section 5 discusses the experimental results and the influence of the limit parameter on them. We make a summary of the paper in Sect. 6.

2   Related Work

Existing machine learning-based Android malware detection techniques are mainly divided into static feature-based approaches and sandbox-based dynamic behavior approaches.

Static feature detection mainly extracts information such as Android application permissions, API, and bytecode as features. Zarni et al. [8] extracted the permission information applied by Android applications through the analysis of the Android application configuration file AndroidManifest.xml and used it as a feature. Aafer et al. [9] proposed DroidAPIMiner, which extracts relevant information about the characteristics of malware behavior captured at the API level, and uses generated feature sets to evaluate the classification effect in different classifiers. Arp et al. [10] proposed Drebin, in which the information such as required hardware, applied permissions, APP components, special APIs, and network addresses are extracted as features embedding matrix through static analysis, and then linear model and corresponding weight of each feature are obtained by SVM(Support Vector Machine) algorithm, the detection accuracy is further improved. Shatnawi et al. [11] proposed a static classification method for malicious APP detection based on android permissions and API calls, and evaluated the method on a new Android malware dataset through three machine-learning algorithms.

Dynamic detection is performed by dynamically monitoring the execution of an application and its interaction with the external environment. Common dynamic features include API call sequences, dynamic data streams, and so on. Mariconti et al. [12] proposed Mamadroid, which constructs a behavior model from the abstract API call sequence based on the Markov chain and uses this behavior model to extract features and classify Android applications. Enck et al. [13] proposed TaintDroid, this method determines whether private data is leaked by tracking whether the private data goes beyond the preset system boundary during the propagation process, to determine whether it is malicious software. Hou et al. [14] proposed Deep4MalDroid, a new dynamic analysis method - component traversal, which can automatically perform each given APP as completely as possible, and a weighted directed graph is built based on extracted Linux kernel system call characteristics, then a deep learning model based on graph features is used to detect unknown Android application.

Some Android malware detection models using deep learning methods are gaining more attention, such as image-based detection technology, which directly converts APK files into images and then detects the images. This method does not require any feature engineering, and can effectively prevent symmetric encryption and malicious code confusion. Zhang et al. [15] combined the visualization features of XML file with the data part of DEX file to create grayscale image datasets and then transferred these images to the time convolutional network [16] (TCN) to detect Android malware. Nisa et al. [17] combined features extracted from the pretrained deep neural networks with image features of malicious code based on segmentation fractal texture analysis (SFTA) to construct a multi-modal representation of malicious code to detect the grayscale image. Zhu et al. [18] proposed a novel convolutional neural network variant, named MADRF-CNN, which converts the significant parts of the Dalvik executable to RGB image and captures the dependent relationships between various sections of the RGB image.

Some detection models using graph neural networks are as follows: Gao et al. [19] proposed GDroid, a novel approach based on the graph convolutional network. The overall thought of this method is to construct a heterogeneous information graph using the APPs and Android APIs, and then the initial question is converted into a node classification task. HinDroid [20] represents Android applications (APPs), associated APIs, and the abundant relations between them as a structured heterogeneous information network [21] (HIN), employing metapaths to describe the semantic correlation between APPs and APIs, and then using multi-kernel learning algorithm to classify APPs. Hei et al. [22] proposed a fast Android malware detection method-HAWK, which is based on heterogeneous graph attention networks. HAWK constructs heterogeneous graphs by extracting static features (permission, API, interface, etc.), and extracts homogeneous graphs based on defined metapaths, then puts them into a graph attention network [23]. All the above models more or less underutilize the node information in the heterogeneous graph. In this paper, we propose a MAAMD method with metapath aggregated to get more comprehensive information.

3   Our Approach

3.1   Feature Engineering

Android applications are stored in an APK (Android Application Package) file, which contains all the data and resource files of Android applications. We decompiled the apk files using Androguard to extract only the two most critical types of information: API and Permission. Permission is a kind of security mechanism, which is mainly used to restrict the use of certain functions within an application and the access of components between applications. API is a predefined function of the Android system, it enables applications and programmers to access a set of routines with no need to access the source code or understand the inner working details. For API extraction, there are tens of thousands of APIs in an APK file, and among these APIs, there are many basic system APIs that are not significant for malware detection. If all APIs are extracted for the experiments, it is time-consuming and inefficient. Therefore, in some static detection, only the top dozens of APIs in sensitivity ranking are used for experiments, or API calls are extracted to construct flowcharts for analysis. In this paper, we first analyze all APK files using Androguard. For each APK file, Androguard will give a list of all the APIs that have occurred, and we make an API dictionary to record the number of occurrences of each API (that is, how many apk files contain this API), and then determine a range of occurrences. For each occurrence in this range, we randomly select n APIs as the final extracted APIs for the experiment.

As for the extraction of permission, there are only a few hundred different permissions extracted from all APK files, so in this paper, we select all of them for the experiment. After extracting all the APIs and permissions, we treat each API and permission as a node (of course, API and permission belong to different types of nodes, respectively), and we also treat each APP as a node. In the following section, we connect all nodes according to the relationship between nodes to construct a heterogeneous graph.

3.2   Constructing Heterogeneous Graph by Metapaths

In our proposed model, we define APP, API, and Permission as three types of nodes, that is APP, API, and P, then define the semantic relationship between them as follows: APP-API: An APP has an API; APP-Permission: An APP has a Permission; API-APP: An API is contained by an APP; Permission-APP: A Permission is contained by an APP. We represent the semantic relationship between nodes by defining two adjacency matrices: A and P, by Eq.(1) and Eq.(2).

$$A_{i,j} = \begin{cases} 1, & \text{if APP } i \text{ has APP } j \\ 0, & \text{otherwise} \end{cases}, \tag{1}$$

$$P_{i,j} = \begin{cases} 1, & \text{if APP } i \text{ has Permission } j \\ 0, & \text{otherwise} \end{cases}. \tag{2}$$

where A represents the adjacency relationship between APP and API, and P represents the adjacency relationship between APP and Permission. Transpose A and P to get A^T and P^T, which represent the adjacency relationship between API and APP, and between Permission and APP, respectively.

We then connect different nodes of the same type by defining metapaths to capture rich semantic information between them. Figure 1 shows the metapath types defined in this article. APP-API-APP means that two APPs have the same API, and they are connected through this API; APP-P-APP means that two APPs have the same permission, and they are connected through this permission. Pathway represents the metapath between two APP nodes. For example, in Fig. 2, there are three pathways under APP-API-APP and two pathways under APP-P-APP for the destination node. Although only two kinds of metapaths are defined in this article, they are sufficient to aggregate the most critical information while saving system resources and improving efficiency.

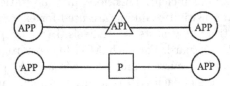

Fig. 1. Metapaths.

In this experiment, we use the onehot matrix to represent the features of APP nodes, denoted as H. First, we standardize A^T and P^T, then $A^T H$ and $P^T H$ are calculated respectively, which serve as the features of API nodes and Permission nodes. So far, the original inputs of the model, namely the features of all kinds of nodes and the adjacency matrix between nodes, have been prepared.

3.3   Metapath Aggregation Android Malware Detection

We divide the embedding of APP nodes into two stages: intra-metapath aggregation and inter-metapath aggregation. Figure 2 shows the overall architecture of MAAMD, the shadow represents the aggregated metapath.

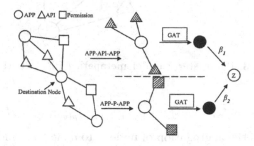

Fig. 2. The overall architecture of MAAMD.

Intra-metapath Aggregation. Unlike the previous method of obtaining relevant information from neighbor nodes, MAAMD aggregates the pathways information related to the destination node to obtain the embeddings of the destination APP node under each kind of metapath. Experiments show that pathways do contain more useful information than neighbor nodes.

In practice, each APP node is connected to hundreds of API and Permission nodes, and these API and Permission nodes may be connected to a large number of other APP nodes. As a result, tens of thousands of pathways will be generated in the experiment, which will bring a lot of system resources and time consumption. Therefore, we limit the number of each kind of metapath by defining the parameter λ, which represents the maximum number of intermediate nodes after selection between two APP nodes under a certain kind of metapath, or the maximum number of pathways after selection between two APP nodes, as shown in Fig. 3. For example, when the value of λ is 5, if the number of metapaths between an APP node and its neighbor node is less than or equal to 5, then all of them are selected; if more than 5, we select 5 pathways at random. In this way, the number of metapath is limited, therefore, system resource consumption is reduced and efficiency is improved while sufficient information can be provided from the selected metapaths.

For the selected pathway, the relational rotation encoder RotatE [24] is used to encode the nodes and edges in turn according to the sequence of the metapath to obtain the aggregation of this pathway. Given a pathway $p_{v,u} = (t_0, t_1, ..., t_n)$, $t_0 = u$, $t_n = v$, v is the destination node to be aggregated, u is the neighbor node based on the metapath, R_i represents the relationship between node t_{i-1} and node t_i , r_i is the relationship vector of R_i. The specific encoding method is as follows:

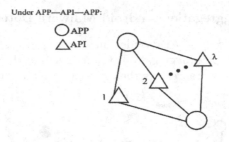

Fig. 3. Parameter λ under metapath APP-API-APP.

$$o_0 = f_{t0} = f_u, \; o_i = f_{ti} + o_{i-1} \odot r_i, \; f_{p(v,u)} = o_n/(n+1), \tag{3}$$

where o_i represents the aggregation of node t_0 to node t_i, f_i indicates the feature of node i, $\odot$ is the element-wise product, $f_{p(v,u)}$ is the aggregation of this pathway. Through this encoding method, not only the features of the neighbor nodes are aggregated, but also the features of the intermediate nodes on the metapath and the features of the destination node itself are included.

After all the metapath aggregations associated with the destination node v under metapath p are calculated, these metapath aggregations and node v form a homogenous graph G^p (Metapath aggregations can be regarded as nodes of the same type as APP nodes). G^p is sent into the graph attention network to obtain the embedding of node v under metapath p by Eq. (4).

$$\alpha_v^p = GAT(G^p). \tag{4}$$

Under each kind of metapath, the intra-metapath aggregation is performed on each APP node, and finally, each APP node will obtain a set of embeddings: $\{\alpha^{p_1}, \alpha^{p_2}, ...\}$, the number of elements in the set is the number of types of metapaths.

Inter-metapath Aggregation. In the intra-metapath aggregation stage, we have aggregated the features of relevant nodes under each kind of metapath. Next, we integrate the feature information from different kinds of metapaths through inter-metapath aggregation to obtain more comprehensive semantic information. Because each kind of metapath represents a semantic relationship, each kind of metapath has a different weight. The final node embedding is obtained by summing the product of the embedding of the APP node under each metapath and the weight of the corresponding metapath. The overall process of MAAMD is shown in Algorithm 1.

The weight of each kind of metapath is calculated by Eq. (5).

$$e_p = \frac{1}{|V_A|} \sum_{v \in V_A} q^T \cdot \tanh\left(M \cdot \alpha_v^p + b\right), \tag{5}$$

Algorithm 1. Metapath Aggregated Android Malware Detection

Input: The heterogeneous graph $G = (V, E)$, Node type $\{APP, API, P\}$, Metapaths $P = \{p_1, p_2\} = \{APP\text{-}API\text{-}APP, APP\text{-}P\text{-}APP\}$, Node features $\{f_v, \forall v \in V\}$.
Output: The APP nodes classification results.
1: **for** metapath $p \in P$ **do**
2:     **for** $v \in V_A$ **do**
3:         Calculate $f_{p(v,u)}$ for all $u \in N_v^p$ using the Rotate encoder function under the constraint of λ;;
4:         Learn the intra-metapath node embedding α_v^p using Eq. (4);
5:     **end for**
6: **end for**
7: Calculate the weight β_p for each kind of metapath $p \in P$ using Eq. (5) and Eq.(6);
8: Get Z_v by fusing the embeddings from different metapaths using Eq. (7);
9: Get Z_v, $v \in V_A$;
10: Feed Z_v into the SVM classifier.

---

where $|V_A|$ represents the number of APP nodes, q is the attention vector, M is the weight matrix, b is the bias vector. Then the metapath weight is normalized by Eq. (6).

$$\beta_p = \frac{\exp(e_p)}{\sum_{i \in P} \exp(e_i)}, \tag{6}$$

where β_p represents the weight of the metapath p.

The embeddings under different metapaths are fused to obtain the final embedding:

$$Z_v = \sum_{p \in P} \beta_p \cdot \alpha_v^p. \tag{7}$$

The final embeddings are fed into a linear SVM classifier for training, and the final classification results are obtained. The loss function of the training process is shown in Eq. (8):

$$L = -\sum_{i=1}^{N} [y_i \cdot \log(p_i) + (1 - y_i) \cdot \log(1 - p_i)], \tag{8}$$

where N represents the number of samples, y_i is the label of sample i, malicious APPs are labeled 1, benign APPs are labeled 0, p_i represents the probability that sample i is predicted to be positive.

4   Experimental Setup

4.1   Environment and Datasets

MAAMD is evaluated on a 10-core Intel(R) Xeon(R) Gold 6148 CPU with 72G RAM and GPU is NVIDIA GeForce RTX 2060. MAAMD is implemented by pytorch v1.2.0, CUDA v10.0, and dgl-cu100 v0.3.1.

To compare with the experimental results of HAWK [22], we selected 2505 malicious APPs and 2431 benign APPs released in 2017, among which malicious APPs were collected from VirusShare. And we download the desired benign APPs from AndroZoo [25] according to the following criteria, that is, date: 2017, max size: 3,000,000 bytes, source: play.google.com.

4.2 Evaluation Metrics

We use Acc and $F1$ to evaluate the effect of this model on Android datasets. Their definitions are the following:

$$F1 = 2 \times \frac{Precision \times Recall}{Precision + Recall}, \tag{9}$$

$$Acc = \frac{TP + TN}{TP + TN + FP + FN}, \tag{10}$$

where

$$Precision = \frac{TP}{TP + FP}, \quad Recall = \frac{TP}{TP + FN},$$

where TP (true positive) indicates the number of malicious APPs that are correctly identified; FP (false positive) indicates the number of benign APPs that are mistakenly identified; FN (false negative) indicates the number of malicious APPs that are mistakenly identified; TN (true negative) indicates the number of benign APPs that are correctly identified.

4.3 Baseline

In this experiment, we compared our experimental results with the state-of-the-art of 6 baselines:

(1) HAN: The importance of metapath-based neighbor nodes is learned by node-level attention and the importance between different metapaths is learned by semantic-level attention. Finally, the final node embedding is generated by fusing them.
(2) HinDroid: A heterogeneous graph is constructed with entities, such as APPs and APIs, and abundant intermediate relationships, then information from different semantic metapaths is aggregated. The embedding representation of the APP is calculated using multi-core learning.
(3) HAWK: The Android entities and behavior relationships are modeled as a HIN, and the embedding representations of APP nodes are obtained through the heterogeneous graph attention network.
(4) RS-GAT [23]: The heterogeneous information graphs are transformed into corresponding homogeneous graphs under different meta-structures, then each homogeneous graph is passed into a graph attention network, and the optimal performance among different graphs is reported.
(5) Metagraph2Vec [26]: A meta-graph is used to direct the creation of random walks and learn the potential embeddings of multiple types of HIN nodes.

(6) DroidEvolver [27]: DroidEvolver uses online learning techniques with evolutionary signature sets and pseudo-tags to make the necessary lightweight updates, and can automatically and continuously update itself during detection.

4.4 Model Parameters

In the API extraction part of this experiment, in the range of occurrence numbers (2 99), three APIs are randomly selected for each occurrence number, and 294 APIs are finally selected from the API dictionary. For permissions, a total of 230 permissions are extracted. In this experiment, the datasets were divided into training, validation, and testing sets whose number is 494(10.01%), 494(10.01%), and 3948(79.98%), respectively. To ensure the fairness, we trained the model 200 times as with HAWK.

5 Experimental Results

In the paper, we use 20%, 40%, 60%, and 80% of datasets respectively for the experiment, and compare the experimental results when $\lambda = 5$ with the results of HAWK (HAWK uses 14,000 benign APPs and 9865 malicious APPs). Table 1 shows the $F1$ values and accuracy of each method.

Table 1. $F1$ and Acc of each model.

Metrics	Approaches	20%	40%	60%	80%
$F1$	HAN [6]	0.9511	0.9617	0.9671	0.9705
	HinDroid [20]	0.9643	0.9669	0.9684	0.9746
	HAWK [22]	0.9857	0.9859	0.9871	0.9878
	RS_GAT [23]	0.9507	0.9631	0.9653	0.9664
	Metagraph2Vec [26]	0.9750	0.9766	0.9764	0.9771
	DroidEvolver [27]	0.9412	0.9517	0.9566	0.9605
	Ours	0.9942	0.9966	0.9969	0.9969
Acc	HAN [6]	0.9521	0.9657	0.9675	0.9699
	HinDroid [20]	0.9688	0.9698	0.9722	0.9764
	HAWK [22]	0.9843	0.9855	0.9867	0.9854
	RS_GAT [23]	0.9486	0.9620	0.9652	0.9664
	Metagraph2Vec [26]	0.9686	0.9698	0.9748	0.9762
	DroidEvolver [27]	0.9329	0.9506	0.9557	0.9623
	Ours	0.9941	0.9965	0.9968	0.9968

As can be seen from Table 1, the detection effect achieved by our model with small-scale samples exceeds that of other methods with large-scale training samples. This is because our model not only aggregates features from neighbor nodes

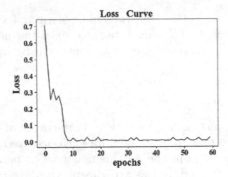

Fig. 4. The loss function curve of a training process.

but also aggregates features of all nodes along the whole metapath, obtaining more useful information and more features helpful to distinguish malware.

The loss function curve in the training process is shown in Fig. 4, which demonstrates that our model can converge quickly during the training process. And the training process is stable.

Table 2. Effect of λ on $F1$ and Accuracy in different data volumes.

λ	F1				Acc			
	20%	40%	60%	80%	20%	40%	60%	80%
1	0.9933	0.9959	0.9964	0.9962	0.9932	0.9958	0.9963	0.9961
3	0.9942	0.9965	0.9969	0.9968	0.9941	0.9964	0.9968	0.9967
5	0.9942	0.9966	0.9969	0.9969	0.9941	0.9965	0.9968	0.9968
7	0.9941	0.9965	0.9969	0.9969	0.9940	0.9964	0.9968	0.9968
None	0.9940	0.9964	0.9968	0.9969	0.9939	0.9963	0.9967	0.9968

We evaluate the effects of the parameters on $F1$ and Acc, and the results are shown in Table 2. As can be seen from Table 2, no matter what the parameter λ is or whether the parameter λ is implemented, the $F1$ value and Acc change slightly and their value is higher than HAWK. The reason is that in this experiment, the features of APIs and Permissions are generated by the features of APPs, so the features of intermediate nodes on the metapath are less important than the features of the destination node itself. Compared with HAWK, our model does not lose any of the original neighbor nodes, but only additionally obtains information of other nodes on the metapath. The added restriction parameter λ also only limits the number of pathways to neighbor nodes. Therefore, our method always outperforms HAWK.

From Fig. 5, we can also see the influence of the parameter λ on the number of pathways. For each kind of metapath, the number of metapath has decreased significantly for involving parameter λ, especially when $\lambda = 1$, the number of

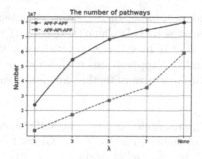

Fig. 5. The variation curve of the number of pathways.

metapath APP-P-APP is cut by about 70%, the number of metapath APP-API-APP is cut by about 89%. Meanwhile, Table 3 also shows the average time that each epoch spends during the training process, we can see that the time consumption is reduced by nearly 7 s when $\lambda = 1$ compared with that without parameter λ. It greatly significantly reduces the consumption resources.

Table 3. Effect of λ on the running time.

λ	1	3	5	7	None
Time (s)	16.55	19.31	20.46	21.79	23.27

6 Conclusions and Future Work

In this paper, we propose a MAAMD model based on metapath aggregated graph neural network to solve the problem of underutilization of the features from other nodes in heterogeneous graphs. The existing methods only aggregate the features of neighbor APP nodes. However, our method aggregates the features of all nodes along the whole metapath. Experiments show that pathway does contain more feature information than simple neighbor nodes. Our method aggregates more information helpful to malware detection. Therefore, our method using small-scale training samples exceeds the results achieved by other methods using large-scale training samples. In addition, we define a limiting parameter λ to solve the problem of an excessive number of pathways in heterogeneous graphs. By involving the parameter λ, the number of each kind of metapath is greatly reduced while very high detection accuracy is still maintained.

However, the randomly selected APIs more or less affect the final experimental results. If a more scientific API extraction algorithm is used, the final experimental results will be more stable and accurate. In addition, we use the parameter λ to limit the number of path-ways in the intra-metapath aggregation module. The limitation of this method is that as the number of applications increases, the corresponding number of APIs and Permissions will increase, and thus the number of metapaths will also keep increasing. Therefore, for large-scale application samples, it is still inevitable to produce a large number of pathways. Moreover, only Android applications with known attributes are tested. In the future, we will explore malware detection for out-of-sample and incremental APPs by our method.

Acknowledgements. This research was funded by NSFC under Grant 61572170, Natural Science Foundation of Hebei Province under Grant F2021205004, Science and Technology Foundation Project of Hebei Normal University under Grant L2021K06, Science Foundation of Returned Overseas of Hebei Province Under Grant C2020342, and Key Science Foundation of Hebei Education Department under Grant ZD2021062.

References

1. Chen, Y., Chen, H., Zhang, Y., Han, M., Siddula, M., Cai, Z.: A survey on blockchain systems: attacks, defenses, and privacy preservation. High-Confidence Comput. **2**(2), 100048 (2022)
2. Zuo, C., Lin, Z., Zhang, Y.: Why does your data leak? Uncovering the data leakage in cloud from mobile apps. In: 2019 IEEE Symposium on Security and Privacy (SP), pp. 1296–1310. IEEE (2019)
3. Dada, E.G., Bassi, J.S., Chiroma, H., Adetunmbi, A.O., Ajibuwa, O.E.: Machine learning for email spam filtering: review, approaches and open research problems. Heliyon **5**(6), e01802 (2019)
4. Zheng, X., Liu, Y., Pan, S., Zhang, M., Jin, D., Yu, P.: Graph neural networks for graphs with heterophily: a survey. arXiv preprint arXiv:2202.07082 (2022)
5. Wu, Z., Pan, S., Chen, F., Long, G., Zhang, C., Philip, S.Y.: A comprehensive survey on graph neural networks. IEEE Trans. Neural Networks Learn. Syst. **32**(1), 4–24 (2020)
6. Wang, X., et al.: Heterogeneous graph attention network. In: The World Wide Web Conference, pp. 2022–2032 (2019)
7. Fu, X., Zhang, J., Meng, Z., King, I.: MAGNN: metapath aggregated graph neural network for heterogeneous graph embedding. In: Proceedings of The Web Conference, vol. 2020, pp. 2331–2341 (2020)
8. Zarni Aung, W.Z.: Permission-based android malware detection. Int. J. Sci. Technol. Res. **2**(3), 228–234 (2013)
9. Aafer, Y., Du, W., Yin, H.: DroidAPIminer: mining API-level features for robust malware detection in android. In: Security and Privacy in Communication Networks: 9th International ICST Conference, pp. 86–103 (2013)
10. Arp, D., Spreitzenbarth, M., Hubner, M., Gascon, H., Rieck, K., Siemens, C. E. R. T.: Drebin: effective and explainable detection of android malware in your pocket. In: Proceedings of the 21st Annual Network and Distributed System Security Symposium (NDSS), vol. 14, pp. 23–26 (2014)
11. Shatnawi, A.S., Yassen, Q., Yateem, A.: An android malware detection approach based on static feature analysis using machine learning algorithms. Procedia Comput. Science **201**, 653–658 (2022)
12. Mariconti, E., Onwuzurike, L., Andriotis, P., De Cristofaro, E., Ross, G., Stringhini, G.: Mamadroid: detecting android malware by building markov chains of behavioral models. arXiv preprint arXiv:1612.04433 (2016)

13. Enck, W., et al.: Taintdroid: an information-flow tracking system for realtime privacy monitoring on smartphones. ACM Trans. Comput. Syst. **32**(2), 1–29 (2014)
14. Hou, S., Saas, A., Chen, L., Ye, Y.: Deep4MalDroid: a deep learning framework for android malware detection based on Linux kernel system call graphs. In: 2016 IEEE/WIC/ACM International Conference on Web Intelligence Workshops (WIW), pp. 104–111 (2016)
15. Zhang, W., Luktarhan, N., Ding, C., Lu, B.: Android malware detection using TCN with bytecode image. Symmetry **13**(7), 1107 (2021)
16. Bai, S., Kolter, J. Z., Koltun, V.: An empirical evaluation of generic convolutional and recurrent networks for sequence modeling. arXiv preprint arXiv:1803.01271 (2018)
17. Nisa, M., et al.: Hybrid malware classification method using segmentation-based fractal texture analysis and deep convolution neural network features. Appl. Sci. **10**(14), 4966 (2020)
18. Zhu, H., Wei, H., Wang, L., Xu, Z., Sheng, V.S.: An effective end-to-end android malware detection method. Expert Syst. Appl. **218**, 119593 (2023)
19. Gao, H., Cheng, S., Zhang, W.: GDroid: android malware detection and classification with graph convolutional network. Comput. Secur. **106**, 102264 (2021)
20. Hou, S., Ye, Y., Song, Y., Abdulhayoglu, M.: Hindroid: an intelligent android malware detection system based on structured heterogeneous information network. In: Proceedings of the 23rd ACM SIGKDD International Conference on Knowledge Discovery and Data Mining, pp. 1507–1515 (2017)
21. Zhao, B., Hu, L., You, Z., Wang, L., Su, X.: HINGRL: predicting drug-disease associations with graph representation learning on heterogeneous information networks. Briefings Bioinf. **23**(1), bbab515 (2022)
22. Hei, Y., Yang, R., Peng, H.: HAWK: rapid android malware detection through heterogeneous graph attention networks. arXiv preprint arXiv:2108.07548 (2021)
23. Veličković, P., Cucurull, G., Casanova, A., Romero, A., Lio, P., Bengio, Y.: Graph attention networks. arXiv preprint arXiv:1710.10903 (2017)
24. Sun, Z., Deng, Z., Nie, J., Tang, J.: Rotate: knowledge graph embedding by relational rotation in complex space. arXiv preprint arXiv:1902.10197 (2019)
25. Allix, K., Bissyandé, T. F., Klein, J., Le Traon, Y.: Androzoo: collecting millions of android APPs for the research community. In: Proceedings of the 13th International Conference on Mining Software Repositories, pp. 468–471 (2016)
26. Zhang, D., Yin, J., Zhu, X., Zhang, C.: MetaGraph2Vec: complex semantic path augmented heterogeneous network embedding. In: Phung, D., Tseng, V.S., Webb, G.I., Ho, B., Ganji, M., Rashidi, L. (eds.) PAKDD 2018. LNCS (LNAI), vol. 10938, pp. 196–208. Springer, Cham (2018). https://doi.org/10.1007/978-3-319-93037-4_16
27. Xu, K., Li, Y., Deng, R., Chen, K., Xu, J.: Droidevolver: self-evolving android malware detection system. In: 2019 IEEE European Symposium on Security and Privacy (EuroS&P), pp. 47–62. IEEE (2019)

A Joint Resource Allocation and Task Offloading Algorithm in Satellite Edge Computing

Zhuoer Chen, Deyu Zhang$^{(\boxtimes)}$, Weijun Cai, Wei Luo, and Yin Tang

School of Computer Science and Engineering, Central South University, Changsha, China
{214712158,zdy876}@csu.edu.cn

Abstract. This paper studies the task offloading problem for ground users in remote areas in satellite edge computing. Each user can offload computation tasks to either the Geosynchronous Earth Orbit (GEO) satellite, forward them to the ground cloud computing center, or offload them to a Low Earth Orbit (LEO) satellite which is constantly moving relative to the ground. To obtain the optimal task offloading plan and resource allocation plan that minimize system computing delay, we formulate this problem as a mixed integer nonlinear programming (MINLP) problem and propose a low complexity solution algorithm for it. Through mathematical derivation, we can organize the MINLP problem into three separate solutions: optimal allocation of computing resources, optimal transmission power control, and optimal offloading plan. In our algorithm, we apply the Lagrange multiplier method and binary search to obtain the optimal allocation of computing resources and optimal transmission power control under a given offloading plan. Then, using our proposed method based on the idea of greedy algorithm, we obtain an approximate optimal solution for task offloading. Compared to other algorithms, our proposed algorithm significantly reduces the system cost with a low computation complexity.

Keywords: Satellite edge computing · Task offloading · Mixed integer nonlinear programming problem · Low complexity · Lagrange multiplier method

1 Introduction

Due to the limited computing performance and battery capacity of terminal devices, executing AI applications with high computational complexity (such as image classification and object detection) on the terminal device often results in huge energy consumption and time cost. In recent years, edge computing [1–4] has been proposed, which pushes cloud services from the network core to the network edge which is closer to IoT devices and data sources. The continuously developing 5G communication technology in recent years provides stronger technical support for the development of IoT, and provides high-reliability and low-latency services in various critical scenarios. However, due to economic cost and

Z. Tari et al. (Eds.): ICA3PP 2023, LNCS 14489, pp. 358–377, 2024.
https://doi.org/10.1007/978-981-97-0798-0_21

technological limitations, conventional land-based networks cannot cover harsh environments such as deserts, forests, plateaus, and oceans, and the ground network infrastructure is also fragile in natural disasters such as floods, earthquakes, and tsunamis.

With the continuous exploration and technological development of future 6G networks [5–7], the internet of everything becomes possible. Satellites have the ability to provide broad coverage and continuous services for ground users, and therefore will become an important part of future 5G/6G communication technology. Combining satellites with mobile edge computing and deploying mobile edge computing (MEC) servers on satellites has become one of the key research directions in edge computing.

Using satellites for edge computing still faces many challenges. Firstly, due to the scattered distribution of terminal devices, it is difficult to maintain communication and share information among them, and it is difficult to aggregate their information and assign task offloading decisions for each user. Secondly, considering the orbital motion of low-earth orbit satellites, it is difficult for ground terminal devices to maintain good communication with low-earth orbit satellites. Furthermore, as the number of users increases, the process of solving the system's optimal solution using a specific computing method requires a large amount of time, and this solving time will cause users to miss the best opportunity to communicate with low-earth orbit satellites and perform task offloading. In order to cope with these challenges, this paper constructs a model that can provide task offloading services for ground users using both low-earth orbit satellites and synchronous satellites. The synchronous satellite can not only provide computing offloading services as an edge server, but also forward more complex computing tasks to cloud computing centers on the ground with stronger computing capabilities, to provide better task offloading services for terminal users. We propose a task offloading algorithm based on the greedy algorithm, which can solve a relatively optimal task offloading plan with extremely low time complexity, thereby reducing the overall computation cost of the system. The main contributions of this paper are as follows:

1. We present a novel task offloading model in satellite edge computing. Users on the ground can offload computing tasks to the stationary GEO satellite or the LEO satellite orbiting around the Earth, or forward tasks to the ground-based cloud computing center through the GEO satellite.
2. We aim to optimize the overall system computation time and formulate a mixed-integer non-linear programming (MINLP) problem. Based on the greedy algorithm, we design a joint resource allocation and task offloading algorithm to solve this problem.
3. Simulation results show that our proposed algorithm can provide flexible offloading schemes to different users in scenarios where tasks and the LEO positions vary, thus reducing system costs. Compared to other algorithms, the ours is more effective and stable. This advantage becomes more apparent as the number of users increases.

The remaining sections of our article are organized as follows: In Sect. 2, we summarize previous related researches. Section 3 presents our system model.

Section 4 introduces our proposed algorithm for joint resource allocation and task offloading. In Sect. 5, we present simulation results and analysis. Finally, Sect. 6 provides a summary of our findings.

2   Related Work

In scenarios where traditional cellular communication systems cannot be relied upon for edge computing, some researches explore the construction of integrated air, space, and ground networks by relying on satellite and unmanned aerial vehicle (UAV) technology to provide communication links or edge computing services to the outside world. In reference [8], satellite MEC is proposed and a collaborative computation offloading model is designed, enabling user devices without nearby MEC servers to also benefit from MEC services through satellite links. Reference [9] provides a comprehensive and detailed analysis of collaborative computation offloading, multi-node task scheduling, and mobility management in Satellite-Terrestrial Integrated network, and points out the technical challenges faced by satellite edge computing. Reference [10] discusses the challenges in maritime communication network and suggests utilizing MEC, space-air-ground-sea integrated network, and blockchain technology to solve these challenges.

References [11,12] utilize drones and LEO satellites to offload computation tasks from ground users, reducing device energy consumption and computational latency. A joint iterative optimization of UAV trajectories, computational resource allocation, and offloading strategies is performed to achieve a locally optimal solution after convergence. Reference [13] investigates an energy-efficient MEC framework for unmanned aerial vehicles (UAV) with non-orthogonal multiple access (NOMA). Through successive convex approximation and quadratic approximation, the originally highly non-convex problem is transformed into two convex problems. A heuristic algorithm is then proposed to iteratively optimize these two sub-problems alternately, significantly reducing the system's computational cost. Reference [14] constructs a dual-edge computing satellite-ground network and proposes a dual-edge computing offloading algorithm to reduce system energy consumption and latency. However, typical LEO satellites cannot maintain constant communication with ground devices due to their inability to remain relatively stationary with respect to the earth. Reference [15] takes into account the communication coverage time of LEO satellites to limit the computation offloading time, and uses the binary variable relaxation method to relax the offloading strategy from a binary 0 or 1 to a decimal value. After obtaining the optimal value, the binary variable is re-obtained as 0 or 1 to obtain the best strategy.

For scenarios with multiple users and MEC servers, such problems are often mixed-integer nonlinear programming problems, which are NP-hard problems, given that multiple users compete for limited computing and communication resources and that a single computing task is often offloaded to a single server. For these problems with many users and offloading options, exhaustive enumeration of all offloading schemes would result in extremely high time complexity.

Reference [16] proposes a curriculum learning-multi-agent deep deterministic policy gradient approach to learn the near-optimal offloading strategy. Reference [17,18] utilize a game-theoretic approach to optimize computation offloading strategy in satellite edge computing and UAV-assisted multi-access edge computing system, respectively. Reference [19] uses a heuristic search algorithm to solve this problem.

However, most of these articles do not consider the possibility that the movement of LEO satellites may prevent task offloading. The impact of this situation, as well as other ways of task offloading, should be worth discussing.

3 System Model

In this section, we first introduce our proposed satellite MEC scenario. Then, we propose our system computation model. The final step is to define the objective function that we want to optimize.

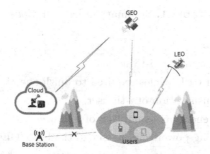

Fig. 1. Satellite MEC system model

As shown in Fig. 1, our satellite MEC model includes some user devices located in remote areas that cannot maintain communication connections with communication infrastructure such as base stations. These users can only choose to communicate with satellites. In this paper, U ground terminal users are dispersed and incapable of communicating with each other. Each user possesses their own CPU with a certain computation capability and their own computing tasks. There is a GEO satellite located about 36,000 km above ground level, which carries a MEC server that can utilize lightweight management platforms like Docker. This enables the satellite to provide computing and content delivery functions, providing task offloading services for ground users. Furthermore, the GEO allows it to maintain real-time communication with the cloud computing center located in ground cities via satellite communication terminals, and ground users can forward their requests to the cloud via the GEO for cloud computing services. Additionally, there is a LEO satellite with a MEC server that orbits around the earth's surface in a periodic circular motion. When the LEO flies above the horizon, it can establish communication connections with ground terminal devices and provide edge computing services. As the GEO satellite can maintain constant communication with users and the cloud, and use satellite links to obtain information on the LEO's status, we allow the GEO to collect users' task information, the LEO's status information, and cloud computing performance information, and make unified decisions to provide the system with task offloading decisions and computing resource allocation plans.

For U terminal users, each user's computing task is represented by $D_i(i = 1, 2, \ldots, U)$ and $C_i(i = 1, 2, \ldots, U)$, where D_i is the size of the task's data and C_i is the number of CPU cycles required to compute each bit of data for the task. Let $\mathbf{X}$ be a $U \times 4$ matrix, where for the i-th row, the corresponding vector is $(x_{i,l}, x_{i,G}, x_{i,GC}, x_{i,L})$. $x_{i,l} = 1$ represents the choice of the i-th user to perform local computing, $x_{i,G} = 1$ indicates offloading the task to the GEO for computing, $x_{i,GC} = 1$ means that the user chooses to offload the task to the ground cloud computing center via the GEO, and $x_{i,L} = 1$ indicates the way by the LEO.

For ease of reference, the key notations used in the article are summarized in Table 1.

Table 1. Summary of key notations

Notation	Description
D_i	Data size of task at i-th user
C_i	Number of CPU cycles required to finish one bit of task at i-th user
E_i^{total}	Energy consumption of i-th user
E_i^{tol}	Tolerated energy consumption of i-th user
P_i^L / P_i^G	Transmitting power of i-th user to the LEO/GEO
$P_i^{\max}$	Maximum transmitting power limitation of i-th user
f_i^l	Computation capability of i-th user
f^C	Computation capability of the Cloud for task at every user
f_i^L / f_i^G	CPU frequency of the LEO/GEO for task at i-th user
f_L / f_G	Computation capability of the LEO/GEO
θ	Position angle of the LEO
R^{GC}	Transmission rate between the GEO and the Cloud
R_i^L / R_i^G	Transmission rate between i-th user and the LEO/GEO
κ	Effective switched capacitance of CPU at users
N_0	Power spectral density of addictive white Gaussian noise of system
B^L / B^G	Uplink bandwidth of the LEO/GEO
h^L / h^G	Channel gain between users and the LEO/GEO

3.1 Geometric Model

Figure 2 shows the geometric relationship among the LEO, the GEO, the cloud, and users on the ground. The radius of the Earth is R, the height above the

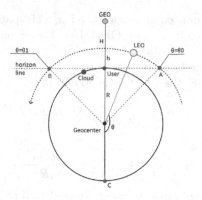

Fig. 2. Geometric relationship

ground for the GEO is H, and the LEO orbits the Earth in a circular orbit at a height of h. The two intersection points between the user's horizon and the LEO orbit are designated as points A and B, while the point symmetric to the user's location with respect to the Earth's center is designated as point C. The angle formed by the LEO, the geocenter, and point C is θ, and the corresponding angles for points A and B are θ_0 and θ_1, respectively, where $\theta_0 = \pi - arccos\frac{R}{R+h}$ and $\theta_1 = \pi + arccos\frac{R}{R+h}$. The LEO can establish communication with the ground users and provide task offloading services only when $\theta_0 < \theta < \theta_1$. Based on the knowledge of celestial motion, we can obtain $\frac{(R+h)3}{T_1^2} = \frac{(R+H)3}{T_0^2}$, where T_1 is the period for the LEO to orbit the Earth, and T_0 is the period for the GEO to orbit the Earth, which is also the period for the Earth's rotation. Thus, the relative angular velocity between the LEO and the rotation of the Earth is $\Delta\omega = \frac{2\pi}{T_1} - \frac{2\pi}{T_0}$, and the relative motion period is $T = \frac{2\pi}{\Delta\omega}$. When the LEO moves in the arc AB section, the relationship between the distance from the user to AB and the angle θ is: $s(\theta) = \sqrt{(R+h)^2 + R^2 - 2(R+h) \cdot R \cdot \cos(\pi - \theta)}$. The time required for the LEO to move above the user's horizon for mission offloading is:

$$
T_i^{upwait}(\theta) = \begin{cases} \dfrac{(\theta_0 - \theta)}{2\pi} * T & 0 \leq \theta \leq \theta_0 \\ 0 & \theta_0 < \theta \leq \theta_1 \\ \dfrac{(2\pi - \theta + \theta_0)}{2\pi} * T & \theta_1 < \theta < 2\pi \end{cases} \tag{1}
$$

3.2 Communication Model

U_l represents the set of users opting for local computing, U_G denotes the set of users offloading tasks to the GEO for computation, while U_{GC} signifies the set of users offloading tasks to the ground cloud computing center. And U_L is the set of users who choose to offload tasks to the LEO. Note that offloading tasks to both the GEO and the Cloud requires transmitting task data to the GEO, therefore $U_{G+GC} = U_G + U_{GC}$.

The wireless communication model used in this paper adopts orthogonal frequency division multiple access (OFDMA). Based on the number of users who need to transmit, the available bandwidth for uploading to the LEO or the GEO is evenly distributed among these users. Therefore, the transmission rate for users to the GEO and the LEO respectively is:

$$R_i^G = \frac{B^G}{U_{G+GC}} \log_2 \left(1 + \frac{U_{G+GC} P_i^G h^G}{N_0 B^G}\right) \tag{2}$$

$$R_i^L = \frac{B^L}{U_L} \log_2 \left(1 + \frac{U_L P_i^L h^L}{N_0 B^L}\right) \tag{3}$$

The explanations of the relevant parameters are listed in Table 1. And the channel gain for the user transmitting data to the GEO or the LEO include path loss and antenna gain.

3.3 Computing Model

Each user has a computing task and can only choose one of the four types of offloading methods. There is no air or cloud in space, and there is no obstruction to the sunlight for the solar panels carried by the GEO and the LEO. The energy produced by the solar panels is sufficient to maintain their own normal energy consumption and provide energy consumption for edge computing for these users. Therefore, we do not consider the energy consumption for computing cost for the GEO and the LEO when providing task offloading services. There is an abundance of power supply in the ground cloud computing center. For cloud computing users, we ignore the energy consumption of the cloud server during the cloud computing process. In addition, since the data size for computing results is generally small, we generally consider ignoring delay and energy consumption generated by the transmission of computing results [16].

Local Computing. f_i^l represents the computing capability of the CPU of the i-th user in cycles/s. The time delay and energy consumption of computing the task are $T_i^l = \frac{D_i C_i}{f_i^l}$ and $E_i^l = \kappa (f_i^l)^2 D_i C_i$ respectively, where κ represents energy consumption and depends on the structure of the CPU chip.

GEO Satellite Computing. If the i-th user chooses to offload the computing task to the GEO, considering the long distance between the user and the GEO and the non-negligible time for electromagnetic wave propagation [9], the computation task delay is composed of three parts, namely, the electromagnetic wave propagation time, data transmission time, and the GEO computation time. The propagation delay is expressed as $t_{i,G}^{pro} = \frac{H}{c}$, where c denotes the propagation speed of electromagnetic waves in vacuum. The transmission delay is represented as $t_{i,G}^{trans} = \frac{D_i}{R_i^G}$, with the transmission rate given by formula (2). The computation delay is $t_{i,G}^{comp} = \frac{D_i C_i}{f_i^G}$, where f_i^G is CPU frequency of the GEO's MEC server for task at i-th user and satisfies $f_i^G < f_G$, where f_G represents the computing capacity of the GEO's MEC server (unit: cycles/s). Therefore, the computation delay is $T_i^G = t_{i,G}^{pro} + t_{i,G}^{trans} + t_{i,G}^{comp}$. Concerning the computation energy consumption, we only consider the energy consumption generated by the user transmitting data during the data upload process, denoted as $E_i^G = \frac{P_i^G D_i}{R_i^G}$, where P_i^G represents the transmission power of i-th user sending data to the GEO.

Cloud Computing. If the i-th user chooses to forward the computing task to the cloud computing center through the GEO, the computational delay will also be divided into the electromagnetic wave propagation time, data transmission time, and cloud computing time. The propagation delay is represented as: $t_{i,GC}^{pro} = \frac{2H}{c}$. Considering that the GEO can maintain stable communication with satellite terminals in cities, we take the transmission rate from the GEO to the Cloud as a fixed value R^{GC}, therefore the transmission delay is represented as: $t_{i,GC}^{trans} = \frac{D_i}{R_i^G} + \frac{D_i}{R^{GC}}$. Considering the large-scale high-performance servers available in the ground cloud computing center, we set a specific computational resource allocation for each user in the center. The computational capability is fixed and relatively high, which we denote as f^C. The calculation delay is: $t_{i,GC}^{comp} = \frac{D_i C_i}{f^C}$. Therefore, the computing delay is $T_i^{GC} = t_{i,GC}^{pro} + t_{i,GC}^{trans} + t_{i,GC}^{comp}$. For computing energy consumption, we only consider the energy consumption generated by users when uploading data, represented as $E_i^{GC} = \frac{P_i^G D_i}{R_i^G}$.

LEO Satellite Computing. If the i-th user chooses to offload the computation task to the LEO, as opposed to the GEO, the offloading process in the LEO needs to take into account the relative position between the LEO and the user, as well as the time required for the LEO to rise, as shown in Eq. (1). The propagation delay is represented as $t_{i,L}^{pro} = \frac{s(\theta)}{c}$. The transmission delay is represented as $t_{i,L}^{trans} = \frac{D_i}{R_i^L}$. The computation delay is represented as $t_{i,L}^{comp} = \frac{D_i C_i}{f_i^L}$, where f_i^L represents the allocation of computing capacity from the LEO to the i-th user's task, and $f_i^L < f_L$, where f_L is the computing capacity of the LEO's MEC server (in cycles/s). Therefore, the total offloading delay is $T_i^L = t_{i,L}^{pro} + t_{i,L}^{trans} + t_{i,L}^{comp} + T_i^{upwait}(\theta)$. The computation energy is represented as $E_i^L = \frac{P_i^L D_i}{R_i^L}$, where P_i^L is the transmit power from i-th user to the LEO.

3.4   Problem Formulation

For each user's computing task, the time and energy consumption from task generation to obtaining the computing result are as follows:

$$T_i^{total} = x_{i,l}T_i^l + x_{i,G}T_i^G + x_{i,GC}T_i^{GC} + x_{i,L}T_i^L \qquad (4)$$

$$E_i^{total} = x_{i,l}E_i^l + x_{i,G}E_i^G + x_{i,GC}E_i^{GC} + x_{i,L}E_i^L \qquad (5)$$

We define the cost of each user's task as follows:

$$J_i = \frac{T_i^{total}}{T_i^l} \tag{6}$$

$$s.t. \quad E_i^{total} < E_i^l \tag{6a}$$

The computation cost is defined as the ratio of the delay when considering the four offloading modes for each user task to the delay when considering only local computing. We use energy consumption as a constraint and require that the energy consumption of considering the four offloading modes should not be higher than the energy consumption of considering only local computing. We define the system cost as the weighted sum of the computation cost of each user, and the goal is to minimize the system cost by optimizing the resource allocation and task offloading strategy. The objective function can be expressed as follows:

$$\min_{\mathbf{X}, \mathbf{P}^L, \mathbf{P}^G, \mathbf{f}^L, \mathbf{f}^G} \sum_{i=1}^{U} \lambda_i J_i \tag{7}$$

$$s.t. \quad x_{i,l}, x_{i,G}, x_{i,GC}, x_{i,L} \in \{0,1\}, \ \forall \, i \in U \tag{7a}$$

$$x_{i,l} + x_{i,G} + x_{i,GC} + x_{i,L} = 1, \ \forall \, i \in U \tag{7b}$$

$$E_i^{total} \le E_i^{tol}, \ \forall \, i \in U \tag{7c}$$

$$0 \le P_i^L, P_i^G \le P_i^{\max}, \ \forall \, i \in U \tag{7d}$$

$$\sum_{i \in U_L} f_i^L \le f_L, \ \sum_{i \in U_G} f_i^G \le f_G \tag{7e}$$

$$f_i^L, f_i^G \ge 0, \ \forall \, i \in U \tag{7f}$$

In the equation, $\mathbf{X}$ represents the offloading plan matrix for all users, $\mathbf{P}^L$ and $\mathbf{P}^G$ represent the transmission power for all user devices to transfer data to the LEO and GEO, $\mathbf{f}^L$ and $\mathbf{f}^G$ represent the allocation of computing capacity of MEC servers carried by the LEO and GEO to users. Constraints (7a–7b) indicate that each user can only choose one computation offloading method, constraint (7c) indicates that the energy consumption of the user cannot exceed the energy consumption of local computation, constraint (7d) represents that the user's transmission power cannot exceed the maximum transmission power, and constraints (7e–7f) represent the allocation restrictions of the LEO and GEO computing capacity to users.

4 Joint Resource Allocation and Task Offloading Algorithm

Formula (7) indicates that our goal is to achieve the minimum system cost by jointly optimizing the users' offloading plan, transmission power for data transmission to the LEO and GEO, and allocation of computation capability of the LEO and GEO. To optimize our objective function, we first rewrite the system cost as follows:

$$
\begin{aligned}
&\sum_{i=1}^{U} \lambda_i J_i \\
&= \sum_{i=1}^{U} \lambda_i \frac{U_L f_i^l x_{i,L}}{C_i B^L \log_2 \left(1 + \frac{U_L p_i^L h^L}{N_0 B^L}\right)} \\
&+ \sum_{i=1}^{U} \lambda_i \frac{U_{G+GC} f_i^l (x_{i,G} + x_{i,GC})}{C_i B^G \log_2 \left(1 + \frac{U_{G+GC} p_i^G h^G}{N_0 B^G}\right)} \\
&+ \sum_{i=1}^{U} \lambda_i \frac{f_i^l x_{i,L}}{f_i^L} + \sum_{i=1}^{U} \lambda_i \frac{f_i^l x_{i,G}}{f_i^G} \\
&+ \sum_{i=1}^{U} \lambda_i \frac{f_i^l}{D_i C_i} T_i^{up_{wait}}(\theta) x_{i,L} + \sum_{i=1}^{U} \lambda_i \frac{f_i^l s(\theta) x_{i,L}}{c D_i C_i} \\
&+ \sum_{i=1}^{U} \lambda_i \frac{f_i^l}{f^C} x_{i,GC} + \sum_{i=1}^{U} \lambda_i \frac{H f_i^l}{c D_i C_i} (x_{i,G} + 2 x_{i,GC}) \\
&+ \sum_{i=1}^{U} \lambda_i x_{i,l} + \sum_{i=1}^{U} \lambda_i \frac{f_i^l x_{i,GC}}{C_i R^{GC}}
\end{aligned}
\tag{8}
$$

We split the various parts of Eq. (8) into the following formulas:

$$
\boldsymbol{\Gamma_1}\left(\mathbf{X,P^L}\right) = \sum_{i \in U_L} \lambda_i \frac{U_L f_i^l}{C_i B^L \log_2 \left(1 + \frac{U_L p_i^L h^L}{N_0 B^L}\right)}
\tag{9}
$$

$$
\boldsymbol{\Gamma_2}\left(\mathbf{X,P^G}\right) = \sum_{i \in U_{G+GC}} \lambda_i \frac{U_{G+GC} f_i^l}{C_i B^G \log_2 \left(1 + \frac{U_{G+GC} p_i^G h^G}{N_0 B^G}\right)}
\tag{10}
$$

$$
\Lambda_1\left(\mathbf{X, f^L}\right) = \sum_{i \in U_L} \lambda_i \frac{f_i^l}{f_i^L}
\tag{11}
$$

$$
\Lambda_2\left(\mathbf{X, f^G}\right) = \sum_{i \in U_G} \lambda_i \frac{f_i^l}{f_i^G}
\tag{12}
$$

$$
\begin{aligned}
\mathbf{L} = &+ \sum_{i \in U_L} \lambda_i \frac{f_i^l}{D_i C_i} T_i^{up_{wait}}(\theta) + \sum_{i \in U_L} \lambda_i \frac{f_i^l s(\theta)}{c D_i C_i} \\
&+ \sum_{i \in U_{GC}} \lambda_i \frac{f_i^l}{f^C} + \sum_{i \in U_{G+GC}} \lambda_i \frac{H f_i^l}{c D_i C_i} + \sum_{i \in U_l} \lambda_i + \sum_{i \in U_{GC}} \lambda_i \frac{f_i^l}{C_i R^{GC}}
\end{aligned}
\tag{13}
$$

The values of Eqs. (9) and (10) are only related to the offloading plan and users' transmission power, while the values of Eqs. (11) and (12) are only related to the offloading plan and the allocation of computing resources in the LEO and the GEO. Equation (13) is only related to the offloading plan. The variables can be decoupled, and for a given offloading plan, some mathematical methods can be used to solve for the optimal computing resource allocation and power control separately.

4.1 Uplink Power Control

For a fixed offloading plan, we can use the monotonicity of the function to find the optimal power. We will now solve Eq. (9) to obtain the optimal power of the user's transmission to the LEO:

$$\min_{\mathbf{P}^L} \sum_{i \in U_L} \frac{U_L \lambda_i f_i^l}{C_i B^L \log_2 \left(1 + \frac{U_L p_i^L h^L}{N_0 B^L}\right)} \tag{14}$$

$$s.t. \quad \frac{p_i^L D_i U_L}{B^L \log_2 \left(1 + \frac{U_L p_i^L h^L}{N_0 B^L}\right)} \le E_i^{tol}, \ \forall i \in U_L \tag{14a}$$

$$0 \le P_i^L \le P_i^{\max}, \ \forall i \in U_L \tag{14b}$$

Letting $\phi_i = U_L \lambda_i f_i^l / C_i B^L$, $\psi_i = U_L h^L / N_0 B^L$, $\varphi_i = D_i U_L / B^L$, we can rewrite Eqs. (13) and (14a) as follows:

$$\Gamma_1 = \sum_{i \in U_L} \frac{\phi_i}{\log_2 \left(1 + \psi_i p_i^L\right)} \tag{15}$$

$$s.t. \quad \frac{\varphi_i p_i^L}{\log_2 \left(1 + \psi_i p_i^L\right)} \le E_i^{tol}, \ \forall i \in U_L \tag{15a}$$

By analyzing the monotonicity of the function through derivative analysis, we can see that each term in Eq. (15) will decrease as p_i^L increases, while the left-hand side of inequality (15a) will increase as p_i^L increases. Therefore, the optimal transmission power should be either at the maximum transmission power limit or at the power that makes the energy consumption equal to the local computation energy consumption. A binary search algorithm can be used to find the optimal energy consumption that satisfies the constraints of Eqs. (14a) and (14b). The method for solving Eq. (10) is the same as before.

4.2 Computing Resource Allocation

For a fixed offloading plan and how to optimize the allocation of computing resources, our work is similar to that of other researchers, where we use the

Lagrange formula for solution [19]. We will now solve Eq. (11):

$$\min_{\mathbf{f}^L} \sum_{i \in U_L} \lambda_i \frac{f_i^l}{f_i^L} \tag{16}$$

$$s.t. \sum_{i \in U_L} f_i^L \leq f_L \tag{16a}$$

$$f_i^L \geq 0, \ \forall \ i \in U_L \tag{16b}$$

Using the Lagrangian multiplier method and the Karush-Kuhn-Tucker (KKT) condition, we can obtain the optimal computing resource allocation of the i-th user for task offloading on the LEO and the minimum value of problem (16) as follows:

$$f_i^{L*} = \frac{f_L \sqrt{\lambda_i f_i^l}}{\sum_{i \in U_L} \sqrt{\lambda_i f_i^l}}, i \in U_L \tag{17}$$

$$\Lambda_1\left(\mathbf{X}, \mathbf{f}^{L*}\right) = \frac{1}{f_L}\left(\sum_{i \in U_L} \sqrt{\eta_i}\ \right)^2 \tag{18}$$

The method for solving Eq. (12) is the same as above.

4.3 Task Offloading

The above describes how to optimize computation resource allocation and user transmission power given a specific task offloading plan. Here, we propose a Greedy-based joint Resource Allocation and Task Offloading (GjRATO) algorithm, which is given in Algorithm 1 to get the best task offloading strategy.

We first initialize by letting all users choose local computing. Then we loop through each user's each offloading option on the premise that other users' offloading decisions remain the same and simultaneously optimize the transmission power and computation resource allocation. In each iteration, if a certain offloading option of user a can get the lowest system cost compared to the offloading options of other users, then the user a selects that option, which is named as x_a, while the other users' offloading options remain unchanged. However, if $x_{a,l} = 1$, which means the user a choose to compute his task locally, this algorithm requires user a to exchange its option with every user b who has previously chosen an offloading decision, in sequential order. The user a will adopt the option chosen by user b, and then user b will select the optimal decision again and update it. We record the order of users who make choices in each iteration, as well as their options. This process continues until all users have made their selections regarding their task offloading plans, at which point the algorithm ends.

The advantage of this design algorithm is that it can choose a better task offloading method for each user with very low time complexity, and reduce the

---

Algorithm 1. Greedy-based joint Resource Allocation and Task Offloading Algorithm

---

Input: Various parameters, the set of users U.
Output: the best $\mathbf{X}, \mathbf{P}^L, \mathbf{P}^G, \mathbf{f}^L, \mathbf{f}^G$, and the optimal system cost $\mathbf{J}_{\min}$.
  1: Initialize: Set $x_{i,l} = 1, \forall\, i \in U$, record the corresponding decision profile $\mathbf{X}^*$,
      a empty queue Q;
  2: **repeat**
  3:    $J = +\infty, a = 0$;
  4:    **for** $i \in U$ **do**
  5:      **if** i is not in Q **then**
  6:        Try four offloading decisions to obtain the best decision x_i;
  7:        Calculate system cost J';
  8:        **if** $J' < J$ **then**
  9:          $J = J', a = i$;
10:        **end if**
11:      **end if**
12:    **end for**
13:    **if** $x_{a,l} \neq 1$ **then**
14:      Update x_i in $\mathbf{X}^*$;Update $\mathbf{P}^L, \mathbf{P}^G, \mathbf{f}^L, \mathbf{f}^G$;
15:    **else**
16:      **for** $b \in Q$ **do**
17:        $x_a = x_b$, update x_a in $\mathbf{X}^*$; Update $\mathbf{P}^L, \mathbf{P}^G, \mathbf{f}^L, \mathbf{f}^G$;
18:        Try four offloading decisions to obtain the best decision x_b;
19:        Update x_b in $\mathbf{X}^*$; Update $\mathbf{P}^L, \mathbf{P}^G, \mathbf{f}^L, \mathbf{f}^G$;
20:      **end for**
21:    **end if**
22:    Store a into Q;
23: **until** every user is in Q;
24: Output $\mathbf{X} = \mathbf{X}^*$, get system cost $\mathbf{J}_{\min}$;

---

computing cost of the system as much as possible. Meanwhile, during the iteration, users who have previously selected the offloading method still have multiple opportunities to choose again. The computation complexity of Algorithm 1 is $O(n^2)$, where n is the number of the users. Compared to the exhaustive method that requires 4^n exploration attempts to explore all task offloading strategies, our algorithm greatly reduces computational complexity and saves solution time.

5   Simulation Results

In this section, we assess the performance of our proposed algorithm through simulation experiments, comparing it with four other algorithms. Our designed approach shines, showcasing its superiority over the alternatives. In our proposed computation model, different task offloading plans are provided to users based on the computing capacities of users and other MEC servers, task data size and computation requirements, as well as the location of the LEO. The GEO

Table 2. Simulation parameters

Parameter	Value	Parameter	Value
P_i^{max}	1 W	N_0	-130 dBm/Hz
f_i^l	0.5 GHz	κ	10^{-27}
f^C	2 GHz	B^L	10 MHz
f_L	20 GHz	B^G	10 MHz
f_G	8 GHz	h^L	-20 dB
R^{GC}	3 Mbits/s	h^G	-30 dB

altitude H is 36,000 km, while the LEO altitude h is 1,000 km. The Earth has a radius R of 6,000 km. All user devices have the same CPU performance, and the data size for each user is between 0.1G bits and 1G bits. The computational requirement for each bit of data is between 200 and 400 cycles per bit. Other relevant simulation parameters are presented in Table 2.

5.1 Algorithm Performance Presentation

Figure 3 shows the task distribution and computing time percentage among four different offloading methods during the LEO's rotation around the Earth when the number of users is 30. It can be observed that ground users change their task offloading plans as the LEO rotates. When the LEO has not yet risen above the horizon and is far away from it, the cost of offloading tasks via the LEO is too high, and users only prefer the other three methods. As the LEO approaches the horizon, users become more willing to wait for the LEO to rise before offloading tasks onto it, and the proportion of local computing or cloud computing accordingly decreases. Since cloud computing and GEO computing compete for the same uplink bandwidth resources, a decrease in the proportion of cloud computing can increase the proportion of GEO computing. Once the LEO has risen completely, the local computing becomes disadvantageous, and users only prefer the other three offloading methods. As the offloading method via LEO saturates, cloud computing becomes more attractive. At this point, the proportion of task offloading and computing time consumption of both LEO and cloud computing dominates, resulting in lower overall cost.

Figure 4 shows the task distribution for four different offloading methods when the number of users is 30 and the LEO rotation angle is 120°. Figure 4(a) indicates that, while the system's total computation requirement is kept constant, larger task data sizes result in increased time for data transmission via task offloading. Therefore, to minimize the overall computation cost of the system, more users will prefer to select the local computing or offload to the LEO, which offers lower computation and transmission costs. As a result, there will be a reduced proportion of tasks offloaded to the Cloud that depend on data transmission twice. It's worth noting that reducing the proportion of Cloud offloading can release some bandwidth pressure for offloading via the GEO, thus increasing the proportion of tasks offloaded to the GEO. From Fig. 4(b), it is observed that when the total computing demands of the system increase while the total task

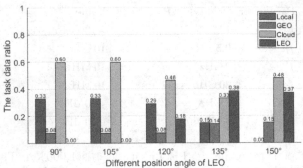

(a) The task data ratio of four different ways of offloading

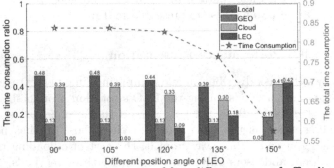

(b) The time consumption ratio of four different ways of offloading

Fig. 3. The task data ratio and time consumption ratio

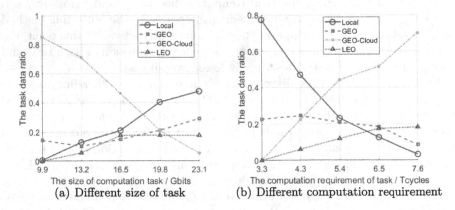

(a) Different size of task   (b) Different computation requirement

Fig. 4. The task data ratio of four different ways of offloading, $\theta=120°$

data size remains constant, the proportion of tasks performed locally will significantly decrease, as the users' local computing power is relatively weak, and performing local calculations would result in considerable time consumption. In such cases, offloading tasks to the powerful cloud computing center or the LEO can considerably reduce computation time. However, it should be noted that since cloud computing only competes for communication resources and not for computation resources, the proportion of tasks using cloud computing will increase significantly. But this will cause competition for communication bandwidth for users who offload tasks to the GEO, leading to higher transmission delays, and thus reduce the proportion of tasks offloaded to the GEO.

5.2   Comparison of Offloading Decisions

To evaluate the performance of our proposed GjRATO algorithm, we will compare it with the following four offloading algorithms. These five algorithms use the same optimization method for uplink transmission power control and computation resource allocation, but differ in their choice of algorithm for task offloading scheme:

(1) All Local (AL) computing decision: all users choose to perform computing locally without offloading tasks.
(2) Random Offloading (RO) decision: all users randomly choose one of the four offloading methods, and the cost is calculated by averaging over multiple experiments.
(3) Game-Theoretic (GT) scheme: every user carries out his current optimal offloading decision to reacts to other users' offloading decisions, which is applied in reference [18]. This scheme can obtain an near-optimal solution to the problem when it achieves Nash equilibrium (NE).
(4) Simulated Annealing-based (SA) solution: initialize the task offloading scheme for all users, and then use a heuristic simulated annealing algorithm to reassign the task offloading scheme for them, which is improved and also used to find the suboptimal solution for task offloading in reference [19]. As the calculation result of the heuristic algorithm has strong randomness, here we take the average value for ten experiments.

Figure 5 presents a comparison of the computational cost of applying these five algorithms to the system in this scenario. The AL decision serves as the benchmark, with the system cost always equaling 1. Figure 5(a) shows the variations in system cost generated by five offloading algorithms as a function of LEO angle when the number of users is 30. In cases where the LEO has not yet risen above the earth's horizon, the LEO offloading way generates long wait time, leading to high system costs with the RO algorithm. SA, GT and our proposed GjRATO can significantly reduce system cost, especially when the LEO satellite is above the horizon and users can offload heavy computational tasks to the powerful LEO satellite without waiting, resulting in a significant reduction in system costs. In this scenario, our proposed GjRATO outperforms SA and GT by 15% in terms of performance improvement. As heuristic SA algorithm

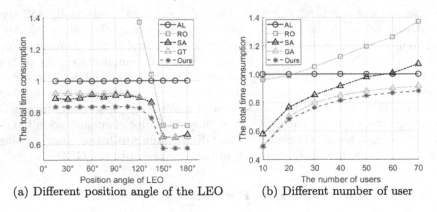

(a) Different position angle of the LEO   (b) Different number of user

Fig. 5. System cost at five different algorithms

is based on continuous attempts to find local optimal solutions, the system will generate different offloading plans each time, thus obtaining different computation costs, even if the LEO is in the same position and user tasks are the same. The solution obtained from the Nash equilibrium state using GT merely allows each user to choose his own optimal solution given the current states of all other users. However, the resulting solution for the entire system may not necessarily be the optimal solution, and it may not even be a relatively good suboptimal solution. Figure 5(b) shows the system cost of five offloading algorithms under different numbers of users when the LEO rotates at an angle of 135°. As the number of users increases, the number of offloading plans for all users exponentially increases and the resource competition intensifies. The SA algorithm that rely on explorations to find solutions will suffer from this exponential growth. It can be seen that when the number of users reaches 50, the SA is almost unable to reduce the system cost further, while our proposed GjRATO can still significantly reduce the system cost.

Figure 6 shows the system cost of three algorithms, SA, GT and GjRATO, under different task data sizes and computing requirements. As depicted in Fig. 6(a), for the same computing requirement, the system cost increases with the task data size due to the increased latency caused by transferring large data. Figure 6(b) demonstrates that for the same task data size, the system cost decreases with the increasing computing requirement. This is because the local computation time becomes significant when the computing requirement is large, while offloading to high computing power MEC servers such as the LEO, the GEO, and the Cloud can significantly reduce the computation time, with the data transmission time remaining constant. These two figures indicate that our proposed GjRATO algorithm outperforms SA by a 15% performance improvement and is even more superior to GT.

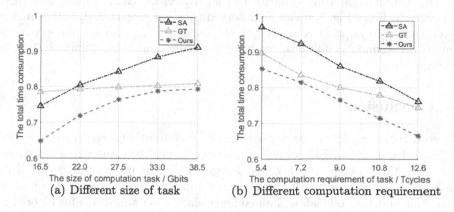

(a) Different size of task   (b) Different computation requirement

Fig. 6. System cost at SA, GT and GjRATO, $\theta=135°$

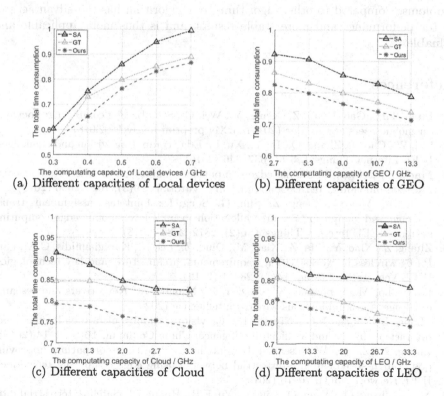

(a) Different capacities of Local devices   (b) Different capacities of GEO

(c) Different capacities of Cloud   (d) Different capacities of LEO

Fig. 7. System cost at SA, GT and GjRATO, $\theta=135°$

Figure 7 shows the system cost of three algorithms, SA, GT and GjRATO, under different computing capacities of local devices, GEO, Cloud, and LEO respectively. From Fig. 7, it can be observed that our proposed algorithm incurs lower time consumption compared to the other two algorithms under various computing resource conditions, resulting in a reduction of approximately 10% in average.

6   Conclusion

We construct a model in this paper where edge computing services are provided to ground users via GEO, Cloud and LEO. To minimize the system cost, we transform the problem into a mixed integer nonlinear programming (MINLP) problem. We design an optimization algorithm for joint computation resource allocation and task offloading, and conduct simulation analysis. This algorithm can adjust task offloading plans based on the location of the LEO satellite and the nature of the computing tasks, leading to a significant reduction in system latency. Our algorithm has a low computation complexity and can quickly obtain excellent results, providing valuable insights for solving similar MINLP problems. Compared to other algorithms, our algorithm has the advantages of better performance and more stable results, and is thus more applicable and valuable.

References

1. Luan, T.H., Gao, L., Li, Z., Xiang, Y., Wei, G., Sun, L.: Fog computing: Focusing on mobile users at the edge (2015). arXiv preprint arXiv:1502.01815
2. Shi, W., Cao, J., Zhang, Q., Li, Y., Xu, L.: Edge computing: vision and challenges. IEEE Internet Things J. 3(5), 637–646 (2016)
3. Mach, P., Becvar, Z.: Mobile edge computing: a survey on architecture and computation offloading. IEEE Commun. Surv. Tutorials 19(3), 1628–1656 (2017)
4. Wang, P., Yao, C., Zheng, Z., Sun, G., Song, L.: Joint task assignment, transmission, and computing resource allocation in multilayer mobile edge computing systems. IEEE Internet Things J. 6(2), 2872–2884 (2018)
5. Zhang, Z., Xiao, Y., Ma, Z., Xiao, M., Ding, Z., Lei, X., Karagiannidis, G.K., Fan, P.: 6g wireless networks: vision, requirements, architecture, and key technologies. IEEE Veh. Technol. Mag. 14(3), 28–41 (2019)
6. Latva-aho, M., Leppänen, K., Clazzer, F., Munari, A.: Key drivers and research challenges for 6g ubiquitous wireless intelligence (2020)
7. Zhang, L., Liang, Y.C., Niyato, D.: 6g visions: mobile ultra-broadband, super internet-of-things, and artificial intelligence. China Commun. 16(8), 1–14 (2019)
8. Zhang, Z., Zhang, W., Tseng, F.H.: Satellite mobile edge computing: improving qos of high-speed satellite-terrestrial networks using edge computing techniques. IEEE Network 33(1), 70–76 (2019)
9. Xie, R., Tang, Q., Wang, Q., Liu, X., Yu, F.R., Huang, T.: Satellite-terrestrial integrated edge computing networks: architecture, challenges, and open issues. IEEE Network 34(3), 224–231 (2020)

10. Pang, Y., Wang, D., Wang, D., Guan, L., Zhang, C., Zhang, M.: A space-air-ground integrated network assisted maritime communication network based on mobile edge computing. In: 2020 IEEE World Congress on Services (SERVICES), pp. 269–274. IEEE (2020)
11. Mao, S., He, S., Wu, J.: Joint uav position optimization and resource scheduling in space-air-ground integrated networks with mixed cloud-edge computing. IEEE Syst. J. **15**(3), 3992–4002 (2020)
12. Liu, M., Wang, Y., Li, Z., Lyu, X., Chen, Y.: Joint optimization of resource allocation and multi-uav trajectory in space-air-ground iort networks. In: 2020 IEEE Wireless Communications and Networking Conference Workshops (WCNCW), pp. 1–6. IEEE (2020)
13. Zhang, X., Zhang, J., Xiong, J., Zhou, L., Wei, J.: Energy-efficient multi-uav-enabled multiaccess edge computing incorporating noma. IEEE Internet Things J. **7**(6), 5613–5627 (2020)
14. Wang, Y., Zhang, J., Zhang, X., Wang, P., Liu, L.: A computation offloading strategy in satellite terrestrial networks with double edge computing. In: 2018 IEEE international conference on communication systems (ICCS), pp. 450–455. IEEE (2018)
15. Tang, Q., Fei, Z., Li, B., Han, Z.: Computation offloading in leo satellite networks with hybrid cloud and edge computing. IEEE Internet Things J. **8**(11), 9164–9176 (2021)
16. Wang, Z., Yu, H., Zhu, S., Yang, B.: Curriculum reinforcement learning-based computation offloading approach in space-air-ground integrated network. In: 2021 13th International Conference on Wireless Communications and Signal Processing (WCSP), pp. 1–6. IEEE (2021)
17. Wang, Y., Yang, J., Guo, X., Qu, Z.: A game-theoretic approach to computation offloading in satellite edge computing. IEEE Access **8**, 12510–12520 (2019)
18. Zhang, K., Gui, X., Ren, D., Li, D.: Energy-latency tradeoff for computation offloading in UAV-assisted multiaccess edge computing system. IEEE Internet Things J. **8**(8), 6709–9719 (2021)
19. Tran, T.X., Pompili, D.: Joint task offloading and resource allocation for multi-server mobile-edge computing networks. IEEE Trans. Veh. Technol. **68**(1), 856–868 (2018)

gGMED: Towards GPU Accelerated Geometric Modeling Evaluation and Derivative Processes

Zhibo Xuan[1], Hailong Yang[1(✉)], Pengbo Wang[1], Xin Sun[1], Jiwei Hao[1], Shenglin Duan[2], Yongfeng Shi[2], Zhongzhi Luan[1], and Depei Qian[1]

[1] School of Computer Science and Engineering, Beihang University, Beijing, China
{xzb,hailong.yang,wangpengbo2022,sxin2206,jiweihao,07680,depeiq}@buaa.edu.cn
[2] Avic Digital Co., Ltd., Beijing, China
{duansl,shiyf002}@avic.com

Abstract. Geometric modeling algorithms serve as the fundamental computation of CAD/CAM software in the field of computer graphics. The evaluation and derivative processes, being an essential component of geometric modeling algorithms, significantly impact their overall performance. However, when dealing with scenarios involving high-precision models or large-scale datasets, the lack of parallel acceleration for geometric modeling computation results in prolonged computation time and low computation efficiency, hindering the satisfactory experience of user interaction. Although the massive parallelism of GPUs has been proved with successful performance acceleration in various application fields, it has not been effectively utilized for accelerating geometric modeling algorithms. In this paper, we propose *gGMED*, a GPU-based approach specifically designed for accelerating the evaluation and derivative processes in geometric modeling. To leverage the massive parallel capability of GPU, our approach provides several optimizations such as data reuse, bank conflict avoidance, and pipeline execution, for effectively improving the performance of evaluation and derivative processes. The experiment results on representative GPUs and various NURBS models demonstrate that our approach can achieve up to $10.18\times$ and $34.56\times$ performance speedup in end-to-end process and kernel computation respectively, compared to the state-of-the-art geometric modeling libraries.

Keywords: Geometric modeling algorithms · Evaluation · Derivative · Parallel optimization · GPU

1 Introduction

Geometric modeling algorithms find extensive applications in the field of computer graphics. Ivan Sutherland created a groundbreaking program called Sketchpad [25], which was the world's first interactive computer graphics system. Charles Lang started researching 3D CAD software and began commercializing it [22]. Based on geometric modeling algorithms, researchers have developed geometric modeling libraries such as *OCCT* (Open CASCADE Technology)

Z. Tari et al. (Eds.): ICA3PP 2023, LNCS 14489, pp. 378–397, 2024.
https://doi.org/10.1007/978-981-97-0798-0_22

[2–4,24], *CGAL* (Computational Geometry Algorithms Library) [5–7], *EGADS* (Extensible Geometry and Discretization Specification) [16], *EGADSlite* [8], etc. Building upon these geometric modeling libraries, examples include CAD (Computer-Aided Design) and CAM (Computer-Aided Manufacturing) software tailored to specific industries have been developed. CAD software, such as *Auto-CAD, SolidWorks, CATIA*, is used for creating, modifying, and optimizing three-dimensional geometric models and design drawings. CAM software, such as *Mastercam, SolidCAM*, and *GibbsCAM*, is used in computer-aided manufacturing processes. These geometric modeling libraries and software are widely employed in industries such as aerospace, architectural design, precision instrument manufacturing, and medical fields, offering enhanced productivity and efficiency in their respective domains.

There are various methods for representing curves and surfaces in geometic modeling, including Bézier Curves and Surfaces, B-Spline, NURBS (Non-Uniform Rational B-Splines) [19,20], Subdivision Surface, Implicit Surface, and Voxel-Based Representations, etc. Ken Versprille proposed NURBS [27], a method for representing geometric models that exhibit natural smoothness and flexible curve inflection points, enabling the precise fitting of curves and surfaces, including both regular and complex freeform shapes. NURBS can be represented as piecewise polynomials with local control characteristics, allowing them to meet various geometric requirements through interpolation and approximation. Besides, NURBS can achieve continuous and high-order smoothness by adjusting weights and knot vectors. Therefore, NURBS representation is widely used, and geometric modeling libraries such as *OCCT, CGAL*, and *EGADSlite*, as well as software like *RhinoCreo*, provide support for NURBS. The standardization of a unified representation format allows models to be exchanged between different libraries and software, enabling the utilization of different tools' advantages for processing purposes.

With the advancement of the manufacturing industry, there is an increasing demand for NURBS with higher precision in model processing across various sectors. Improving precision leads to a significant increase in computational workload for model calculations. This places a greater burden on geometric modeling libraries in terms of computational requirements. However, many basic geometric modeling implementations lack inherent parallelism in their algorithms, such as evaluation and derivative, which further exacerbate the challenge of handling increased computational demands. Consequently, excessive computation times become unacceptable. Prolonged execution times prevent users from receiving timely responses and observing real-time visual feedback of CAD and CAM software, thereby failing to meet the requirements for interactive usage.

In contrast, in recent years, Graphics Processing Units (GPUs) have witnessed rapid advancements, providing new possibilities for accelerating geometric modeling algorithms. Compared to CPUs, GPUs feature a higher number of computational units, faster memory access speeds, and specialized high-speed computing units. As a result, GPUs possess superior parallel computing capabilities, faster data transfer rates, and stronger processing speeds for specific computation patterns. However, there has been limited work in leveraging GPU acceleration in existing open-source geometric modeling libraries. Consequently, the increasing computational demands of NURBS in geometric modeling cannot fully utilize the computational power improvements brought about by the rapid development of GPUs. This issue hampers the further enhancement of computational speeds in geometric modeling algorithms.

Geometric modeling algorithms encompass a diverse set of tasks, including curve and surface fitting, surface reconstruction, curve and surface subdivision, boolean operations, voxelization, and more. Among these tasks, surface intersection and distance analysis stand as fundamental algorithms that provide essential support. At the core of these geometric modeling algorithms lies the evaluation process for NURBS, which is responsible for calculating the model-space (Cartesian multi-dimensional space) coordinates based on the parameter-space coordinates. Geometric modeling libraries heavily rely on the evaluation function and surface derivative function to obtain surface curvature and perform subsequent calculations. As precision requirements increase, the number of function calls, particularly for the evaluation and derivative functions, grows exponentially. Unfortunately, these frequently called functions have not undergone sufficient optimization for parallel execution. Consequently, this limitation leads to inefficient performance of functions that depend on the evaluation process, thereby impacting the responsiveness and interactive usage that users demand.

To address the aforementioned issues regarding the lack of systematic optimization in existing open-source solutions, we proposed *gGMED*, a GPU-based acceleration approach built upon the *EGADSlite* open-source geometric modeling library. *gGMED* focuses on parallel optimization of the evaluation and derivative processes utilizing GPUs. It combines the computation of the two functions and fully exploits the evaluation and derivative characteristics of NURBS, aiming to utilize the computational and bandwidth resources of the GPU effectively and improve computational efficiency.

Specifically, this paper makes the following contributions:

- We propose a parallel optimization approach for the evaluation and derivative processes, addressing the limitations of existing geometric modeling libraries. By employing this approach, the evaluation and derivative processes can be efficiently accelerated through the utilization of GPU.
- We further introduce three techniques to improve the performance of the evaluation and derivative kernel, which enable the evaluation and derivative processes fully leverage the computational resources of the GPU and reduces unnecessary computations and memory access.
- Through experiments, we demonstrate the performance improvements achieved through parallel optimization by comparing with state-of-the-art open-source geometric modeling libraries. Additionally, we validate the effectiveness of the parallel approach and optimization methods.

The rest of this paper is organized as follows. Section 2 details the computation of evaluation and derivative processes, respectively. Section 3 introduces the related works about geometric modeling algorithms accelerating. Section 4 discusses the detailed design and implementation of our parallel approaches based on the *EGADSlite* library. Section 5 provides the evaluation results and the roofline analysis. We conclude this paper and discuss the future work in Sect. 6.

2   Background

2.1   Evaluation Process

A NURBS surface can be defined by a control point grid and two knot vectors that in the u and v directions of the parameter space. Evaluation refers to the process of mapping NURBS surfaces from parameter space (u, v) to model space (x, y, z). The NURBS model is definited as Eq. 1, where P denotes the control points, W represents the control point weights, and N_i^p represents the basis functions of a NURBS model with degree p, which is detailed in Eq. 2 and Eq. 3 [21]. The basis functions are recursive, where u_i and v_i represent the i-th component in the u or v direction of the knot vector. The overall evaluation process of NURBS surfaces is shown in Fig. 1, which is important for further geometric modeling algorithms.

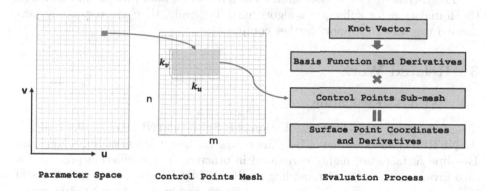

Fig. 1. The evaluation process of NURBS.

$$S(u,v) = \frac{\sum_{i=0}^{n} \sum_{j=0}^{m} N_i^p(u) N_j^p(v) w_{ij} P_{ij}}{\sum_{i=0}^{n} \sum_{j=0}^{m} N_i^p(u) N_j^p(v) w_{ij}} \tag{1}$$

$$N_i^p(u) = \frac{u - u_i}{u_{i+p} - u_i} N_i^{p-1}(u) + \frac{u_{i+p+1} - u}{u_{i+p+1} - u_{i+1}} N_{i+1}^{p-1}(u) \tag{2}$$

$$N_i^0(u) = \begin{cases} 1 \, if \, u_i \le u \le u_{i+1}, \\ 0 \, otherwise. \end{cases} \tag{3}$$

2.2 Derivative Process

Taking derivatives of specific coordinates in NURBS is crucial and widely used in rendering, surface intersection, curvature calculation, and other geometric modeling algorithms. For simplicity, we use B-spline surface as an illustrative example for derivative computation. Equation 4 defines a B-Spline surface, and the corresponding basis function is detailed in Eq. 5. The derivative of a specific point can be obtained by multiplying the control points P with the derivatives of the basis functions, as shown in Eq. 6 and Eq. 7.

$$S(u, v) = \sum_{i=0}^{n} \sum_{j=0}^{m} N_i^p(u) N_j^p(v) P_{ij} \tag{4}$$

$$\frac{\partial N_i^p(u)}{\partial u} = \frac{p-1}{u_{i+p} - u_i} N_i^{p-1}(u) - \frac{p-1}{u_{i+p+1} - u_{i+1}} N_{i+1}^{p-1}(u) \tag{5}$$

$$\frac{\partial S(u, v)}{\partial u} = \sum_{i=0}^{n} \sum_{j=0}^{m} \frac{\partial N_i^p(u)}{\partial u} N_j^p(v) P_{ij} \tag{6}$$

$$\frac{\partial S(u, v)}{\partial v} = \sum_{i=0}^{n} \sum_{j=0}^{m} N_i^p(u) \frac{\partial N_j^p(v)}{\partial v} P_{ij} \tag{7}$$

The derivative calculation often follows the evaluation process and serves as the foundation for subsequent algorithms. Typically, they are commonly performed together to support further computations.

3 Related Work

3.1 NURBS Evaluation

There are many works in the evaluation of B-Spline and NURBS [9, 10, 15, 16, 20]. As NURBS is an extension of B-spline representation, the works on NURBS and B-spline surfaces are highly correlated in practice. The evaluation process can also give an intuitive understanding of NURBS by directly seeing the effects of different control point positions, weight vectors, and knot vectors. NURBS evaluation algorithms are essentially numerical computational methods that compute model-space points on a NURBS curve or surface from control points, weights, knot vectors and other given parameter values. To represent NURBS in the model space, the spline should be evaluated at multiple parameters u, where $0 \leq u \leq 1$, and the basis functions are needed as well. Several NURBS evaluation algorithms are detailed in this section.

The most famous solution for B-spline evaluation is De Boor's algorithm which utilizes Cox-de Boor recursive formulas [20]. De Boor's algorithm is a generalization of de Casteljau's algorithm for B-spline and can execute the NURBS evaluation techniques based on the observation that curve point $C(u)$ is positioned at the location of the control point P_{k-p}, when $u = u_k$, and the knot multiplicity at u equals p. The two-stage Cox-de-Boor technique avoids repeated calculations of the same components or intermediate values by calculating and caching them [15]. The values of nonzero u basis functions and their derivatives can be computed simultaneously in a single loop for each u point $u_{m,k}$ by applying the Cox-de Boor recursive formulas to B-splines that are nonzero at $u_{m,k}$. This process requires many costly divisions and many of the denominators (the difference of u knots) are the same. Therefore, it requires the computation of a large number of duplicates for higher-order B-splines and the computation consumption would be very expensive. The Cox-de Boor recursive formula is prone to the problems of rounding error and numerical instability, especially when there are repeated knots in the knots vector.

Lee et al. [13] proposed an indirect approach that converts a B-spline basis representation to a power basis representation, performs multiplication by convolving coefficients, and then converts back to a B-spline basis representation via the de Boor-Fix formula. As the whole process is computationally expensive, Lee developed a scheme to evaluate the coefficients of the product B-spline a group at a time by computing a chain of blossoms. B-spline blossoming provides another direct approach that can be straightforwardly translated from NURBS expression to model space points. Ueda et al. [26] reported a direct approach for B-spline multiplication based on a blossom representation of B-splines, and proved its equivalence to Morken's earlier discrete B-spline approach. However, observing that computing the product B-spline coefficients directly from the blossom representation of product B-spline is inefficient.

3.2 GPU-Based Parallel Optimization

Many academic studies and research works [9,11,12,14,17,21,23] heavily rely on the performance of evaluation and derivative calculations in the context of geometry modeling. Numerous algorithms commonly employed in geometric modeling, such as surface intersection [12,14], distance analysis [9], and rendering [23], necessitate the utilization of evaluation for solving model space coordinates. Moreover, these algorithms often require derivative computations for evaluating curvature and related data. The accurate determination of points on surfaces through evaluation and the precise calculation of derivatives play pivotal roles in ensuring the robustness and effectiveness of these geometric modeling techniques.

McMains et al. [17] accelerate modeling operations such as ray intersections and surface-surface intersections by constructing multi-level axis-aligned bounding-boxes that can enclose the surface, which reduce the unnecessary computation and provide sufficient parallelism at a finer resolution. Adarsh et al. [10] parallelize the de Boor evaluation algorithm and perform NURBS evaluation modeling operations using the programmable fragment processor on the GPU. The GPU-based NURBS evaluator that evaluates NURBS surfaces is extended to compute exact normals for either standard or rational B-spline surfaces for usage in rendering and geometric modeling. However, few GPU optimization techniques have been applied to this work. McMains et al. [9,11] provide parallelizd minimum distance and clearance queries algorithms based on the existing work [10], demonstrating the significant role of parallelization in the evaluation and derivative processes of geometric modeling algorithms. Horner's Scheme is the NURBS evaluation technique that does not require a curve's derivative. Horner's Scheme exploits the recursive derivation of the binomial coefficients and evaluates a Bezier curve of degree n in $O(n)$ steps by using nested multiplications [18]. The small and constant number of registers required by Horner's scheme enables the evaluation of arbitrary degree curves on GPUs and Horner's scheme is suitable for NURBS curve evaluation for finite elements.

Therefore, the optimization and enhancement of evaluation and derivative processes significantly contribute to the advancement and refinement of academic research in the field of geometric modeling. However, there is a lack of recent work conducted on state-of-the-art GPUs, which significantly differs from previous studies.

3.3   Mainstream Geometry Modeling Libraries

Geometric Modeling Libraries are the software libraries for CAD, CAM, Computer Aided Engineering (CAE), Virtual Reality (VR), and Computer-generated Imagery (CGI). The Geometric Modeling Libraries provide a range of functions and data structures that allow developers to easily create, edit, and manipulate 2D and 3D geometric models. Some mainstream open-source geometric modeling libraries include *OCCT* [2–4,24], *CGAL* [5–7], *EGADS* [16], *EGADSlite* [8], etc.

OCCT is an open-source C++ geometry modeling library developed by Matra Datavision company [2], which is designed to assist the rapid development of complex domain-specific CAD/CAM/CAE applications. *OCCT* provides 3D curves, surface and entity modeling, CAD data exchange, and visualization functionality. *OCCT* also allows the development of 2D or 3D geometry modeling for simulation applications or illustration tools. Besides, *OCCT* contains a geometric characterization of all 2D and 3D geometry elements in *Geom2d* and *Geom* directories. The data structure of the boundary representation for *OCCT* is defined in the *BRep* directory, and the *TopoDS* directory defines the topology of Opencascade's 3D geometry. In addition to providing rich model support at the geometric model level, *OCCT* also provides code in the *src* directory to support the function of geometric model intersection and distance. Many algorithms in geometric modelings, such as surface curve intersection and surface curve distance query, all depend on the modeling results of NURBS evaluation. The *Geom_BSplineCurve* and *Geom_BSplineSurface* classes of *OCCT* provide methods for evaluating NURBS curves and surfaces, respectively.

CGAL [5–7] is Computational Geometric Modeling Library designed and supported by the INRIA organization. *CGAL* provides a large number of computational geometry algorithms for geometric entities such as points, lines, surfaces, polygons, convex hulls, and triangulated meshes, and offers efficient, reliable, and convenient implementations of these algorithms in the form of a C++ library. However, *CGAL* does not provide native support for NURBS evaluation. To perform NURBS evaluations, *CGAL* can be used together with other libraries that specialize in NURBS, such as the OpenNURBS library [1].

EGADS [16] is a geometric modeling library designed to handle complex 3D geometric models. It is designed for scientific computing and engineering simulation applications and provides a general and extensible way to describe, manipulate, and analyze geometric data. *EGADSlite* [8] is the lite version of *EGADS*, which has a smaller footprint and is designed for applications with limited resources or less demanding requirements for geometric operations and analysis. *EGADSlite* retains basic geometric manipulation and analysis capabilities while offering a more streamlined and lightweight solution.

In summary, the existing works to accelerate geometric modeling libraries primarily focus on algorithmic innovations and domain-specific optimizations, with limited work dedicated to GPU acceleration. GPU acceleration work is scarce, and the few existing implementations do not fully harness the potential of GPU acceleration.

4   Methodology

4.1   Parallel Implementation

Due to the independency of evaluation between different parameter pair (u, v), we can parallelize the evaluation among different compute units intuitively. We transfer the information required for evaluation to the device memory (e.g., degree of u and v direction, control points, (u, v), etc.). Specifically, gGMED partitions the computation into 2-dimensional blocks and threads for further optimization convenience, which is detailed in Sect. 4.2. The process of evaluation (shown in Fig. 1) is as follows: *1)* Identify the sub-grid of control points corresponding to the target point; *2)* Calculate the first-order derivatives, second-order derivatives, and the non-zero basis functions in the u and v directions for the target point; *3)* Multiply the values and derivative values of the non-zero basis functions with their corresponding control points and sum up the results.

However, in this situation, the implementation provided by EGADSlite will introduce a huge memory copy overhead including both device-to-host and host-to-device. The reason is frequent memory page replacement, namely *cold-start*. To address this issue, considering the interdependence between this process and subsequent geometric modeling algorithms, we employ locked page memory to mitigate it.

Additionally, we aggregate memory operations to minimize stride-based memory access patterns and strive to store data in registers as much as possible to accelerate memory access speed. Meanwhile, we move the generation of (u, v) from CPU to GPU to further reduce the data transfmission.

4.2   Data Reuse Optimization

Based on the method proposed in Sect. 4.1, a thread block of size (m, n) needs to perform evaluation according to the sub-grid $\{(u_i, v_j), i \in [0, m], j \in [0, n]\}$ in parameter space. In this process, each thread needs to compute two sets of basis

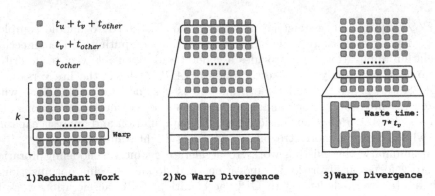

Fig. 2. The warp divergence.

function values for u_i and v_j, $N_m^p(u_i), ..., N_{m+p}^p(u_i)$ and $N_n^q(v_j), ..., N_{n+q}^q(v_j)$, where m and n are integer values determined by u_i, v_j and the knots vector. p, q represent degrees along u and v directions. The computation of basis function values takes a large portion of the evaluation process. However, in the context of evaluation over a grid of parameters, there are only m unique values along u direction and n unique values along v direction in each sub-grid, making it possible to share results inside a thread block to speed up the evaluation. A clear approach is to assign the computation of basis function values of u_i/v_j to the thread responsible for the evaluation of $(u_i, v_0), i \in [0, m]/(u_0, v_j), j \in [0, m]$, and store results to shared memory, where all threads in a thread block can access with low latency.

However, such an intuitive reuse strategy doesn't perform well due to the load imbalance within a warp reducing the parallelism of the kernel, as shown in Fig. 2 **(1)**. Assuming there are k warps in a thread block, the computation of basis function values $N_i^p(u)$ and $N_j^q(v)$ takes t_u and t_v respectively, and other operation takes t_{other}. Ideally, we would expect only 2 warps to execute for $t_u + t_v + t_{other}$, and other warps to execute only for t_{other}, making the overall execution time of the thread block $t_u + t_v + k \cdot t_{other}$, like Fig. 2 **(2)**. However, threads in the same wrap can be fully parallelized only if they share the same control flow, otherwise, the execution of different control flows will be serialized, namely the thread divergence. As shown in Fig. 2 **(3)**, thread divergence explains the poor performance of our naive approach. The basis function values computation along u direction is scattered across warps, so every warp must execute at least for $t_v + t_{other}$, leading to the overall execution time of $t_u + k \cdot (t_v + t_{other})$. Additionally, as shown in Fig. 3 **(2)**, the current 2D indexing scheme will also lead to wasted parallelism, impacting evaluation performance.

Thus, we should assign the computation of basis function values to threads in the same warp to get optimal execution time, as shown in Fig. 2 **(2)**. However, it requires complex control logic under the current 2D thread block indexing scheme. For example, thread blocks for the marginal sub-grid of a parameter may need more than one warp to compute the basis function value in a certain

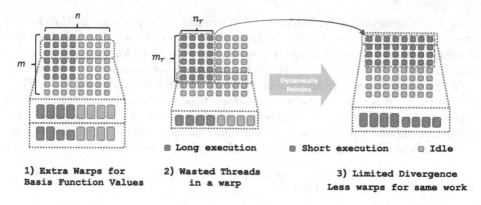

Fig. 3. Reindex Scheme: 1) wasted parallelism, 2) redundant warps and 3) reindex solution.

parametric direction, as shown in Fig. 3 (**1**). A solution is to dynamically reindex threads in a thread block, as shown in Fig. 3 (**3**). For a 2D thread block of size (m, n), we ues (m_r, n_r) to denote the actual size of the sub-grid, where $m_r < m, n_r < n$. For every thread in the warp, we set its scope id as $sid = warp\_id * 32 + lane\_id$. A thread process for the evaluation of NURBS surface on $(sid/n_r, sid\%n_r)$ provided that $sid < m_r \cdot n_r$. In this way, there will be at most 1 non-full warp in a thread block, limiting the waste of parallelism. The basis function value computation can also be easily assigned to different warps, minimizing warps divergence and leading to theoretically fast evaluation (a.k.a., $t_u + t_v + k \cdot t_{other}$).

4.3 Bank Conflict Avoidance

		n_1	n_2	n_3	n_4
		i_1	i_1	i_1	i_1
k_1		i_2	i_2	i_2	i_2
		i_3	i_3	i_3	i_3
		i_4	i_4	i_4	i_4
		i_1	i_1	i_1	i_1
k_2		i_2	i_2	i_2	i_2
		i_3	i_3	i_3	i_3
		i_4	i_4	i_4	i_4

		i_1	i_2	i_3	i_4
	n_1	i_1	i_2	i_3	i_4
k_1	n_2	i_1	i_2	i_3	i_4
	n_3	i_1	i_2	i_3	i_4
	n_4	i_1	i_2	i_3	i_4
	n_1	i_1	i_2	i_3	i_4
k_2	n_2	i_1	i_2	i_3	i_4
	n_3	i_1	i_2	i_3	i_4
	n_4	i_1	i_2	i_3	i_4

Fig. 4. Bank Conflict Avoidance: before and after.

To maximize the throughput of shared memory access, we employ a carefully chosen data storage pattern to avoid bank conflict. On typical platforms, shared memory in CUDA is implemented by 32 equally sized memory banks that can be accessed simultaneously. However, when threads in warp access multiple locations that map to the same bank, all these access have to be serialized. So instead of using multi-dimensional arrays where the basis functions for the same parameter are continuous (left of Fig. 4), we use an interleaved memory layout where basis function values for different threads are continuous (right of Fig. 4). For the n-th basis function value of param u_i/v_i whose degree is $p-k, k \in [0,2]$, the shared memory index is calculated with $k \cdot block\_dim * (p+1) + n * block\_dim + i$. When loading and storing basis function values along u and v directions, threads in a wrap will access data in different banks without conflict.

4.4   Pipeline Optimization

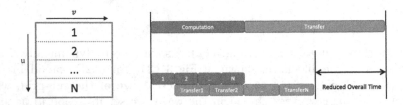

Fig. 5. Pipeline optimization: work partitioning and overlapping.

Although the optimized evaluation kernel exhibits commendable performance improvement, the data transfer rate between DRAM and GPU VRAM is relatively limited, thus incurring a significant overhead for overall time. To mitigate this problem, we introduced the pipeline optimization technique which enables the overlap of computation and data transfer using the CUDA stream. CUDA stream is a sequence of operations that run in order on the GPU, where computation and data transfer operations in different streams can be executed concurrently. We employed a one-dimensional workload partitioning scheme on the target parameter space, as shown in the left of Fig. 5. The evaluation over the whole parameter space is partitioned into N tasks in u direction, each task evaluates points in its slice of parameter space and then transfers output data back to the host. By partitioning the parameter space in the u direction only, we ensured that all data transfer operations process a continuous area in memory. Tasks are then executed in different streams, so the computation and data transfer process in different tasks can overlap effectively, as shown in the right of Fig. 5, significantly reducing the overall evaluation time.

5   Evaluation

5.1   Experimental Setup

We evaluate $gGMED$ on three workstations to create an environment for regular usage of geometric modeling algorithms. Each workstation has 52 cores with two

Intel Xeon Gold 6230R processors and 384GB of memory. To conduct performance testing of different accelerators, we use non-GPU, Tesla V100, and RTX 3090 on three workstations, respectively. The detailed hardware and software configurations of experiment environments are listed in Table 1.

Table 1. The hardware and software specifications.

Environment	W1	W2	W3
Processor	Intel(R) Xeon(R) Gold 6230R CPU @ 2.10 GHz		
Memory	384 GB	384 GB	384 GB
Accelerator	-	Tesla V100-PCIE-32 GB	GeForce RTX 3090
Device Memory	-	32 GB	24 GB
Software	gcc 7.5.0, CUDA 11.3		
System	Ubuntu 18.04.6 LTS		

For comparison, we choose the *OCCT* (*version 7.6.0*), and *EGADSlite* (*version 1.22*) as the state-of-the-art geometric modeling libraries and for both libraries we choose the evaluation and derivative processes as target. For evaluation, we create three NURBS models with different *degree* and numbers of *control points*, as shown in Table 2. The three models have different computational and memory access requirements due to the difference in their parameters. The corresponding figures of the three models are shown in Fig. 6.

For performance evaluation, we compared the execution times of different geometric modeling libraries for various NURBS models across three different configurations. For scalability evaluation on a specific GPU, we compared the speedup ratios of the same GPU across different parallel scales. For illustrating the effectiveness of optimization, we conducted a series of ablation experiments to provide a visual demonstration of the performance improvements achieved by different optimizations.

Table 2. The configurations of three NURBS models.

Case	U-degree	V-degree	U-control points	V-control points
Case1	3	3	155	150
Case2	6	6	123	123
Case3	9	9	213	213

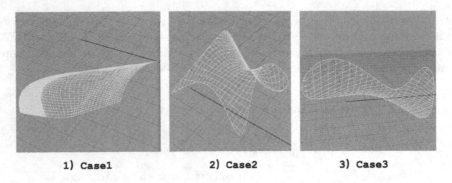

1) Case1   **2) Case2**   **3) Case3**

Fig. 6. Three NURBS models.

5.2   Performance Comparison

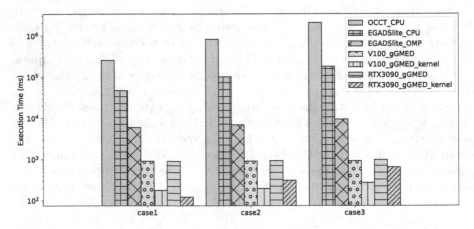

Fig. 7. Time comparison between different settings with a scale of 9000 × 9000.

For *OCCT*, we call the *D2 function* of *Geom_BSplineSurface Class* to perform evaluation and derivative processes. For *EGADSlite*, we choose the *EG_evaluate function* as the comparison target. For *gGMED*, we provide both the end-to-end computation time and kernel computation time on different experiment specifications, as the memory access time can be reduced in future work. Figure 7 presents the time data of the evaluated cases on different experiment specifications. We have observed a significant disparity in execution times between *gGMED* and *OCCT_CPU* and *EGADSlite_CPU*, spanning multiple orders of magnitude. This discrepancy renders a fair comparison impractical. Therefore, for the purpose of calculating the speedup ratio, we have opted to consider the Naive parallel version of *EGADSlite* utilizating *OpenMP* as our baseline (a.k.a, *EGADSlite_OMP*). Based on the time data, we further calculate the speedup

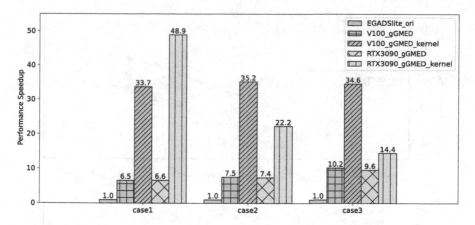

Fig. 8. Performance comparison between different settings with a scale of 9000×9000.

of the computation, as shown in Fig. 8. On such specifications, *gGMED* incurs an optimization effect up to $10.18\times$ and $9.57\times$ on V100 and RTX 3090 GPU, respectively. The performance of *gGMED* on V100 is slightly higher than that on RTX 3090 because the higher performance of float64 computation. Moreover, if we consider only the kernel time, *gGMED* achieves a $34.56\times$ and $14.43\times$ speedup in computation compared to *EGADSlite_OMP*, respectively, whose potential is likely to be utilized in other algorithms that rely on evaluation and derivative processes. And the results demonstrate that the memory access speed is still the bottleneck.

5.3 Scalability

To demonstrate the scalability potential towards a large amount of computation, we evaluate *gGMED* using varying degrees of parallelism. We executed the same computational tasks on the GPU using different parallel scales. We carefully selected a range of parallel scales, ranging from 10×10 to 9000×9000, to cover a wide spectrum of parallelization scenarios.

For each parallel scale, we measured the execution time of the computational tasks and compared them to the execution time of the *EGADSlite* sequential implementation on the same GPU. The speedup ratio was then calculated as the ratio of the sequential execution time to the parallel execution time. The results of our comparison showed varying speedup ratios across different parallel scales. Generally, we observed that as the parallel scale increased, the speedup ratio improved, indicating a higher level of performance enhancement achieved through parallelization, as well as the scalability, as shown in Fig. 9.

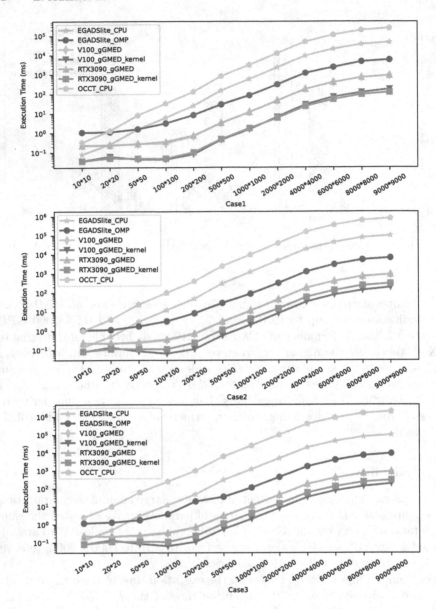

Fig. 9. Overall ablation results, where the y-axis indicates the speedup normalized to *EGADSlite*: **1)** *Case1*, **2)** *Case2* and **3)** *Case3*.

5.4   Ablation Study

Overall Ablation Experiment. As mentioned earlier, both the end-to-end computation performance and the kernel computation performance have real potential for utilization. We conducted two ablation experiments to evaluate both the effect of end-to-end optimization and kernel optimization separately, as shown in Fig. 10 and Fig. 11. Note that the kernel computation time includes both the kernel computation time and the shared memory access time, but it does not include the time spent on cudaMemcpy operations. We choose **W2** as the baseline environment for all cases at a 4000 × 4000 scale. The performance data is normalized using the method described in Sect. 5.2.

End-to-End Performance. There are four optimized implementations, as shown in Fig. 10. We use p to denote the parallel optimization detailed in Sect. 4.1, m to denote the bank conflict avoidance detailed in Sect. 4.3, r to denote the data reuse detailed in Sect. 4.2 and l to denote the pipeline optimization detailed in Sect. 4.4. The best optimization effect is from *Case 3*, which has the highest computational workload. Besides, *naive* denotes the naive GPU version of *EGADS_OMP*.

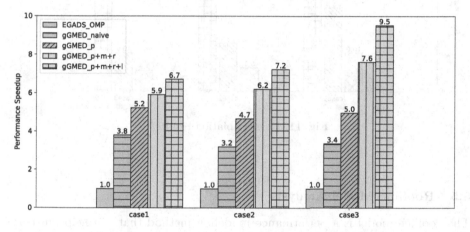

Fig. 10. End-to-End ablation results, where we take into account the memory access overhead.

Parallelizing the evaluation and derivative processes using GPUs provided a significant speedup of 3.36×, which is the most notable improvement among all optimizations. When performing the optimization detailed in Sect. 4.1, the end-to-end speedup increases to 4.97×. This is because that cold-start of memory significantly impacts the speed of cudaMemcpy operations. When applying data reuse and bank conflict avoidance optimization, the end-to-end speedup increases to 7.61×. Due to the aforementioned optimizations, the memory access time and kernel computation time are within the same order of magnitude. Thus pipeline optimization further increases the speedup to 9.52×.

Kernel Computing Performance. There are three optimized implementations in kernel computing performance experiments, too. The configuration in the Fig. 11 is consistent with the Fig. 10, but it lacks the pipeline optimization because this optimization is only applicable to the end-to-end scenario. Parallelizing the computation process using a naive way results in a 10.01× speed up. After perfroming the parallelization in depth, the speedup has further increased to 21.84×. Thanks to the characteristics of NURBS model evaluation, the optimization of data reuse and bank conflict avoidance based on the distribution features of u and v has further increased the speed up to 42.49×.

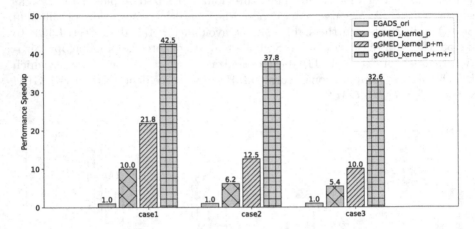

Fig. 11. Kernel ablation results.

5.5   Roofline Model Analysis

The roofline model is a performance modeling method that can help illustrate the optimization and limitations of specific CUDA kernels. To gain a deeper understanding of the effectiveness of our proposed optimization methods, we plotted the roofline model before and after the optimization with *Case3* on *V100*.

In the roofline model analysis, we only present the experiment result of *Case2* for illustration. As shown in Fig. 12, the compute intensity is the primary metric we use to measure the effectiveness of optimizations. And we can observe and draw the following conclusions: *1)* The operational intensity of *EGADSlite-ori* is only *0.51 Flops/Byte*. *2)* After applying parallel optimization, the compute intensity improves from *0.51 Flops/Byte* to *0.93 Flops/Byte*. This is due to the higher memory access efficiency, which means we can take full advantage of more abundant resources, i.e. computational resources. *3)* With the incorporation of data reuse optimization, the compute intensity improves to *4.82 Flops/Byte*. In more intuitive terms, it means that the knee point of the roofline model is closer to the upper-right corner. *4)* After applying all the optimizations, the compute intensity improves to *4.87 Flops/Byte*. At this point, as shown in Fig. 12, we can observe that the roofline model closely aligns with the

GPU's Memory Bound and Compute Bound, indicating the effectiveness of our optimizations. This demonstrates that we have achieved a good balance between memory operations and compute operations, maximizing the utilization of the GPU's resources.

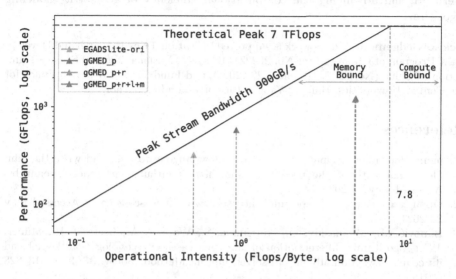

Fig. 12. The roofline model of *gGMED* in V100.

In summary, the roofline model of *gGMED* reveals that when applying all the optimization techniques, *1240.1 Gflops* calculations should be performed. Compared to the theoretical performance limit, we believe that improving computational intensity and memory efficiency remains crucial for optimizing this process.

6 Conclusion

In this paper, we proposed *gGMED*, a novel GPU-based parallel optimization approach for the evaluation and derivative processes in geometric modeling libraries. Our approach exploits the inherent independence and parallelism of the evaluation and derivative processes and leverages the computational power of GPUs to achieve accelerated computations for large-scale evaluation and derivative tasks. Furthermore, we employ efficient data reuse techniques between evaluations and derivatives, thereby improving the utilization of shared memory. Through extensive experimentation on three distinct cases across multiple GPUs, we demonstrate the remarkable computational speed improvements achieved by our proposed method. Specifically, we observe a *10.18×* enhancement in end-to-end processing speed and a significant *34.56×* acceleration in kernel execution speed. These remarkable kernel acceleration results exemplify the promising potential of our approach for future research endeavors.

Based on the foundational contributions presented in this paper, our future research endeavors will focus on further parallel optimization of geometric modeling algorithms that heavily rely on the evaluation and derivative processes. Our objectives include enhancing data reuse strategies, reducing memory access overhead, and advancing the computational efficiency of geometric modeling algorithms.

Acknowledgements. This work is supported by National Key Research and Development Program of China (Grant No. 2022ZD0117805), National Natural Science Foundation of China (No. 62072018 and U22A2028), and Fundamental Research Funds for the Central Universities. Hailong Yang is the corresponding author.

References

1. mcneel/opennurbs: Opennurbs libraries allow anyone to read and write the 3dm file format without the need for rhino. https://github.com/mcneel/opennurbs. Accessed May 24 2023
2. Open cascade, part of capgemini. https://www.opencascade.com/ Accessed May 24 2023
3. Banović, M., Mykhaskiv, O., Auriemma, S., Walther, A., Legrand, H., Müller, J.D.: Algorithmic differentiation of the open cascade technology cad kernel and its coupling with an adjoint cfd solver. Optim. Methods Softw. **33**(4–6), 813–828 (2018)
4. Bedaka, A.K., Lin, C.Y.: Cad-based robot path planning and simulation using open cascade. Pro. Comput. Sci. **133**, 779–785 (2018)
5. Boissonnat, J.D., Devillers, O., Teillaud, M., Yvinec, M.: Triangulations in cgal. In: Proceedings of the Sixteenth Annual Symposium on Computational Geometry, pp. 11–18 (2000)
6. Fabri, A., Giezeman, G.J., Kettner, L., Schirra, S., Schönherr, S.: On the design of cgal a computational geometry algorithms library. Softw.: Pract. Exp. **30**(11), 1167–1202 (2000)
7. Fabri, A., Pion, S.: Cgal: the computational geometry algorithms library. In: Proceedings of the 17th ACM SIGSPATIAL International Conference on Advances in Geographic Information Systems, pp. 538–539 (2009)
8. Haimes, R., Dannenhoffer, J.: Egadslite: a lightweight geometry kernel for hpc. In: 2018 AIAA Aerospace Sciences Meeting. p. 1401 (2018)
9. Krishnamurthy, A., Khardekar, R., McMains, S.: Optimized gpu evaluation of arbitrary degree nurbs curves and surfaces. Comput. Aided Des. **41**(12), 971–980 (2009)
10. Krishnamurthy, A., Khardekar, R., McMains, S., Haller, K., Elber, G.: Performing efficient nurbs modeling operations on the gpu. In: Proceedings of the 2008 ACM symposium on Solid and physical modeling, pp. 257–268 (2008)
11. Krishnamurthy, A., McMains, S., Halle, K.: Accelerating geometric queries using the gpu. In: 2009 SIAM/ACM Joint Conference on Geometric and Physical Modeling, pp. 199–210 (2009)
12. Krishnamurthy, A., McMains, S., Haller, K.: Gpu-accelerated minimum distance and clearance queries. IEEE Trans. Visual Comput. Graphics **17**(6), 729–742 (2011)
13. Lee, E.: Computing a chain of blossoms, with application to products of splines. Comput. Aided Geomet. Design **11**(6), 597–620 (1994)

14. Lin, H., Qin, Y., Liao, H., Xiong, Y.: Affine arithmetic-based b-spline surface inter-section with gpu acceleration. IEEE Trans. Visual Comput. Graphics **20**(2), 172–181 (2013)
15. Luken, W.L., Cheng, F.: Comparison of surface and derivative evaluation methods for the rendering of nurb surfaces. ACM Trans. Graph. (TOG) **15**(2), 153–178 (1996)
16. Marti, L., et al.: Evaluation of gadolinium's action on water cherenkov detector systems with egads. Nucl. Instrum. Methods Phys. Res., Sect. A **959**, 163549 (2020)
17. McMains, S., Krishnamurthy, A.: Parallel gpu algorithms for interactive cad/cam operations
18. Pavlidis, T.: Algorithms for graphics and image processing. Springer Science & Business Media (2012)
19. Piegl, L.: On nurbs: a survey. IEEE Comput. Graphics Appl. **11**(01), 55–71 (1991)
20. Piegl, L., Tiller, W.: The NURBS book. Springer Science & Business Media (1996)
21. Prasad, A.D., Balu, A., Shah, H., Sarkar, S., Hegde, C., Krishnamurthy, A.: Nurbs-diff: a differentiable programming module for nurbs. Comput. Aided Des. **146**, 103199 (2022)
22. Requicha, A.A.: Mathematical models of rigid solids. Tech. Memo28, Production Automation Project. University of Rochester (1977)
23. Schollmeyer, A., Froehlich, B.: Efficient and anti-aliased trimming for render-ing large nurbs models. IEEE Trans. Visual Comput. Graphics **25**(3), 1489–1498 (2018)
24. Slyadnev, S., Malyshev, A., Turlapov, V.: Cad model inspection utility and proto-typing framework based on opencascade. In: Conference Paper: GraphiCon (2017)
25. Sutherland, I.E.: Sketch pad a man-machine graphical communication system. In: Proceedings of the SHARE Design Automation Workshop, pp. 6–329 (1964)
26. Ueda, K.: Multiplication as a general operation for splines. Curves and Surfaces in Geometric Design, pp. 475–482 (1994)
27. Versprille, K.J.: Computer-aided design applications of the rational b-spline approximation form. Syracuse University (1975)

Parallelized ADMM with General Objectives for Deep Learning

Yanqi Shi, Yu Tang, Hao Zheng, Zhigang Kan, and Linbo Qiao[✉]

National University of Defense Technology, Changsha 410073, China
linboqiao@nudt.edu.cn

Abstract. While considerable efforts have been dedicated to improving models that employ regularized functions, the direct solution of non-convex models using most stochastic gradient optimization algorithms poses significant challenges due to their inherent non-convex nature. The Alternating Direction Method of Multipliers (ADMM) has emerged as a promising approach for addressing both convex and non-convex problems, boasting rapid convergence and effective constraint-handling capabilities. However, ADMM has not yet achieved significant advancements in the realm of non-convex regularized deep learning, and the development of parallelized ADMM techniques for non-convex objectives remains lacking. To address these challenges, this paper proposes the implementation of ADMM as a solution for solving general (non-convex regularized) deep learning tasks and presents a comprehensive analysis of its convergence properties. Furthermore, a parallelized framework for ADMM is proposed to address the absence of such advancements for general objectives. Experimental results reveal the stable convergence properties of ADMM when applied to non-convex objectives, demonstrating superior performance compared to ADMM with convex objectives. Additionally, we evaluate the computational efficiency of our proposed parallelized framework for ADMM.

Keywords: General objectives · Stochastic ADMM · Parallel computing

1 Introduction

Optimizing complex deep neural networks (DNNs) has been a major challenge in the field of machine learning, and various optimization algorithms have been proposed to address this issue [20,27,33]. One such algorithm is the Alternating Direction Method of Multipliers (ADMM) [9], which has been utilized to solve convex objective functions in deep neural networks.

While ADMM has shown promising results for convex objective function in DNNs, applying it to general (non-convex) objective function presents significant challenges. In some application scenarios, ADMM with non-convex objectives outperforms ADMM with convex ones [30]. However, unlike convex problems,

Z. Tari et al. (Eds.): ICA3PP 2023, LNCS 14489, pp. 398–410, 2024.
https://doi.org/10.1007/978-981-97-0798-0_23

non-convex optimizations based on ADMM are much more difficult, and the behaviour of ADMM for non-convex problems in deep learning has been largely a mystery [41].

We wonder whether it is convergent for non-convex problems in deep learning and whether it is efficient in solving these problems via ADMM. This has motivated us to explore and respond to this question: how ADMM performs in dealing with non-convex objective deep learning tasks, and what conditions could drive this problem into convergence? Can ADMM solve non-convex problems in parallel? To simplify our terminology, we will henceforth refer to Parallelized ADMM with non-convex objectives for Deep Learning as *Pn-ADMM*.

In our experiment, we trained a simple deep neural network on Fashion MNIST [39] to compare the performance of non-convex and convex objectives. The neural network contained three computation layers and two activation layers. We used cross-entropy as the loss function, which we expressed as:

$$\min -\frac{1}{N} \sum_{i=1}^{N} \log\left(\frac{e^{h_j}}{\sum_{j=1}^{K} e^{h_j}}\right) + \frac{\mu}{2} \sum_{i=1}^{N} \|x_i - c_{y_i}\|_2^2,$$

where N is the number of data samples, K is the number of all classes, h_j is prediction for the $j-th$ class, c_{y_i} is the "center" of the class y_i and μ is a hyperparameter. These results are summarized in Table 1 and Table 2. In Table 1, "CEL" represents the "cross-entropy loss", "CL" represents the "center-loss", and "SUM" means the sum of the cross-entropy loss and the center-loss. Non-convex objectives achieve higher accuracy and lower loss than convex ones.

Table 1. The losses for non-convex and convex objectives on Fashion MNIST.

	non-convex			convex
	CEL	CL	SUM	(CEL)
training loss	0.44136	0.00591	**0.44727**	0.5182
test loss	0.50093	0.00588	**0.50681**	0.6531

Table 2. The accuracy for non-convex and convex objectives on Fashion MNIST.

	non-convex	convex
training accuracy	**0.87**	0.867
test accuracy	**0.847**	0.845

The comparisons we conducted regarding loss and accuracy indicate that non-convex objectives in deep learning produce more satisfactory results than convex objectives. Therefore, non-convex objectives could be more effective in the realm of deep learning.

We summarize our contributions as follows:

- We propose parallel and distributed ADMM, named Pn-ADMM, to solve deep neural network with general objective functions. We introduce Pn-ADMM, which includes SPn-ADMM and APn-ADMM, to enhance the efficiency of the algorithm. It can efficiently train neural networks in a multi-worker environment.
- The convergence analysis of proposed algorithms is presented under more realistic assumptions with both convex and non-convex objective functions.
- Numerical experiments are conducted on handcrafted neural networks, and the experiment results demonstrate the training efficiency of the parallel ADMM scheme. Additionally, these results also show that ADMM produces a small enough generation gap in deep learning.

2   Related Work

ADMM was first introduced in [9]. Its convergence was established in [8,10]. Since ADMM can decompose large problems with constraints into several small ones, it has been one of the most powerful optimization frameworks. It shows a multitude of well-performed properties in plenty of fields, such as machine learning [2], signal processing [30] and tensor decomposition [12] and modal decomposition [25].

Significant theoretical and practical work on ADMM has been done for convex cases [3,11,15,26,42]. Recent progress has also been made on non-convex problems. For instance, [21] used ADMM to solve the tensor decomposition problem and achieved better results. In 2016, [32] proposed a new method for training neural networks without using gradients by utilizing the Alternating Direction Method of Multipliers. [37] was the first to apply ADMM to deep learning for convex objective functions and achieved remarkable results. They proposed a framework named dlADMM. The concept of dlADMM also appeared for the first time in this paper. In our paper, we also use dlADMM to represent training deep neural networks via ADMM.

Recently, some work has been put forward to solve the non-convex ADMM problem [13,22]. It has been proved that ADMM could be a powerful tool to solve non-convex optimization problems [31]. [5] used ADMM to solve the group sparsity problems and developed an ADMM algorithm for finding representations for group sparsity using non-convex regularization. [1] applied ADMM for certain non-convex quadratic problems. In [30], Sun & Jiang considered solving a class of non-convex and non-smooth optimization problems via ADMM and got satisfying results in signal processing and machine learning. Besides, the non-convex stochastic ADMMs [17,43] produced a good performance in solving big data problems. What's more, ADMM shows its great characteristic in some other non-convex problems, such as matrix completion sensing [28,29,40] and tensor factorization [24]. Wang et al. proposed a method to solve the deep learning problem of non-convex objectives from another perspective [35,36].

Besides, since it is much simple to achieve ADMM in parallel [2], there is also much work related to the application of ADMM in parallel. In 2016, Chang et al. proposed Asynchronous Distributed ADMM (AD-ADMM) for large-scale optimization, including its algorithms and convergence analysis [4]. They also provided a linear convergence analysis for large-scale optimization in [4]. In [19], these authors used a dynamic scheduling strategy in the asynchronous ADMM algorithm for distributed optimization, and this strategy improved the convergence speed and communication efficiency of ADMM in large-scale clusters. In [14], they provided an ADMM-based framework for parallel deep learning training. In 2020, Wang et al. proposed a model parallelism for deep neural network based on gradient-free ADMM framework [34]. Compared with traditional data parallelism and model parallelism, ADMM in parallel is not only simple to implement but also does not suffer from massive communication in data parallelism.

3   Problem Formulation

Considering a deep neural network with L layers, we reformulate a typical neural network as the following minimization problem:

$$
\min_{W_l, b_l, x_l, o_l} R(W_l, b_l, x_l, o_l; y) + \sum_{l=1}^{L} \Omega(W_l),
$$

$$
\text{s.t.} \quad o_l = W_l x_{l-1} + b_l, x_l = f_l(o_l) \quad (l = 1, \cdots, L), \tag{1}
$$

where $R(W_l, b_l, x_l, o_l; y)$ is the loss function and $\Omega(W_l)$ is a regularization term, y is the labels of the training samples, $W_l, b_l, x_l, o_l \in \mathbb{R}^d$, $l \in [1, 2, \cdots, L]$, d is the number of feature dimensions, W_l and b_l denotes its weight and bias respectively, while x_l and o_l denotes the input and output of this layer. Here, $o_l = W_l \cdot x_l + b_l$, f_l represents the activation layer in deep neural networks.

The problem could be solved with the ADMM framework, and the Augmented Lagrangian function is formulated as:

$$
\mathcal{L}_\rho(W_l, b_l, x_l, o_l, \lambda) = R(W_l, b_l, x_l, o_l; y) + \sum_{l=1}^{L} \Omega(W_l)
$$

$$
+ \frac{\rho}{2} \| o_L - W_L x_{L-1} - b_L + \frac{\lambda}{\rho} \|_2^2
$$

$$
+ \frac{\nu}{2} \sum_{l=1}^{L-1} \left(\| o_l - W_l x_{l-1} - b_l \|_2^2 + \| x_l - f_l(o_l) \|_2^2 \right) \tag{2}
$$

where λ is the Lagrange multiplier, ρ is a penalty parameter and ν is a tuning parameter. Mathematically, Eq.(2) could be solved through an iterative way listed in Eq.(3).

$$\begin{cases} W_l^{t+1} \leftarrow \min\limits_{W_l,b_l,x_l,o_l} \mathcal{L}_\rho(W_l^t,b_l^t,x_l^t,o_l^t,\lambda^t), \\[2mm] b_l^{t+1} \leftarrow \min\limits_{W_l,b_l,x_l,o_l} \mathcal{L}_\rho(W_l^{t+1},b_l^t,x_l^t,o_l^t,\lambda^t), \\[2mm] o_l^{t+1} \leftarrow \min\limits_{W_l,b_l,x_l,o_l} \mathcal{L}_\rho(W_l^{t+1},b_l^{t+1},x_l^t,o_l^t,\lambda^t), \\[2mm] x_l^{t+1} \leftarrow \min\limits_{W_l,b_l,x_l,o_l} \mathcal{L}_\rho(W_l^{t+1},b_l^{t+1},x_l^t,o_l^{t+1},\lambda^t), \\[2mm] \lambda^{t+1} \leftarrow \lambda^t + \rho(o_L^{t+1} - W_L^{t+1}x_{L-1}^{t+1} - b_L^{t+1}). \end{cases} \tag{3}$$

3.1 Convergence Analysis

First, we give some mild assumptions as follows:

Assumption 1. *The gradient of the objective function R is H-Lipschitz continuous.*

Under Assumption 1, we will have two following properties:

Property 1. If $\rho > H$, $\mathcal{L}_\rho(W_l,b_l,x_l,o_l,\lambda)$ is lower bounded.

Property 2. If $\rho > 2H$ and $C_1 = \rho/2 - H/2 - H^2/\rho > 0$, there will exist $C_2 = \min(\rho/2, C_1)$ satisfying

$$\mathcal{L}_\rho(W_l^t,b_l^t,x_l^t,o_l^t,\lambda^t) - \mathcal{L}_\rho(W_l^{t+1},b_l^{t+1},x_l^{t+1},o_l^{t+1},\lambda^{t+1})$$
$$\geq C_2\|(o_L - W_L x_{L-1} - b_L)^{t+1} - (o_L - W_L x_{L-1} - b_L)^t\|_2^2,$$

where t is the current iteration.

Assumption 2. *We have the feasible set $\mathcal{F} := \{(W_l,b_l,x_l,o_l,\lambda) \in \mathbb{R}^{5 \times d}|o_L = W_L x_{L-1} + b_L\}$. The objective function $R(W_l,b_l,x_l,o_l,\lambda)$ is coercive over this set, that is to say $R(W_l,b_l,x_l,o_l,\lambda) \rightarrow \infty$ if $(W_l,b_l,x_l,o_l,\lambda) \in \mathcal{F}$ and $\|(W_l,b_l,x_l,o_l,\lambda)\| \rightarrow \infty$.*

If the sequence $(W_l,b_l,x_l,o_l,\lambda)$ in the feasible set is bounded, then Assumption 2 holds trivially for any continuous objective function.

We refer [38] as a more detailed explanation. Based on these assumptions, when the parameters update, $\mathcal{L}_\rho(W_l,b_l,x_l,o_l,\lambda)$ has a trend of decent and as $t \rightarrow \infty$, $\mathcal{L}_\rho(W_l,b_l,x_l,o_l,\lambda)$ converges in the unique limit point $(W_l^*,b_l^*,x_l^*,o_l^*)$. As far as the convergence rate, we have the following lemmas [16] firstly:

Lemma 1. *Suppose the sequence $\{W_l^t,b_l^t,x_l^t,o_l^t,\lambda^t\}$ is generated by Eq.(3). Then we will have the following inequality:*

$$\mathbb{E}\|\lambda^{k+1} - \lambda^k\|^2 \leq \zeta\|z_L^k - z_L^{k-1}\|^2 + \zeta_1\mathbb{E}\|z_L^{k+1} - z_L^k\|^2 + \frac{10\sigma^2}{M},$$

where $\zeta = \dfrac{5(L^2\eta^2 + (\phi_{max}^H)^2)}{\eta^2}$ and $\zeta_1 = \dfrac{5(\phi_{max}^H)^2}{\eta^2}$.

Lemma 1 provides the upper bound of $\mathbb{E}\|\lambda^{k+1} - \lambda^k\|^2$. then we define a useful sequence $\{\Psi^k\}_{k=1}^T$ as follows:

$$\psi^k = \mathbb{E}[\mathcal{L}_\rho(W_l^t, b_l^t, x_l^t, o_l^t, \lambda^t) + \frac{5(L^2\eta^2 + (\phi_{max}^H)^2)}{\rho\eta^2}\|z_L^k - z_L^{k-1}\|^2].$$

For notational simplicity, let $\tilde{L} = L + 1$, $\phi^H = (\phi_{min}^H)^2 + 20(\phi_{max}^H)^2$ and $\varphi = (\tilde{L} + 10L^2/(\rho)) - \rho$.

Lemma 2. *Suppose that the sequence $\{W_l^t, b_l^t, x_l^t, o_l^t, \lambda^t\}$ is generated by Eq.(3).* Let $\rho_* = \frac{\tilde{L} + \sqrt{40L^2 + \tilde{L}^2}}{2}$, $\triangle = (\phi_{min}^H)^2 + \frac{20(\phi_{max}^H)^2}{\rho}(\rho - (\tilde{L} + \frac{10L^2}{\rho}))$, and

$$\rho_0 = \frac{10\phi_{max}^H(\tilde{L}\phi_{max}^H + \sqrt{\tilde{L}^2(\phi_{max}^H)^2 + 2L^2\phi^H})}{\phi^H} \tag{4}$$

and suppose the parameters ρ and η, respectively, satisfy

$$\begin{cases} \eta \in (\frac{\phi_{min}^H - \sqrt{\triangle}}{\varphi}, \frac{\phi_{min}^H + \sqrt{\triangle}}{\varphi}), & \rho \in (\rho_0, \rho_*); \\ \eta \in (\frac{10(\phi_{max}^H)^2}{\rho\phi_{min}^H}, \frac{r-1}{\rho}], & \rho = \rho_*; \\ \eta \in (\frac{\phi_{min}^H - \sqrt{\triangle}}{\varphi}, \frac{r-1}{\rho}], & \rho \in (\rho_*, +\infty). \end{cases}$$

Then we have $\gamma = \frac{\phi_{min}^H}{\eta} + \frac{\rho}{2} - \frac{\tilde{L}}{2} - \frac{5(L^2\eta^2 + 2(\phi_{max}^H)^2)}{\rho\eta^2} > 0$, and it holds that

$$\frac{1}{T}\sum_{k=0}^{T-1}\mathbb{E}\|z_L^k - z_L^{k+1}\|^2 \le \frac{\Psi_0 - \Psi^*}{\gamma T} + \frac{(\rho + 20)\sigma^2}{2\gamma\rho M}, \tag{5}$$

where Ψ^* is a lower bound of sequence $\{\Psi^k\}_{k=1}^T$.

Given these conditions and let $\kappa_1 = 3(L^2 + \frac{(\phi_{max}^H)^2}{\eta^2})$, $\kappa_2 = \frac{\zeta}{\rho^2}$, $\kappa_3 = \rho^2\|B\|^2\|A\|^2$, and $\kappa_4 = \frac{\rho+20}{2\rho}$. Let

$$T = \frac{\max\{\kappa_1, \kappa_2, \kappa_3\}}{\epsilon\gamma}(\Psi^1 - \Psi^*),$$

where Ψ^* is a lower bound of the sequence $\{\Psi^k\}_{k=1}^T$. Thus if assuming the optimal solution of Eq.(2) is $(W_l^*, b_l^*, x_l^*, o_l^*)$, we will have:

$$\mathbb{E}[R(W_l, b_l, x_l, o_l; y) + \sum_{l=1}^L \Omega(W_l) - R(W_l^*, b_l^*, x_l^*, o_l^*; y)$$

$$- \sum_{l=1}^L \Omega(W_l^*) + \rho\|o_L - W_L x_{L-1} - b_L\|] = O(1/T),$$

which means its convergence rate is $O(1/T)$.

4   Paralleled Algorithm

In this section, we introduce the distributed training algorithms in ADMM, including Asynchronous Parallel ADMM (APn-ADMM) and Synchronous Parallel ADMM (SPn-ADMM).

Data parallelism and model parallelism [7] have their own advantages and disadvantages. Data parallelism splits the training data into several subsets and is much simpler to implement than model parallelism. However, it consumes a lot of memory and suffers from communication overheads. The communication between workers in model parallelism is significantly less than in data parallelism, but traditional model parallelism faces backpropagation issues and suffers from staleness, resulting in instability [6].

In our distributed algorithms, we adopt a "*master-worker*" method. The *master* is responsible for initializing parameters and managing them, while the *worker* performs all the calculations in ADMM.

4.1   SPn-ADMM

Synchronous Parallel ADMM is slightly different from traditional synchronous algorithms. In our Synchronous Parallel ADMM, we divide the training process into n blocks, denoted as $B_1, B_2, \cdots, B_n$. These blocks can be considered as workers in a distributed system, where all the workers calculate parts of updates for the parameters at the same time. At iteration k, worker B_m, where $m = 1, 2, \cdots, n$ sends the parameters from the last iteration to B_{m+1}. The parameters are not updated until all of these parameters are computed. Once all of these parameters are updated, they will be sent to the former block to continue the training process. This approach is inspired by [18]. To illustrate this process, we provide an example in Fig. 1.

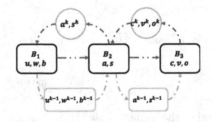

Fig. 1. An example of Synchronous Parallel ADMM.

There are three workers in this example, B_1, B_2, and B_3. These three workers are responsible for different parameter updates. At iteration $k \leq 1$, B_1 sends u^{k-1}, w^{k-1} and b^{k-1} to B_2. At the mean time, B_2 sends a^{k-1} and s^{k-1} to B_3. B_1, B_2, and B_3 update their parameters at the same time. After the updates, the relative parameters will be sent to their workers to continue the process.

4.2 APn-ADMM

SPn-ADMM performs better than vanilla ADMM, but it is still limited by the slowest workers, especially when the workers have different computation and communication delays. Therefore, we proposed Asynchronous Parallel ADMM (APn-ADMM), which will maintain a send queue Q_s, a receive queue Q_r. Q_r receives hyperparameters from the master. The send queue Q_s in APn-ADMM sends the updated parameters immediately, regardless of the source worker. This way, convergence can speed up significantly.

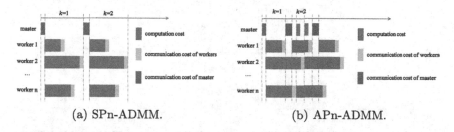

(a) SPn-ADMM.   (b) APn-ADMM.

Fig. 2. Illustration of Synchronous and Asynchronous Paralleled ADMM.

Figure 2 illustrates the detailed process of SPn-ADMM and APn-ADMM. In SPn-ADMM, the training time is limited by the longest computation time. The other workers need to wait for the most time-consuming worker. As a result, the master takes a lot of time waiting for the parameters in each iteration. However, in APn-ADMM, the parameters can be sent to the master as soon as they are updated, saving a lot of unnecessary time. The master is much busier in APn-ADMM than in SPn-ADMM.

5 Experiments

5.1 Results on MNIST and Fashion MNIST

Figures 3(a) and 3(b) summarize our experimental results of Pn-ADMM on MNIST [23]. The curve trends in these two figures indicate that the training and testing processes converge well with Pn-ADMM. Additionally, we observe that the curves have a similar form, which implies that the generation gap is small enough to ignore. Therefore, Pn-ADMM is a good method to reduce the generation gap resulting from the training process, especially when non-convex objective functions in deep learning are more prone to the generation gap. In our experiments, we achieved a maximum training accuracy of 0.966 and the best testing accuracy of 0.954, while the minimum losses for training and testing were 0.208 and 0.233, respectively.

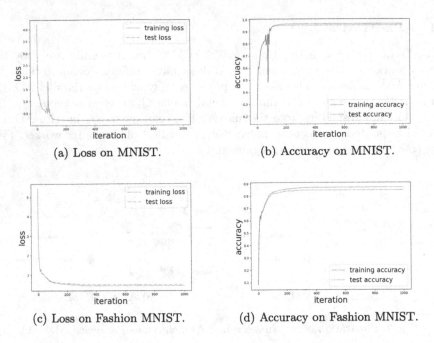

(a) Loss on MNIST.   (b) Accuracy on MNIST.

(c) Loss on Fashion MNIST.   (d) Accuracy on Fashion MNIST.

Fig. 3. Experiment on the MNIST and Fashion MNIST datasets.

However, in Fig. 3(a) and 3(b), we observe a jitter when the iteration number is around 150. This jitter caused a larger loss or a smaller accuracy. Our analysis suggests that this jitter is caused by the non-convex function and the use of subgradients of the objective in the parameter updates. During the process of solving non-convex objective functions with the training and testing sets, the derivative of the objective function will proceed into saddle points.

We conducted further experiments to test the performance of Pn-ADMM on Fashion MNIST, which are presented in Fig. 3(c) and 3(d), respectively. For Fashion MNIST, the best training accuracy is 0.870, and the best testing accuracy is 0.847. As for the loss, the training loss and the testing loss are 0.447 and 0.507, respectively.

We find that Pn-ADMM also achieves convergent results on Fashion MNIST. Compared with results on MNIST, it shows a smoother trend, especially when the iteration is around 150. In these two figures, there is still some jitter, but it is much smaller than that seen in Fig. 3(a) and 3(b). However, the generation gap is similarly small.

5.2   Paralleled Experiments

We conducted experiments on MNIST to compare the performance of vanilla ADMM, SPn-ADMM, and APn-ADMM. The experimental results are presented in Fig. 4(a) and 4(b), respectively. In each test, both SPn-ADMM and APn-ADMM had three workers. The results clearly demonstrate that SPn-ADMM and APn-ADMM are faster than vanilla ADMM.

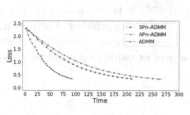
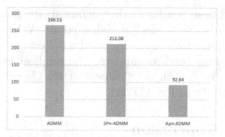

(a) Convergence speed.   (b) Training time within 50 iterations.

Fig. 4. The Comparison of vanilla ADMM, SPn-ADMM and APn-ADMM.

In our experiment, vanilla ADMM required 266.53 s to complete 50 iterations, which is approximately one minute longer than the time taken by SPn-ADMM. Therefore, not only can SPn-ADMM solve large-scale models that cannot be solved on a single machine through distributed algorithms, but it also saves valuable time.

Comparing APn-ADMM with vanilla ADMM and SPn-ADMM, we observe that it takes the least amount of time among these three methods, which is approximately one-third of the time taken by vanilla ADMM. This represents a significant speedup given the presence of 3 workers in the system. Furthermore, it is worth noting that APn-ADMM converges faster than both ADMM and SPn-ADMM.

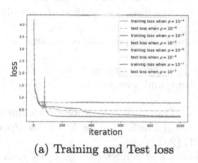
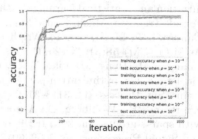

(a) Training and Test loss   (b) Training and Test accuracy

Fig. 5. Experiment with different values of ρ on MNIST.

5.3 Varying ρ

It is worth noting that values of ρ are crucial in vanilla ADMM. We plan to test the performance of different initial values of ρ on MNIST and explore whether ADMM is robust in deep learning for non-convex objectives. We will present our results in Fig. 5(a) and 5(b), respectively. For each experiment, we set ρ $10^{-4}, 10^{-5}, 10^{-6}$ and 10^{-7}.

These results further demonstrate that a large ρ leads to faster convergence but also to a less satisfying result. When $\rho = 10^{-4}$, it is the first to converge but has the largest loss and smallest accuracy. As ρ gets smaller, the loss and accuracy tend to improve. The smallest loss and best accuracy are achieved with $\rho = 10^{-6}$, with the sharpest jitter. Therefore, a proper value of ρ must be chosen carefully in Pn-ADMM.

6   Conclusion

In this paper, we have proposed and analyzed a solution to non-convex regularized deep learning tasks using ADMM. Our approach addresses the current limitations in ADMM for non-convex objectives and presents a comprehensive analysis of its convergence properties. We also introduce a parallelized framework to improve computational efficiency.

Our experimental results demonstrate the stable convergence properties of ADMM for non-convex objectives, outperforming ADMM with convex objectives. Additionally, we have evaluated the computational efficiency of our proposed parallelized framework. Our contributions provide a promising path forward for solving complex deep-learning tasks with non-convex objectives.

References

1. Ames, B.P., Hong, M.: Alternating direction method of multipliers for penalized zero-variance discriminant analysis. Comput. Optim. Appl. **64**(3), 725–754 (2016)
2. Boyd, S., Parikh, N., Chu, E., Peleato, B., Eckstein, J., et al.: Distributed optimization and statistical learning via the alternating direction method of multipliers. Found. Trends® Mach. Learn. **3**(1), 1–122 (2011)
3. Candès, E.J., Li, X., Ma, Y., Wright, J.: Robust principal component analysis? J. ACM (JACM) **58**(3), 11 (2011)
4. Chang, T.H., Hong, M., Liao, W.C., Wang, X.: Asynchronous distributed ADMM for large-scale optimization-part I: algorithm and convergence analysis. IEEE Trans. Sig. Process. **64**, 3118–3130 (2015)
5. Chartrand, R., Wohlberg, B.: A nonconvex ADMM algorithm for group sparsity with sparse groups. In: 2013 IEEE International Conference on Acoustics, Speech and Signal Processing, pp. 6009–6013. IEEE (2013)
6. Chen, C.C., Yang, C.L., Cheng, H.Y.: Efficient and robust parallel DNN training through model parallelism on multi-GPU platform. arXiv abs/1809.02839 (2018)
7. Dean, J., et al.: Large scale distributed deep networks. In: NIPS (2012)
8. Fortin, M., Glowinski, R.: Augmented Lagrangian methods: applications to the numerical solution of boundary-value problems (1983)
9. Gabay, D., Mercier, B.: A dual algorithm for the solution of nonlinear variational problems via finite element approximation. Comput. Math. Appl. **2**(1), 17–40 (1976)
10. Glowinski, R., Tallec, P.L.: Augmented Lagrangian and operator-splitting methods in nonlinear mechanics (1987)
11. Goldfarb, D., Ma, S., Scheinberg, K.: Fast alternating linearization methods for minimizing the sum of two convex functions. Math. Program. **141**(1–2), 349–382 (2013)

12. Goldfarb, D., Qin, Z.: Robust low-rank tensor recovery: models and algorithms. SIAM J. Matrix Anal. Appl. **35**(1), 225–253 (2014)
13. Guan, L., et al.: An efficient parallel and distributed solution to nonconvex penalized linear SVMs. Front. Inf. Technol. Electron. Eng. **21**(4), 17 (2020)
14. Guan, L., Yang, Z., Li, D., Lu, X.: pdlADMM: an ADMM-based framework for parallel deep learning training with efficiency. Neurocomputing **435**, 264–272 (2021). https://doi.org/10.1016/j.neucom.2020.09.029
15. He, B., Yuan, X.: On the o(1/n) convergence rate of the Douglas-Rachford alternating direction method. SIAM J. Numer. Anal. **50**(2), 700–709 (2012)
16. Huang, F., Chen, S.: Mini-batch stochastic ADMMs for nonconvex nonsmooth optimization. arXiv preprint arXiv:1802.03284 (2018)
17. Huang, F., Chen, S., Lu, Z.: Stochastic alternating direction method of multipliers with variance reduction for nonconvex optimization. arXiv preprint arXiv:1610.02758 (2016)
18. Huo, Z., Gu, B., Yang, Q., Huang, H.: Decoupled parallel backpropagation with convergence guarantee. arXiv abs/1804.10574 (2018)
19. Jiang, S., Mei Lei, Y., Wang, S., Wang, D.: An asynchronous ADMM algorithm for distributed optimization with dynamic scheduling strategy. 2019 IEEE 21st International Conference on High Performance Computing and Communications; IEEE 17th International Conference on Smart City; IEEE 5th International Conference on Data Science and Systems (HPCC/SmartCity/DSS), pp. 1–8 (2019)
20. Kingma, D., Ba, J.: Adam: a method for stochastic optimization. Comput. Sci. (2014)
21. Kolda, T.G., Bader, B.W.: Tensor decompositions and applications. SIAM Rev. **51**(3), 455–500 (2009)
22. Lan, Q., Qiao, L.B., Wang, Y.J.: Stochastic extra-gradient based alternating direction methods for graph-guided regularized minimization. Front. Inf. Technol. Electron. Eng. (006), 019 (2018)
23. LeCun, Y., Bottou, L., Bengio, Y., Haffner, P., et al.: Gradient-based learning applied to document recognition. Proc. IEEE **86**(11), 2278–2324 (1998)
24. Liavas, A.P., Sidiropoulos, N.D.: Parallel algorithms for constrained tensor factorization via alternating direction method of multipliers. IEEE Trans. Sig. Process. **63**(20), 5450–5463 (2015)
25. Masuyama, Y., Kusano, T., Yatabe, K., Oikawa, Y.: Modal decomposition of musical instrument sound via alternating direction method of multipliers. In: 2018 IEEE International Conference on Acoustics, Speech and Signal Processing (ICASSP), pp. 631–635. IEEE (2018)
26. Monteiro, R.D., Svaiter, B.F.: Iteration-complexity of block-decomposition algorithms and the alternating minimization augmented Lagrangian method. Manuscript, School of Industrial and Systems Engineering, Georgia Institute of Technology, Atlanta, GA, pp. 30332–0205 (2010)
27. Robbins, H.E.: A stochastic approximation method. Ann. Math. Stat. **22**, 400–407 (1951)
28. Shen, Y., Wen, Z., Zhang, Y.: Augmented Lagrangian alternating direction method for matrix separation based on low-rank factorization. Optim. Methods Softw. **29**(2), 239–263 (2014)
29. Sun, D.L., Fevotte, C.: Alternating direction method of multipliers for non-negative matrix factorization with the beta-divergence. In: 2014 IEEE International Conference on Acoustics, Speech and Signal Processing (ICASSP), pp. 6201–6205. IEEE (2014)

30. Sun, T., Jiang, H., Cheng, L., Zhu, W.: Iteratively linearized reweighted alternating direction method of multipliers for a class of nonconvex problems. IEEE Trans. Sig. Process. **66**(20), 5380–5391 (2018)

31. Suzuki, T.: Dual averaging and proximal gradient descent for online alternating direction multiplier method. In: International Conference on Machine Learning, pp. 392–400 (2013)

32. Taylor, G., Burmeister, R., Xu, Z., Singh, B., Patel, A., Goldstein, T.: Training neural networks without gradients: A scalable ADMM approach. In: International Conference on Machine Learning, pp. 2722–2731 (2016)

33. Tieleman, T., Hinton, G.: Lecture 6.5-rmsprop, coursera: neural networks for machine learning. Technical report, University of Toronto (2012)

34. Wang, J., Chai, Z., Cheng, Y., Zhao, L.: Toward model parallelism for deep neural network based on gradient-free ADMM framework. In: Proceedings - IEEE International Conference on Data Mining, ICDM 2020, November, pp. 591–600 (2020). https://doi.org/10.1109/ICDM50108.2020.00068

35. Wang, J., Li, H., Zhao, L.: Accelerated gradient-free neural network training by multi-convex alternating optimization. Neurocomputing **487**, 130–143 (2022). https://doi.org/10.1016/j.neucom.2022.02.039

36. Wang, J., Zhao, L.: Convergence and applications of ADMM on the multi-convex problems. In: Gama, J., Li, T., Yu, Y., Chen, E., Zheng, Y., Teng, F. (eds.) PAKDD 2022. LNCS (LNAI), vol. 13281, pp. 30–43. Springer, Cham (2022). https://doi.org/10.1007/978-3-031-05936-0_3

37. Wang, J., Yu, F., Chen, X., Zhao, L.: ADMM for efficient deep learning with global convergence. arXiv preprint arXiv:1905.13611 (2019)

38. Wang, Y., Yin, W., Zeng, J.: Global convergence of ADMM in nonconvex nonsmooth optimization. J. Sci. Comput. **78**(1), 29–63 (2019)

39. Xiao, H., Rasul, K., Vollgraf, R.: Fashion-mNIST: a novel image dataset for benchmarking machine learning algorithms. arXiv preprint arXiv:1708.07747 (2017)

40. Xu, Y., Yin, W., Wen, Z., Zhang, Y.: An alternating direction algorithm for matrix completion with nonnegative factors. Front. Math. China **7**(2), 365–384 (2012)

41. Xu, Z., De, S., Figueiredo, M., Studer, C., Goldstein, T.: An empirical study of ADMM for nonconvex problems. arXiv preprint arXiv:1612.03349 (2016)

42. Yang, J., Zhang, Y.: Alternating direction algorithms for \ell_1-problems in compressive sensing. SIAM J. Sci. Comput. **33**(1), 250–278 (2011)

43. Zheng, S., Kwok, J.T.: Stochastic variance-reduced ADMM. arXiv preprint arXiv:1604.07070 (2016)

Multi-view Neighbor-Enriched Contrastive Learning Framework for Bundle Recommendation

Yuhang Chen[1], Sheng Liang[1], and Songwen Pei[1,2,3]($\boxtimes$)

[1] School of Optical-Electrical and Computer Engineering, University of Shanghai for Science and Technology, Shanghai 200093, China
swpei@usst.edu.cn
[2] State Key Laboratory of Computer Architecture, Institute of Computing Technology, Chinese Academy of Sciences, Beijing 100190, China
[3] Engineering Research Center of Software/Hardware Co-design Technology and Application, Ministry of Education (East China Normal University), Shanghai 200062, China

Abstract. Bundle recommendation aims to recommend a group of items with a similar theme to users. The previous methods devoted to alleviating the data sparsity problem. However, they either modeled the intuitive interaction between users and items (bundles) or randomly sampled the negative samples during the training process. It is far from enough to learn the user and bundle embeddings because of insufficient modeling of collaborative information. We propose a Multi-view Neighbor-enriched Contrastive learning framework for Bundle Recommendation (MNCBR). MNCBR learns representations of users and bundles from two separate views (*i.e.* item and bundle view). Meanwhile, different contrastive learning strategies are applied to each view respectively. Specifically, the item-view contrastive mechanism jointly learns the high-order relations of users and bundles, and obtains the global preferences of users. The bundle-view contrastive mechanism explores the collaborative information via structural neighbors on the interaction graph. Extensive experiments on two public datasets show the proposed MNCBR outperforms the state-of-the-art methods.

Keywords: Bundle Recommendation · Contrastive Learning · Graph Neural Network

1 Introduction

Bundle recommendation (BR) differs from the traditional recommendation that recommends individual item. BR aims to recommend to users a package of multiple items with a single theme. BR [1,3,5] involves multiple types of nodes and relations, which naturally leads to complex structures. Recent works [5,21] focus on leveraging graph neural networks (GNN) to model these interactions, design different message passing mechanisms to learn node embeddings. These graph-based methods need sufficient training data to stimulate their advantages. However, the interaction data between users, items and bundles is severely sparse in BR.

Contrastive learning [20] has grown rapidly in recent years. It can alleviate the data sparsity problem by extracting self-supervised signals from the data itself without using additional labels [15]. This technique maximizes the consistency of positive pairs while minimizing the consistency of negative pairs by extracting positive and negative samples from the data. Some recent approaches [6] generate positive and negative pairs by removing nodes and edges from the graphs. These supervised signals are all derived from low-level observed signals, which may not fully capture the intricacies of the underlying graph structures and could limit the model's capacity to generalize effectively.

To address the above issues, we propose a Multi-view Neighbor-enriched Contrastive learning framework for Bundle Recommendation (MNCBR). In order to learn self-supervised signals, we construct two graph views with different contrastive learning mechanism to learn the feature representations. In the item view, MNCBR performs information propagation between user-item graphs and bundle-item graphs with conventional contrastive learning paradigm. In the bundle view, MNCBR applies a structural neighbor contrastive mechanism for the problem of sparse user and bundle interactions.

The main contributions are summarized as below:

- We combine the user-item and bundle-item graphs, design a message propagation method to learn higher-order information and applies contrastive learning to alleviate the data sparsity.
- We use neighbor contrastive mechanism to enrich the collaborative information to aggregate user/bundle embeddings. F
- We conduct extensive experiments on two real-world datasets to evaluate the performance of MNCBR and the effectiveness of each components.

2　Related Work

2.1　Graph-Based Bundle Recommendation

Compared with traditional recommendations, there are multiple relationships should be considered in BR. Previous works attempt to fuse different types of user and item interactions using matrix factorization [10]. Subsequent research [2] focuses on the relationship between bundles and items, and incorporates these relations into joint training. With the rapid development of GNN [9,14] and the remarkable ability to process graph data, more works integrate GNNs into their models. BGCN [3] combines the user-item-bundle relationship into a heterogeneous graph to learn the representation of users and bundles. MIDGN [21] decouples user interests into global and local views by classifying each item in the bundle. CrossCBR [12] uses cross-view contrastive learning, the representations under the two views are mutually enhanced. However, due to the sparse interaction between users and bundles, conventional data augmentation methods (*i.e.* node/edge dropout) will make it lose more raw structural information.

2.2 Contrastive Learning in Recommendation

Contrastive learning was initially active in the field of computer vision [7,8]. Recently, it makes progress in the exploration of recommendations [19,22], such as conventional graph-based recommendations [16], hypergraph-based recommendations [17,18], *etc.* Its key aim is to construct positive and negative sample pairs to facilitate representation learning. S^3-Rec [22] uses multiple pre-training to enhance the representation ability of data. It deploys auxiliary task of self-supervised learning to improve the performance of sequence recommendation through random masking of node attributes in the sequence. DHCN [18] uses a dual-channel hypergraph convolution and contrastive learning to process the two views to extract feature embeddings separately. In this work, we construct self-supervised signals from two perspectives, and jointly captures the collaborative relationship under the two view.

3 Preliminaries

3.1 Problem Formulation

We use $\mathcal{U} = \{u_1, u_2, \ldots, u_M\}$ as the user set, $\mathcal{B} = \{b_1, b_2, \ldots, b_N\}$ as the bundle set, and $\mathcal{I} = \{i_1, i_2, \ldots, i_O\}$ as the item set, where M, N and O denote the respective numbers of users, bundles, and items respectively. At the same time, we define user-bundle interactions, user-item interactions and bundle-item interactions as $X_{M \times N} = \{x_{ub} | u \in \mathcal{U}, b \in \mathcal{B}\}$, $Y_{M \times O} = \{y_{ui} | u \in \mathcal{U}, i \in \mathcal{I}\}$ and $Z_{N \times O} = \{z_{bi} | b \in \mathcal{B}, i \in \mathcal{I}\}$. Taking X as an example, $x_{ub} = 1$ means that user $u \in \mathcal{U}$ interacts with bundle $b \in \mathcal{B}$, otherwise it is 0, and Y and Z satisfy the same definition.

According to the above definition, the problem is: through the historical interaction sets X, Y and Z, predict the possibility that user u may interact with bundle b that never interacted before.

3.2 Graph Encoder Layer

Graph-based methods learn user and item representations by applying the propagation and readout function on the graph. In MNCBR, we use LightGCN as the kernel of graph encoder. Following LightGCN, we also remove the spin of the nodes on the graph and the nonlinear layer in the information propagation process. The propagation function as follows:

$$z_u^{(k)} = GraphEncoder(u, c) = \sum_{c \in N_u} \frac{1}{\sqrt{|\mathcal{N}_u||\mathcal{N}_c|}} z_c^{(k-1)} \tag{1}$$

$$z_c^{(k)} = GraphEncoder(c, u) = \sum_{u \in N_c} \frac{1}{\sqrt{|\mathcal{N}_c| \, |\mathcal{N}_u|}} z_u^{(k-1)} \qquad (2)$$

where $c = \{b, i\}$ is a uniform placeholder for bundle symbol b and item symbol i.

4   Model

In this section, three parts of the MNCBR will be presented: 1) *Item-view constrastive module*. It obtains user and bundle features from user-item graphs and bundle-item graphs, and then applies a contrastive learning; 2) *Bundle-view contrastive module*. It directly learns user preferences from user-bundle graphs, and then applies a constrastive learning with neighbors. 3) *Joint Learning*. It optimizes the prediction loss, item-view and bundle-view contrastive loss to optimize the model. In the following section, we present the details of MNCBR (Fig. 1).

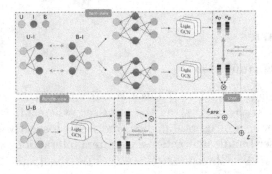

Fig. 1. An overview of the proposed framework.

4.1   Item-View Contrastive Module

Item-View Representation Learning. This module uses the user-item graphs to encode and learn representations of the user and item, and then obtains the bundle representations by aggregating item representations. In order to accelerate representation learning, we use LightGCN as the way of information propagation. The representations of the user and item are obtained by performing average pooling, and the obtained user representations are item-view user representations. The information propagation of layer k can be expressed as:

$$e_u^{I(k)} = GraphEncoder(u, i), \; e_i^{I(k)} = GraphEncoder(i, u) \qquad (3)$$

where $e_u^{I(k)}$ and $e_i^{I(k)}$ are the representations of user u and item i through k-th layer's information propagation; $e_u^{I(0)}$ and $e_i^{I(0)}$ are randomly initialized.

We encode the information in the bundle-item interaction graphs in the same way, and the obtained item representations are embeddings containing bundle semantic information. After the k-th layer's information propagation, the following is obtained:

$$e_{i'}^{I(k)} = GraphEncoder\,(i', b) \tag{4}$$

where $e_{i'}^{I(k)}$ is the representation of item i' propagated through k-th layer's information under the guidance of item-bundle graph.

Then, add the two item representations guided by different interaction graphs as the final item representation. Finally, the results of k-th layer's information propagation are aggregated to obtain the final representation under the item view:

$$e_{i*}^{I(k)} = e_{i}^{I(k)} + e_{i'}^{I(k)} \tag{5}$$

$$c_{u}^{\mathcal{G}_1} = aggregate(e_{u}^{I(k)}),\, e_{i}^{\mathcal{G}_1} = aggregate(e_{i*}^{I(k)}) \tag{6}$$

Among them, $\mathcal{G}_1$ represents user-item and bundle-item graphs are original; $e_{u}^{\mathcal{G}_1}$, $e_{i}^{\mathcal{G}_1}$ and $e_{b}^{\mathcal{G}_1}$ are the representations of users, items and bundles in the item view. We use simple sum function as aggregate function, and $\mathcal{N}_b$ represents the set of items contained in a bundle ðÍŚŘ.

$$e_{b}^{\mathcal{G}_1} = \frac{1}{|\mathcal{N}_b|} \sum_{i \in \mathcal{N}_b} e_{i}^{\mathcal{G}_1} \tag{7}$$

Item-View Contrastive Learning. We destroy the original graph by randomly removing edges to get a new graph $\mathcal{G}_2$. With the same learning process in $\mathcal{G}_1$, we can obtain representations of the user and bundle as $e_{u}^{\mathcal{G}_2}$ and $e_{u}^{\mathcal{G}_2}$. Then, the representations of two graphs are optimized by contrastive learning, enforcing that the encoded embeddings guided by different perspectives of graphs are consistent with each other and can be distinguished from embeddings of other nodes. We use the commonly InfoNCE loss which makes the same user and bundle in different views as similar as possible. The loss equation is as follows:

$$\mathcal{L}_{U}^{I} = \frac{1}{|\mathcal{U}|} \sum_{u \in \mathcal{U}} -\log \frac{\exp\left(\theta(e_{u}^{\mathcal{G}_1}, e_{u}^{\mathcal{G}_2})/\tau\right)}{\Sigma_{u^- \in \mathcal{U}} \exp\left(\theta(e_{u}^{\mathcal{G}_1}, e_{u^-}^{\mathcal{G}_2})/\tau\right)} \tag{8}$$

$$\mathcal{L}_{B}^{I} = \frac{1}{|\mathcal{B}|} \sum_{b \in \mathcal{B}} -\log \frac{\exp\left(\theta(e_{b}^{\mathcal{G}_1}, e_{b}^{\mathcal{G}_2})/\tau\right)}{\Sigma_{b^- \in \mathcal{B}} \exp\left(\theta(e_{b}^{\mathcal{G}_1}, e_{b^-}^{\mathcal{G}_2})/\tau\right)} \tag{9}$$

Among them, $\mathcal{L}_{U}^{I}$ and $\mathcal{L}_{B}^{I}$ represent contrastive learning for users and bundles; $\theta(\cdot)$ is the cosine similarity function; τ is the temperature parameter. Averaging the contrastive losses of the two views obtains the final item-view contrastive loss $\mathcal{L}^{I}$:

$$\mathcal{L}^{I} = \frac{1}{2}(\mathcal{L}_{U}^{I} + \mathcal{L}_{B}^{I}) \tag{10}$$

4.2   Bundle-View Contrastive Module

Bundle-View Representation Learning. In order to learn feature representations of the user and bundle in the bundle view, we construct the interaction between the user and bundle into a bipartite graph. The message propagation process is as follows:

$$e_u^{B(k)} = GraphEncoder(u, b), e_b^{B(k)} = GraphEncoder(b, u) \qquad (11)$$

Among them, $e_u^{B(k)}$ and $e_b^{B(k)}$ represent users and bundles that propagate through k-th layer message propagation. Likewise, spins and non-linear layers are removed.

Finally, the final expression after k-th layer message propagation is:

$$e_u^B = aggregate(e_u^{B(k)}), e_b^B = aggregate(e_b^{B(k)}) \qquad (12)$$

where e_u^B and e_b^B are the bundle-view user and bundle representations respectively, $e_b^{B(0)}$ are randomly initialized.

Bundle-View Contrastive Learning. The contrastive learning based on graph enhancement focuses on nodes and edges to model the implicit relationships between nodes. However, using structural dropout for graph enhancement may lead to loss of raw information. This module applies the structural contrastive loss inspired by NCL [11]. By this way, the distance between positive pairs can be minimized:

$$\mathcal{L}_U^{\mathcal{B}} = \sum_{u \in \mathcal{U}} -\log \frac{\exp\left(\left(e_u^{(k)} \cdot e_u^{(0)}/\tau\right)\right)}{\Sigma_{u- \in \mathcal{U}} \exp\left(\left(e_u^{(k)} \cdot e_{u-}^{(0)}/\tau\right)\right)} \qquad (13)$$

$$\mathcal{L}_B^{\mathcal{B}} = \sum_{b \in \mathcal{B}} -\log \frac{\exp\left(\left(e_b^{(k)} \cdot e_b^{(0)}/\tau\right)\right)}{\Sigma_{b- \in \mathcal{B}} \exp\left(\left(e_b^{(k)} \cdot e_{b-}^{(0)}/\tau\right)\right)} \qquad (14)$$

Among them, k is an even number; τ is the temperature parameter. Finally, the average of the above two losses is used as the bundle-view contrastive loss:

$$\mathcal{L}^B = \frac{1}{2}(\mathcal{L}_U^{\mathcal{B}} + \mathcal{L}_B^{\mathcal{B}}) \qquad (15)$$

4.3   Prediction and Optimization

After getting the final representation, first combine the inner product of e_u^I and e_b^I under the item view and the inner product of e_u^B and e_b^B under the bundle view as the final prediction.

$$y_{u,b} = e_u^{I*\top} e_b^{I*} + e_u^{B*\top} e_b^{B*} \qquad (16)$$

We use the conventional Bayesian Personalized Ranking (BPR) loss [13] as the main loss:

$$\mathcal{L}^{BPR} = \sum_{(u,b,b')\in\mathcal{Q}} -\ln \sigma(y_{u,b} - y_{u,b'}) \tag{17}$$

where the set $\mathcal{Q} = \{(u, b, b')|u \in \mathcal{U}, b, b' \in \mathcal{B}, x_{ub} = 1, x_{ub'} = 0\}$, $\sigma(\cdot)$ is the sigmoid function.

Then, the BPR loss $\mathcal{L}^{BPR}$, the item-view contrastive loss $\mathcal{L}^I$ and the bundle-view contrastive loss $\mathcal{L}^B$ are summed to obtain the final optimization objective:

$$\mathcal{L} = \mathcal{L}^{BPR} + \alpha\mathcal{L}^I + \beta\mathcal{L}^B \tag{18}$$

where α and β are hyper-parameters. We optimize these hyper-parameters jointly throughout the training process.

5 Experiments

5.1 Experimental Settings

Table 1. Dataset Statistics.

Dataset	U	B	I	U-B	B-I	U-I
Youshu	8039	4771	32770	51337	176667	138515
NetEase	18528	22864	123628	302303	1778838	1128065

To evaluate the performance of MNCBR, we use two public datasets: (1) Youshu [4]: a dataset provided by Youshu.com, which contains users' individual preferences for a single book and multiple book packages. (2) NetEase [2]: a NetEase Cloud music dataset provided by NetEase, which recommends a single music and a list of at least 15 music tracks to each user every day. The data of the two datasets are shown in Table 1. We use R@K and NDCG@K as evaluation metrics to judge the merits of ranked lists. R@K indicates the proportion of test bundles in the top-k ranked list. NDCG@K assigns higher scores to click in higher positions in the top-k ranked list.

For all methods, the embedding dimension is set to 64, the model is optimized using the Adam optimizer with the learning rate ranged from {0.001, 0.002, 0.003}, and the batch size is set to 2048. For our method, the number of layers is selected from {1, 2, 3}, the temperature parameter τ is selected from {0.1, 0.15, 0.2, 0.25, 0.3}. The trade-off parameters {α, β} are tuned with the range of {0.01, 0.025, 0.05, 0.1, 0.25, 0.5, 0.75, 1}. Following the previous works, we set the message dropout within {0, 0.1, 0.2, 0.3}.

5.2   Baselines

The proposed MNCBR model is compared with other benchmark experiments to verify the superiority of MNCBR.

- MFBPR [13] optimizes matrix decomposition by BPR loss.
- DAM [4] uses a multitasking framework for capturing cooperative signals between bundles using an attention mechanism.
- LightGCN [9] is a recommendation model based on GNN and collaborative filtering utilizing a lightweight graph learning kernel.
- BundleNet [5] uses GCN and multi-task learning to learn the three-part graph of user-bundle-goods.
- BGCN [3] decouples the relationship of the three into a bundle view and a commodity view, and uses GCN to learn the representation of both.
- MIDGN [21] captures the diversity of goods in a bundle and learns user preferences at a more granular level.
- CrossCBR [12] models collaborative associations between two views via cross-view contrastive learning.

Table 2. Performance comparisons on two real-world datasets.

Dataset	Models	R@20	NDCG@20	R@40	NDCG@40
Youshu	MFBPR	0.1959	0.1117	0.2735	0.1320
	DAM	0.2082	0.1198	0.2890	0.1418
	LightGCN	0.2286	0.1344	0.3190	0.1592
	BundleNet	0.1895	0.1125	0.2675	0.1335
	BGCN	0.2347	0.1345	0.3248	0.1593
	MIDGN	0.2682	0.1527	0.3712	0.1808
	CrossCBR	0.2813	0.1668	0.3785	0.1938
	MNCBR	0.2882	0.1703	0.3893	0.1983
	%Improv	2.45	2.10	2.85	2.32
NetEase	MFBPR	0.0355	0.0181	0.0600	0.0246
	DAM	0.0411	0.0210	0.0690	0.0281
	LightGCN	0.0496	0.0254	0.0795	0.0334
	BundleNet	0.0391	0.0201	0.0690	0.0281
	BGCN	0.0491	0.0258	0.0829	0.0346
	MIDGN	0.0678	0.0343	0.1085	0.0451
	CrossCBR	0.0842	0.0457	0.1264	0.0569
	MNCBR	0.0860	0.0464	0.1306	0.0582
	%Improv	2.14	1.53	3.32	2.28

5.3   Performance Comparison

As shown in Table 2, first comparing the recommendation effects of all benchmark models with MNCBR on 2 datasets, MNCBR performs better than other benchmark models. Among baseline models, we note that the metrics of BGCN and MIDGN achieve superior metrics compared to the rest of the benchmark models. This can be attributed to the incorporation of negative sample sampling in training and intention-oriented contrastive learning modules, respectively. CrossCBR performs best among baselines due to the use of cross-view contrastive learning, enabling each view to extract cooperative information from the other and they enhanced mutually. MNCBR demonstrates the effectiveness of incorporating the neighbors and multi-view contrastive mechanism, achieving significant improvement over all multi-view learning benchmarks for bundle recommendation.

5.4   Ablation Study

We continue to explore three ablation models for MNCBR: MNCBR-B represents removing bundle-view contrastive mechanism. MNCBR-I represents removing item-view contrastive mechanism. MNCBR-BI represents removes the both contrastive mechanism and uses only BPR loss optimization. From the data in Fig. 2, it is observed that:

Removing the contrastive learning module for both views make MNCBR significantly less effective (MNCBR-BI), which implies that contrastive learning plays a key role in capturing self-supervised signals and mitigating the data sparsity problem.

The relatively poor effect of MNCBR-B compared to MNCBR is due to deficiently to obtain supervised signals from the user-bundle view, the result shows the effectiveness of neighbors.

The performance of MNCBR-I is suboptimal compared to the full MNCBR. This implies the effectiveness of items as a collaborative signal for constructing users and bundles, and thus the global semantic information learned is robust.

5.5   Hyper-parameters Analysis

The Effect of the Parameter α. The trade-off parameter α controls the proportion of item-view contrastive learning during the training process in Eq. 18. As shown in Fig. 3, the performance of MNCBR is worst when α is set to 0.01, which illustrates the critical role of the item-view contrastive learning module. In addition, when α is set to 1, the performance drops sharply, which is due to the overly global feature being weighted too heavily and ignoring the fine grained features. From the results, the best performance of MNCBR is obtained when $\alpha = 0.5$ on Youshu and $\alpha = 0.25$ on NetEase.

The Effect of the Parameter β. The trade-off parameter β controls the bundle-view contrastive mechanism. From the results in Fig. 4, MNCBR obtains better performance when β was increased from 0 to 0.1 on Youshu and when β was increased from 0 to 0.5 on NetEase. This demonstrates the important effectiveness of neighbor guided contrastive learning. In addition, the effectiveness of

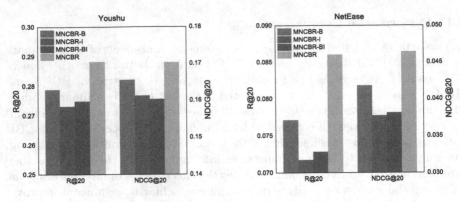

Fig. 2. Ablation study on two datasets.

the model decreases significantly when β is set to 1. It indicates that over focus on the collaborative signals of users and bundles reduces the performance of the model.

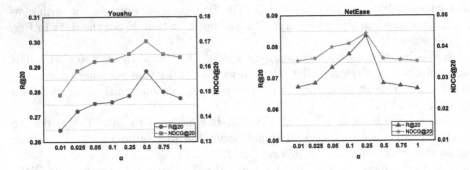

Fig. 3. Sensitive of parameter α

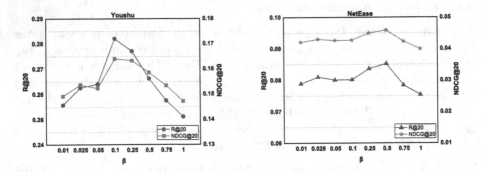

Fig. 4. Sensitive of parameter β

6   Conclusion

In this work, we propose a novel model named MNCBR. In contrast to previous approaches, MNCBR designs two views (*i.e.*, item and bundle view) and uses different contrastive learning to obtain user and bundle representations under these two views. The item-view contrastive module propagates information jointly guided by the user-item and bundle-item graphs and performs the traditional contrastive learning paradigm. And the bundle-view contrastive module leverages neighbors between users and bundles to regularize embeddings. Comprehensive experiments demonstrate that our method performs significantly better than the baseline on both datasets, indicating its superior performance.

Acknowledgments. The authors would like to thank anonymous reviewers for their invaluable comments. This work was partially funded by the National Natural Science Foundation of China under Grant No. 61975124, Shanghai Natural Science Foundation (20ZR1438500), State Key Laboratory of Computer Architecture (ICT, CAS) under Grant No. CARCHA202111, and Engineering Research Center of Software/Hardware Co-design Technology and Application, Ministry of Education, East China Normal University under Grant No. OP202202. Any opinions, findings and conclusions expressed in this paper are those of the authors and do not necessarily reflect the views of the sponsors.

References

1. Bai, J., et al.: Personalized bundle list recommendation. In: The World Wide Web Conference, pp. 60–71 (2019)
2. Cao, D., Nie, L., He, X., Wei, X., Zhu, S., Chua, T.S.: Embedding factorization models for jointly recommending items and user generated lists. In: Proceedings of the 40th International ACM SIGIR Conference on Research and Development in Information Retrieval, pp. 585–594 (2017)
3. Chang, J., Gao, C., He, X., Jin, D., Li, Y.: Bundle recommendation with graph convolutional networks. In: Proceedings of the 43rd international ACM SIGIR conference on Research and development in Information Retrieval, pp. 1673–1676 (2020)
4. Chen, L., Liu, Y., He, X., Gao, L., Zheng, Z.: Matching user with item set: Collaborative bundle recommendation with deep attention network. In: IJCAI, pp. 2095–2101 (2019)
5. Deng, Q., Wang, K., Zhao, M., Zou, Z., Wu, R., Tao, J., Fan, C., Chen, L.: Personalized bundle recommendation in online games. In: Proceedings of the 29th ACM International Conference on Information & Knowledge Management, pp. 2381–2388 (2020)
6. Hassani, K., Khasahmadi, A.H.: Contrastive multi-view representation learning on graphs. In: International Conference on Machine Learning, pp. 4116–4126. PMLR (2020)
7. He, K., Chen, X., Xie, S., Li, Y., Dollár, P., Girshick, R.: Masked autoencoders are scalable vision learners. In: Proceedings of the IEEE/CVF Conference on Computer Vision and Pattern Recognition, pp. 16000–16009 (2022)

8. He, K., Fan, H., Wu, Y., Xie, S., Girshick, R.: Momentum contrast for unsupervised visual representation learning. In: Proceedings of the IEEE/CVF Conference on Computer Vision and Pattern Recognition, pp. 9729–9738 (2020)

9. He, X., Deng, K., Wang, X., Li, Y., Zhang, Y., Wang, M.: Lightgcn: Simplifying and powering graph convolution network for recommendation. In: Proceedings of the 43rd International ACM SIGIR Conference on Research and Development in Information Retrieval, pp. 639–648 (2020)

10. Le, D.T., Lauw, H.W., Fang, Y.: Basket-sensitive personalized item recommendation. IJCAI (2017)

11. Lin, Z., Tian, C., Hou, Y., Zhao, W.X.: Improving graph collaborative filtering with neighborhood-enriched contrastive learning. In: Proceedings of the ACM Web Conference 2022, pp. 2320–2329 (2022)

12. Ma, Y., He, Y., Zhang, A., Wang, X., Chua, T.S.: Crosscbr: cross-view contrastive learning for bundle recommendation. In: Proceedings of the 28th ACM SIGKDD Conference on Knowledge Discovery and Data Mining, pp. 1233–1241 (2022)

13. Rendle, S., Freudenthaler, C., Gantner, Z., Schmidt-Thieme, L.: Bpr: Bayesian personalized ranking from implicit feedback. arXiv preprint arXiv:1205.2618 (2012)

14. Wang, X., Jin, H., Zhang, A., He, X., Xu, T., Chua, T.S.: Disentangled graph collaborative filtering. In: Proceedings of the 43rd International ACM SIGIR Conference on Research and Development in Information Retrieval, pp. 1001–1010 (2020)

15. Wang, X., Liu, N., Han, H., Shi, C.: Self-supervised heterogeneous graph neural network with co-contrastive learning. In: Proceedings of the 27th ACM SIGKDD Conference on Knowledge Discovery & Data Mining, pp. 1726–1736 (2021)

16. Wu, J., Wang, X., Feng, F., He, X., Chen, L., Lian, J., Xie, X.: Self-supervised graph learning for recommendation. In: Proceedings of the 44th International ACM SIGIR Conference on Research and Development in Information Retrieval, pp. 726–735 (2021)

17. Xia, L., Huang, C., Xu, Y., Zhao, J., Yin, D., Huang, J.: Hypergraph contrastive collaborative filtering. In: Proceedings of the 45th International ACM SIGIR Conference on Research and Development in Information Retrieval, pp. 70–79 (2022)

18. Xia, X., Yin, H., Yu, J., Wang, Q., Cui, L., Zhang, X.: Self-supervised hypergraph convolutional networks for session-based recommendation. In: Proceedings of the AAAI Conference on Artificial Intelligence, vol. 35, pp. 4503–4511 (2021)

19. Xie, X., et al.: Contrastive learning for sequential recommendation. In: 2022 IEEE 38th International Conference on Data Engineering (ICDE), pp. 1259–1273. IEEE (2022)

20. You, Y., Chen, T., Sui, Y., Chen, T., Wang, Z., Shen, Y.: Graph contrastive learning with augmentations. Adv. Neural. Inf. Process. Syst. **33**, 5812–5823 (2020)

21. Zhao, S., Wei, W., Zou, D., Mao, X.: Multi-view intent disentangle graph networks for bundle recommendation. In: Proceedings of the AAAI Conference on Artificial Intelligence, vol. 36, pp. 4379–4387 (2022)

22. Zhou, K., et al.: S3-rec: Self-supervised learning for sequential recommendation with mutual information maximization. In: Proceedings of the 29th ACM International Conference on Information & Knowledge Management, pp. 1893–1902 (2020)

Efficient Respiration Rate Estimation Based on MIMO mmWave Radar

Zhicheng Xu, Ling Deng, Biyun Sheng, Linqing Gui$^{(\boxtimes)}$, and Fu Xiao

Nanjing University of Posts and Telecommunications, Nanjing 210023, China
{b19031820,1022040919,biyunsheng,guilq,xiaof}@njupt.edu.cn

Abstract. Respiration rate provides important reference for disease diagnosis and rehabilitation treatment. MmWave-based respiration rate estimation has increasingly attracted attention in recent years due to its non-intrusiveness and cost-effectiveness. However, when the subject locates far away from the mmWave radar and also deviated from it, the low accuracy of respiration rate estimation becomes a major concern. This paper presents MMRes, a mmWave-based respiration monitoring system that can achieve high accuracy of respiration rate estimation even in the scenario of long range and large deviation angle. In order to extract desired information from weak signals, MMRes finds the most sensitive channel from all channels and extract respiration rate by adaptively decomposing the signals in the channel. Specifically, MMRes first selects diverse channels by a coarse screening method, then finds the best one with the maximum total energy by an optimal channel selection method, and finally decompose the signals of the best channel by a WOA-VMD based algorithm with optimized parameters. The computing time of MMRes is reduced by data selection in the coarse screening and parallel computation in the optimal channel selection. Our extensive experiments show that MMRes can improve the accuracy of respiration rate estimation by up to 60.5%, when the subject is 4 m away from the radar and the deviation angle is 60°C. The fast coarse screening can reduce the computing time by up to 76.2%, while the optimal channel selection with 8 cores can reduce the time by up to 33.7% than that with 4 cores.

Keywords: millimeter-wave radio · wireless sensing · respiration monitoring · non-contact sensing

1 Introduction

Vital sign signals are important indicators in modern health care and medical applications, which can provide important reference for disease diagnosis and rehabilitation treatment. Conventionally, continuous vital sign monitoring (e.g. respiration rate monitoring) demands cumbersome wearable sensors such as respiratory belt that should be tightly wrapped around the chest [7]. However, it is cumbersome and uncomfortable to wear these dedicated sensors and may

© The Author(s), under exclusive license to Springer Nature Singapore Pte Ltd. 2024
Z. Tari et al. (Eds.): ICA3PP 2023, LNCS 14489, pp. 423–442, 2024.
https://doi.org/10.1007/978-981-97-0798-0_25

influence the effect of respiratory surveillance [2]. To overcome the drawbacks of contact sensing and achieve ubiquitous vital sign monitoring, contact-free sensing has attracted increasing attention in both academia and industry.

Non-contact respiration monitoring systems are capable of detect respiration rate of human subject without contacting the body [9]. The detection can be accomplished based on different kinds of signals, such as sound [17], visible light [10], infrared [6] and electromagnetic waves [4, 18–20]. Compared to other types of signals, electromagnetic waves have distinct advantages on both privacy protection and detecting accuracy. A variety of systems use electromagnetic waves to detect respiration rate, such as RFID [18], WiFi [4], UWB radar [20] and millimeter wave (mmWave) radar [19]. Among them, mmWave radar can achieve much better respiration detection accuracy due to its shorter wavelength, larger bandwidth and much finer-grained sensing capability [11]. So mmWave radar has become a promising solution for accurate and non-contact detection of human respiration.

Many mmWave radar-based vital sign monitoring systems have been proposed these years. For example, a band-pass filter and FFT were adopted in [14] to process the data from mmWave radar and estimate vital signs of the subject. In [13], respiration signals were reconstructed by jointly optimizing the decomposition of all the extracted compound vital signals over different range-azimuth bins. In [16], a hierarchy variational mode decomposition (VMD)-based approach was designed to extract vital signs from mmWave radar data. The work in [8] leveraged commodity mmWave Radar to contactlessly measure the vital signs of humans in real-time. Vital sign waveforms were recovered in [1, 3, 15] based on Support Vector Machine (SVM), differentiate and cross-multiply (DACM) and sequential deep learning network, respectively. Most of the existing methods monitor vital signs at close range. For example, each subject in [8, 13, 16] was no more than 1 m away from the radar. In addition, the subject was directly facing the radar. But in practice, when the subject locates far away from the mmWave radar and also deviated from it, the accuracy of vital sign estimation cannot be guaranteed by the above existing methods.

In order to accurately estimate respiration rate in the scenarios of long range and large deviation angle, this paper proposes MMRes, a new respiration rate estimation system based on MIMO mmWave radar. The new system aims to address the problem of low estimation accuracy in long-range and large-deviation scenarios. There are mainly two root causes of low estimation accuracy. One is that when the subject is far away from the radar, the signal reflected by the subject is too weak at the receiver for accurate respiratory estimation. The other is that when the deviation angle between the subject and the radar is so large that goes beyond the field of view of the radar, the signal reflected by the subject is also gets too weak at the receiver.

For solving the above problem, our system first quickly screens the most diverse channels, then selects the best channel from the screened channels, finally estimates respiration rate by adaptively decomposing the signal of the best channel. The main contributions of this paper are summarized as follows.

1) In order to effectively improve the accuracy of respiration rate estimation in the scenarios of long range and large deviation angle, MMRes finds the most sensitive channel from all channels and extract respiration rate by adaptively decomposing the signals in the channel. Specifically, MMRes first selects diverse channels by a coarse screening method, then finds the best one with the maximum total energy by an optimal channel selection method, and finally decompose the signals of the best channel by a WOA-VMD based algorithm with optimized parameters.

2) In order to reduce computing time, the coarse screening and the optimal channel selection are designed in a way of fast processing and parallel processing, respectively. In coarse screening, part of data in each channel are selected, which improves the efficiency since radars are having increasing number of channels. Then the data of screened channels are processed parallelly in the optimal channel selection, since each screened channel shares the same processing procedure.

3) We evaluate the performance of the whole system as well as each module in the system. Experimental results show that show that MMRes can improve the accuracy of respiration rate estimation by up to 60.5%, when the subject is 4 m away from the radar and the deviation angle is 60°C. In the same scenario, the fast coarse screening can reduce the computing time by up to 76.2%, while the optimal channel selection with 8 cores can reduce the time by up to 33.7% than that with 4 cores.

The rest of this paper is organized as follows. Section 2 models respiration estimation based on MIMO mmWave radar signal. Section 3 introduces the design and implementation of our system in detail. Section 4 evaluates the whole system and each module in the system. The conclusion is presented in Sect. 5.

2 Model Respiration Estimation Based on MIMO MmWave Radar Signal

In this section, we theoretically analyse the feasibility of our respiration estimation scheme by building a corresponding respiration estimation model. To this end, we first model how human respiration influences the propagation of MIMO radar signal. Then based on the propagation model, we further build a respiration estimation model by integrating our basic radar signal processing scheme.

2.1 Model Radar Signal Propagation with the Influence of Respiration

As shown in Fig. 1, the millimeter wave radar has P transmitting antenna and Q receiving antenna. Each transmitting antenna continuously transmits a FMCW signal. A chirp signal transmitted by the p-th antenna at time t can be expressed as

$$s_p(t) = a_p(t)e^{j(2\pi f_0 t + \pi B_c t^2 / T_c)}, \quad 0 \le t \le T_c, \tag{1}$$

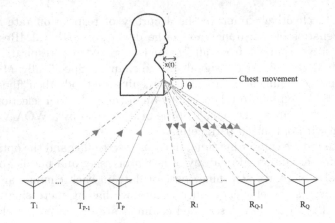

Fig. 1. MIMO radar signal propagation with the influence of respiration

where $a_p(t)$ represents the amplitude of the signal, f_0 is the start frequency, B_c is the bandwidth of the chirp signal, and T_c is the duration of the chirp signal.

During propagation, the chirp signal is reflected by the subject before arriving at each receiving antenna. When the subject is motionless, its chest still rises and falls regularly with respiration. As the subject breaths, the displacement of the chest can be theoretically formulated as

$$x(t) = A_m cos 2\pi f_b t, \tag{2}$$

where A_m represents the maximum displacement of the chest and f_b represents respiration rate.

The whole distance of signal propagation from the p-th transmitting antenna to the q-th receiving antenna, including the reflection at the chest, can be formulated as

$$d_{p,q}(t) = \bar{d}_{p,q} - x(t) cos\theta, \tag{3}$$

where $\bar{d}_{p,q}$ represents the distance of the chest when it is at the end of exhalation, and θ represents the angle between the reflected path and the direction of chest movement.

According to the signal propagation distance, the propagation time delay is given by

$$\tau_{p,q}(t) = d_{p,q}(t)/c, \tag{4}$$

where c is the speed of radar signal.

Then the signal from the p-th transmitting antenna at the q-th receiving antenna is formulated as

$$r_{p,q}(t) = a_q(t)e^{j(2\pi f_0(t-\tau_{p,q}(t))+\pi B_c(t-\tau_{p,q}(t))^2/T_c)}, \tag{5}$$

where $a_q(t)$ is the amplitude of the received signal. Substituting (2), (3) and (4) into (5), we can have the phase of $r_{p,q}(t)$ as

$$\angle r_{p,q}(t) = 2\pi f_0(t - \bar{d}_{p,q} - \frac{A_m cos\theta cos2\pi f_b t}{c}) + \frac{\pi B_c(t - \bar{d}_{p,q} - \frac{A_m cos\theta cos2\pi f_b t}{c})^2}{T_c}. \quad (6)$$

2.2   Model Respiration Estimation Based on Radar Signal Processing

Mixing the transmitted signal and the received signal, we can have an intermediate frequency (IF) signal output as

$$y_{p,q}(t) = a_p(t)a_q(t)e^{j(2\pi d_{p,q}(t)/\lambda_0 + 2\pi B_c d_{p,q}(t)t/(cT_c) - 2\pi B_c d_{p,q}^2(t)/(c^2 T_c))}, \quad (7)$$

where $\lambda = c/f_0$. In home settings, $d_{p,q}(t)$ is normally less than 5 m, so the last term in the phase of the above signal is much smaller than the first two terms. Neglecting the last term, we can rewrite (7) to

$$y_{p,q}(t) = a_p(t)a_q(t)e^{j(2\pi d_{p,q}(t)/\lambda_0 + 2\pi B_c d_{p,q}(t)t/(cT_c))}. \quad (8)$$

When processing the signal, the radar turns it into discrete version. t can be discretized as $(K(m-1) + k)\Delta t$, where m represents the m-th chirp, K is the number of samples per chirp, k represents the k-th sample in a chirp, Δt represents sample interval and equals T_c/K. In addition, since the duration of one chirp is very short (e.g. 42.6 μs in our setting), $d_{p,q}(t)$, $a_p(t)$ and $a_p(t)$ can be regarded to be constant during one chirp. So in a chirp, the discrete version of $y_{p,q}(t)$ can be written as

$$y_{p,q,m}(k) = a_{p,m}a_{q,m}e^{j(2\pi d_{p,q,m}/\lambda_0 + 2\pi B_c d_{p,q,m}k/(cT_c))}. \quad (9)$$

The Fourier transform of (9) is given by

$$Y_{p,q,m}(n) = \sum_{k=0}^{K} y_{p,q,m}(k)e^{-j2\pi kn/N}, \qquad 1 \leq n \leq N, \quad (10)$$

where n represents the n-th frequency point. Substituting (9) into (10), we can have

$$Y_{p,q,m}(n) = \sum_{k=0}^{K} a_{p,m}a_{q,m}e^{j(2\pi d_{p,q,m}/\lambda_0 + 2\pi B_c d_{p,q,m}k/(cT_c) - 2\pi kn/N)}. \quad (11)$$

At each frequency point, the sum of the amplitude of $Y_{p,q,m}(n)$ over all chirps is

$$X_{p,q,n} = \sum_{m=1}^{M} |Y_{p,q,m}(n)|. \quad (12)$$

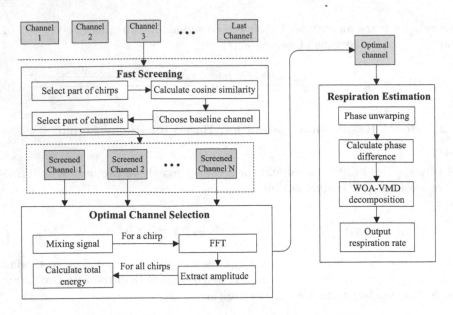

Fig. 2. An overview of MMRes architecture

Theoretically when n equals $2B_c d_{p,q,m}/(cT_c)$, $X_{p,q,n}$ reaches its maximum. Then the optimal frequency point n^* corresponds to the maximum $X_{p,q,n}$, i.e. $n^* = \arg\max_{(n)} X_{p,q,n}$.

Among all the $P*Q$ channels, we can find the maximum X_{p,q,n^*}. The optimal p and q are given by

$$(p^*, q^*) = \arg\max_{p,q} X_{p,q,n^*}. \tag{13}$$

Based on (n^*, p^*, q^*) and (11), the optimal frequency-domain chirp sequence is given by

$$Y_{p^*,q^*,m} = K a_{p^*,m} a_{q^*,m} e^{j(2\pi d_{p^*,q^*,m}/\lambda_0)}. \tag{14}$$

Extracting the phase sequence of (14), we can estimate the respiration rate.

3   Design

3.1   System Overview

We design a mmWave-based respiration monitoring system, MMRes. Figure 2 shows the architecture of MMRes. The whole system consists of three parts: Fast Screening, Optimal Channel Selection and WOA-VMD Based Respiration Estimation. In Fast Screening, we select part of data out of all data in each channel, calculate the difference between the selected data of any two channels, and find the diverse channels which have relatively large difference between each other. In Optimal Channel Selection, we parallelly process the data of all screened channels. We perform FFT on each chirp of IF signal, extract the amplitude of the range-domain signal, accumulate the square of amplitude over all chirps, and finally select the channel with maximum total energy. In WOA-VMD Based Respiration Estimation, we use WOA algorithm to adaptively optimize the VMD parameters and estimate the respiration rate by decomposing the data of the optimal channel.

3.2 Fast Screening

Since each channel in MIMO radar corresponds to one pair of transmitting antenna and receiving antenna, the number of channels in a MIMO radar is $P * Q$. So at the same time, the radar receives signals of as many as $P * Q$ channels. Although the transmitting and receiving antennas are compactly placed in the radar, their positions are still different from each other. So the signal from each transmitting antenna propagates toward the chest of the subject at a slightly different angle, and arrives at each receiving antenna through different reflection paths. Hence, the signal of each channel experiences a different propagation path. Then the received signals of $P * Q$ channels should be different from each other.

In general, the difference between the signals of two channels is related to the relative positions of the corresponding transmitting and receiving antennas. For example, if the i-th transmitting antenna T_i is placed closer to the p-th transmitting antenna T_p, while the position of the j-th receiving antenna R_j remains unchanged, then the difference between the signal of channel (T_i, R_j) and that of channel (T_p, R_j) should become smaller. Motivated by the above basic principle, our fast screening method will select N_{fs} channels out of all $P * Q$ channels, while the signals of those N_{fs} channels should be more different from each other compared to the signals of other channels. The detail of the fast screening method is illustrated as follows.

Step 1: In order to reduce time cost of screening, we select a part of data out of the whole data in the signal of each channel. The selected number of chirps can be denoted as αM, where M is the total number of chirps in the received signal and α is the proportion of selected chirps. Though theoretically any αM chirps in the received signal can be selected, without loss of generality, we choose the αM chirps in the middle of the received signal for fast screening.

Step 2: We calculate the difference between the received data of any two channels during the period of the selected chirps. The difference between the data of two channels can be measured by cosine similarity. More specifically, we first calculate the cosine similarity between the data of two channel as:

$$cos((T_i, R_j), (T_p, R_q)) = \frac{\sum_{l=1}^{\alpha M K}(|y_{i,j}(l)||y_{p,q}(l)|)}{\sqrt{\sum_{l=1}^{\alpha M K}|y_{i,j}(l)|^2 \sum_{l=1}^{\alpha M K}|y_{p,q}(l)|^2}}, \qquad (15)$$

where $i, p \in [1, P]$, $j, q \in [1, Q]$, (T_i, R_j) represents the channel between the i-th transmitting antenna T_i and the j-th receiving antenna R_j.

Step 3: The baseline channel will be found in this step. For each channel (T_p, R_q) , since we have calculated all the similarity values between the data in

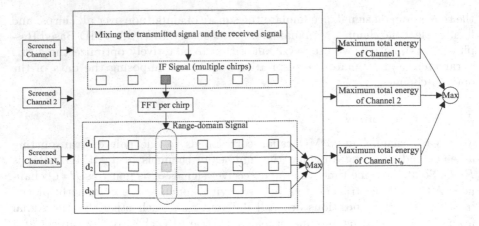

Fig. 3. Optimal Channel Selection

(T_p, R_q) and that in every other channel, among them we can find the maximum similarity denoted as $maxSim_{(T_p, R_q)}$. Then among all the $maxSim_{(T_p, R_q)}$, we can find the minimum one and its corresponding channel is the baseline channel that we suppose to find. This step can be formulated as :

$$(\hat{p}, \hat{q}) = \arg\min_{p,q}(\max_{i,j}(cos((T_p, R_q), (T_i, R_j)))), \tag{16}$$

where $(T_{\hat{p}}, R_{\hat{q}})$ represents the baseline channel.

Step 4: Using the above channel as the baseline, we select $N_{fs} - 1$ channels whose similarities to the baseline channel are smaller than other channels. The $N_{fs} - 1$ channels together with the baseline channel are the results of our fast screening.

3.3   Optimal Channel Selection

The N_{fs} channels selected by fast screening have differences in space, so the radar data of the N_{fs} channels varies in sensing performance. In order to obtain the best respiration estimation performance, in this section we process the data of all N_{fs} channels to select the best one, as shown in Fig. 3.

In order to reduce data processing time, the multi-channel signals after fast screening are supposed to be processed in parallel by a multi-core CPU. The number of cores in the CPU is denoted as N_{cr}. If N_{cr} is always larger than N_{fs}, the signals of N_{fs} channels can be easily processed in parallel. However, in practice, many existing CPUs have limited number of cores. N_{cr} is often less than N_{fs}, especially with the increased number of antennas in MIMO radars. For example, we use a TI radar with 3 transmitting antennas and 4 receiving antennas, so the total number of channels is 12. After fast screening, 8 out of 12 channels are supposed to be selected, i.e. $N_{fs} = 8$. But the CPU in a computer may have only 4 cores. In this case, since N_{cr} is less than N_{fs}, the data of N_{fs}

channels cannot be processed in parallel simultaneously. Instead, every time the data in N_{cr} out of N_{fs} channels are processed parallelly. The parallel processing should conducted N_{fs}/N_{cr} times.

The scheme how we process the data of each channel is expressed as follows.

Step 1: We first mix the transmitted signal and the received signal to get the IF signal, then we perform FFT on each chirp of the IF signal. According to (9) and (10), the distance between the subject and the radar is proportional to the frequency of the signal, so the signal's frequency domain is actually equivalent to its range domain. Thus we can obtain the range-domain signal at each range point by performing FFT on each chirp. Then we extract the amplitude of the range-domain signal.

Step 2: At each range point, we accumulate the square of amplitude of the range-domain signal over all chirps. So we obtain the sum of the amplitude square of the signal at the range point over all chirps, which is regarded as the total energy at the range point. The range point with the maximum total energy corresponds to the accurate position of subject. The reason why we perform the summation on all chirps rather than a single chirp is as follows: one chirp has very short duration and is vulnerable to interference, so the calculated range point with the maximum total energy is probably not the accurate position of the subject. Instead we can effectively reduce the errors of range estimation by calculating the total energy over the chirps.

Step 3: We perform the above steps in parallel for each of the N_{cr} channels and then find the optimal range point and the optimal channel with the maximum total energy.

3.4 WOA-VMD Based Respiration Estimation

The weak chest movement caused by respiration can be detected by FMCW radar signal. The corresponding signal has non-stationary characteristics. Moreover, the signal is easily interfered by the subject's own slight movement such as slight shaking the body. To extract vital signs from non-stationary radar signal, VMD has been demonstrated to be a very useful method [5]. Among the parameters in VMD, the number of decomposition layers and penalty factor have great influence on the performance of VMD decomposition. However, these two parameters are set manually in traditional VMD method, directly affecting the effect of respiration waveform construction. In order to achieve the best effect of respiration waveform construction, we adopt WOA algorithm to adaptively optimize the parameters of VMD decomposition.

VMD-Based Signal Decomposition. Variational mode decomposition(VMD) can decompose the range-domain signal $Y_{p*,q*,n*}(t)$ into G IMF components u_g ($g = 1, 2, 3, \ldots, G$). The center frequency of each component u_g is denoted as ω_g. The bandwidth of u_g is optimized to be as small as possible, so the mode aliasing problem in other decomposition methods such as EMD can be effectively suppressed. Hence, VMD is modeled as a constrained variational problem whose objective is to find the minimum bandwidth of each IMF component, i.e.

$$
\min_{\{u_g\},\{\omega_g\}} \& \left\{ \sum_g \left\| \partial_t \left[(\delta(t) + \frac{j}{\pi t}) * u_g(t) \right] e^{j\omega_g t} \right\|_2^2 \right\}, \& \text{s.t.} \quad \sum_g u_g(t) = Y_{p*,q*,n*}(t),
$$

(17)

where $u_g \in \{u_1, u_2, \ldots, u_G\}$, $\omega_g \in \{\omega_1, \omega_2, \ldots, \omega_G\}$, $(\delta(t) + \frac{j}{\pi t}) * u_g(t)$ is Hilbert transform, $*$ represents convolution.

In order to solve (17), the quadratic penalty factor β and the Lagrange multiplication operator $\lambda(t)$ are introduced. The solution of (17), namely the saddle point $(\{u_g\}, \{\omega_g\}, \lambda)$, is obtained by the alternating direction multiplier method, which updates u_g, ω_g and λ iteratively in the frequency domain. Then the G IMF components are obtained.

The steps of VMD decomposition are illustrated as follow. In the first step, the number of decomposition layers G and the penalty factor β are set manually. In the second step, $\{u_g^1\}$ and, $\{\omega_g^1\}$ are randomly initialized, while λ^1 is initialized to 0. In the third step, IMF is updated according to

$$
\hat{u}_g^{n+1}(\omega) = \frac{\hat{Y}_{p*,q*,n*}(\omega) - \sum_{i=1, i \neq g}^{G} \hat{u}_i^n(\omega) + \frac{\hat{\lambda}^n(\omega)}{2}}{1 + 2\alpha(\omega - \omega_g^n)^2},
$$

(18)

while the central frequency of each IMF component is updated according to

$$
\omega_g^{n+1}(\omega) = \frac{\int_0^\infty \omega \left| \hat{u}_g^{n+1}(\omega) \right|^2 d\omega}{\int_0^\infty \left| \hat{u}_g^{n+1}(\omega) \right|^2 d\omega},
$$

(19)

then the Lagrange multiplier is updated as

$$
\hat{\lambda}^{n+1}(\omega) = \hat{\lambda}^n(\omega) + \tau [\hat{Y}_{p*,q*,n*}(\omega) - \sum_{g=1}^{G} \hat{u}_g^{n+1}(\omega)],
$$

(20)

where τ can be set to 0 in order to obtain a good noise reduction effect.

In the fourth step, we repeat the third step until the following termination condition is satisfied, i.e.

$$
\frac{\sum_g \left\| \hat{u}_g^{n+1} - \hat{u}_g^n \right\|_2^2}{\left\| \hat{u}_g^n \right\|_2^2} < \varepsilon,
$$

(21)

where ε is the discriminant accuracy. Finally, the IMF components are obtained according to the newest $\{u_g\}$ and $\{\omega_g\}$.

VMD Parameter Optimization Based on WOA. Since the number of decomposition layers G and penalty factor β have great influence on the performance of VMD, we need to find the optimal G and β to achieve sufficient decomposition of the radar signal. WOA algorithm is adopted to adaptively optimize the VMD parameters. Compared to traditional optimization algorithms such as PSO and GSA, WOA can achieve better local optima avoidance and higher convergence speed simultaneously during iterations [12]. The procedure of optimizing the VMD parameters based on WOA is illustrated as follow.

In the first step, we initialize the parameters $[G, \beta]$ as $[G^0, \beta^0]$, decompose the radar signal into multiple IMF components based on VMD with $[G^0, \beta^0]$, and calculate the fitness function of the IMF components based on envelope entropy. The envelope entropy is usually used to measure dynamic information in a signal. A smaller envelope entropy value indicates more feature information in the signal. The envelope entropy of radar signal can be calculated as

$$\begin{cases} E_p = \sum_{i=1}^{N} p_i \lg p_i \\ p_i = a_i / \sum_{i=1}^{N} a_i \end{cases}, \qquad (22)$$

where a_i is the envelope signal of the G components decomposed by VMD after Hilbert demodulation, p_i is the probability distribution obtained by normalizing a_i, N is the number of sampling points.

The calculated fitness function value will be used to update the best fitness function value, if it is less than the current best value.

In the second step, we iteratively updates the parameters $[G, \beta]$ based on WOA. The algorithm has 3 hunting stages: encircling prey, bubble-net attacking, searching for prey. In the n-th iteration, we first maps the parameters $[G, \beta]$ to the positions of whales $X(n)$. Then we decide the stage of hunting according to variable p and A. p is a random number between $[0, 1]$, indicating that the whales have 50% chance to use the round-up prey or the bubble-net prey. A is a coefficient controlling the way the whale swims and is calculated as

$$A = 2ar_a a, \qquad (23)$$

where r_a is a random vector in $[0, 1]$, $a = 22(n/N_{max})$, N_{max} is the maximum number of iterations. When $p \geq 0.5$, the hunting stage should be bubble-net attacking, which is formulated as

$$X(n+1) = D' \cdot e^{bl} \cdot cos(2\pi l) + X^*(n), \qquad (24)$$

where $D = |X^*(n) - X(n)|$, $X^*(n)$ is the position of the best whale, b is a constant usually set to 1, and l is a random number between $[-1, 1]$.

When $p < 0.5$ and $|A| <= 1$, the hunting stage should be encircling prey, which is formulated as

$$X(n+1) = X^*(n) - A \cdot D, \qquad (25)$$

where $D = |2r_b \cdot X^*(n) - X(n)|$, r_b is a random vector in $[0, 1]$.

When $p < 0.5$ and $|A| > 1$, the hunting stage is searching for prey and the whales move randomly for global exploration. It is formulated as

$$X(n+1) = X_{rand} - A \cdot D, \tag{26}$$

where $D = |C \cdot X_{rand} - X(n)|$, $X_{rand(t)}$ is the position of a whale randomly selected in current population.

Finally, when the number of iterations reach N_{max}, the iteration terminates. The optimal parameters $[G^*, \beta^*]$ are obtained according to $X^*(N_{max})$. Based on $[G^*, \beta^*]$, the signal is decomposed into multiple IMFs. The IMF inside respiration rate range is extracted and then the respiration rate is obtained.

4   Evaluation

Experimental Setup. We adopt a MIMO mmWave radar as the sensing front end and a laptop as the data processing back end. The mmWave radar is equipped with three on-board transmitter antennas and four receiver antennas. The parameters corresponding to the mmWave radar setting are listed in Table 1. During the experiment, the subject sits in a chair in front of the mmWave radar. The distances between the mmWave radar and the subject is 0.5 m, 1 m, 1 m, 2 m and 4 m. The subject's chest is at basically the same height with the mmWave radar. And the angle of the subject corresponding to the mmWave radar is in the range of $[0°, 60°]$. The subject wears respiratory zone to capture the ground truth of human respiration.

Table 1. Main Parameters of mmWave Radar

Parameters	Value
Starting Frequency	60 GHz
Bandwidth	2.986 GHz
Chirp Duration	42.667 μs
Sampling Rate	3 MHz
Chirp Sampling Number	128

4.1   Effectiveness of Our Fast Screening Method

To evaluate how N_{fs} in the fast screening method influences respiration rate estimation accuracy, we set N_{fs} first to 8 and then to 4. In both cases of N_{fs}, the distance between the subject and the radar changes from 0.5 m to 4 m. The angle between them is 0°C, meaning the subject is directly facing the radar. Meanwhile, we also estimated the respiration rate without using the fast screening method, i.e., N_{fs} equals 12. The experiment results are shown in Fig. 4.

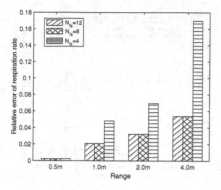
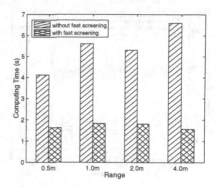

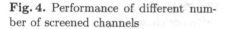

Fig. 4. Performance of different number of screened channels

Fig. 5. Computing time with/without fast screening

When N_{fs} reduces from 8 to 4, the error of respiration rate estimation gets larger at each distance. The reason is that when fewer channels are selected, the dimension as well as the amount of available data for the subsequent modules is also reduced, then it is more likely to lose high-quality data. On the other hand, in Fig. 4 we can also observe that when N_{fs} reduces from 12 to 8, respiration rate estimation error does not get worse in all cases of distance. The reason is that our fast screening method selects the 8 channels with the largest diversity from all 12 channels. Since the data of so many diverse channels are retained, the data with high sensing quality should also be retained. Therefore, selecting the data of N_{fs} diverse channels (e.g. $N_{fs}=8$ out of 12 channels), we can obtain the same respiration estimation performance as that of using all channels.

To evaluate whether our fast screening method can effectively reduce the computing time, we obtain the computing time in two cases, i.e. with and without the fast screening method. The number of selected channels in the fast screening method, i.e. N_{fs}, is set to 8. The experiment results are shown in Fig. 5. We can observe that when our fast screening method reduces the number of channels from 12 to 8, the computing time is significantly reduced. For example, with our fast screening method, in all cases of different distances, it always takes less than 2 s to process 2-minute radar data. On the contrary, if the fast screening method is not used, the computing time is increased significantly. For example, when the range is 4 m, the fast screening can reduce the computing time by 76.2%.

To verify the effect of our fast screening method, the estimated respiration rate by our fast screening is compared with the result by random screening. Here, random screening means randomly selecting N_{fs} channels from all $P*Q$ channels. Since the channels are randomly selected, accordingly the respiration rate estimation accuracy of random screening also changes in a random manner. So we will present both the best and the worse results of random screening, and compare them with the result of our fast screening. As shown in Fig. 6, when the subject is only 0.5 m away from the radar, random screening and our fast

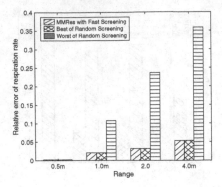
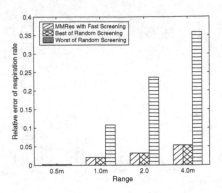

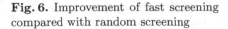

Fig. 6. Improvement of fast screening compared with random screening

Fig. 7. Maximum and minimum similarity per channel

screening can both achieve very accurate respiration rate estimation, as their estimation errors are all below 0.3%. As the subject gets further away from the radar, the estimation errors of fast screening and random screening also increase. But no matter how far the subject is away from the radar, our fast screening can always achieve the same accuracy of respiration rate estimation as the best result of random screening, with the respiration rate estimation error always below 5%. On the contrary, the worst result of random screening shows very low estimation accuracy, especially when the subject is far away from the radar. The above results demonstrate that our fast screening method can select the best channels and achieve the best respiration rate estimation accuracy, compared to random screening.

As to each channel, its data similarity with every other channel can be calculated, so 11 similarity values can be obtained for each channel. Among them the maximum and the minimum similarity are shown in Fig. 7 as per each channel. The subject is 1 m away from the radar and directly facing the radar. Observed from Fig. 7, the minimum similarity for all channels is always less than 85%, showing the inherent diversity between the data of any two channels. Thus good performance requires to retain as many channels as possible. It can also be observed that the maximum similarity could be as high as 91%, e.g. channel 6. The channel with such a high similarity should be screened out for improving the efficiency of the system.

When comparing data similarity between channels, we select only part of the data in each channel, so as to improve the efficiency of fast screening. The proportion has been denoted as α. Figure 8 shows the accuracy of respiration estimation under different setting of α. It can be observed that the respiration rate estimation error increases with the reduction of α. The reason is that the similarity calculated based on a smaller part of the data in each channel should deviate farther from the true similarity. The above deviation induced by a tiny α could worsen the accuracy of respiration rate estimation significantly, especially

when the subject is far away from the radar. For example, when the subject is
4 m away from the radar and α is 0.01, the respiration rate estimation error is
as high as 27.8%. On the contrary, under the same distance, when α is set to
a higher value such as 0.5 and 0.8, the respiration rate estimation error is only
5.36%. So it is recommended to set an appropriate value for α.

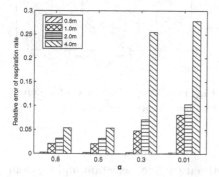

Fig. 8. Performance of different data proportions

Fig. 9. Computing time of different data proportions

Figure 9 shows how the proportion α influences the computing time of the
fast screening method. It should be noted that before fast screening the amount
of data in each channel has already been downsampled. Otherwise, the amount
of data would be too huge for timely processing. The downsampling rate is set
to either 1/50 or 1/100. From the figure, it can be observed that the computing
time of the fast screening method increases with the increase of α. For reducing
the computing time, a lower α is always preferable. But α cannot be too small.
Otherwise, the respiration rate estimation accuracy would be too bad. Therefore,
setting an appropriate α should take into account both estimation accuracy and
computing time.

4.2   Effectiveness of Our Optimal Channel Selection Method

To verify whether our optimal channel selection method can obtain the best
channel for the best estimation accuracy, the estimated respiration rate by our
optimal channel selection method is compared with the result by random selection. Here, random selection means randomly selecting one channel from N_{fs}
channels. So we will present both the best and the worse results of random
selection, and compare them with the result of our optimal channel selection
method.

As shown in Fig. 10, when the subject is only 0.5 m away from the radar,
random selection and our optimal channel selection method can both achieve
very accurate respiration rate estimation. When the subject gets further away

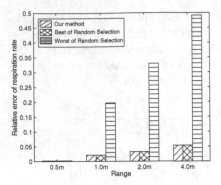

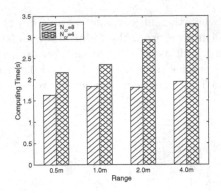

Fig. 10. Improvement of optimal channel selection over random selection

Fig. 11. Computing time of different number of cores

from the radar, the estimation errors of optimal channel selection and random selection both increase. Nevertheless, our optimal channel selection can always achieve the same accuracy as the best random selection. Its corresponding estimation error is always below 6%, even when the subject is far (e.g. 4 m) away from the radar. It is much smaller than the worst result of random selection. When the subject is 4 m away from the radar, the worst random selection has an estimation error as high as 49.1%. The above results demonstrate that our optimal channel selection method can achieve much better respiration rate estimation accuracy than random channel selection.

Our optimal channel selection processes the data of N_{fs} fast-screened channels in parallel. To evaluate how the number of CPU cores influences the computing time of optimal channel selection method, the number of CPU cores used for data processing, denoted as N_{cr} is first set to 8 and then to 4. In both cases, we obtain the computing time of optimal channel selection method. The experiment results are shown in Fig. 11. It can be observed that in the case of $N_{cr} = 4$ the optimal channel selection method needs more computing time than that in the case of $N_{cr} = 8$. The optimal channel selection with 8 cores can reduce the time by up to 33.7% than that with 4 cores. So more CPU cores are always preferable to reduce the computing time of optimal channel selection.

In previous results, the subject is directly facing the radar, i.e. the angle between the subject and the radar is 0 degree. To further evaluate how the angle between the subject and the radar influences the performance of optimal channel selection method, in the following we compare the estimation accuracy of optimal channel selection with that of random channel selection when the angle varies from 0 to 60°C. The distance between the subject and the radar is set to 2 m. The experiment results are shown in Fig. 12. As shown in the figure, when the angle between the subject and the radar increases, the estimation errors of optimal channel selection and random selection both increase. Our optimal channel selection still achieve the same accuracy as the best random selection, with the estimation error always below 5%, even when the angle between the subject and the radar is as large as 60°C. On the contrary, the worst random selection always has much larger estimation errors in all the angle settings. The above results demonstrate the good performance of our optimal channel selection method when the subject is not directly facing the radar.

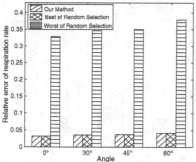

Fig. 12. Improvement of optimal channel selection in different angles (fixed range)

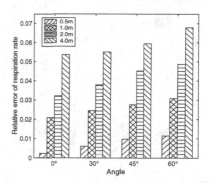

Fig. 13. Performance of optimal channel selection in different angles and ranges

Different from the above setting which fixes the distance between the subject and the radar, we now change the distance and the angle together to evaluate how the two factors jointly influence the performance of our system. The experiment results are shown in Fig. 13. It can be observed that the distance and angle both influence the accuracy of the respiration rate estimation, but the distance has stronger influence than the angle. Under the same angle setting (e.g. 60°C), the error of respiration rate estimation in the distance of 0.5 m is 1.15%, but the error in the distance of 4 m is significantly increased to 6.79%. On the contrary, under the same distance setting (e.g. 4 m), the estimation error in 0°C is 5.36%, while the error in 60°C is slightly increased to 6.79%. In addition, we can observe that the errors of respiration rate estimation are all below 7%, demonstrating the good performance of our system even when the subject is far away and deviated from the radar.

4.3   Effectiveness of WOA-VMD Based Respiration Waveform Reconstruction

To evaluate whether VMD can improve the estimation accuracy, the respiration rate estimated based on the VMD algorithm in [16] is compared with the result without VMD. In the case without VMD, an alternative and also traditional solution is the combination of bandpass filter and fast Fourier transform (FFT). More specifically, the alternative solution first filters the data of the optimal channel by a bandpass filter, then transforms the filtered time-series data into the frequency domain by FFT, and finally estimated the respiration rate based on the amplitude peak in the frequency-domain data. The errors of respiration rate estimation with/without VMD are shown in Fig. 14.

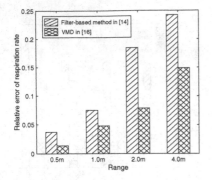

Fig. 14. Improvement of VMD over filter-based method

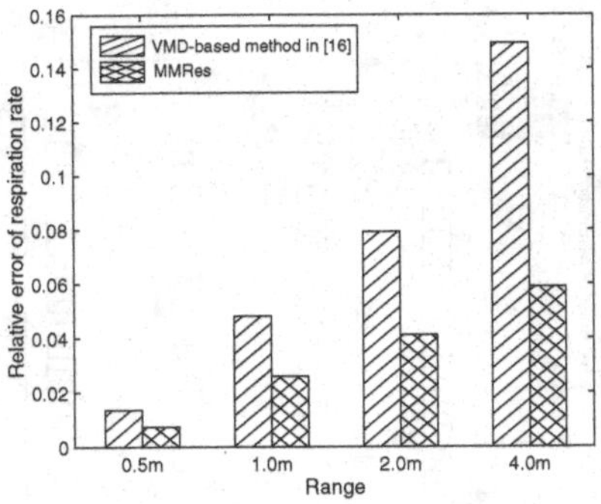

Fig. 15. Improvement of our system over VMD-based method

It can be observed that the error of respiration rate estimation increases with the distance between the subject and the radar, whether VMD is used or not. Nevertheless, in all cases of the distance, the accuracy of respiration rate estimation with VMD is always better than that without VMD. The advantage of VMD over the alternative solution is obvious when the subject is far away from the radar. For example, when the subject is 4 m away from the radar, the estimation error of the alternative solution is as high as 24.3%, but the estimation error by VMD is reduced to 14.9%. So the experiment results demonstrate that VMD-based method can achieve much better accuracy of respiration rate estimation than filter-based method.

Although the above results show the advantage of VMD, the estimation error of the VMD-based method often stays at a high level, especially when the subject is far away from the radar. We have used the WOA algorithm to optimize the key parameters in VMD and improve the performance of VMD. To evaluate the improvement by the WOA-VMD algorithm, the accuracy of respiration rate estimation based on our system MMRes is compared with the VMD-based method. The experiment results are shown in Fig. 15.

It can be observed that when the distance between the subject and the radar gets larger, the error of respiration rate estimation by our MMRes also increases. However, in all distance settings, the estimation accuracy of our system is always better than the accuracy of the VMD-based method. It demonstrates the superiority of our system. Moreover, when the distance is large, e.g. 4 m, our MMRes can significantly improve the accuracy of respiration rate estimation. In this far distance scenario, our MMRes successfully reduces the error of respiration rate estimation from 14.9% to no more than 6%. It demonstrates the importance of optimizing the parameters in VMD. It should also be noted that in both Fig. 14 and Fig. 15, the estimation results under each distance setting are already been averaged in terms of the angles between the subject and the radar.

5 Conclusion

Non-contact sensing based on mmWave radar has been an efficient approach to non-intrusively monitor human respiration. However, current mmWave-based respiration monitoring systems still suffer from low accuracy of respiration rate estimation when the subject locates far away from the radar and also deviated from it. This paper presents MMRes, a mmWave-based respiration monitoring system that can achieve high accuracy of respiration rate estimation even in the scenario of long range and large deviation angle. MMRes can find the most sensitive channel from all available channels and extract respiration rate by adaptively decomposing the signals from the channel. To this end, MMRes first selects the most diverse channels by a coarse screening method, then finds the best one with the maximum total energy by an optimal channel selection method, and finally decompose the signals of the best channel by a WOA-VMD based algorithm with optimized parameters. The computing time of MMRes is reduced by data selection in the coarse screening and parallel computation in the optimal channel selection. Our extensive experiments show that MMRes can improve the accuracy of respiration rate estimation by up to 60.5%, when the subject is 4 m away from the radar and the deviation angle is 60°C. The fast coarse screening can reduce the computing time by up to 76.2%, while the optimal channel selection with 8 cores can reduce the time by up to 33.7% than that with 4 cores.

Acknowledgements. This work is supported jointly by the National Natural Science Foundation of China (Grant 62125203, 61932013, 61972201).

References

1. Abedi, H., et al.: Non-visual and contactless wellness monitoring for long term care facilities using mm-wave radar sensors. In: 2022 IEEE Sensors, pp. 1–4 (2022)
2. Chen, A., et al.: Machine-learning enabled wireless wearable sensors to study individuality of respiratory behaviors. Biosens. Bioelectron. **173**, 112799 (2021)
3. Dong, S., Li, Y., Lu, J., Zhang, Z., Gu, C., Mao, J.: Accurate detection of doppler cardiograms with a parameterized respiratory filter technique using a k-band radar sensor. IEEE Trans. Microw. Theory Tech. **71**(1), 71–82 (2023)
4. Guo, Z., Yuan, W., Gui, L., Sheng, B., Xiao, F.: Breatheband: a fine-grained and robust respiration monitor system using wifi signals. ACM Trans. Sen. Netw. **19**(4) (2023)
5. Li, T., Qian, Z., Deng, W., Zhang, D., Lu, H., Wang, S.: Forecasting crude oil prices based on variational mode decomposition and random sparse bayesian learning. Appl. Soft Comput. **113**, 108032 (2021)

6. Maurya, L., Mahapatra, P., Chawla, D.: Non-contact breathing monitoring by integrating rgb and thermal imaging via rgb-thermal image registration. Biocybernet. Biomed. Eng. **41**(3), 1107–1122 (2021)

7. Nassi, T.E., et al.: Automated scoring of respiratory events in sleep with a single effort belt and deep neural networks. IEEE Trans. Biomed. Eng. **69**(6), 2094–2104 (2022)

8. Salman, M., Noh, Y.: Contactless vital signs tracking with mmwave radar in real-time. In: 2023 IEEE International Conference on Big Data and Smart Computing (BigComp), pp. 389–390 (2023)

9. Scebba, G., Da Poian, G., Karlen, W.: Multispectral video fusion for non-contact monitoring of respiratory rate and apnea. IEEE Trans. Biomed. Eng. **68**(1), 350–359 (2021)

10. Shokouhmand, A., Eckstrom, S., Gholami, B., Tavassolian, N.: Camera-augmented non-contact vital sign monitoring in real time. IEEE Sens. J. **22**(12), 11965–11978 (2022)

11. Singh, A., Rehman, S.U., Yongchareon, S., Chong, P.H.J.: Multi-resident non-contact vital sign monitoring using radar: a review. IEEE Sens. J. **21**(4), 4061–4084 (2021)

12. Sun, Y., Chen, Y.: Multi-population improved whale optimization algorithm for high dimensional optimization. Appl. Soft Comput. **112**, 107854 (2021)

13. Wang, F., Zeng, X., Wu, C., Wang, B., Liu, K.J.R.: Driver vital signs monitoring using millimeter wave radio. IEEE Internet Things J. **9**(13), 11283–11298 (2022)

14. Wang, F., Zhang, F., Wu, C., Wang, B., Liu, K.J.R.: Vimo: multiperson vital sign monitoring using commodity millimeter-wave radio. IEEE Internet Things J. **8**(3), 1294–1307 (2021)

15. Wang, Y., Gu, T., Luan, T.H., Yu, Y.: Your breath doesn't lie: multi-user authentication by sensing respiration using mmwave radar. In: 2022 19th Annual IEEE International Conference on Sensing, Communication, and Networking (SECON), pp. 64–72 (2022)

16. Xu, X., et al.: mmecg: monitoring human cardiac cycle in driving environments leveraging millimeter wave. In: IEEE INFOCOM 2022 - IEEE Conference on Computer Communications, pp. 90–99 (2022)

17. Xue, B., Deng, B., Hong, H., Wang, Z., Zhu, X., Feng, D.D.: Non-contact sleep stage detection using canonical correlation analysis of respiratory sound. IEEE J. Biomed. Health Inform. **24**(2), 614–625 (2020)

18. Yang, C., Wang, X., Mao, S.: Respiration monitoring with rfid in driving environments. IEEE J. Sel. Areas Commun. **39**(2), 500–512 (2021)

19. Zhai, Q., Han, X., Han, Y., Yi, J., Wang, S., Liu, T.: A contactless on-bed radar system for human respiration monitoring. IEEE Trans. Instrum. Meas. **71**, 1–10 (2022)

20. Zhang, S., Zheng, T., Wang, H., Chen, Z., Luo, J.: Quantifying the physical separability of rf-based multi-person respiration monitoring via sinr. In: Proceedings of the 20th ACM Conference on Embedded Networked Sensor Systems, SenSys 2022, pp. 47–60, Association for Computing Machinery, New York (2023)

Running Serverless Function on Resource Fragments in Data Center

Yukang Chu, Wenhao Huang, and Laiping Zhao[✉]

College of Intelligence and Computing (CIC), Tianjin University, Tianjin, China
{chuyukang,hwh,laiping}@tju.edu.cn

Abstract. Repetitive scheduling of resources in data centers leads to a large number of transient resource fragments. Serverless computing can decompose long-period mixed deployment applications into small-volume and short-period functions. Fine-grained functions can effectively respond to the transient fluctuation behavior of fragments. However, the fluctuation characteristics and predictability of resource fragments in the cloud environment vary greatly, which may still interfere with the execution process of functions.

In this paper, we introduce Tianxuan, a function-level resource scheduler. It classifies the resource fragments in the cloud environment according to their usage forms into unallocated and unused fragments and extracts the fluctuation characteristics from open-source datasets of each type of fragment. Then, it builds a fragment feature classification-aware spatiotemporal prediction model based on these characteristics and schedules functions on spatiotemporally complementary fragments to achieve resource efficiency by reducing the probability of resource contention. Experimental results show that Tianxuan improves the prediction accuracy of resource fragments by 15% compared to existing techniques and increases server CPU utilization by 16%–25% and function throughput by 17%–37%.

Keywords: transient resource · serverless · spatiotemporal prediction

1 Introduction

Background. Serverless computing is a popular application development and hosting model that divides traditional applications into multiple functions [16,18], enabling better scalability and fault tolerance mechanisms. Based on real-time user loads, this model automates the scaling of function replicas and reduces overall operational costs through pay-as-you-go billing. By simplifying infrastructure management and providing a scalable and fault-tolerant application development process, Serverless computing is changing the way developers build and deploy applications in the cloud.

The rapid development of Serverless has changed the form of business on existing infrastructures. User applications are gradually migrating from

---

Y. Chu and W. Huang—Contributed equally to this work.

Z. Tari et al. (Eds.): ICA3PP 2023, LNCS 14489, pp. 443–462, 2024.
https://doi.org/10.1007/978-981-97-0798-0_26

large-grained virtual machines to small-grained Serverless functions, and infrastructures are transitioning from a single virtualization architecture to a hybrid architecture of virtual machines and containers [28,31]. Despite no changes to the given infrastructure configuration, resource inefficiency in data centers remains a problem to be solved. According to data provided by some cloud service providers, their internal resource utilization rates remain at 25%-60% [1,3,4], and the large amount of idle resources in the cloud environment not only harms the resource utilization rate of data centers but also results in unnecessary physical energy consumption.

Moviation. Idle "unallocated resource fragments" and "allocated unused resource fragments" are the direct cause of resource inefficiencies in the data centre. Unallocated resource fragments refer to resources that are not sold in a timely manner, as well as resources that are actively set aside to provide services such as resilient scalability and disaster recovery to users. Shared allocated unused resource fragments refer to resources that are allocated to users but not fully used by them. Users often over-purchase resources to ensure the quality of service for their business. According to data from Unicloud [2] and Azure Cloud [3], the CPU allocation rate of Unicloud is only 61.29%, and the average CPU usage rate allocated to users is 24.125%; about half of the servers in Azure Cloud have more than 20% unallocated resources [31], and the average CPU usage rate of allocated servers is only 15.134% [13]. With large clusters of servers running in data centres for long periods of time, any small amount of recycling of unused resources can have a huge economic impact.

Related Work. Cloud service providers often use hybrid deployment techniques to improve data center resource utilization [23,25,26,30,34]. However, resource contention and interference between hybrid applications can cause performance degradation in cloud services, especially for latency-sensitive interactive services with significantly increased tail latency [11,14,20,22,24,27]. Serverless computing can decompose long-lived hybrid part applications into several short-lived and small-sized functions, and short-lived and small-sized functions can effectively cope with the transient fluctuation phenomenon of resource fragments. Recently, the function hybrid technique can be mainly summarized below:

Scavenger [15] calculated the resource reuse value of memory, CPU, and network bandwidth by observing the average and standard deviation of the time window for virtual machine service resource usage, monitored the IPC of virtual machines to provide feedback for adjusting resource allocation values. ServerMore [28] replaced IPC with the Cohen coefficient of IPC (the ratio of the average difference to the standard deviation), which can better evaluate CPU performance degradation based on Scavenger. However, these approaches still suffer from the problem of lower priority for secondary citizens, and the slow feedback may affect tail-latency-sensitive function services. Harvestless [31] builds a load balancing algorithm internally, calculates the ratio of CPU usage to CPU harvestable on nodes, decides the degree of resource shortage and whether to place functions, and reduces cold start by establishing the minimum working set of functions.

Our Work. We have implemented a function-level resource scheduling system called Tianxuan based on the open-source Serverless framework Open-FaaS [5]. The core modules of Tianxuan are the temporal-spatial predictor of resource fragments and the temporal-spatial scheduler of functions. Based on the temporal-spatial fluctuation characteristics of resource fragments, we designed a KNN model with feature characterization and an LSTM model with waveform recombination for unallocated and allocated unused resource fragments. The temporal-spatial scheduler schedules functions on the resource fragments that satisfy their temporal and spatial resource requirements based on the prediction results of the model, ensuring that the resource requirements of the functions during execution are met.

Our contributions can be summarised as follows:

1. We have analysed the fragmentation dataset of the public cloud, derived universal temporal and spatial fluctuation characteristics of unallocated resource fragments and allocated unused resource fragments, and built a high-precision fragmentation prediction model.
2. We design a two-dimensional function scheduling strategy in time and space to schedule functions on resource fragments that are satisfied by temporal and spatial resources, guaranteeing function service quality and achieving full resource recovery.
3. We implemented Tianxuan based on OpenFaaS, an open source FaaS platform, and demonstrated its significant improvement in resource efficiency and cloud service quality from real public cloud traces.

2   Characterization

This section undertook a comprehensive analysis of fragmented datasets derived from public clouds. It entailed examining the temporal and spatial fluctuation attributes exhibited by both unallocated resource fragments and allocated but unused resource fragments. Through meticulous investigation, we substantiated the universality of these distinctive features.

2.1   Analysis of the 'Time' Characteristics of Unallocated Resource Fragments

Fluctuations in unallocated resource fragments primarily stem from two situations: the creation and removal of virtual machines. When functions are deployed on these fragments, only the creation of new virtual machines affects the reduction of unallocated resources, which can lead to interference caused by resource competition. Consequently, this section provides a summary of the fluctuation characteristics observed in unallocated resource fragments by specifically analyzing the creation of virtual machines.

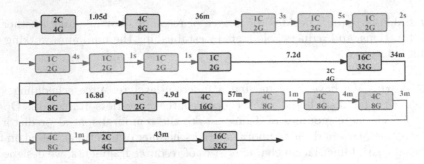

Fig. 1. Virtual machine request interval diagram

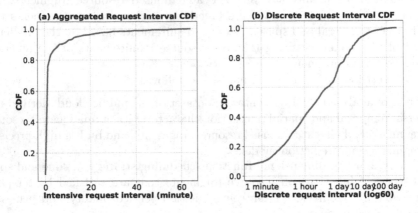

Fig. 2. Dense discrete request interval CDF diagram

Figure 1 provides a glimpse into the temporal trajectory of user-created virtual machines on server 27 within the Unicloud Beijing 3 cluster. The second black trajectory demonstrates the creation of merely five virtual machines over a span of 29 d, indicating a consistently sparse fluctuation pattern over an extended duration. In contrast, the first pink trajectory illustrates the creation of six virtual machines within a brief 16-second interval. Comparatively, when compared to the fluctuation frequency observed in the black trajectory of Fig. 1, the pink trajectories exhibit locally dense fluctuations confined to a short timeframe.

In order to investigate the nature of the dense requests depicted in Fig. 1, a thorough analysis was conducted on the 12,273 user virtual machine purchase requests within the entire Beijing 3 cluster. These requests were categorized into two distinct groups: dense and discrete. Dense request segments were identified as those with continuous time intervals falling within the range of 0.9–1.1 times and shorter than one hour, or entirely less than one minute, indicating a strong temporal correlation. On the other hand, the remaining requests were classified as discrete request segments, characterized by a weaker temporal correlation. Based on these classifications, the cumulative distribution function (CDF) diagrams illustrating the time intervals for dense requests (Fig. 2(a)) and discrete requests (Fig. 2(b)) were plotted using the obtained results.

The statistical findings are as follows: among the 12,273 virtual machine creation records, a total of 3,989 dense requests were identified, with an average interval time of 1.72 min per request and a mode of 3 s per request. Additionally, there were 9,489 discrete requests, exhibiting an average interval time of 3.12 d per request, with a mode of 1.67 d per request. Dense requests constitute a certain proportion within the cluster, indicating that their occurrence is not a random phenomenon.

2.2   Analysis of the 'Space' Characteristics of Unallocated Resource Fragments

Within the depicted dense request segments of Fig. 1, each virtual machine (VM) is configured with 1 CPU and 2G of memory, while another set of VMs in the same segments are equipped with 4 CPUs and 8G of memory. It is worth noting that VMs within each dense request segment share the same CPU and memory specifications. Conversely, in the discrete request segment, the VM sizes vary, ranging from 1 CPU and 2G of memory to 16 CPUs and 32G of memory, with adjacent VMs having different sizes. During the "Locally Dense" time period, there is typically consistency in VM sizes, whereas in the "Stable Sparse" time period, VM sizes exhibit greater randomness.

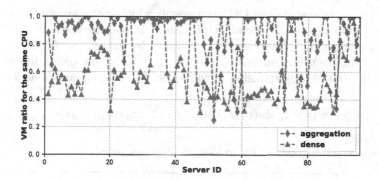

Fig. 3. Node CPU Consistency Diagram

In order to validate the universality of the observed feature across server nodes, the probabilities of CPU and size consistency were calculated for VM creation data from 96 servers within the cluster. Servers with less than 100 samples were excluded from the analysis. Figure 3 presents a probability diagram showcasing CPU consistency for each node. Blue dots represent the ratio of consistent CPUs within the dense request segments for each respective server, while red dots indicate the ratio of consistent CPUs within the discrete request segments. The results indicate that on 90.62% of the servers, the CPU consistency probability within dense request segments exceeds that within discrete request segments. Specifically, the CPU consistency probability within dense request segments is 88.01%, compared to 58.39% within the discrete request segments. Thus, the robust size consistency feature within dense segments holds true universally, both from a cluster-wide and individual server perspective.

2.3   Analysis of the "Temporal" Characteristics of Allocated Unused Resource Fragments

In order to accurately predict changes in fragment resource size, it is first necessary to distinguish which virtual machine applications exhibit periodicity, as the resource utilization of periodic virtual machines often has higher predictive accuracy.

The Fourier transform can decompose any periodic function into a combination of sine and cosine functions. In this section, the Fourier transform is derived to establish the relationship between the amplitude in the frequency spectrum and periodic and non-periodic signals. This enables the identification of periodicity in virtual machine resource fragmentation through the frequency spectrum of virtual machine utilization rates. Let $f(x)$ be the fitted function of virtual machine utilization rates, and expand $f(x)$ through Fourier transform to obtain:

$$f(x) = z_0 + \sum_{n=1}^{\infty} z_n \cos(n\omega x + \varphi_n) \tag{1}$$

In Eq. 1, z_0 represents the DC component, which is the average value of the signal and is a constant independent of time. It is also the value at $x = 0$ in the frequency spectrum. z_n represents the amplitude, $n\omega$ represents the frequency of each cosine function, and φ_n represents the phase of the cosine function.

By using the cosine expansion formula, Eq. 1 can be expanded as follows:

$$f(x) = z_0 - z_n \sin(n\omega x) \sin(\varphi_n) + \sum_{n=1}^{\infty} z_n \cos(n\omega x) \cos(\varphi_n) \tag{2}$$

After doing the integration of 2 in $[-T/2, T/2]$, we get:

$$z_n = \sqrt{(\frac{2}{T} \int_{-\frac{T}{2}}^{\frac{T}{2}} \cos(n\omega x) \cdot f(x))^2 + (\frac{2}{T} \int_{-\frac{T}{2}}^{\frac{T}{2}} \sin(n\omega x) \cdot f(x) dx)^2} \tag{3}$$

According to Eq. 3, the period T directly determines the amplitude z_n, that is, directly determines the position of the point in the frequency spectrum.

If the utilization rate of the application is periodic, then T is a constant $T \in \mathbb{C}$. Substituting the constant T into Eq. 3, we obtain:

$$z_n \in C \tag{4}$$

Therefore, the amplitude z_n is a constant, that is, **the frequency spectrum of the periodic virtual machine utilization rate through the Fourier** transform has at least one outlier point (constant z_n), and the rest of the points in the spectrum converge to and tend towards zero.

If the utilization rate is not periodic, T is considered as infinite $T \longrightarrow +\infty$ (that is, the entire domain of $f(x)$ is considered as a period), and substituting T into Eq. 3, we get:

$$z_n \longrightarrow 0 \tag{5}$$

Therefore, the amplitude z_n tends to zero, that is, **the frequency spectrum of the non-periodic virtual machine utilization rate through the Fourier transform has no outlier points.**

3 Design

In this chapter, we describe the specific workflow of the function-level resource scheduling system Tianxuan in a public cloud environment, and provide detailed descriptions of the implementation of the prediction module and scheduling module.

3.1 System Architecture

Figure 4 presents the overall architecture of the Tianxuan function-level resource scheduling system in public cloud environments, which consists of four main components: the Resource Profiler, the Resource Predictor, the Time Space, and the Function Scheduler. These four modules work together to predict resource fragmentation and schedule functions accordingly. Specifically, the Resource Profiler collects resource utilization and creation time intervals of virtual machines at runtime. Based on the collected data, the Temporal-Spatial Predictor classifies resource fragmentation into four categories and predicts their temporal and spatial resource requirements, which are then stored in the Time Space. The Function Profiler characterizes the temporal and spatial resource demands of functions, and the Temporal-Spatial Scheduler schedules functions on the available resource fragments that can satisfy their requirements. Through these steps, Tianxuan strives to ensure sufficient resources for function execution, minimize the chance of resource contention, and maximize the recycling of resource fragments.

3.2 Temporal-Spatial Predictor

Based on the "temporal-spatial" characteristics of resource fragmentation, this section designs two prediction models for unallocated resource fragments and idle resource fragments, respectively, collectively referred to as the Temporal-Spatial Predictor.

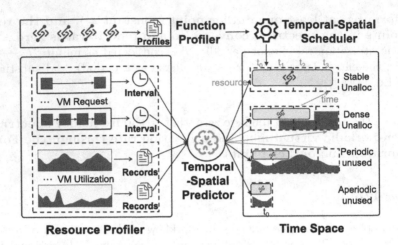

Fig. 4. Tianxuan design

3.2.1 Unallocated Resource Prediction Model

We design the "Unallocated Resource Statistical Prediction Model" based on the characteristics of unallocated resource fragments being "stable sparse, locally dense" and the "strong consistency of dense request size". Rule-based virtual machine schedulers will produce similar or identical results when receiving the same input. Therefore, the creation time of a new virtual machine on the same node is related to the creation time of the same category of requests in the historical record. The statistical classification method is suitable for predicting the creation time of this type of virtual machine.

We define $vm_i(t, c, p, s, b = 0, e = 0)$ as the information of the ith virtual machine created on the server, including the creation time t, the number of physical cores c, the scheduling node p, the remaining resources s, and the initialized Boolean dense feature b and size feature e with a default value of 0.

We first calculate the "stable sparse, locally dense" feature of each virtual machine. A dense request segment is defined as a virtual machine creation time interval that does not exceed the previous and next α times and is less than β minutes, or the time interval between the previous and next times is less than γ minutes. Based on production experience, we set $\alpha = 0.1, \beta = 60, \gamma = 1$. The Boolean value $vm_k.d$ is 1 if the virtual machine is in a locally dense time segment and 0 if it is in a stable and sparse time segment.

$$vm_k.d = ((ABS(vm_{k+1}.t - vm_k.t) < \alpha \times vm_k.t) \wedge (vm_k.t < \beta \wedge vm_{k+1}.t < \beta))$$
$$\vee (vm_k.t < \gamma \wedge vm_{k+1}.t < \gamma)$$

$$(6)$$

According to the "dense request size strong consistency" feature, we determine if the CPUs of consecutive virtual machines are equal. The Boolean value $vm_k.e$ is 1 if the virtual machine meets the feature of dense size consistency, and 0 otherwise:

$$vm_k.e = (vm_k.c == vm_{k+1}.c) \tag{7}$$

The virtual machines in the sequence that satisfy the dense features 6 and 7 are divided into a dense request sequence $X^{u:v}$, which means that the dense features $d = 1, e = 1$ are all present from the u-th virtual machine to the v-th virtual machine.

$$\prod_{k=u}^{v} vm_k.d \times vm_k.e = 1 \tag{8}$$

The dense request data structure $dense$ is created by transforming the dense request sequence $X^{u:v}$, where $d = 1, e = 1$. The $dense$ structure contains the average time interval t_{mean} between virtual machines in the sequence, the minimum interval t_{min}, the maximum interval t_{max}, the average resource consumption r_{mean} of the sequence, the average remaining resources r_{spare} of the node, and the node p where the virtual machines are scheduled.

$$
\begin{aligned}
dense.t_{mean} &= \left(\frac{\sum_{k=u}^{v}(vm_{k+1}.t - vm_k.t)}{v - u + 1} \right) \\
dense.t_{max} &= MAX(\{vm_{k+1}.t - vm_k.t | v >= k >= u\}) \\
dense.t_{min} &= MIN(\{vm_{k+1}.t - vm_k.t | v >= k >= u\}) \\
dense.r_{mean} &= \left(\frac{\sum_{k=u}^{v} vm_k.r}{v - u + 1} \right) \\
dense.r_{spare} &= \left(\frac{\sum_{k=u}^{v} vm_k.s}{v - u + 1} \right)
\end{aligned}
\tag{9}
$$

The dense data structure, which contains dense requests across all nodes in the cluster, is input to the KNN model with higher accuracy, is shown in steps 1 and 2 of Fig. 5. The features used are the average time interval between virtual machine creations t_{mean} and the remaining resources on the node r_{spare}. The object is composed of the minimum time interval t_{min}, maximum time interval t_{max}, mean resource quantity r_{mean}, and average remaining resources on the node r_{spare}. The KNN model is initialized based on these features, and the model matches the feature vector with the shortest Euclidean distance to the average time interval and remaining resources on the node.

The prediction process is shown in steps 3 and 4 of Fig. 5. First, the information of the two most recent VM creations on the node where the model resides is collected. The remaining resources on the dense segment closest to the current node $dense.rspare$ are chosen as the predicted result.

3.2.2 Resource Regression Prediction Model for Fragmented and Unused Resources

Due to the fact that Fourier Transform can distinguish whether a fragment has periodicity only based on resource utilization in black box scenarios, and the utilization curve of periodicity has similar numerical values of the dependent variable utilization rate corresponding to the independent variable time every

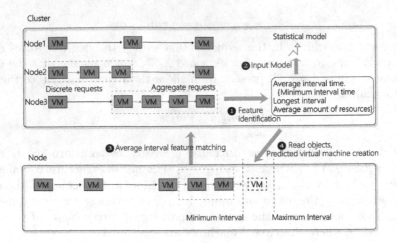

Fig. 5. Unallocated resources statistical forecasting model workflow

period, it is suitable to use a regression model for prediction. Based on the spatiotemporal characteristics of the fragmented and unused resources, this section designs a fragmented and unused resource regression prediction model.

Given historical virtual machine CPU utilization sequence Y_1^0, perform Fourier transform FT to obtain all amplitude points z_n^0.

$$z_n^0 = FT(Y_1^0) \tag{10}$$

As the periodic virtual machine utilization rate has at least one outlier point in the frequency spectrum after being transformed by Fourier Transform and the other points in the frequency spectrum converge and tend to zero, a given periodic amplitude z_n^1 contains a point greater than the other points by δ times and is of the same order of magnitude at points in the domain $[\varepsilon * n : n]$. According to production experience, $\delta = 10, \varepsilon = 3/4$ are proposed.

$$z_n^1 = (\exists z_i > \delta \times z_t, |, z_t \in z_n - z_i)$$
$$\wedge (\ z_p/z_q < \delta, |, p, q \in [\varepsilon * n : n]) \tag{11}$$

The Fourier inverse transform (IFT) is performed on the periodic amplitude to obtain the periodic utilization rate Y_1^1:

$$Y_1^1 = IFT(z_n^1) \tag{12}$$

After screening out the periodic utilization rate Y_1^1, this section selects the median filter as the filter for the model, which has the relatively best prediction accuracy and convergence speed. A median filter with a filtering amplitude of 3 is performed on the periodic utilization rate Y_1^1 to filter out noise points, obtaining a new utilization rate Y_1^2:

$$Y_1^2 = (Median(fi - 1, fi, f1 + 1)|fi - 1, fi, f1 + 1 \in Y_1^1) \tag{13}$$

After completing operations such as waveform filtering and noise reduction, the utilization rate generated is usually difficult to maintain consistently high across different periods, resulting in fluctuations. The LSTM model is chosen as the regression model. We train a multi-step LSTM model on sample Y_1^2 and input the time scale of the next m seconds $[t+1 : t+m]$ into the model to obtain the spatio-temporal prediction results of periodic allocated unused resource fragments $R_{periodic}$. The LSTM model for waveform reconstruction is constructed as follows:

$$Rperiodic = LSTM_{Y_1^2}(t + 1 : t + m) \qquad (14)$$

As for the aperiodic allocated unused resource fragments $R_{aperiodic}$, which do not possess periodicity and exhibit severe fluctuations, we collect their spatial resource size every second. The time interval for each collection of such resources is one second.

3.3   Temporal-Spatial Scheduler

The temporal scheduler sets different scheduling priorities for the functions depending on their execution time and available resource fragments. The scheduling process is as follows: first, the function is executed in a black box and its resource consumption and execution time are recorded (Algorithm 1, line 1). The variance of the phase space of all nodes and the current free resources of the nodes are calculated and the top K nodes with optimal stability are selected (Algorithm 1, lines 4 to 8).

When the function execution time is less than one second, it is a short function and is scheduled on small-volume and highly volatile resource fragments first. The algorithm checks whether the capacity of a non-periodic partitioned and unused resource segment, a periodic partitioned and unused resource segment, or an unallocated resource segment meets the function's resource requirement.(Algorithm 1, lines 9 to 23). When the function execution time is greater than one second, the algorithm checks whether the capacity of an unallocated resource segment array, a periodic partitioned but unused segment array, or the interval corresponding to their sum is greater than the function resource array. (Algorithm 1, lines 24 to 40).

In summary, the temporal scheduler uses short functions that should recover resource fragments with small capacity and high volatility, and ensures that long functions prioritize the consumption of resource fragments with stable volatility, i.e., safeguarding function performance and fully recovering resources.

4   Implementation

The Tianxuan system is designed based on the open-source serverless computing framework OpenFaaS. The system reused many modules from OpenFaaS, primarily modifying the function scheduling logic and adding a resource view module and a prediction module for function scheduling services. The system consists of approximately 5000 lines of Go code and 400 lines of Python code. The module

---

Algorithm 1. Schedule function

Require: input function f
Ensure: output result

```
 1: f_cpu, f_time = Charactor(f);
 2: Phase_Space[] = Uuallocated[] + Periodic_Allocated[] + Aperiodci_Allocated;
 3: index = Time.Now() - Phase.Start_Time();
 4: for node in Global_Node do
 5:     node.variance = get_variance(node.Phase_Space[]);
 6:     node.resource = node.Phase_Space[index];
 7: end for
 8: node = Select top K of Global_Node.variance where Global_Node.resource >f_cpu
    ;
 9: if f_time <1 then
10:     if f_cpu <Aperiodci_Allocated then
11:         Aperiodci_Allocated -= f_cpu;
12:     else
13:         if f_cpu <Periodic_Allocated[index] then
14:             Periodic_Allocated[index] -= f_cpu;
15:         else
16:             if f_cpu <Uuallocated[index] then
17:                 Uuallocated[index] -= f_cpu ;
18:             else
19:                 return False;
20:             end if
21:         end if
22:     end if
23: else
24:     f_arrey = f[f_time];
25:     if Uuallocated[index : index+f_time] - f_arrey[] >0 then
26:         Uuallocated[index : index+f_time] -= f_arrey[];
27:     else
28:         if Periodic_Allocated[index : index+f_time] - f_arrey[] >0 then
29:             Periodic_Allocated[index : index+f_time] -= f_arrey[];
30:         else
31:             if Uuallocated[index : index+f_time] + Periodic_Allocated[index :
    index+f_time] - f_arrey[] >0 then
32:                 Uuallocated[index : index+f_time] = 0 ;
33:                 difference[] = f_arrey[] - Uuallocated[index : index+f_time];
34:                 Periodic_Allocated[index : index+f_time] -= difference[];
35:             else
36:                 return False;
37:             end if
38:         end if
39:     end if
40: end if
41: return True;
```

---

connections are as follows: the fragment collector collects runtime data from virtual machines and sends it to the spatiotemporal prediction model, which then sends the predicted results to the main node's spatiotemporal resource view for storage. User function requests are transmitted through the OpenFaaS API gateway and are received by the spatiotemporal scheduler. The scheduler selects the appropriate scheduling node based on the spatiotemporal resource views of all nodes in the cluster and sends the function's configuration file to Kubernetes via an instance creator, ultimately creating the function on the corresponding worker node.

5 Evaluation

5.1 Methodology

Testbed. We conducted our experiments on five servers, each equipped with 24 CPUs and 96 GB of memory. The CPU model used in the servers is the Intel(R) Xeon(R) Gold 5118 CPU, with a clock frequency of 2.30 GHz, 32K of level-1 cache, 4M of level-2 cache, and 16M of level-3 cache. The CPU architecture is x86_64, and the operating system used is CentOS Linux release 7.3.1611. We employed KVM as the virtual machine hypervisor, and Kubernetes and OpenFaaS as the function management frameworks.

Table 1. Function loads

Function Name	Pressure	SLA	CPU Consumption
Floating Point Operation	CPU	0.003 s	0.2
Model Training	CPU	9.8 s	14.58
K Nearest Neighbors	CPU	4.25 s	2.134
Naive Bayes	CPU	1.38 s	2.563
Random Forest	CPU	2.80 s	2.112
Support Vector Machine	CPU	1.37 s	1.975
Disk Read/Write	Memory	0.045	0.457
Compression	Memory	0.063 s	0.311
Network Speed Test	Network	9.56 s	0.0125
JSON Dump	Network	0.204 s	0.0012

Workload Selection. We selected virtual machine applications from Cloudsuite Bench [12], including Web Serving, Graph Analytics, Data Analytics, as well as Redis [7], Mysql [6], and LSTM [8]. Each virtual machine was configured with 4 cores and 8 GB of memory.

The functions used in this experiment include six functions from Function Bench [17] and four machine learning functions: KNN, RF, SVM, and NB. The function workloads are shown in Table 1.

Evaluation Metrics. The evaluation metrics include resource utilization, function throughput (throughput = number of jobs executed/resource consumption/task time), short function 99th percentile tail latency, LC service 99th percentile tail latency, and long function tail latency.

Comparison Method. We compare Tianxuan system with two existing systems, Servermore [28] and Harvestless [31], as follows:

(1) Servermore [28]: Servermore adopts a feedback scheduling algorithm, which evaluates the standard deviation and variance of the IPC (the number of instructions executed by the CPU per clock cycle) of the application at each moment. When the IPC does not meet the expected value, the system adjusts the resource allocation for the next moment through feedback.
(2) Harvestless [31]: Harvestless employs a load balancing algorithm. When a function request arrives, the controller selects its working set based on the function type and sends the request to the scheduling program in the node through Kafka. To maintain load balance, the scheduling program calculates whether the ratio of the current node's CPU usage to its harvestable CPU is greater than a threshold. If so, the function execution is delayed; otherwise, it is executed immediately.

5.2 Model Prediction Accuracy Evaluation

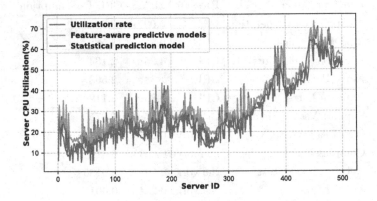

Fig. 6. Comparison of model prediction results

The feature-aware prediction model compares Servermore [29]'s statistical prediction model for experiments to predict the future amount of idle resources by counting the CPU and IPC mean and standard deviation of the moving window.

This section simulates 5-minute virtual machine request record from server 168 of the Unicloud Beijing cluster was used. Virtual machines with 4CPU8G were deployed at 0 s, 90 s, 170 s, 380 s, 440 s, and 480 s. The applications within the virtual machine are in order Web Serving, Graph Analytics, Data Analytics [12], Redis [7], MySQL [6], and LSTM [8].

The resource utilization is shown in Fig. 6. When the acceptable threshold for successful prediction is set to an error between predicted value and real value of less than 1 CPU (4.16% of 24 cores), the Servermore statistical prediction model successfully predicted 114 points, with a prediction accuracy of only 22.80%, while the feature-aware prediction model designed in this chapter successfully predicted 346 points, with a prediction accuracy of 69.20%. When successful prediction is defined as an error of less than 2 CPU (8.32%), the Servermore statistical prediction model successfully predicted 371 points, with a prediction accuracy of only 74.20%, while the feature-aware prediction model designed in this chapter successfully predicted 1,023 points, with a prediction accuracy of 89.20%.

5.3 Mixing Effect Evaluation

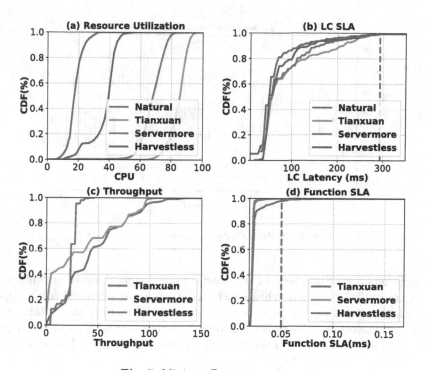

Fig. 7. Mixing effect comparison

This section evaluates the results of the hybrid deployment of Tianxuan and Servermore, Harvestless. Each scenario deploys six 4C8G VMs running for 20 min to simulate a hybrid workload in a cloud environment. Before executing the hybrid plan, six VM services are run in a non-disruptive environment and the CPU utilization of the server and the tail latency of the LC service are recorded as a control group. Then run Tianxuan, Servermore and Harvestless separately and record the CPU utilization of the server, the tail latency of the LC service, the throughput of the function and the tail latency of the function service.

The results are shown in Fig. 7.(a) Graph showing the average utilization of 84.32%, 67.74%, and 36.91% for Tianxuan, Servermore, Harvestless, and Natural, respectively. (b) Plots showing the p99th percentile latencies of 293ms, 307ms, 297ms and 280ms for Natural, Tianxuan, Servermore and Harvestless, respectively.(c) Plots showing the average throughput of 44.05, 32.77 and 21.68 for Tianxuan, Servermore and Harvestless, respectively.(d) Plots shows that 98%, 99% and 100% of function requests of Tianxuan, Servermore and Harvestless are completed within SLA 0.05 s, respectively. In summary Tianxuan can effectively improve the throughput with guaranteed service quality.

5.4   System Overhead Evaluation

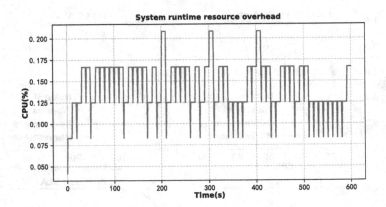

Fig. 8. System runtime resource overhead

Figure 8 displays the overhead of the master node during the system runtime for ten minutes. The system overhead includes the running overhead of the Open-FaaS platform and the model inference overhead of reading the database per second. The average CPU utilization of the system is 0.145%, and the highest CPU utilization is 0.247%. Therefore, the system overhead is low and can be neglected.

6 Related Work

Recent hybrid deployment techniques fall into the following four main types:

(1)Resource Allocation Technology Based on SLO Feedback. Heracles [19] uses a software and hardware collaboration mechanism to isolate workloads and uses the gradient descent method to select the optimal resource allocation for BE jobs. Rhythm [33] packages and deploys more BE jobs on LC services with strong anti-interference ability by characterizing the anti-interference ability between LC components. PARTIES [10] tracks various SLOs to jointly locate multiple LC workloads and manages resources based on the degree of proximity between each workload and its SLO. ICE [21] reduces the number of requests on servers with poor SLOs through load balancing algorithms. Q-Clouds [25] builds a multi-input multi-output model using feedback to capture interference sources.

(2)Resource Allocation Technology Based on Offline Characterization of Interference Perception. Bubble-Up [23] characterizes the curve model of the QoS of applications fluctuating with pressure to jointly locate the performance interference between applications. Bubble-flux [30] accurately predicts the degree of influence of mixed business on the QoS of the original business by dynamically detecting the sequential pressure of shared hardware on the server. DeepDive [26] determines whether VMs are affected by interference by comparing them with offline-characterized underlying indicators. Dirigent [34] dynamically predicts the completion time of tasks in the service execution process through a lightweight runtime and completion time predictor.

(3)Online predictive load balancing technology. Zhang [32] predicted the resources required for VMs using the historical utilization of CPU and disk. Scavenger [15] calculated the resource reuse value of memory, CPU, and network bandwidth by observing the average and standard deviation of the time window for virtual machine service resource usage, monitored the IPC of virtual machines to provide feedback for adjusting resource allocation values. Server-More [28] replaced IPC with the Cohen coefficient of IPC (the ratio of the average difference to the standard deviation), which can better evaluate CPU performance degradation based on Scavenger. Harvest VM [9] predicts the survival time of HVM by inputting the number of servers and racks in the cluster, hardware information, and the total amount of CPU memory into a machine learning model. Harvestless [31] builds a load balancing algorithm internally, calculates the ratio of CPU usage to CPU harvestable on nodes, decides the degree of resource shortage and whether to place functions, and reduces cold start by establishing the minimum working set of functions. MHVM [13] recovers memory resources from unallocated resources, reduces the time for recovering memory using batch processing and pre-recovery mechanisms, and reduces the possibility of NUMA crossing by lowering memory pages using unallocated memory buffer.

7   Conclusion

This paper first describes the existence of unallocated resources and allocated unused resources in public cloud environments, analyzes the fluctuation characteristics of these two types of resources, and establishes statistical and regression prediction models based on their different characteristics and the relationships between them and functions. We propose "Tianxuan", which is based on spatio-temporal prediction and scheduling to determine whether the resource demands of functions during their lifecycles can be satisfied, thus safely scheduling the execution of functions. Our research results show that Tianxuan can maximize resource reclamation while ensuring the service quality of both the original service and function services.

8   Declarations

8.1   Ethical Approval

Not applicable.

8.2   Competing Interests

Not applicable.

8.3   Authors' Contributions

Not applicable.

8.4   Funding

This work is supported by the National Key Research and Development Program of China No.2022YFB4500702; project ZR2022LZH018 supported by Shandong Provincial Natural Science Foundation; the National Natural Science Foundation of China under grant 62141218, 62372322 and the open project of Zhejiang Lab (2021DA0AM01/003).

Author contributions. Yukang Chu and Wenhao Huang : contributed equally to this work.

References

1. Alibaba. [cp/ol]. https://github.com/alibaba/clusterdata (2018).
2. Unicloud. [cp/ol]. https://www.unicloud.com/ (2018)
3. Azure. [cp/ol]. https://github.com/Azure/AzurePublicDataset (2019)
4. Google. [cp/ol]. https://github.com/google/cluster-data (2019)
5. Openfaas. openfaas [cp/ol]. https://www.openfaas.com/ (2021)
6. Mysql. [cp/ol]. https://www.mysql.com/ (2022)

7. Redis. [cp/ol]. https://redis.com/ (2022)
8. Abadi, M., et al.: {TensorFlow}: a system for {Large-Scale} machine learning. In: 12th USENIX Symposium on Operating Systems Design and Implementation (OSDI 16), pp. 265–283 (2016)
9. Ambati, P., et al.: Providing {SLOs} for {Resource-Harvesting}{VMs} in cloud platforms. In: 14th USENIX Symposium on Operating Systems Design and Implementation (OSDI 20), pp. 735–751 (2020)
10. Chen, S., Delimitrou, C., Martínez, J.F.: Parties: Qos-aware resource partitioning for multiple interactive services. In: Proceedings of the Twenty-Fourth International Conference on Architectural Support for Programming Languages and Operating Systems, pp. 107–120 (2019)
11. Cook, H., Moreto, M., Bird, S., Dao, K., Patterson, D.A., Asanovic, K.: A hardware evaluation of cache partitioning to improve utilization and energy-efficiency while preserving responsiveness. ACM SIGARCH Comput. Architect. News **41**(3), 308–319 (2013)
12. Ferdman, M., et al.: Clearing the clouds: a study of emerging scale-out workloads on modern hardware (2012). http://infoscience.epfl.ch/record/173764
13. Fuerst, A., et al.: Memory-harvesting vms in cloud platforms. In: Proceedings of the 27th ACM International Conference on Architectural Support for Programming Languages and Operating Systems. pp. 583–594 (2022)
14. Govindan, S., Liu, J., Kansal, A., Sivasubramaniam, A.: Cuanta: quantifying effects of shared on-chip resource interference for consolidated virtual machines. In: Proceedings of the 2nd ACM Symposium on Cloud Computing, pp. 1–14 (2011)
15. Javadi, S.A., Suresh, A., Wajahat, M., Gandhi, A.: Scavenger: A black-box batch workload resource manager for improving utilization in cloud environments. In: Proceedings of the ACM Symposium on Cloud Computing, pp. 272–285 (2019)
16. Jonas, E., et al.: Cloud programming simplified: a berkeley view on serverless computing. arXiv preprint arXiv:1902.03383 (2019)
17. Kim, J., Lee, K.: Practical cloud workloads for serverless faas. In: Proceedings of the ACM Symposium on Cloud Computing., pp. 477–477 (2019)
18. Li, Z., Guo, L., Cheng, J., Chen, Q., He, B., Guo, M.: The serverless computing survey: a technical primer for design architecture. ACM Comput. Surv. (CSUR) **54**(10s), 1–34 (2022)
19. Lo, D., Cheng, L., Govindaraju, R., Ranganathan, P., Kozyrakis, C.: Heracles: Improving resource efficiency at scale. In: Proceedings of the 42nd Annual International Symposium on Computer Architecture, pp. 450–462 (2015)
20. Machina, J., Sodan, A.: Predicting cache needs and cache sensitivity for applications in cloud computing on cmp servers with configurable caches. In: 2009 IEEE International Symposium on Parallel & Distributed Processing, pp. 1–8. IEEE (2009)
21. Maji, A.K., Mitra, S., Bagchi, S.: Ice: An integrated configuration engine for interference mitigation in cloud services. In: 2015 IEEE International Conference on Autonomic Computing, pp. 91–100. IEEE (2015)
22. Manikantan, R., Rajan, K., Govindarajan, R.: Probabilistic shared cache management (prism). In: 2012 39th Annual International Symposium on Computer Architecture (ISCA), pp. 428–439. IEEE (2012)
23. Mars, J., Tang, L., Hundt, R., Skadron, K., Soffa, M.L.: Bubble-up: Increasing utilization in modern warehouse scale computers via sensible co-locations. In: Proceedings of the 44th annual IEEE/ACM International Symposium on Microarchitecture, pp. 248–259 (2011)

24. Mei, Y., Liu, L., Pu, X., Sivathanu, S., Dong, X.: Performance analysis of network i/o workloads in virtualized data centers. IEEE Trans. Serv. Comput. **6**(1), 48–63 (2011)
25. Nathuji, R., Kansal, A., Ghaffarkhah, A.: Q-clouds: managing performance interference effects for qos-aware clouds. In: Proceedings of the 5th European Conference on Computer Systems, pp. 237–250 (2010)
26. Novaković, D., Vasić, N., Novaković, S., Kostić, D., Bianchini, R.: {DeepDive}: Transparently identifying and managing performance interference in virtualized environments. In: 2013 USENIX Annual Technical Conference (USENIX ATC 13), pp. 219–230 (2013)
27. Srikantaiah, S., Kandemir, M., Wang, Q.: Sharp control: controlled shared cache management in chip multiprocessors. In: Proceedings of the 42nd Annual IEEE/ACM International Symposium on Microarchitecture, pp. 517–528 (2009)
28. Suresh, A., Gandhi, A.: Servermore: Opportunistic execution of serverless functions in the cloud. In: Proceedings of the ACM Symposium on Cloud Computing, pp. 570–584 (2021)
29. Wang, Y., et al.: Smartharvest: harvesting idle cpus safely and efficiently in the cloud. In: Proceedings of the Sixteenth European Conference on Computer Systems, pp. 1–16 (2021)
30. Yang, H., Breslow, A., Mars, J., Tang, L.: Bubble-flux: Precise online qos management for increased utilization in warehouse scale computers. ACM SIGARCH Comput. Arch. News **41**(3), 607–618 (2013)
31. Zhang, Y., et al.: Faster and cheaper serverless computing on harvested resources. In: Proceedings of the ACM SIGOPS 28th Symposium on Operating Systems Principles, pp. 724–739 (2021)
32. Zhang, Y., Prekas, G., Fumarola, G.M., Fontoura, M., Goiri, Í., Bianchini, R.: {History-Based} harvesting of spare cycles and storage in {Large-Scale} datacenters. In: 12th USENIX Symposium on Operating Systems Design and Implementation (OSDI 16), pp. 755–770 (2016)
33. Zhao, L., et al.: Rhythm: component-distinguishable workload deployment in datacenters. In: Proceedings of the Fifteenth European Conference on Computer Systems, pp. 1–17 (2020)
34. Zhu, H., Erez, M.: Dirigent: enforcing qos for latency-critical tasks on shared multicore systems. In: Proceedings of the Twenty-first International Conference on Architectural Support for Programming Languages and Operating Systems, pp. 33–47 (2016)

IKE: Threshold Key Escrow Service with Intermediary Encryption

Yang Yang[1], Bingyu Li[1(✉)], Shihong Xiong[1], Bo Qin[3], Yan Zhu[1], Haibin Zheng[2], and Qianhong Wu[1]

[1] School of Cyber Science and Technology, Beihang University, Beijing 100191, China
{y_yang,libingyu,sy2239215,zhuyan,qianhong.wu}@buaa.edu.cn
[2] Hangzhou Innovation Institute, Beihang University, Hangzhou 310051, China
zhenghaibin29@buaa.edu.cn
[3] School of Information, Renmin University of China, Beijing 100191, China
bo.qin@ruc.edu.cn

Abstract. Blockchain has gained significant attention for its potential to revolutionize various fields. The security of the blockchain system heavily relies on private key management, and traditional key management schemes pose a higher risk of key compromise or loss. A general key management solution with security supervision is needed to overcome the problems in the traditional key management scenario and ensure the long-term development of the blockchain system. Drawing upon the aforementioned principles, this paper proposes a threshold key escrow scheme with intermediaries that affords secure, generic and convenient distributed key escrow services. The scheme utilizes Paillier encryption and intermediary encryption to provide two layers of encryption protection for the escrow key. Moreover, we consider the recovery needs of committee member changes and key share losses in practical scenarios and propose secure methods to obtain new shares. Finally, we prove the correctness and semantic security of the scheme; and through theoretical and the practicability and efficiency of the scheme in practical applications are verified through theoretical and experimental analysis.

Keywords: Key Escrow · Intermediary Encryption · Threshold Cryptography · Blockchain

1 Introduction

Blockchain technology has gained widespread attention due to its unique features of transparency, immutability, and decentralization [1]. The security of blockchain systems heavily relies on the management of private keys, which are

Supported by Beijing Natural Science Foundation M21031; National Key R & D Program of China 2022YFB2702901; National Natural Science Foundation of China U21A20467, 61932011, 61972019, 62002011; Youth Top Talent Support Program of Beihang University under Grant YWF-22-L-1272.

used to access and control digital assets [2]. In blockchain's distributed scenario, private keys determine the ownership of UTXO or account, making their protection and attack a lingering concern since the inception of blockchain [3].

Traditional key management schemes for blockchain wallets involve storing private keys on a local device, such as a computer, or on a centralized server, such as a cryptocurrency exchange [4]. However, these schemes are vulnerable to various attacks, including phishing, malware, physical theft, and server hacking. Furthermore, this traditional storage method is not user-friendly. For example, the cold wallet [5] is completely stored offline, which brings great inconvenience to the use and people may lose or damage their keys due to their own mistakes, potentially resulting in significant losses [6]. Most current key management solutions are tailored to specific scenarios, such as transportation [7] or supply chain [8], which lack of universality. Key management solutions designed for large-scale systems often integrate specific system requirements, making them non-portable. Therefore, these solutions are not user-friendly for individual users.

Moreover, supervision is a significant challenge for blockchain projects to be implemented on a large scale. For example, the U.S. government requires encryption products to comply with encryption standards that support key recovery mechanisms for law enforcement agencies when needed [9]. The Chinese government also stated that for the criminal acts of blockchain information services, it is necessary to implement supervision, quickly identify the responsible person, and help the victims [10]. A robust key escrow scheme ensures that individuals do not need to fully disclose their private keys to the power departments, but through the form of key share encryption escrow, which guarantees adequate supervision of blockchain systems without compromising the security of the key.

The key management scheme is an important support for the development of blockchain [11]. However, the form of local storage of the private key brings great inconvenience to use and cannot be used for the supervision of blockchain projects. There is a lack of universality and convenience in traditional key escrow schemes, leading to neglecting individual users' or group systems' key escrow requirements. Thus, a more general, secure, and flexible key escrow solution is needed to cater to diverse users' needs.

To address the limitations of traditional key escrow schemes, this paper proposes a threshold key escrow service with intermediary encryption based on the blockchain. We have built a secure key escrow service in the blockchain scenario by providing services through committees. Moreover, we consider the change of committee members or shares and provide a new key escrow share without compromising the shares of the remaining members. We build a threshold key escrow service solution with strong universality and high security.

In terms of scheme design, our proposal begins with a basic data encryption scheme with intermediaries, utilizing the Paillier and Elgamal schemes. Building upon this foundation, we incorporate threshold cryptography technology to establish a committee as intermediaries, providing security protection. By combining this with blockchain, we achieve distributed escrow and verification of key shares, ultimately providing secure, generic and convenient distributed key escrow services for blockchain-related key scenarios.

We implement and test the data encryption scheme with intermediaries and our IKE scheme using Python 3.9. For the protocol mentioned in the paper, we have analyzed and proved its correctness and semantic security. Regarding the IKE protocol, we conducte a time test on each stage of the protocol operation under the condition of different numbers of nodes and different thresholds, and the test results show that the efficiency of the protocol reached expectations.

In summary, the contributions of this paper are as follows:

1. We propose a secure data encryption scheme with intermediaries and the scheme can be extended in terms of threshold and homomorphism, which can be of independent interests.
2. On the basis of the basic encryption scheme, a complete threshold key escrow scheme with intermediate encryption is designed, which provides practical and secure key management services, and fully considers special circumstances such as replacement of members or loss of shares.
3. We also analyze the correctness and semantic security of the scheme, and perform performance tests on the prototype system to prove its feasibility.

2   Related Work and Preliminaries

2.1   Related Work

In 1992, Micali [12] proposed the first key escrow scheme, which laid the groundwork for developing this system. However, the scheme was not practical enough due to its inefficiency in selecting the number of trustees. In the following years, subsequent research by experts such as Kilian [13] clarified the concept and attributes of key escrow. Early research on key escrow systems focused mainly on ensuring the security, with less attention given to additional attributes.

To further improve the security of the key escrow scheme, Micali et al. [14] later proposed the GPKE system, which includes enhanced verification to ensure the authenticity of the trustee's possession of the key fragment. Bellare et al. [15] added the publicly verifiable property to the scheme, which improves efficiency and guarantees correctness in establishing partially escrowed keys.

The increasing popularity of blockchain technology has led to heightened demand for secure private key protection and expanded the application scenarios of key escrow solutions. Lei et al. [7] proposed a dynamic key management scheme for heterogeneous intelligent transportation systems based on blockchain, which can enhance the scheme's efficiency and robustness. Ma et al. [16] designed a distributed key management scheme based on the blockchain-based IoT layered architecture that offers good scalability in privacy protection but comes with high overhead. Cha et al. [8] optimized the security of the long-life cycle supply chain by designing a framework for the blockchain and key escrow encryption system based on the weapon supply chain system's application requirements. In response to the challenge of key custody in traditional public key encryption systems, Mwitende et al. [17] proposed a pairing-based certificateless scheme to solve the key escrow problem. However, due to the limited scenarios and inconvenient use of these schemes, there is a need a general, secure, and flexible key escrow scheme that can cater to diverse users' needs.

2.2 Carmichael Function

Carmichael function $\lambda(n)$ of a positive integer n is the smallest positive integer that holds for every integer a coprime to n.

Typically, $a^{\lambda(n)} \equiv 1 \bmod n$.

2.3 Paillier Cryptosystem

The Paillier scheme [18] is a probabilistic public-key encryption algorithm based on the complicated problem of decisional composite residuosity assumption.

Definition 1. *Decisional Composite Residuosity Assumption. The decisional composite residuosity assumption (DCRA) is a mathematical assumption commonly used in cryptography. Given a composite integer n and an integer z, it is difficult to determine whether z is an n-residue modulo n^2, meaning whether there exists an integer y such that $z \equiv y^n \;(\mathrm{mod}\, n^2)$.*

Key Generation: Randomly select two prime numbers p and q, and compute $N = pq$. The $lcm()$ function means to find the least common multiple of two numbers. For convenience, we directly define $\lambda = lcm(p - 1, q - 1)$.

Select random integer g where $g \in \mathbb{Z}_{N^2}^*$, while ensure $\mu = \left(L\left(g^\lambda \bmod N^2\right)\right)^{-1}$ exists. Function $L(x)$ is defined as $L(x) = \frac{x-1}{N}$. Since $g = N + 1$ satisfies the condition, $g = N + 1$ is directly taken in the scheme. Then select a random seed b to compute another generator $h = b^N$ while ensure $ord_{N^2}(h) = \varphi(N)$. Finally, the system public key is (N, g, h) and the private key is (λ, μ).

Encryption: For any plaintext message $m \in \mathbb{Z}_N$, choose a random number $r \in \mathbb{Z}_N^*$ and calculate the ciphertext $c \equiv g^m h^r \bmod N^2$.

Decryption: Calculate the plaintex m as: $m \equiv L\left(c^\lambda \bmod N^2\right) * \mu \bmod N \equiv \frac{L\left(c^\lambda \bmod N^2\right)}{L\left(g^\lambda \bmod N^2\right)} \bmod N$.

2.4 Proof of Equality of Discrete Logarithms

We use a variant of non-interactive Chaum-Pedersen protocol using the Fait-Shamir heuristic. The prover commits to $P = g_1^a h_1^b h_2^d$ and $Q = g_2^a$ publicly and proves that a is equal and the prover knows b and d.

1. Randomly select r_1, r_2, r_3, compute $t_1 = g_1^{r_1} h_1^{r_2} h_2^{r_3}$ and $t_2 = g_2^{r_1}$. Given a hash function H(), compute challenge value $c = \mathrm{H}(g_1, h_1, h_2, P, Q, t_1, t_2)$.
2. Then compute $s_1 = r_1 + a \cdot c$, $s_2 = r_2 + b \cdot c$, $s_3 = r_3 + d \cdot c$, and output the proof pair $\pi = (t_1, t_2, s_1, s_2, s_3)$ of discrete equality.

In the verification phase, the verifier evaluates whether $g_1^{s_1} \cdot h_1^{s_2} \cdot g_2^{s_1}$ and $P^c \cdot t_1$, $g_2^{s_1}$ and $Q^c \cdot t_2$ are equal.

3   Warm-Up: A Basic Data Encryption Scheme with Intermediaries

3.1   Overview

The warm-up scheme in this section implements the basic encryption scheme with intermediaries, laying the foundation for the IKE key escrow scheme in Sect. 4. In traditional encryption scenarios, there may be vulnerable to man-in-the-middle attacks. To mitigate this issue, the intermediary encryption method is used, which adds a trusted intermediary between the sender and receiver and establishes a double public key encryption system. The scheme utilizes the reliable Paillier encryption scheme in combination with ElGamal to form a secure intermediary encryption system. Before decrypting a message, the decryption requester must first obtain partial decryption of the message from the intermediary, improving the overall security of the traditional encryption scheme.

3.2   Description

In this section, we will provide a detailed explanation of the overall scheme.

System Identity Initialization: Everyone needs to initialize their Paillier public-private key pair. Bob, the message receiver, follows the steps in 2.3 to complete the initialization. Then, the public key of Bob is (N, g, h) and the Bob's private key is (λ, μ). In addition, An intermediary randomly selects $\alpha \in \mathbb{Z}_{N^2}^*$ and generates its own independent encryption public-private key pair $(x, y = \alpha^x)$.

Message Encryption: Alice, the message sender, selects random element r and s from $\mathbb{Z}_{N^2}^*$ and sends the ciphertext pair $CN = (CN_1, CN_2)$ to Bob, where $CN_1 \equiv g^M h^r y^s \bmod N^2, CN_2 \equiv \alpha^s \bmod N^2$.

Intermediary Decryption: Bob sends the ciphertext CN_2 to the intermediary. Then the intermediary uses the private key x to perform a preliminary decryption operation on the ciphertext CN_2, and returns the decrypted message $E \equiv CN_2^x \equiv \alpha^{sx} \bmod N^2$ to Bob.

Message Decryption: Bob uses the partially decrypted message E to calculate the ciphertext $C \equiv CN_1/E \equiv g^M h^r \bmod N^2$ encrypted with his own public key. After obtaining C, Bob runs the Paillier decryption scheme to decrypt the encrypted plaintext message $M \equiv L\left(C^\lambda \bmod N^2\right) * u \bmod N$.

3.3   Proof of Correctness and Security

Correctness. According to the definition of Carmichael theorem mentioned in 2.2, $\lambda\left(N^2\right) = lcm\left(\lambda\left(q^2\right), \lambda\left(p^2\right)\right) = lcm\left(\phi\left(q^2\right), \phi\left(p^2\right)\right) = lcm(q(q-1), p(p-1)) = pq(lcm(p-1, q-1)) = N\lambda$.

Then, given any $w \in \mathbb{Z}_{N^2}^*$, $w^{N\lambda} \equiv 1 \bmod N^2$. For instance, $h = b^N$, then $h^{r\lambda} \equiv b^{Nr\lambda} \equiv 1 \bmod N^2$.

$C^\lambda \equiv g^{M\lambda}h^{r\lambda} \equiv g^{M\lambda} \equiv (1+N)^{M\lambda} \equiv 1+NM\lambda \bmod N^2$, and $g^\lambda \equiv (1+N)^\lambda \equiv 1 + \lambda N \bmod N^2$.

Therefore, $M \equiv \frac{L(C^\lambda \bmod N^2)}{L(g^\lambda \bmod N^2)} \equiv \frac{M\lambda \bmod N^2}{\lambda \bmod N^2} \bmod N$.

Security. The ElGamal's security is based on the decisional Diffie-Hellman problem. In the scheme, the intermediary only has access to the ciphertext CN_2, and hence, cannot obtain any information about the plaintext M. Similarly, the receiver cannot obtain any information about the intermediary's private key x from the message E. Additionally, the final Paillier encryption of the scheme satisfies the standard security definition of encryption schemes: semantic security, i.e., ciphertext indistinguishability under chosen plaintext attack (IND-CPA). Through security analysis, the scheme's security can finally be reduced to the Decisional Composite Residuosity Assumption mentioned in 2.3.

4   IKE: Threshold Key Escrow Service with Intermediary Encryption

4.1   Overview

The IKE scheme based on threshold Paillier scheme is an extension of the warm-up scheme in previous section. In this scheme, system members hold threshold Paillier distributed key shares and encrypt their own shares through a secure encryption mode with a middleman, which is then escrowed on the blockchain. The shares of the committee members are all escrowed on the chain and support verification. Members who meet the threshold can recover the key to decrypt the message together. A multi-party distributed key escrow service model can be constructed by linking the Paillier key with the private key of the blockchain system. Our threshold key escrow service also considers the change of committee members of the custody shares or the loss of members' private key shares and designs a safe new share recovery method.

The scheme provides a key escrow service with intermediary encryption, including two key escrow modes: partial key escrow and complete key escrow, depending on whether the service party has created a committee or not. And we will directly introduce the core key escrow process after creating the committee, as shown in Fig. 1.

4.2   Description

In this section, we will provide a detailed explanation of the overall threshold key escrow service scheme with intermediary encryption.

System Identity Initialization: After determining the holder of the key share, the system dealer first initializes the system parameters.

- The dealer selects two safe primes $p = 2p' + 1$, and $q = 2q' + 1$ where p' and q' are also prime. Then use p and q to compute $N = pq$.

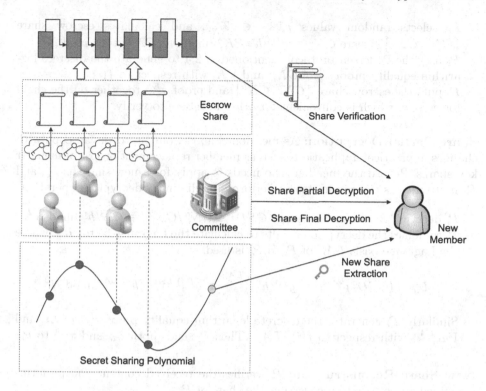

Fig. 1. The basic process overview of the IKE scheme.

- The dealer randomly picks up $\beta \in \mathbb{Z}_N^*$ and computes $m = p'q'$, $\theta \equiv m\beta \bmod N$, $\Delta = n!$ where n is the number of members.
- Similarly, the dealer directly sets $g = N + 1$ and selects a random seed b to compute another generator $h = b^N$ while ensure $ord_{N^2}(h) = \varphi(N)$. The system public key is (N, g, h, θ), and the private key is βm.

Threshold Key Initialization: The committee recorded as $P_1, P_2, \ldots, P_n$ start initialization operations such as the distribution of threshold keys.

- The dealer picks up random values a_i from $\{0, 1, \ldots, Nm - 1\}$ to construct polynomial $f(x) = \beta m + a_1 x + a_2 x^2 + \cdots + a_t x^t \bmod Nm$. Send $f(i)$ to the corresponding member P_i to complete the allocation of threshold shares.
- The dealer randomly selects $r \in \mathbb{Z}_{N^2}^*$ and discloses the verification keys VK and VK_i for P_i as $VK = v = r^2 \bmod N^2$, and $VK_i = v^{\Delta f(i)} \bmod N^2$.
- The dealer selects $\alpha \in \mathbb{Z}_N^*$, and each member P_i construct independent public-private key pair $(PK_i = \alpha^{x_i}, SK_i = x_i)$, where x_i is the random private key chosen in $\mathbb{Z}_{N^2}^*$.

Key Share Escrow: Committee members need to encrypt their own Paillier threshold key share as a message. For each P_i:

- P_i selects random values $r_i, s_i \in \mathbb{Z}_{N^2}^*$, and calculates escrow share $\{CN_i^a, CN_i^b\}$, where $CN_i^a \equiv g^{f(i)} h^{r_i} PK_i^{s_i} \bmod N^2, CN_i^b \equiv \alpha^{s_i} \bmod N^2$.
- P_i uses the Pedersen protocol mentioned in 2.4 to generate the discrete logarithm equality proof π_i^1 of CN_i^a and VK_i with respect to $f(i)$.
- P_i puts the escrow share $\{CN_i^a, CN_i^b\}$ and proof π_i^{ES} together on the chain for escrow, which is convenient for other members to verify.

Share Partial Decryption: As mentioned above, our protocol can cope with changes in practical applications such as member replacement or loss of member key shares. Record the member who needs to apply for a new share as P_e, and P_e first requests the committee to perform a preliminary decryption operation.

- P_i decrypt his own escrow share $CSD_i^a \equiv CN_i^a / CN_i^{b^{x_i}} \equiv g^{f(i)} h^{r_i} \bmod N^2$.
- P_i combines the decryption results to calculate the Lagrange factor L_{ie}, where the Lagrange basis LA_{ie} of P_i to P_e is used.

$$L_{ie} \equiv (CSD_i^a)^{LA_{ie}} \equiv \left(g^{f(i)} h^{r_i} \right)^{LA_{ie}} \equiv g^{f(i) \cdot LA_{ie}} \cdot h^{r_i \cdot LA_{ie}} \bmod N^2$$

- Similarly, P_i generates the discrete logarithm equality proof π_i^{LA} of L_{ie} and $VK_i^{LA_{ie}}$ with respect to $f(i) \cdot LA_{ie}$. Then P_i sends both L_{ie} and π_i^{LA} to P_e.

New Share Reconstruction: P_e verifies the received content and performs combined operations to reconstruct the share of P_e.

- P_e verifies the correctness after receiving the Lagrange factors of each node, and adds the correct ones to the set UQ_e.
- After the Lagrange factors whose number exceeds the threshold value are collected, the combination operation of shares can be performed.

$$L_e \equiv \prod_{i \in UQ_e} \left(g^{f(i) \cdot LA_{ie}} \cdot h^{r_i \cdot LA_{ie}} \right) \equiv g^{\sum_{i \in UQ_e} f(i) \cdot LA_{ie}} \cdot \sum_{i \in UQ_e} h^{r_i \cdot LA_{ie}}$$
$$\equiv g^{f(e)} \cdot \sum_{i \in UQ_e} h^{r_i \cdot LA_{ie}} \equiv g^{f(e)} \cdot h^R \bmod N^2$$

Note that the above steps combine the received Lagrange factors to obtain the corresponding encrypted content containing the private key share f_e. The random r_i combination of each node in the corresponding formula can be regarded as a randomly selected R, which is the Paillier encryption of secret f_e.

Share Final Decryption: P_e sends the reconstruction share L_e to the committee members to request the committee's threshold Paillier decryption.

- Each committee member P_i calculates threshold decryption share c_i by using its secret share for the received L_e, which $c_i \equiv L_e^{2\Delta f(i)} \bmod N^2$.
- Similarly, P_i publishes the discrete logarithm equality proof π_i^{De} of c_i and VK_i with respect to $f(i)$. Then P_i sends both c_i and π_i^{De} to P_e.

New Share Extraction: P_e verifies the correctness of each node's decryption share. After receiving the c_i that meets the threshold value, they will be reconstructed in combination with the Lagrange interpolation method to obtain the secret $f(e)$ of the new share.

$$f(e) = L \left(\prod_{i \in S} c_i^{2 \cdot \Delta \cdot LA_{i0}} \bmod N^2 \right) \times \frac{1}{4\Delta^2 \theta} \bmod N$$

where LA_{i0} is the Lagrange basis of P_i to P_0 and $L(x) = \frac{x-1}{N}$.

4.3 Proof of Correctness and Security

Correctness. This threshold key escrow scheme with intermediary encryption uses ElGamal encryption to increase the security of escrowed shares, whose correctness has been explained in Sect. 3.3.

Moreover, our scheme uses the threshold Paillier scheme to recover the encrypted form of the new key share, whose correctness is described as follows.

$$\prod_{i \in S} c_i^{2\Delta LA_{i0}} \equiv c^{4\Delta^2 \sum_{i \in S} f(i) LA_{i0}} \equiv c^{4\Delta^2 m\beta} \equiv \left(g^{f(e)} h^R \right)^{4\Delta^2 m\beta}$$

$$\equiv g^{4\Delta^2 m\beta f(e)} \equiv 1 + 4\Delta^2 m\beta f(e) N \bmod N^2$$

Since $ord_{N^2}(h) = \varphi(N)$ and $m = p'q'$, $h^{4R\Delta^2 m\beta} \equiv 1 \bmod N^2$.
Since $g = N + 1$, $g^{4\Delta^2 m\beta f(e)} \equiv 1 + 4\Delta^2 m\beta f(e) N \bmod N^2$.
$L \left(\prod_{i \in S} c_i^{2 \cdot \Delta \cdot LA_{i0}} \bmod N^2 \right) \times \frac{1}{4\Delta^2 \theta} \equiv 4\Delta^2 m\beta f(e) \times \frac{1}{4\Delta^2 \theta} \equiv f(e) \bmod N$.

Theorem 1. *The proposed threshold key escrow scheme with intermediary encryption is semantically secure if the underlying threshold Paillier encryption scheme is semantically secure.*

Proof. We aim to prove semantic security for the IKE scheme. The proof is based on the idea that if an adversary can break the semantic security of the IKE scheme, the adversary can also break the security of the threshold Paillier encryption scheme in the system. We assume that an adversary can compromise at most t servers, where t is the threshold of the Paillier scheme. In the following simulation game, we consider only non-adaptive adversaries, and the set of servers to be compromised is determined before the shares are distributed.

We assume that if there exists an adversary $\mathcal{A}$ that can break the semantic security of our IKE scheme, then there exists an adversary $\mathcal{A}_{TP}$ that can use $\mathcal{A}$ to break the semantic security of the threshold Paillier encryption scheme with a similar (polynomially related) advantage. Let us consider the following Game.

Setup Phase: Challenger $\mathcal{C}$ generates public parameters of IKE and sends to $\mathcal{A}$. For the nodes $P = \{P_1, P_2, \ldots, P_n\}$, $\mathcal{A}$ selects a set of at most t nodes and corrupts them. Without loss of generality $\mathcal{A}$ corrupts the first t nodes $T = \{P_1, P_2, \ldots, P_t\}$. Then $\mathcal{A}$ sends public parameters and key shares of T to $\mathcal{A}_{TP}$.

Query Phase 1: $\mathcal{A}$ interacts with the challenger $\mathcal{C}$ who actually operates our protocol to encrypt and escrow the set of public and private keys. The adversary $\mathcal{A}$ selects a node i from T, and claims that the share of node i is lost, and interacts with challenger $\mathcal{C}$ to request committee members to run the protocol to help restore i's key share. Notice that we set the node i as P_e of our scheme.

Let P_{-T} denote the set $\{P_{t+1}, \ldots, P_n\}$. Then $\mathcal{A}$ will receive the Lagrange factor L_{ie} and decryption share c_i along with proofs π_i^{LA} and π_i^{De} where $i \in P_{-T}$. $\mathcal{A}_{TP}$ receives the corresponding parameters from $\mathcal{A}$ and submits them to the decryption and reconstruction oracle $\mathcal{O}$ of threshold Paillier scheme. And $\mathcal{A}$ will interpolate the new share of i. This step can be repeated as often as $\mathcal{A}$ wishes.

Challenge Phase: Adversary $\mathcal{A}$ chooses two equivalent points (i, i'), and at the same time initiates a request to challenger $\mathcal{C}$ for new shares. Then the challenger $\mathcal{C}$ tosses a random coin $b \in \{0, 1\}$, if $b = 0$, challenger $\mathcal{C}$ chooses to generate the encryption of the new share of the node i, and if $b = 1$, challenger $\mathcal{C}$ chooses to generate the encryption of the new share of the node i'. Challenger $\mathcal{C}$ send Lagrange factors to $\mathcal{A}$.

Query Phase 2: Adversary $\mathcal{A}$ starts the oracle $\mathcal{O}$ of the threshold Paillier scheme for query interaction. Adversary $\mathcal{A}$ adaptively interacts with the uncorrupted nodes and reconstruct the encrypted form of a new share. However, $\mathcal{A}$ is not allowed to proceed to share the final decryption phase of our IKE scheme to request the decryption of encrypted shares.

Guess Phase: Adversary $\mathcal{A}$ finally outputs a random bit $b' \in \{0, 1\}$ according to the query. If the output random bit b' is equal to b selected by challenger $\mathcal{C}$, it means the attack of $\mathcal{A}$ is successful. If the adversary $\mathcal{A}$ can attack the semantic security of our IKE scheme, then adversary $\mathcal{A}_{TP}$ can exploit adversary $\mathcal{A}$ to win the IND-CPA game of the threshold Paillier scheme. Assuming that the probability of adversary $\mathcal{A}$ winning is $Adv_\mathcal{A}$, and the probability of adversary $\mathcal{A}_{TP}$ winning is $Adv_{\mathcal{A}_{TP}}^{IND-CPA}$, they have the following relationship.

$$
\begin{aligned}
Adv_\mathcal{A} &= \Pr\left[b = b'\right] = \Pr\left[\mathcal{A}(\text{ guess }) = 1 \mid b = 1\right] \cdot \Pr\left[b = 1\right] \\
&\quad + \Pr\left[\mathcal{A}(\text{ guess }) = 0 \mid b = 0\right] \cdot \Pr\left[b = 0\right] \\
&= \frac{1}{2} \cdot \Pr\left[\begin{matrix}\mathcal{A}_{TP}(\text{ guess }) = 1, \\ b' = 1\end{matrix} \mid b = 1\right] + \frac{1}{2} \cdot \Pr\left[\begin{matrix}\mathcal{A}_{TP}(\text{ guess }) = 0, \\ b' = 0\end{matrix} \mid b = 0\right] \\
&= \frac{1}{4} \cdot \{\Pr\left[\mathcal{A}_{TP}(\text{ guess }) = 1 \mid b = 1\right] + \Pr\left[\mathcal{A}_{TP}(\text{ guess }) = 0 \mid b = 0\right]\} \\
&= \frac{1}{4} \cdot \Pr\left[Adv_{\mathcal{A}_{TP}}^{IND-CPA}\right] \leq \frac{1}{2} + \epsilon
\end{aligned}
$$

where ϵ is a negligible function. $\square$

5 Implementation and Evaluation

We implement the prototype system based on python 3.9, which is evaluated on a laptop machine of Windows 10, equipped with an Intel(R) Core(TM) i5-7300HQ CPU with 16 GB of RAM.

The IKE scheme provides a threshold key escrow service where multiple nodes form committees. We analyzed the scheme's performance under different nodes and conducted performance tests for four scenarios with 32, 64, 128 and 256 committee nodes, each with varying threshold values. We evaluated the protocol's performance under member key share loss and request recovery scenarios using threshold values of $t = n - 1$ and $t = n/2$.

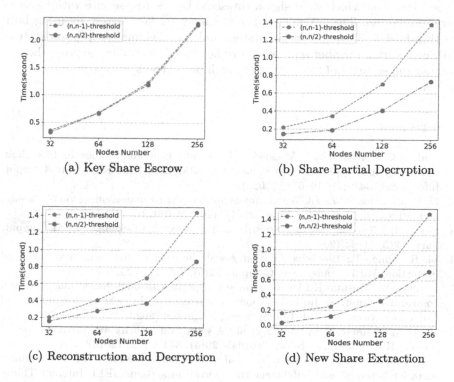

(a) Key Share Escrow   (b) Share Partial Decryption

(c) Reconstruction and Decryption   (d) New Share Extraction

Fig. 2. The time observed from the escrow to decryption process of IKE scheme.

Figure 2 shows the results of the performance test. In practice, the time required for hosting on-chain, proof and verification can vary depending on the blockchain carrier used and the specific service mode. Therefore, in this test, we deploy nodes locally to simulate the actual situation, with the primary objective being to measure the time overhead in the encryption and decryption process of the test scheme. We tested the time of the four main stages.

From the results of the performance test, it can be seen that the time for hosting encryption is not affected by the number of nodes. The reason is that each node completes the hosting operation locally. The time for share reconstruction, decryption and other links will be extended as the number of nodes increases. When the number of nodes increases by 50%, the time consumed increases by 50%-60%, which is in the linear growth range. Under the same number of committee nodes, the lower the threshold value, the shorter the completion time.

The decryption time is at the millisecond level. The time performance of the scheme is good, and the key escrow service can be well completed.

6   Conclusion

In this paper, we first introduce a basic data encryption system with intermediaries, based on which we design a threshold key escrow service with intermediary encryption. The scheme implements a secure key escrow by using both the threshold Paillier encryption system and the ElGamal encryption system, while considering member replacement or key loss. Eventually, we prove the correctness and security of the scheme through prototype system test and security analysis.

References

1. Gad, A.G., Mosa, D.T., Abualigah, L., et al.: Emerging trends in blockchain technology and applications: a review and outlook. J. King Saud Univ.-Comput. Inform. Sci. **34**(9), 6719–6742 (2022)
2. Li, W.J., Meng, W.Z.: BCTrustFrame: enhancing trust management via blockchain and IPFS in 6G era. IEEE Netw. **36**(4), 120–125 (2022)
3. Zhang, R., Xue, R., Liu, L.: Security and privacy on blockchain. ACM Comput. Surv. **52**(3), 1–34 (2019)
4. Li, R., Song, T.: Blockchain for large-scale internet of things data storage and protection. IEEE Trans. Serv. Comput. **12**(5), 762–771 (2018)
5. Khanum S., Mustafa K.: Exposure of sensitive data through blockchain wallets: a comparative analysis. In: International Conference on Innovative Computing and Communications, pp. 161–169. Springer, Singapore (2022)
6. Conti, M., Kumar, E.S., Lal, C., et al.: A survey on security and privacy issues of bitcoin. IEEE Commun. Surv. Tutorials **20**(4), 3416–3452 (2018)
7. Lei, A., Cruickshank, H., Cao, Y., et al.: Blockchain-based dynamic key management for heterogeneous intelligent transportation systems. IEEE Internet Things J. **4**(6), 1832–1843 (2017)
8. Cha, S.Y., Baek, S., Kim, S.: Blockchain based sensitive data management by using key escrow encryption system from the perspective of supply chain. IEEE Access **8**, 154269–154280 (2020)
9. Barker E.: Recommendation for key management, part 2: best practices for key management organization. National Institute of Standards and Technology (2018)
10. Lu, W., Wu, L., Zhao, R., et al.: Blockchain technology for governmental supervision of construction work: learning from digital currency electronic payment systems. J. Constr. Eng. Manag. **147**(10), 04021122 (2021)
11. Zhao, H., Bai, P., Peng, Y., et al.: Efficient key management scheme for health blockchain. CAAI Trans. Intell. Technol. **3**(2), 114–118 (2018)
12. Micali S.: Fair cryptosystems. Massachusetts Institute of Technology (1993)
13. Kilian, J., Leighton, T.: Fair cryptosystems, revisited. In: Coppersmith, D. (ed.) CRYPTO 1995. LNCS, vol. 963, pp. 208–221. Springer, Heidelberg (1995). https://doi.org/10.1007/3-540-44750-4_17
14. Micali S.: Guaranteed partial key escrow. Technical Report Technical Memo 537, MIT Laboratory for Computer Science (1995)

15. Bellare M., Goldwasser S.: Verifiable partial key escrow. In: 4th ACM Conference on Computer and Communications Security, pp. 78–91. ACM, USA (1997)
16. Ma, M.X., Shi, G.Z., Li, F.G.: Privacy-oriented blockchain-based distributed key management architecture for hierarchical access control in the IoT scenario. IEEE Access **7**, 34045–34059 (2019)
17. Mwitende, G., Ye, Y.L., Ali, I., et al.: Certificateless authenticated key agreement for blockchain-based WBANs. J. Syst. Architect. **110**, 101777 (2020)
18. Paillier, P.: Public-key cryptosystems based on composite degree residuosity classes. In: Stern, J. (ed.) EUROCRYPT 1999. LNCS, vol. 1592, pp. 223–238. Springer, Heidelberg (1999). https://doi.org/10.1007/3-540-48910-X_16

A Constructive Method for Data Reduction and Imbalanced Sampling

Fei Liu and Yuanting Yan[(✉)] [ID]

Artificial Intelligence Institute, School of Computer Science and Technology,
Anhui University, Hefei, Anhui 230601, People's Republic of China
ytyan@ahu.edu.cn

Abstract. A large number of training data lead to high computational cost in instanced-based classification. Currently, one of the mainstream methods to reduce data size is to select a representative subset of samples based on spatial partitioning. However, how to select a representative subset while maintaining the overall potential distribution structure of the dataset remains a challenge. Therefore, this paper proposes a constructive data reduction method called Constructive Covering Sampling (CCS) for classification problems. The CCS does not rely on any relevant parameters. It iteratively partitions the original data space into a group of data subspaces, which contains several samples of the same class, and then it selects representative samples from the data subspaces. This not only maintains the original data distribution structure and reduces data size but also reduces problem complexity and improves the learning efficiency of the classifier. Furthermore, CCS can also be extended as an effective undersampling method (CCUS) to address class imbalance issues. Experiments on 18 KEEL and UCI datasets demonstrate that the proposed method outperforms other sampling methods in terms of F-measure, G-mean, AUC and Accuracy.

Keywords: constructive covering algorithm · data reduction · undersampling · class imbalance

1 Introduction

Data resampling and data reduction techniques are important data processing technique in machine learning, data mining and other fields. Data reduction can effectively reduce data storage space by reducing the size of the data, which improves the learning efficiency of downstream data mining and machine learning algorithms. Data resampling of imbalanced data in learning refers to the rebalancing of imbalanced training data, including oversampling, undersampling, and hybrid sampling [20].

Undersampling technique is widely used in data reduction and imbalanced classification. Its main goals are: 1) to reduce the data size and improve learning

This work was supported in part by the National Natural Science Foundation of China under Grant 62376002.

efficiency by selecting key samples that preserve the original data distribution structure and overall information to the maximum extent possible. This type of method has obvious advantages when the training data is too large. 2) to reduce the degree of imbalance in the dataset by selecting representative majority class samples and avoid potential learning bias issues. Traditional undersampling methods, such as RUS [7], EasyEnsemble [15], ENN [11], etc., focus on solving the problem of class imbalance at the data level by selecting an representative data subset of majority class. However, after the sampling of these methods, the data size of the samples is mostly relatively fixed (that is, determined by the size of a few types of samples), and the data size after sampling cannot be flexibly set.

As for data reduction which focus on selecting representative subsets for all the training data with different classes. Zhao et al. [23] proposed an adaptive sampling process that iteratively evaluates the probability of mislabeled classes, gradually reducing the risk of selecting incorrectly labeled samples for model training. This approach can construct highly generalizable models when there are a large number of mislabeled samples in the data. Therefore, Xia et al. [19] proposed RSDS which divides the data into boundary points, interior points, and noise points by introducing the idea of random space division. The boundary points are then extracted as the final sampling result to reduce the data size and the negative impact of label noise. They further [18] proposed the Granular Ball Sampling (GBS) method, which utilizes K-means clustering to adaptively generate hyper-spheres that cover the original sample space. And then they proposed a GBS-based data reduction method and a GBS-based undersampling method. However, these methods are all rely on relevant parameters, which leads to weaker adaptability of the algorithm.

To this end, this paper proposes a parameter-free method that can be used for data reduction and imbalanced data classification. The method can not only cover the entire data space but also maintain the original spatial distribution structure well. The overall process of the algorithm does not rely on any parameters, and has strong adaptability. The main contributions of this paper are as follows:

- This paper proposes a parameter-free data reduction method (CCS) based on constructive neural networks for large-scale data reduction. It also introduces a parameter-free undersampling method (CCUS) for imbalanced data.
- CCS and CCUS utilize a constructive process to learn the spatial neighborhood distribution information of the data, and then design data reduction and sample sampling methods based on spatial neighborhood distribution.
- The effectiveness of the proposed methods are verified by comparing with GBS, RSDS, RUS, SMOTE, and IGBS on 18 datasets using DT classifiers.

The rest of this paper is organized as follows: Sect. 2 introduces related work. Section 3 describes the proposed methods CCS and CCUS in detail. Section 4 introduces and analyzes the experimental results. Finally, Sect. 5 presents the conclusion of the study.

2   Related Work

In the past few decades, how to efficiently process large-scale data has been widely studied and of great interest to both academia and industry [6]. Scholars have proposed various methods to reduce the size of data by selecting key samples, such as spatial distribution-based sampling and spatial partition-based reduction [1,8,22]. These methods aim to reduce the computational complexity of algorithms by deleting redundant samples from the original dataset and selecting a representative subset without affecting classification performance [22]. The algorithm that reduces data size through spatial partitioning is one of the most popular and effective prototype generation (PG) algorithms at present. It partitions the original data space into several subspaces, with each subspace containing samples of the same class. It generates prototypes by replacing all samples in a subspace with the mean value of the samples, achieving the goal of reducing data size [6]. Most existing methods initialize prototypes with a small portion of training samples and modify it to maximize the overall classification performance of the entire training set. Furthermore, prototypes can be selected or generated based on the given data [5,10]. Xia et al. [19] used random forests to divide and sample the entire data space, filter noise samples, reduce data size, and improve robustness to noise.

The "granular-ball computing classifiers" was proposed by [17], which generates several spheres by clustering. These classifiers use granular spheres instead of the input of original samples in the mathematical model of the classifier, reducing data size and thereby improving robustness, efficiency, and scalability. Li et al. [9] proposed a sample selection method based on local geometry and statistical information, which removes redundant samples that do not significantly affect the classification results and selects margin and boundary samples as the final sampling results, thus achieving the goal of reducing data size and improving classification performance.

Stefanos et al. [13] proposed a simple and fast variant method of RSP3, which avoids the problem of high computational complexity in partitioning same-class clusters and the problem caused by noise samples in the training set. Aslani et al. [2] proposed the BPLSH method, which extracts boundary points as the final sampling results and eliminates redundant samples that do not significantly help classification, thereby speeding up the training process of the classifier without significantly reducing the original classification accuracy. Ougiaroglou et al. [12] proposed a clustering-based isomorphic clustering reduction (RHC) algorithm, which can quickly construct isomorphic clusters so that each sample in each cluster belongs to the same class. RHC initially treats the entire training set as a nonisomorphic cluster and calculates an average sample for each class in the cluster as the initial seed sample for the cluster. The algorithm is recursively executed until all samples in all clusters belong to the same class, and the centroids of each cluster are used as the final result [14]. Wiharto et al. [16] proposed an improved intrusion detection system model combining improved DBSCAN clustering and classification regression trees, which reduces data using an improved DBSCAN algorithm while overcoming differences in data homogeneity. To some extent, this method can achieve better clustering accuracy, but it cannot find clusters of arbitrary types.

3 Proposed Method

3.1 Constructive Data Partitioning

Given a d-dimensional dataset $D = \{(x_1, y_1), (x_1, y_2), \ldots, (x_n, y_n)\}$, where $x_i = \{x_i^1, x_i^2, \cdots, x_i^d\}$, y_i denotes the label of the i-th sample. CCA partitions the original dataset space into several sub-regions iteratively until all samples are labeled. At this point, a set of cover $C = \{C_1^1, \ldots, C_1^{n_1}, C_2^1, \ldots, C_2^{n_2}, \ldots, C_m^1, \ldots, C_m^{n_m}\}$ are obtained, where C_i^j denotes the j-th cover of the i-th class. And each cover contains samples belonging to the same class. The steps are as follows:

Step 1: Randomly select an uncovered sample x_i as the cover center.

Step 2: Calculate the Euclidean distance d_1 between the different-class samples closest to x_i according to Eq. (1).

$$d_1 = \arg\min_{y_i \neq y_j}\{dist\{x_i, x_j\}\} \tag{1}$$

where $dist\{x_i, x_j\}$ denotes the distance between sample x_i and sample x_j.

Step 3: Calculate the distance d_2 between the samples of the same-class that are farthest from x_i in the range d_1 according to Eq. (2).

$$d_2 = \arg\max_{y_i = y_j}\{dist\{x_i, x_j\}\}, s.t. d_2 < d_1 \tag{2}$$

Step 4: Take x_i as the cover center and d_2 as the cover radius to form a cover, label these samples as learned samples.

Step 5: Repeat steps 1 to 4 until all samples have been learned.

3.2 Constructive Data Partitioning-Based Data Reduction

According to Sect. 3.1, to show the cover learning process of CCA more clearly, we artificially created an original data distribution (as shown in Fig. 1(a)), and learned the cover of the data distribution, and multiple covers can be obtained, as shown in Fig. 1(b). Depending on the number of samples in each cover, different strategies are used in this paper. Specifically, we divide the cover into three different cases based on $2*d$ as a benchmark: single sample, with sample number between $(1, 2*d]$, and with more than $2*d$ samples.

(1) For cover with single sample, we determine whether it is a noise as the following rules: 1) We calculate the average radius R of all the covers with more than $2*d$ samples; 2) We calculate the proportion of same-class samples and different-class samples within the spatial area centered on the sample with radius R. If the proportion of same-class samples is less than that of different-class samples, the sample is considered as a noise sample and directly removed. Otherwise, the sample is retained. As shown in Fig. 1(e), S_1 and S_2 are noise samples. We take the average cover radius of all covers with more than $2*d$ samples (C' in Fig. 1(e)), and within the range of C', there are more different-class samples than same-class samples, indicating that these two samples are noise samples and should be removed.

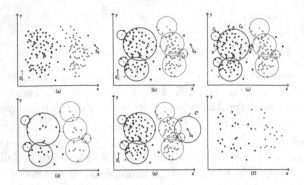

Fig. 1. Concept of CCS. (a) The original dataset; (b) Generated CCs(Constructive covers); (c) 2*d intersection points a, b, c and d. Closest samples in the CC to the intersection points a', b', c' and d'; (d) Sampling of the original data with the CCs; (e) The judgment of noise sample; (f) Final sampling result.

(2) For covers with sample number between $(1, 2*d]$, we directly retain the samples in the cover without sampling. For example, in cover C_1 shown in Fig. 1(c), which contains only three samples, we directly retain them as the sampling result.

(3) For covers containing more than $2*d$ samples, the cover center point c is moved a distance of d_2 along a certain determined coordinate axis, and two intersection points are obtained on each coordinate axis, with one in the positive direction and the other in the negative direction. Specifically, given a d-dimensional dataset D, the i-th cover generated on the dataset D is denoted by c_i, its cover center vector is $c_i = \left(c_i^1, c_i^2, \cdots, c_i^j, \cdots, c_i^d\right)$, the radius is r_i, and the two intersection points in the positive and negative directions on the j-th coordinate axis are represented as follows:

$$a = \left(c_i^1, c_i^2, \cdots, c_i^j + r, \cdots, c_i^d\right), b = \left(c_i^1, c_i^2, \cdots, c_i^j - r, \cdots, c_i^d\right) \quad (3)$$

After the $2*d$ intersections are generated for each cover, the nearest same-class samples to each intersection are selected as the sampling results, as shown in Fig. 1(c). For example, for cover C_2, its intersections are a, b, c, and d, and the nearest same-class samples are a', b', c', and d'. They are the best samples to keep covering the overall structure.

However, to prevent the above process from sampling, the majority class and minority class compression ratios IR_{maj} and IR_{min} are calculated according to Eq. (4). Then, samples are selected from the class with low compression ratio so that the compression ratios of the two classes are the same.

$$IR_{maj} = \frac{S_{maj}}{M}, IR_{min} = \frac{S_{min}}{N} \quad (4)$$

$$n_{min} = \left[S_{min} \times \left(\frac{S_{maj} \times N}{S_{min} \times M} - 1\right)\right], n_{maj} = \left[S_{maj} \times \left(\frac{S_{min} \times M}{S_{maj} \times N} - 1\right)\right] \quad (5)$$

where S_{maj} (S_{min}) represent the number of majority (minority) class samples after sampling; $M(N)$ represent the number of majority (minority) class samples before sampling; $n_{maj}(n_{min})$ represent the number of majority (minority) class samples that need to be supplemented; $[\cdot]$ represents rounding.

The selection rule is as follows: after sampling from covers more than $2*d$ samples, select samples from the class with lower compression ratio. According to Eq. (5), calculate the number of samples $n_{maj}(n_{min})$ to be selected from a class of data with low sampling rate. If $n_{maj}(n_{min})$ is an even multiple of d and more than or equal to $2*d$, select the $2*d$ samples of the same-class that are next closest to the intersection point from the cover with the largest number of samples, and select the remaining samples from the cover with the second most samples until all the required samples are selected. If $n_{maj}(n_{min})$ is less than $2*d$, select the nearest sample to the center point of the cover. If $n_{maj}(n_{min})$ is more than $2*d$ and is an odd multiple of d, select the sample that is next closest to the intersection point first, and then select the remaining samples that are less than $2*d$ from the nearest sample to the center point of the cover until all the required samples are selected. At this point, both compression ratios are the same. This is illustrated in Fig. 1(d). Based on the above description, The pseudo-code for CCS is as follows Algorithm 1.

Algorithm 1. CCS Algorithm

Input: Dataset D;
Output: The sampled results D';
1: Initialize $C=\{\}$;
2: **for** a random sample $x_i \in D$ **do**
3:    Construct a cover C_i on x_i; // (1), (2), x_i is the center of C_i;
4:    $C \leftarrow C_i$;
5: **end for**
6: **for** $c \in C$ **do**
7:    **if** len(c) == 1 **then**
8:       **if** x_i is a noise sample **then** remove
9:       **else** Add x_i to D';
10:   **else if** $1 < $ len(c) $\leq 2*d$ **then**
11:      Add all samples to D';
12:   **else**
13:      Find the corresponding cover radius r_i and center $c_i = \left(c_i^1, c_i^2, \cdots\cdots, c_i^d\right)^T$;
14:      **for** $j = 1, 2, \ldots, d$ **do**
15:         $a = \left(c_i^1, c_i^2, \cdots, c_i^j + r_i, \cdots, c_i^d\right)^T$, $b = \left(c_i^1, c_i^2, \cdots, c_i^j - r_i, \cdots, c_i^d\right)^T$;
16:         Find the samples a' and b' in D, respectively.
17:         Add a' and b' to D';
18:      **end for**
19:   **end if**
20: **end for**
21: Calculate the number of samples to be selected Eqs. (4)~ (5).
22: According to Section 3.2(3), select representative samples and adding them to D';

3.3 Undersampling Method for Imbalanced Data Based on Constructive Data Partitioning

This section mainly introduces how the CCS method deals with data imbalance problem. Imbalanced problems is a major challenge in the field of machine learning. In recent years, researchers have proposed many methods to solve the problems caused by imbalanced data [4], all of which aim to balance the samples of different classes by reducing the majority class or increasing the minority class. Therefore, as an undersampling method (CCUS) to balance different class samples, CCUS can be used to deal with data imbalance problems, thereby improving classification accuracy and learning efficiency.

Figure 2 shows a two-dimensional illustration of imbalanced data, where black dots represent the majority class (negative samples) and red dots represent the minority class (positive samples).

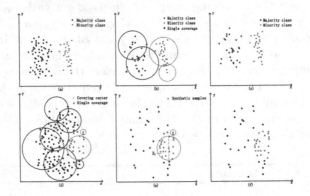

Fig. 2. Sampling process using CCUS on a imbalanced dataset; (a) The original dataset; (b) Generated CCs; (c) Sampling result when CCS is directly used; (d) Generated CCs on a imbalanced dataset; (e) The minority samples synthesis results; (f) Final sampling result.

The CCUS method only samples from covers larger than $2*d$, according to the sampling rules described in Sect. 3.2. At this point, the data is still imbalanced. We may encounter the following three situations after sampling: (1) the number of majority class samples is less than that of minority class samples; (2) the number of majority class samples is equal to that of minority class samples; (3) the number of majority class samples is more than that of minority class samples. To address this issue, we adopt the following approach: The difference is that CCUS directly retains all minority class samples to address data imbalance.

For situation (1), where the number of majority class samples after sampling is less than the number of minority class samples. First, calculate the number of samples by which the majority class is less than the minority class. Next, calculate the proportion of the sample size of the cover with the most samples to the total number of samples in covers larger than $2*d$. Finally, multiply the number of additional majority class samples needed by the corresponding cover to obtain the number of samples to be selected. The rules are as follows: first, select the majority class sample closest to the center of the cover with the largest number of samples, then select from the cover with the next largest number of samples. Repeat the above process until the number of majority class and minority class samples is balanced. This process is illustrated in Fig. 2(b)~(c).

For situation (2), no further action is required. For situation (3), synthetic minority class samples are generated to achieve balance between the two classes. Based on the cover set C obtained from Sect. 3.1, the number of minority class samples to be synthesized is calculated using Eq. (6):

$$\Delta N_{\min} = S_{maj} - N_{min} \tag{6}$$

A few classes of samples are synthesized as follows. For $\forall C_{min}, C_{maj} \in C$, find the corresponding cover center based on Eq. (7):

$$\begin{cases} c_{min} \leftarrow \arg\max |C_{min}| \\ c_{maj} \leftarrow \arg\max |C_{maj}| \end{cases} \tag{7}$$

where c_{min} and c_{maj} are the cover centers of minority and majority classes with the largest sample numbers respectively, C_{min} and C_{maj} are the minority class and majority class cover with the largest number of samples, respectively. According to Eq. (8), the distance d_k between the two cover centers and the radius Z_i of the synthesis area are calculated as follows:

$$d_k = dist\left(c_{min}, c_{maj}\right), Z_i = d_k - r_j \tag{8}$$

where r_j represents the radius of the majority class cover with the most samples. As shown in Fig. 2(e), the synthetic minority class samples will be generated within the Z_1 region, and overlapping samples between the synthetic and original minority class will be removed. Repeat the above process until the number of majority and minority class samples is equal, as shown in Fig. 2(f). The pseudo-code of CCUS is shown below Algorithm 2.

Algorithm 2. CCUS Algorithm

Input: Dataset D;
Output: The sampled results D';
1: Based on steps 2-5 in Algorithm 1, the set of coverings C can be obtained;
2: Add all minority class samples to the set D';
3: **for** $C' \in C$ **do** //cover of greater than or equal to 2*d samples
4:    **if** $C' \in$ majority classes **then**
5:       Find the corresponding cover radius r_i and center $c_i = \left(c_i^1, c_i^2, \cdots\cdots, c_i^d\right)^T$;
6:       **for** $j = 1, 2, \ldots, d$ **do**
7:          $a = \left(c_i^1, c_i^2, \cdots, c_i^j + r_i, \cdots, c_i^d\right)^T, b = \left(c_i^1, c_i^2, \cdots, c_i^j - r_i, \cdots, c_i^d\right)^T$;
8:          Find the samples a' and b' in c that are closest to a and b, respectively;
9:          Add a' and b' to D';
10:       **end for**
11:    **end if**
12: **end for**
13: Check if the number of majority class and minority class samples in set D' is balanced;
14: **if** $S_{maj} < S_{min}$ **then**
15:    Select the majority class samples and add them to D'; // Section 3.3 situation(1)
16: **end if**
17: **if** $S_{maj} > S_{min}$ **then**
18:    Synthesize minority class samples according to Eqs.(6)~(8) and add them to D';
19: **end if**

---

4   Experiments and Analysis

4.1   Datasets Selection and Evaluation Metric

In this paper, we conducted experiments on binary classification datasets selected from the KEEL and UCI databases. For the sake of convenience in demonstration and comparative analysis, we randomly selected 18 binary classification datasets for experimental analysis, as shown in Table 1. Where named, Bean and letter are shorthand for Data-for-UCI-named, Dry-Bean-Dataset and letter-recognition datasets respectively. The experiment used five-fold cross-validation on the datasets, and each dataset was randomly divided into two parts: training set and test set. The process was repeated ten times, and the average was used as the measure of performance, with the highest accuracy highlighted in bold. This paper uses DT classifier to verify the effectiveness of this method in four metrics of F-measure, G-mean, AUC and ACC. And a specific definition is given in reference [21]. For the sake of simplicity, they are not listed one by one here.

4.2   Analysis of CCS Experiment Results

As the method proposed in this study, CCS involves sampling reduction, therefore we compared it with GBS [18], RSDS [19]. Table 2 and Table 3 provide the comparative performance of CCS and these methods on Accuracy and AUC evaluation metrics in the DT classifier. From the tables, it can be seen that CCS has better performance compared to GBS, RSDS, and Original. For example, the

Table 1. Information of the Datasets selected from UCI and KEEL

Datasets	Instances	Attributes
banana	5300	2
penbased	10992	16
phoneme	5404	5
ring	7400	20
satimage	6435	20
segment0	2308	19
shuttle	57999	9
texture	5500	40
thyroid	7200	21
basketball	12567	7
codrna	59535	8
named	10000	13
Bean	13611	15
ijcnn1	141691	22
letter	20000	16
mushrooms	8124	112
svmguide1	7085	4
twonorm	7400	20

Table 2. Results of CCS on Accuracy by DT

Datasets	CCS	GBS	RSDS	Original
banana	0.8762	**0.8885**	0.8575	0.8666
basketball	**0.9718**	0.9316	0.9645	0.8763
codrna	**0.9389**	0.9317	0.9320	0.8563
named	**0.9998**	0.9562	0.9998	0.9998
Bean	**0.9613**	0.9562	0.9254	0.7378
ijcnn1	**0.9804**	0.9654	0.9673	0.9710
letter	0.9376	0.9123	0.9214	**0.9408**
mushrooms	**1.0000**	**1.0000**	0.9956	**1.0000**
penbased	0.9783	0.8914	0.9778	**0.9797**
phoneme	0.8660	0.8658	0.8029	**0.8888**
ring	**0.9299**	0.8718	0.8689	0.8757
satimage	**0.9254**	0.8871	0.8639	0.9092
segment0	**0.9935**	0.8796	0.9786	0.9923
shuttle	**0.9996**	0.8869	0.9987	0.9985
svmguide1	**0.9659**	0.9651	0.9327	0.9566
texture	**0.9520**	0.9081	0.9488	0.6077
thyroid	0.9881	0.9144	**0.9943**	0.9944
twonorm	0.8524	**0.8943**	0.8339	0.8419
Average	**0.9510**	0.9170	0.9313	0.9052

classification performance of the ijcnn1 dataset in CCS is slightly better than the other three methods, while the classification performance of the ring dataset in CCS is significantly improved compared to other methods. Although some datasets have a significant difference in performance compared to other methods in the classifier (such as the phoneme and thyroid datasets), most datasets have higher rankings in CCS's classification performance compared to other methods (such as the penbased dataset), and the overall average is the highest, especially with a significant improvement in the Accuracy evaluation metric.

Although the CCS method may lose some sample information after sampling, these lost samples do not destroy the distribution structure of the original data. That is, compared with the original data, the space structure of the data before and after sampling has not changed significantly, and it will not have a significant impact on the final classification results. Moreover, the CCS method has no parameters in the whole process, which makes the algorithm more adaptable, while other methods have set related parameters. Because the size and spatial distribution of the dataset will affect the classification performance of the classifier, the testing accuracy of the traditional classifier is worse than that of the CCS method.

4.3   Experiments on Imbalanced Datasets

In this section, we mainly introduce the classification performance of CCUS method on the problem of class imbalance. Currently, there are many solutions proposed for class imbalance problem, which can be roughly divided into two categories: algorithm-level and data-level. Algorithm-level solutions mainly modify traditional algorithms or design specific algorithms to address class imbalance issues, while data-level solutions mainly adjust the size of the training set to address class imbalance problems, such as SMOTE [3], RUS [7], etc.

In this paper, SMOTE, RUS, and IGBS [18] are used as comparison methods. Table 4, Table 6 and Table 7 respectively show the classification performance of CCUS and other three methods on DT classifiers, using F-measure, G-mean and AUC as the evaluation metric. It can be seen from the tables that the CCUS method is superior to other methods in terms of average values, especially in the basketball and texture datasets, where its performance is significantly higher than that of other methods. This is because the CCUS method can not only maintain the distribution information and structure of the dataset very well, but also has no parameter setting throughout the whole process, and has strong adaptability of the algorithm. In addition, it can also reduce the size of the dataset, greatly improving the learning performance of the algorithm.

Table 3. AUC results of CCS on KEEL and UCI datasets obtained by DT

Datasets	CCS	GBS	RSDS	Original
banana	**0.8744**	0.8709	0.8576	0.8656
basketball	**0.9710**	0.9063	0.9630	0.8733
codrna	**0.9296**	0.9142	0.9245	0.8095
named	**0.9998**	0.9434	**0.9998**	**0.9998**
Bean	**0.9612**	0.9458	0.9237	0.7460
ijcnn1	0.9349	0.9410	**0.9423**	0.9211
letter	0.9262	0.8875	0.9131	**0.9307**
mushrooms	**1.0000**	**1.0000**	0.9955	**1.0000**
penbased	**0.9784**	0.8638	0.9778	0.9797
phoneme	0.8368	0.8289	0.7880	**0.8548**
ring	**0.9298**	0.8298	0.8689	0.8756
satimage	**0.9243**	0.8366	0.8695	0.9081
segment0	**0.9860**	0.8461	0.9756	0.9848
shuttle	**0.9994**	0.8556	0.9971	0.9977
svmguide1	**0.9649**	0.9473	0.9315	0.9562
texture	**0.9512**	0.8843	0.9485	0.5822
thyroid	0.9341	0.8913	**0.9799**	0.9794
twonnorm	0.8524	**0.8654**	0.8339	0.8419
Average	**0.9419**	0.8921	0.9272	0.8948

Table 4. F-measure results on KEEL and UCI datasets obtained by DT

Datasets	CCUS	RUS	SMOTE	IGBS
banana	0.8612	0.8676	**0.8705**	0.8631
basketball	**0.9688**	0.8926	0.8673	0.9129
codrna	0.8988	**0.9217**	0.8815	0.8359
named	**0.9999**	0.9998	0.9998	0.8769
Bean	**0.9654**	0.7381	0.7483	0.8915
ijcnn1	0.9037	**0.9653**	0.9196	0.8603
letter	0.9008	0.9064	**0.9298**	0.8666
mushrooms	**1.0000**	0.9962	0.9963	0.9987
penbased	**0.9795**	0.9787	0.9791	0.8371
phoneme	0.7828	**0.8762**	0.8407	0.8114
ring	**0.8791**	0.8741	0.8739	0.8133
satimage	**0.9113**	0.9047	0.9071	0.8192
segment0	0.9789	**0.9887**	0.9834	0.8288
shuttle	0.9961	**0.9998**	0.9998	0.9987
svmguide1	0.9541	0.9625	0.9583	**0.9726**
texture	**0.9622**	0.6966	0.5325	0.9603
thyroid	0.9730	**0.9952**	0.9809	0.9658
twonnorm	0.8487	0.8406	0.8412	**0.9340**
Average	**0.9308**	0.9114	0.8950	0.8915

4.4 Statistical Analysis

In order to evaluate whether there is a significant difference between the proposed algorithm and the comparison algorithms, this section uses the non-parametric Friedman test and performs corresponding post-hoc tests to further detect performance differences between the algorithms [21]. Table 5 shows the average ranks among the comparison algorithms, the Friedman statistic (χ_F^2) with (k-1) degrees of freedom, and the F-distribution with (k-1) and (N-1) degrees of freedom, where k represents the number of algorithms and N represents the number of datasets. When the significance level is set 0.05, the critical values are $\chi_F^2(3)$ is 7.8147 and $F(3, 17)$ is 3.20. When the test statistic is larger than the corresponding critical value, the null hypothesis is rejected.

Figure 3 shows the results of the Friedman test for all algorithms at a significance level of $\alpha = 0.05$. The top row corresponds to the critical range CD of the mean rankings, and the coordinates on the axis represent the mean rankings of each method, where the lower the ranking, the further to the left. Methods connected by the same line indicate no significant difference between them. From Fig. 3, it can be seen that the results of G-mean and AUC are similar. There is no significant difference between CCS and Original, but there are significant differences between them and the other three methods. Moreover, in four evaluation metrics, the proposed CCUS method achieved the best rankings compared to the other methods.

Table 5. Statistical Test of 4 Methods on Three Evaluation Metrics by DT

F-measure	CCS	GBS	RSDS	Original	χ_F^2	F_F
	1.8056	2.3333	2.6389	3.2222	11.3500	4.5240
G-mean	CCS	GBS	RSDS	Original	χ_F^2	F_F
	1.4722	2.4167	2.8889	3.2222	18.7500	9.0426
AUC	CCS	GBS	RSDS	Original	χ_F^2	F_F
	1.5556	2.5000	2.8333	3.1111	14.8667	6.4583
ACC	CCS	GBS	RSDS	Original	χ_F^2	F_F
	1.5000	2.4444	3.0000	3.0556	16.8667	7.7217

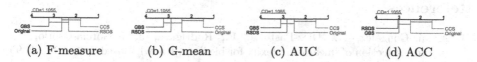

(a) F-measure   (b) G-mean   (c) AUC   (d) ACC

Fig. 3. Comparison of the algorithms on three evaluation metrics. Groups of the algorithms without significant difference are connected.

Table 6. G-mean results on KEEL and UCI datasets obtained by DT

Datasets	CCUS	RUS	SMOTE	IGBS
banana	**0.8739**	0.8688	0.8710	0.8655
basketball	**0.9762**	0.8597	0.8668	0.9159
codrna	**0.9182**	0.8675	0.8665	0.8775
named	**0.9999**	**0.9999**	0.9998	0.9081
Bean	**0.9601**	0.7492	0.7667	0.9171
ijcnn1	**0.9441**	0.9352	0.9349	0.8795
letter	0.9290	0.9192	**0.9329**	0.8840
mushrooms	**1.0000**	0.9963	0.9964	0.9989
penbased	**0.9796**	0.9785	0.9791	0.8614
phoneme	**0.8482**	0.8346	0.8454	0.8326
ring	**0.8815**	0.8741	0.8739	0.8332
satimage	**0.9206**	0.9062	0.9073	0.8385
segment0	**0.9888**	0.9812	0.0862	0.8472
shuttle	0.9989	0.9997	**0.9998**	0.9992
svmguide1	0.9595	0.9578	0.9586	**0.9728**
texture	**0.9653**	0.4692	0.5726	0.9615
thyroid	0.9874	**0.9917**	0.9834	0.9661
twonnorm	0.8469	0.8406	0.8412	**0.9390**
Average	**0.9432**	0.8905	0.8990	0.9054

Table 7. AUC results on KEEL and UCI datasets obtained by DT

Datasets	CCUS	RUS	SMOTE	IGBS
banana	**0.8740**	0.8688	0.8710	0.8655
basketball	**0.9762**	0.8668	0.8668	0.9159
codrna	0.9186	0.8708	0.8665	**0.9206**
named	**0.9999**	**0.9999**	0.9998	0.9404
Bean	**0.9602**	0.7825	0.7667	0.9430
ijcnn1	**0.9443**	0.9353	**0.9349**	0.8980
letter	0.9293	0.9192	0.9329	0.9002
mushrooms	**1.0000**	0.9963	0.9964	0.9989
penbased	**0.9796**	0.9785	0.9791	0.8775
phoneme	**0.8495**	0.8347	0.8454	0.8464
ring	**0.8816**	0.8741	0.8739	0.8461
satimage	**0.9207**	0.9062	0.9073	0.8506
segment0	**0.9888**	0.9813	0.9862	0.8586
shuttle	0.9989	0.9997	**0.9998**	0.9992
svmguide1	0.9595	0.9578	0.9586	**0.9728**
texture	**0.9653**	0.5907	0.5726	0.9615
thyroid	0.9875	**0.9917**	0.9834	0.9661
twonnorm	0.8470	0.8407	0.8412	**0.9390**
Average	**0.9434**	0.8997	0.8990	0.9167

5   Conclusion

This paper proposes a parameter-free data reduction method (CCS) and extends this method to resampling learning with imbalanced data, introducing a resampling method (CCUS) that is suitable for imbalanced data. The method learns the spatial distribution of the data through a supervised process, obtains a set of data partition subsets based on the spatial distribution, and then designs a data reduction method for large-scale data and a data resampling method for imbalanced data based on the local distribution information within the data subsets. Compared with the GBS and RSDS data reduction methods and the SMOTE, RUS, and IGBS imbalanced data sampling methods on 18 datasets and a typical classifiers (DT), the proposed method shows better performance, demonstrating its effectiveness. It should be noted that this method is mainly focused on binary classification problems, and how to extend it to multiclass data is the next issue to be addressed.

References

1. Arnaiz-González, Á., Díez-Pastor, J.F., Rodríguez, J.J., García-Osorio, C.: Instance selection of linear complexity for big data. Knowl.-Based Syst. **107**, 83–95 (2016)
2. Aslani, M., Seipel, S.: Efficient and decision boundary aware instance selection for support vector machines. Inf. Sci. **577**, 579–598 (2021)
3. Chawla, N.V., Bowyer, K.W., Hall, L.O., Kegelmeyer, W.P.: Smote: synthetic minority over-sampling technique. J. Artif. Intell. Res. **16**, 321–357 (2002)
4. Chen, X.W., Wasikowski, M.: Fast: a ROC-based feature selection metric for small samples and imbalanced data classification problems. In: Proceedings of the 14th ACM SIGKDD International Conference on Knowledge Discovery and Data Mining, pp. 124–132 (2008)
5. Escalante, H.J., Graff, M., Morales-Reyes, A.: PGGP: prototype generation via genetic programming. Appl. Soft Comput. **40**, 569–580 (2016)
6. Giorginis, T., Ougiaroglou, S., Evangelidis, G., Dervos, D.A.: Fast data reduction by space partitioning via convex hull and MBR computation. Pattern Recogn. **126**, 108553 (2022)
7. Hasanin, T., Khoshgoftaar, T.: The effects of random undersampling with simulated class imbalance for big data. In: 2018 IEEE International Conference on Information Reuse and Integration (IRI), pp. 70–79. IEEE (2018)
8. Leyva, E., González, A., Pérez, R.: Knowledge-based instance selection: a compromise between efficiency and versatility. Knowl.-Based Syst. **47**, 65–76 (2013)
9. Li, Y., Maguire, L.: Selecting critical patterns based on local geometrical and statistical information. IEEE Trans. Pattern Anal. Mach. Intell. **33**(6), 1189–1201 (2010)
10. Lozano, M., Sotoca, J.M., Sánchez, J.S., Pla, F., Pękalska, E., Duin, R.P.: Experimental study on prototype optimisation algorithms for prototype-based classification in vector spaces. Pattern Recogn. **39**(10), 1827–1838 (2006)
11. Marchiori, E.: Class conditional nearest neighbor for large margin instance selection. IEEE Trans. Pattern Anal. Mach. Intell. **32**(2), 364–370 (2009)
12. Ougiaroglou, S., Evangelidis, G.: RHC: a non-parametric cluster-based data reduction for efficient k-NN classification. Pattern Anal. Appl. **19**(1), 93–109 (2016)
13. Ougiaroglou, S., Mastromanolis, T., Evangelidis, G., Margaris, D.: Fast training set size reduction using simple space partitioning algorithms. Information **13**(12), 572 (2022)
14. Valero-Mas, J.J., Castellanos, F.J.: Data reduction in the string space for efficient KNN classification through space partitioning. Appl. Sci. **10**(10), 3356 (2020)

15. Wang, T., Lu, C., Ju, W., Liu, C.: Imbalanced heartbeat classification using easyensemble technique and global heartbeat information. Biomed. Sig. Process. Control **71**, 103105 (2022)
16. Wicaksana, A.K., Cahyani, D.E., et al.: Modification of a density-based spatial clustering algorithm for applications with noise for data reduction in intrusion detection systems. Int. J. Fuzzy Logic Intell. Syst. **21**(2), 189–203 (2021)
17. Xia, S., Liu, Y., Ding, X., Wang, G., Yu, H., Luo, Y.: Granular ball computing classifiers for efficient, scalable and robust learning. Inf. Sci. **483**, 136–152 (2019)
18. Xia, S., Zheng, S., Wang, G., Gao, X., Wang, B.: Granular ball sampling for noisy label classification or imbalanced classification. IEEE Trans. Neural Netw. Learn. Syst. 2144–2155 (2021)
19. Xia, S., Zheng, Y., Wang, G., He, P., Li, H., Chen, Z.: Random space division sampling for label-noisy classification or imbalanced classification. IEEE Trans. Cybern. **52**(10), 10444–10457 (2021)
20. Yan, Y., Jiang, Y., Zheng, Z., Yu, C., Zhang, Y., Zhang, Y.: LDAS: local density-based adaptive sampling for imbalanced data classification. Expert Syst. Appl. **191**, 116213 (2022)
21. Yan, Y., Zhu, Y., Liu, R., Zhang, Y., Zhang, Y., Zhang, L.: Spatial distribution-based imbalanced undersampling. IEEE Trans. Knowl. Data Eng. **35**, 6376–6391 (2023)
22. Zhai, J., Wang, X., Pang, X.: Voting-based instance selection from large data sets with mapreduce and random weight networks. Inf. Sci. **367**, 1066–1077 (2016)
23. Zhao, Z., Chu, L., Tao, D., Pei, J.: Classification with label noise: a Markov chain sampling framework. Data Min. Knowl. Disc. **33**, 1468–1504 (2019)

Period Extraction for Traffic Flow Prediction

Qingyuan Wang[1], Chen Chen[2], Long Zhang[3], Xiaoxuan Song[1], Honggang Li[1], Qingjie Zhao[3], Bingxin Niu[2], and Junhua Gu[2(✉)]

[1] Hebei Xiongan Rongwu Expressway Co., Ltd., Hebei 071700, China
[2] School of Artificial Intelligence and Hebei Province Key Laboratory of Big Data Calculation, Hebei University of Technology, Tianjin, China
jhgu_hebut@163.com
[3] Hebei Provincial Transportation Planning and Design Institute, Hebei 050000, China

Abstract. Due to the particularity of "Tourist chartered Buses, Liner Buses and Dangerous Goods Transport Vehicles" ("TLD Vehicles"), traffic accidents will bring serious losses. Therefore, traffic flow prediction for "TLD Vehicles" has become an urgent need for traffic management departments. Different from the ordinary traffic flow prediction problem, the traffic flow for "TLD Vehicles" has the characteristics of sparsity in the spatial dimension. The ordinary spatial feature extraction method will capture useless node information and affect the prediction accuracy. The traffic data of "TLD Vehicles" has significant periodic characteristics in the time dimension. Most of the traditional traffic prediction methods extract time characteristics through artificially set cycles, which has certain limitations. In this paper, a Period Extraction model for Traffic Flow Prediction is proposed to solve the problem of data sparsity and insufficient periodic feature extraction ability of traffic flow prediction. The model uses sparse graph convolution combined with Transformer to extract spatial features, uses sequence decomposition and auto-correlation attention mechanism to extract temporal features, and obtains prediction results through stacked spatio-temporal modules. The experimental results show that the proposed algorithm can extract the traffic f low data feature information with stable characteristics more effectively and improve the accuracy of the network model for traffic prediction.

Keywords: Traffic Prediction · Sparse Graph Convolution · Periodic Information Analysis · Spatio-temporal Feature Extraction

1 Introduction

The transportation industry is an important cornerstone of China's social and economic development. With the rapid development of China's transportation industry, the number of "TLD Vehicles" is also increasing. The potential danger caused by the gathering of "TLD Vehicles" has attracted more and more

© The Author(s), under exclusive license to Springer Nature Singapore Pte Ltd. 2024
Z. Tari et al. (Eds.): ICA3PP 2023, LNCS 14489, pp. 490–501, 2024.
https://doi.org/10.1007/978-981-97-0798-0_29

attention. The traffic flow prediction method for "TLD Vehicles" studied in this paper is compared with the traditional traffic flow prediction method. In theory, the traffic flow of "TLD Vehicles" changes smoothly and the cycle information is obvious. The overall traffic flow change trend is more affected by holidays, epidemics, weather and other factors, and the overall traffic flow is larger and the traffic trend is more unstable. As shown in Fig. 1, the traffic flow data of the week from January 2,2021 to January 8,2021 is selected to show the characteristics of the change of traffic flow for "TLD Vehicles" and the overall traffic f low. Compared with the overall traffic flow, the traffic flow of "TLD Vehicles" has stronger periodicity and regularity. The overall traffic flow is also affected by the morning peak and evening peak, while the traffic flow of "TLD Vehicles" is less affected by the morning and evening peak. In summary, compared with the overall traffic flow, the change trend of traffic flow of "TLD Vehicles" in this week is relatively less affected by whether it is a working day, and the change law is more significant and the fluctuation is more stable.

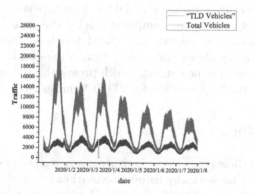

Fig. 1. "Tourist chartered Buses, Liner Buses and Dangerous Goods Transport Vehicles" traffic flow and the overall traffic flow trend chart.

In the field of traffic flow prediction, most scholars' research on spatial feature extraction methods focuses on how to better extract the static global features inherent in space and the dynamic local features that change with time. The classical ASTGCN model [1] captures the correlation between adjacent nodes of the road network more accurately by considering the attention mechanism combined with the graph convolution structure to extract spatial features. However, it does not take into account that the spatial correlation will change with time. The extraction of spatial relationships in the model only depends on the road network adjacency matrix defined in advance at the beginning, and does not take into account the dynamic changes of space. There are also models that take into account the dynamics of the spatial relationship of the road network, such as the DCRNN model [2], the DGCN model, and the DMSTGCN model [3]. However, according to the analysis of the characteristics of the traffic data for "TLD Vehicles" studied in this paper, the spatial relationship of the real "TLD

Vehicles" traffic flow data set used in this paper has the characteristics of sparsity. The method of dynamic spatial feature capture is not very suitable for the data set studied in this paper.

In order to solve the above problems and better predict the spatial sparsity and time periodic data characteristics of "TLD Vehicles", this paper proposes a PETFP model. The main innovations of this paper are as follows:

- A sparse graph convolution combined with Transformer module is proposed to extract the spatial features of traffic data. Different from the traditional graph convolution method, sparse graph convolution can extract useful information between spatial nodes, suppress the extraction of spatial information of invalid nodes, capture spatial features more accurately, and improve the accuracy of feature extraction between network nodes.
- The periodic feature extraction module is proposed to extract the time features of traffic data. The module learns the periodic characteristics of traffic f low data by using the idea of sequence decomposition and autocorrelation mechanism, rather than using artificially set fixed period and step size.
- Using multiple stacked spatio-temporal blocks, the spatio-temporal features of different dimensions are extracted, and the final prediction results are obtained. The model shows good prediction results on the traffic dataset of "TLD Vehicles" and public dataset, which proves the accuracy and versatility of the model in traffic prediction for "TLD Vehicles".

2   Related Work

In the field of traffic flow prediction, there are many studies to capture the spatial correlation in traffic networks by using convolutional neural networks from two dimensional spatio-temporal traffic data [4]. However, the traffic network is very complex and dynamic, and it is difficult to describe it with a two-dimensional matrix, which has certain limitations. Later, graph convolutional neural net works, which are more suitable for highway network, emerged to extract spatial features. GCN generally has two methods, namely, spectrum-based convolution method and space-based convolution method. GCN can better extract the spatial relationship in complex road network structure.

In the field of traffic prediction, RNN and its deformation are usually used to deal with the characteristics of time series data, but when modeling long sequences, these models will experience gradient explosion. Later, a model using convolutional neural networks to process time features appeared. STGCN [5] applied pure convolution structure to extract spatio-temporal features simultaneously from time series data with graphical structure. The Gman model [6] uses extended causal convolution as the time convolution layer of its model to capture the time trend of nodes.

As a standard method for time series analysis, time series decomposition [7] decomposes a time series into several components, each representing a basic class of more predictable patterns. It is mainly used to explore the history over time. In forecasting tasks, decomposition is always used as a preprocessing of historical series before predicting future series [8,9], such as Prophet [10] using trend-seasonality decomposition, N-BEATS [11] using basis expansion, and DeepGLO [12] using matrix decomposition. However, this preprocessing is limited by the plain decomposition effect of the historical series and ignores the hierarchical interactions between the underlying patterns of the long-term future series.

3 Proposed Model

In this section, we present our proposed model in detail. The proposed model is mainly to solve the problem of insufficient periodic information extraction in the time dimension and sparse data in the spatial dimension. The general framework of our proposed model is shown in Fig. 2. The overall architecture of the model contains two main parts: the spatial-temporal block and the prediction layer. Each section is described in detail next.

The model initially separates the periodic sequence term and the trend sequence term by decomposing the input sequence. Then, the autocorrelation mechanism module is used to further advance the periodic features of the periodic items, and the final obtained periodic features and trend items are fused to obtain the feature relationship of the time dimension. Next, the spatial features of the intermediate variables are extracted by using sparse graph convolution combined with Transformer, and the data is extracted by multiple modules at different scales. Features, and finally output prediction results through the prediction layer.

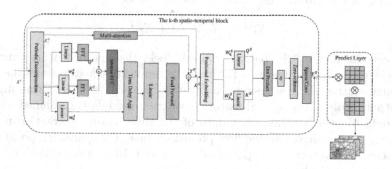

Fig. 2. PETFP model architecture diagram.

3.1 Problem Formulation

The content of this paper is traffic flow prediction. The main research content is to predict the traffic flow information of each gantry node in the next period of time through the traffic flow of all sections in the past period of time. That is, given the traffic flow data of P historical steps, the traffic flow of Q steps in the future is predicted by learning model f.

$$\hat{X}_{t+1:t+Q} = f\left(X_{t-P+1:t}\right) \tag{1}$$

3.2   Input Layer

Before the original data passes through the spatio-temporal block for the first time, the initial feature aggregation of the input original data is first performed through a fully connected convolutional layer, and the number of channels is expanded by convolution to achieve the dimension increase of the number of channels.

The initially aggregated features are used as the input of the first layer of spatio-temporal blocks. The spatio-temporal features are jointly learned by spatio-temporal blocks in the dynamically changing traffic conditions, and different levels of spatio-temporal features are alternately captured by stacked spatio-temporal blocks to extract more complex spatio-temporal relationships.

3.3   Spatial-Temporal Blocks

As we all know, the future traffic state of a node is affected by the traffic conditions of adjacent nodes. Therefore, this paper develops a spatio-temporal block to integrate the feature extraction module of space and time, so as to jointly model the space and time models in the traffic network to achieve accurate prediction. In general, the input of the $l - th$ spatio-temporal block is obtained by the output of the $l - 1$ spatio-temporal block combined with the residual connection of the input. The formula is expressed as:

$$X_l^{ST} = H\left(Y_{l-1}^{ST}\right) + X_{l-1}^{ST} \tag{2}$$

In general, in the l-th spatio-temporal block, the spatial feature extraction module extracts spatial features Y_l^S from the input node features and the graph adjacency matrix A. The residual connection is used for stable training, and the output result of the spatial feature extraction module is combined with the input data to generate the input data of the time feature extraction module. According to the current traffic flow prediction task, the spatio-temporal blocks are stacked to improve the capacity of the model. In summary, the formula of spatio-temporal block can be expressed as follows:

$$Y_l^S = S\left(X_l^S, A\right) \tag{3}$$

$$Y_l^T = T\left(X_l^T\right) \tag{4}$$

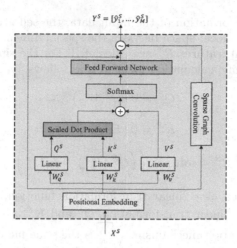

Fig. 3. Spatial feature extraction module structure diagram.

3.4 Prediction Layer

The prediction layer uses two classical convolutional layers to perform multi step prediction based on the spatio-temporal features of the last spatio-temporal block. The input X^{ST} of the prediction layer is composed of the spatio-temporal features of N nodes of the historical l time slices. The multi-step prediction of "TLD Vehicles" in the future T time slices is expressed as:

$$Y = \mathrm{Conv}\left(\mathrm{Conv}\left(X^{ST}\right)\right) \tag{5}$$

The model is trained by mean absolute loss. The formula is expressed as:

$$L = \left\| Y - Y^{\mathrm{truth}} \right\|_1 \tag{6}$$

3.5 Spatial Extraction Module

As shown in Fig. 3, the spatial dependency is captured by combining the spatial Transformer structure of sparse convolution.

For the l-th spatial feature extraction module, the input $X_l^S \in R^{MN \times N}$ is obtained by connecting the output of the previous time feature extraction module with the residual of the input. The formula is as follows:

$$X_l^S = X_l^T + Y_l^T \tag{7}$$

Position embedding coding: For the spatial feature extraction module, position coding is a linear transformation through input data. The formula is as follows:

$$E^S = \emptyset\left(X^S, W_E^S\right) \tag{8}$$

where $\emptyset$ expressed as a linear transformation function, W_E^S is the weight of the linear transformation.

After linear transformation of the input data, the self-attention mechanism is used to extract the local space between the spatial dimension nodes, and the local dynamic spatial features R^S are calculated. For the single-head attention mechanism, the specific calculation formula is as follows:

$$Q^S = \emptyset \left(E^S, W_Q^S \right) \tag{9}$$

$$K^S = \emptyset \left(E^S, W_K^S \right) \tag{10}$$

$$R^S = \text{softmax} \left(\left(Q^S \left(K^S \right)^T \right) / \sqrt{d_k^S} \right) \tag{11}$$

where, $\emptyset()$ is expressed as a linear transformation function, and W_Q^S and W_K^S are the weight of the linear transformation. Q^S and K^S are the query and key value of the self-attention mechanism. $\sqrt{d_k^S}$ is the scale factors that ensure the numerical stability and prevent the value from being too large.

The spatial relationship obtained by multi-head attention is the dynamic spatial feature R_M^S obtained by aggregating the results of each head and 1×1 convolution aggregation results in the time dimension. The result of each time step is the extracted partial spatial feature, and the value represents the influence between the nodes. The rows and columns of R_M^S matrix are combined to obtain a higher dimensional spatial feature F. For the l-th spatio-temporal block, the formula is expressed as follows:

$$F_l^{\text{row}} = \text{Conv} \left(F_{l-1}, K_{1 \times S}^{\text{row}} \right) \tag{12}$$

$$F_l^{col} = \text{Conv} \left(F_{l-1}, K_{S \times 1}^{col} \right) \tag{13}$$

$$F_l = \delta \left(F_l^{\text{row}} + F_l^{col} \right) \tag{14}$$

where F_l^{row} and F_l^{col} are row-based and column-based spatial feature maps in the $l - th$ space-time block, and F_l is the feature mapping result calculated by the nonlinear activation function. $K_{1 \times S}^{row}$ and $K_{S \times 1}^{col}$ are convolution kernels of size $1 \times S$ and $S \times 1$, respectively. For the first spatio-temporal block, F_0 is R_M^S, the convolution operation is filled with a value of 0 to ensure that the input and output are consistent.

The spatial feature threshold is set to generate a mask M_S with sparse effect, and the dynamic sparse adjacency matrix is generated according to the mask. The formula is as follows:

$$M_S = II\{\text{sigmoid}(F) \geq \zeta\} \tag{15}$$

$$A_S = (M_S + I) \odot R_M^S \tag{16}$$

where, if the inequality in function $II\{\}$ holds, then output 1, otherwise output 0. A_S is a sparse adjacency matrix, $\odot$ representing the corresponding elements by phase multiplication.

The obtained sparse adjacency matrix is normalized. In this paper, the adjacency matrix is normalized by the Zero-softmax function to maintain the sparsity of the matrix. The specific formula is as follows :

$$\text{Zero} - \text{softmax}(x_i) = (\exp(x_i) - 1)^2 / \left(\sum_{j}^{K} (\exp(x_i) - 1)^2 + \varepsilon \right) \qquad (17)$$

$$\hat{A}_S = Zero - \text{softmax}(X^S, A_S) \qquad (18)$$

where ε is a constant that can be ignored, the role is to ensure the stability of the value, K is the dimension of the input vector, on this basis, the adjacency matrix is normalized.

Finally, the input of the spatial feature extraction module and the obtained adjacency matrix are convoluted to obtain the extracted spatial features. The formula is expressed as:

$$Y^S = \text{graphConv}(X^S, A_S) \qquad (19)$$

3.6 Temporal Extraction Module

Figure 4 below is the module architecture of time feature extraction proposed in this paper. Figure 4(a) is the overall structure of the time feature extraction module, and Fig. 4(b) is the specific structure of the autocorrelation attention mechanism module in the model. In the extraction of time correlation information, the traffic data is first divided into trend term and periodic term data. Different from the prediction of traffic flow by using the time series decomposition method alone, this paper initially extracts the periodic information in the traffic data through the periodic feature extraction method. After that, the periodic features and time correlation in the data are further extracted through the autocorrelation attention mechanism, and the results of the time feature extraction module are obtained after the neural network processing results.

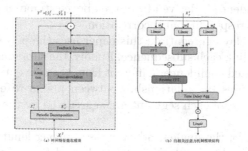

Fig. 4. Temporal extraction module structure diagram.

Periodic Decomposition. In this paper, the idea of time series decomposition is used to learn the complex patterns in time series. By dividing the sequence into long-term trend part and periodic trend part, the periodic features and trend features in the input data are preliminarily extracted. In this paper, the additive decomposition method is used to separate the trend term, and the rest is the periodic term. The input sequence X is P in length and d in dimension. For the input sequence $X \in R^{P \times d}$ with length P, the specific calculation process can be expressed as:

$$X = X_S + X_T \tag{20}$$

$$X_T = \text{LOESS}(\text{Padding}(X)) \tag{21}$$

where, X_S is the periodic term, X_T is the trend term. In this paper, LOESS() is used for local weighted regression, and the filling operation is performed to keep the variable consistent with the length of the input sequence.

Auto-correlation Mechanism. As shown in Fig. 4(b), this paper expands the utilization of periodic information by using autocorrelation combined with attention mechanism. The autocorrelation mechanism discovers the period-based dependency by calculating the autocorrelation coefficient of the sequence, and aggregates similar subsequences through time delay aggregation to further ex tract the periodic characteristics of the data.

In the traffic flow of "TLD Vehicles" change trend map, it is learned that the change trend of traffic flow with time shows similar traffic trend subsequences in the same phase position between each cycle. Therefore, inspired by the idea of stochastic process theory, for a real discrete-time process X_t, this paper can obtain the autocorrelation coefficient by the following formula:

$$R_{xx}(\tau) = \lim_{L \to \infty} \left(1/L \sum_{t=1}^{L} X_t X_{t-\tau} \right) \tag{22}$$

where, $R_{xx}(\tau)$ reflects the time-delay similarity between X_t and its delay sequence $X_{t-\tau}$.

In this paper, autocorrelation is used as the non-standardized confidence level for estimating the period length. Then, k cycle lengths are selected. The period-based dependence is derived from the above estimated period and can be weighted by the corresponding autocorrelation.

By using time delay aggregation, this paper uses the period obtained by cor relation analysis to link each subsequence. Finally, the subsequence is aggregated by softmax normalized confidence. Different from the conventional self-attention mechanism, which uses the relationship between points to extract and aggre-gate, the subsequences of each segment are aggregated by the delay aggregation method.

In simple terms, for the single head case, the sequence length of the time series X is L. After the weight calculation, the Q, K and V values can be obtained, which can replace the self-attention. The calculation formula of the autocorrelation attention mechanism is as follows:

$$\tau_1, \ldots, \tau_k = \underset{\tau \in \{1,\ldots,L\}}{\arg \text{Topk}} (R_{Q,K}(\tau)) \tag{23}$$

$$\hat{R}_{Q,K}(\tau_1), \ldots, \hat{R}_{Q,K}(\tau_k) = \text{soft max}(R_{Q,K}(\tau_1), \ldots, R_{Q,K}(\tau_k)) \qquad (24)$$

$$\text{Auto-correlation } (Q, K, V) = \sum_{i=1}^{k} \text{Roll}(V, \tau_i)\, \hat{R}_{Q,K}(\tau_i) \qquad (25)$$

where, argTopk() is the independent variable of TOPK autocorrelation, $k = \lfloor c \times logL \rfloor$ and c is a hyperparameter. $R_{Q,K}$ is the autocorrelation coefficient between sequences Q and K. $Roll(V, \tau)$ represents an operation on V with a time delay τ, during which the element removed from the first position is reintroduced at the last position.

4   Experiment

4.1   Experimental Results

In Table 1, this paper lists the traffic flow prediction results of the comparison model and the model on a provincial highway dataset. This paper measures the prediction effect of the model by using a unified evaluation index (MAE, MAPE, RMSE). Compared with the extensive baseline, the model in this paper has a certain improvement in traffic prediction, which proves that the model in this paper is effective for the prediction of stable and periodic spatio-temporal data.

Table 1. Comparison of prediction errors of different models under HBDataSet.

Model	HBDataSet								
	96			192			288		
	MAE	MAPE /%	RMSE	MAE	MAPE /%	RMSE	MAE	MAPE /%	RMSE
ARIMA [13]	2.53	5.70	5.18	2.88	7.00	6.15	3.30	8.70	7.20
LSTM [14]	2.05	4.19	4.80	2.20	4.55	5.20	3.27	4.96	5.7
STGCN	1.36	2.96	2.90	1.81	4.27	4.17	3.19	5.69	5.79
T-GCN [15]	2.17	3.94	3.58	2.21	4.36	4.05	3.07	6.37	6.18
STTN	1.93	3.54	3.29	2.16	4.15	3.86	3.24	6.12	5.81
PETFP	1.32	2.75	2.78	1.75	3.95	3.66	2.32	5.51	5.41

From Table 1, it can be observed that the PETFP model exhibits better predictive performance compared to the baseline models across various metrics and prediction windows. Particularly, the PETFP model shows the most significant improvement in prediction compared to the classical ARIMA model. The LSTMmodel demonstrates good predictive performance in long-term traffic flow forecasting compared to the ARIMA model. The PETFP model achieves an average error reduction of 1.28 in root mean square error (RMSE) compared to the LSTM model.

From Table 2 comparing the errors of the PETFP model and other contrast models across various prediction time windows and evaluation metrics, we find that the errors are generally higher than the results shown in Table 1 for the models trained on the HBDataSet. This suggests that the characteristics of traffic f low data for "TLD Vehicles" allow the models to predict future traffic flow more consistently. Furthermore, the PETFP model exhibits a greater improvement in predictive performance compared to the contrast models in the HBDataSet. This indicates that the proposed model is more suitable for traffic flow datasets with prominent periodic features.

Table 2. Comparison of prediction errors of different models under PEMSD7

Model	PEMSD7								
	96			192			288		
	MAE	MAPE /%	RMSE	MAE	MAPE /%	RMSE	MAE	MAPE /%	RMSE
ARIMA	2.96	8.41	5.67	3.42	11.8	7.77	5.21	12.93	9.57
LSTM	2.80	7.94	5.73	3.63	11.27	7.73	5.16	12.38	9.51
STGCN	2.65	6.50	5.12	3.50	10.27	7.19	3.62	10.70	7.45
T-GCN	2.88	7.62	5.74	3.47	9.57	7.24	4.59	11.97	9.40
STTN	2.47	6.18	4.63	2.92	7.03	5.21	3.04	10.14	7.07
PETFP	2.46	6.30	4.50	2.87	6.80	5.42	2.32	5.51	5.41

5   Conclusion

Aiming at the sparse characteristics of traffic flow data in the spatial dimension, this paper proposes PETFP model. Transformer and sparse graph convolution are used to extract the spatial characteristics of traffic flow data. In view of the significant periodicity of the traffic flow data in the time dimension, the sequence decomposition and autocorrelation mechanism are used in the time dimension to further extract the periodic characteristics of the time dimension, so as to better capture the time correlation in the traffic flow data. For "TLD Vehicles", the HBDataSet is used to verify the prediction effect of traffic flow. Compared with the overall traffic flow prediction results, the prediction effect of traffic flow for "TLD Vehicles" is better. In the field of overall traffic flow prediction, the prediction effect of the prediction model proposed in this paper in the direction of traffic flow prediction is verified by using the PEMSD7 Dataset.

Acknowledgements. This work was supported in part by the Hebei Province Innovation Capability Enhancement Plan Project under Grant 22567603H, in part by S&T Program of Hebei under Grant 20310801D, in part by research project (the fourth batch) of KT12 section of the new line of Rongwu Expressway.

References

1. Guo, S., Lin, Y., Feng, N., Song, C., Wan, H.: Attention based spatial-temporal graph convolutional networks for traffic flow forecasting. In: Proceedings of the AAAI Conference on Artificial Intelligence, vol. 33, pp. 922–929 (2019)
2. Li, Y., Yu, R., Shahabi, C., Liu, Y.: Diffusion convolutional recurrent neural network: Data-driven traffic forecasting. arXiv preprint arXiv:1707.01926 (2017)
3. Han, L., Du, B., Sun, L., Fu, Y., Lv, Y., Xiong, H.: Dynamic and multi-faceted spatio-temporal deep learning for traffic speed forecasting. In: Proceedings of the 27th ACM SIGKDD Conference on Knowledge Discovery & Data Mining, pp. 547–555 (2021)
4. Li, Y., Shahabi, C.: A brief overview of machine learning methods for short-term traffic forecasting and future directions. Sigspatial Special 10(1), 3–9 (2018)
5. Yu, B., Yin, H., Zhu, Z.: Spatio-temporal graph convolutional networks: a deep learning framework for traffic forecasting (2017). arXiv preprint arXiv:1709.04875
6. Yao, H., Liu, Y., Wei, Y., Tang, X., Li, Z.: Learning from multiple cities: a meta-learning approach for spatial-temporal prediction. In: The World Wide Web Conference, pp. 2181–2191 (2019)
7. Polson, N., Sokolov, V.: Bayesian particle tracking of traffic flows. IEEE Trans. Intell. Transp. Syst. 19(2), 345–356 (2017)
8. Moreira-Matias, L., Gama, J., Ferreira, M., Mendes-Moreira, J., Damas, L.: Predicting taxi-passenger demand using streaming data. IEEE Trans. Intell. Transp. Syst. 14(3), 1393–1402 (2013)
9. Ren, S., Yang, B., Zhang, L., Li, Z.: Traffic speed prediction with convolutional neural network adapted for non-linear spatio-temporal dynamics. In: Proceedings of the 7th ACM SIGSPATIAL International Workshop on Analytics for Big Geospatial Data, pp. 32–41 (2018)
10. Gong, Y., Li, Z., Zhang, J., Liu, W., Yi, J.: Potential passenger flow prediction: a novel study for urban transportation development. In: Proceedings of the AAAI Conference on Artificial Intelligence, vol. 34, pp. 4020–4027 (2020)
11. Tan, H., Wu, Y., Shen, B., Jin, P.J., Ran, B.: Short-term traffic prediction based on dynamic tensor completion. IEEE Trans. Intell. Transp. Syst. 17(8), 2123–2133 (2016)
12. Yu, H.F., Rao, N., Dhillon, I.S.: Temporal regularized matrix factorization for high-dimensional time series prediction. Advances in neural information processing systems 29 (2016)
13. Williams, B.M., Hoel, L.A.: Modeling and forecasting vehicular traffic flow as a seasonal arima process: theoretical basis and empirical results. J. Transp. Eng. 129(6), 664–672 (2003)
14. Cho, K., Van Merriënboer, B., Gulcehre, C., Bahdanau, D., Bougares, F., Schwenk, H., Bengio, Y.: Learning phrase representations using rnn encoder-decoder for statistical machine translation. arXiv preprint arXiv:1406.1078 (2014)
15. Zhao, L., et al.: T-GCN: a temporal graph convolutional network for traffic prediction. IEEE Trans. Intell. Transp. Syst. 21(9), 3848–3858 (2019)

Author Index

Z. Tari et al. (Eds.): ICA3PP 2023, LNCS 14489, pp. 503–504, 2024.
https://doi.org/10.1007/978-981-97-0798-0

Printed in the United States
by Baker & Taylor Publisher Services